VITAMINS AND HORMONES

VOLUME 58

VITAMINS AND HORMONES

ADVANCES IN RESEARCH AND APPLICATIONS

Editor-in-Chief

GERALD LITWACK
Department of Biochemistry and Molecular Pharmacology
Jefferson Medical College
Thomas Jefferson University
Philadelphia, Pennsylvania

Volume 58

ACADEMIC PRESS

A Harcourt Science and Technology Company

San Diego San Francisco New York
Boston London Sydney Tokyo

This book is printed on acid-free paper. ♾

Academic Press
A Harcourt Science and Technology Company
525 B Street, Suite 1900, San Diego, California 92101-4495, USA
http://www.apnet.com

Academic Press
24-28 Oval Road, London NW1 7DX, UK
http://www.hbuk.co.uk/ap/

International Standard Book Number: 0-12-709858-5

PRINTED IN THE UNITED STATES OF AMERICA
99 00 01 02 03 04 BB 9 8 7 6 5 4 3 2 1

Contents

Protein Folding and Deficiencies Caused by Dominant-Negative Mutants of Hormones

PRISCILLA S. DANNIES

Structural Aspects of the G-Protein Receptor, Rhodopsin

ARLENE D. ALBERT AND PHILIP L. YEAGLE

Molecular Modeling of Mammalian Cytochromes *P*450: Application to Study Enzyme Function

GRAZYNA D. SZKLARZ, SANDRA E. GRAHAM, AND MARK D. PAULSEN

Prostaglandin D Synthase: Structure and Function

YOSHIHIRO URADE AND OSAMU HAYAISHI

Steroid Dehydrogenase Structures, Mechanism of Action, and Disease

WILLIAM L. DUAX, DEBASHIS GHOSH, AND VLADIMIR PLETNEV

Structure-Based Inhibitor Design

SYDNEY P. CRAIG III AND ANN E. EAKIN

Protein Folding Using Contact Maps

MICHELE VENDRUSCOLO AND EYTAN DOMANY

HIV Protease: Enzyme Function and Drug Resistance

SERGEI GULNIK, JOHN W. ERICKSON, AND DONG XIE

Ultrastructure of Cells Undergoing Apoptosis

Kenji Kimura, Hironobu Sasano, Tooru Shimosegawa, Shizue Mochizuki, Hiroshi Nagura, and Takayoshi Toyota

Role of Cytoskeleton in Apoptosis

R. Atencia, A. Asumendi, and M. García-Sanz

Dietary Antioxidants and Cardiovascular Disease

J. Michael Gaziano

Melatonin Receptors and Ligands

Paula A. Witt-Enderby and Pui-Kai Li

Vitamin K-Dependent Proteins

Gary L. Nelsestuen, Amit M. Shah, and Stephen B. Harvey

Steroid/Nuclear Receptor Coactivators

J. Don Chen

Thyroid Hormone Receptor, v-ErbA, and Chromatin

ALAN P. WOLFFE, TREVOR N. COLLINGWOOD, QIAO LI, JANET YEE, FYODOR URNOV, AND YUN-BO SHI

Preface

This volume of *Vitamins and Hormones* may be divided into three groups of contributions. The first set deals with aspects of structure and function, especially those concerning biologically active proteins. The second set contains two contributions on aspects of apoptosis, and then a larger section on specific proteins and receptors follows.

The first paper by P. Dannies concerns protein folding and hormone deficiencies. This is followed by a discussion of the structural aspects of rhodopsin by A. Albert and P. Yeagle. Enzyme function through modeling of cytochromes *P*450 is considered by G. Szklarz, S. Graham, and M. Paulsen. Structure–function of prostaglandin D synthase is reviewed by Y. Urade and O. Hayaishi. Structure–function of steroid dehydrogenase is reviewed by W. Duax, D. Ghosh, and V. Pletnev. Structure-based inhibitor design is the subject of a contribution by S. Craig and A. Eakin. M. Vendruscolo and E. Domany introduce the subject of protein folding using contact maps, and S. Gulnik, J. Erickson, and D. Xie round out this section with a review on HIV protease from the point of view of enzyme function and drug resistance.

Two papers follow on apoptosis. The first, by K. Kimura, H. Sasano, T. Shimosegawa, S. Mochizuki, H. Nagura, and T. Toyota, deals with ultrastructure of cells undergoing apoptosis. The second deals with the cytoskeleton in apoptosis and is by R. Atencia, A. Asumendi, and M. García-Sanz. The next three contributions cover antioxidants, melatonin receptors, and vitamin K-dependent proteins. J. M. Gaziano discusses dietary antioxidants and cardiovascular disease, and P. A. Witt-Enderby and P.-K. Li summarize melatonin receptors and ligands. G. Nelsestuen, A. Shah, and S. Harvey describe vitamin K-dependent proteins. J. D. Chen reviews steroid/nuclear receptor coactivators. Thyroid hormone receptor, v-ErbA, and chromatin are topics written about by A. Wolffe, T. Collingwood, Q. Li, J. Yee, F. Urnov, and Y.-B. Shi to round out the last section.

Academic Press continues its invaluable cooperation in this endeavor. As stated previously, the enlarged scope will allow many diverse reviews in the future to satisfy an increasing audience for this serial.

Gerald Litwack

VITAMINS AND HORMONES, VOL. 58

Protein Folding and Deficiencies Caused by Dominant-Negative Mutants of Hormones

PRISCILLA S. DANNIES

Department of Pharmacology, Yale School of Medicine, New Haven, Connecticut 06510

Protein folding and transport in the secretory pathway of cells is a controlled process, facilitated by chaperones. Proteins that do not fold well elicit several different programmed responses from the cells. A comparison of mutants of growth hormone that result in growth hormone deficiency suggests that cells do not respond in the same way to all growth hormone mutants that cannot fold, because some mutants are dominant and some are recessive. Causes for autosomal dominant hormone deficiencies include accumulation of toxic or dysfunctional forms, competition for chaperones important for folding or transport, induction of protein degradation in the endoplasmic reticulum, or long-term responses of the cells to synthesis of proteins that do not fold that decrease hormone synthesis or cell viability. © 2000 Academic Press.

0083-6729/00 $30.00

I. Introduction

All cells have a secretory pathway consisting of a series of membrane-bound compartments through which pass transmembrane proteins and proteins that are bound for secretion or lysosomes (Fig. 1). The synthesis of proteins destined for the secretory pathway starts in the cytoplasm, during which most of them are transported into the lumen of the endoplasmic reticulum. Transport is through a pore in the endoplasmic reticulum membrane, and regardless of whether the proteins are transported during or after synthesis, they must be unfolded to be transported through the membrane. Folding occurs in the endoplasmic reticulum. After correct folding, and in some cases, after carbohydrate addition, proteins are transported to the *cis*-Golgi complex in small irregularly shaped vesicles. Carbohydrate moieties on the proteins are modified, and other processing, such as sulfation, may occur as proteins proceed through the stacks of membranes forming the Golgi complex. At the *trans*-Golgi side, specific vesicles bear off proteins to their destinations: the plasma membrane, endosomes, or back to the layers of the Golgi complex. Neuroendocrine cells, as well as several other cell types, including mast cells and neurons, have a specialized part of the secretory pathway in which protein hormones or other proteins are stored in large amounts in a concentrated form to be released by exocytosis when cells are stimulated.

The process of protein folding in cells has been an area of intense investigation, and several recent reviews have discussed the specific aspects of protein folding in the secretory pathway (Ruddon and Bedows, 1997; Kim and Arvan, 1998; Kuznetsov and Nigam, 1998). This review focuses on what happens when proteins do not fold well, and how these events may cause some of the cases of isolated hormone deficiencies.

II. Functions of the Endoplasmic Reticulum Related to Protein Folding

The endoplasmic reticulum serves other functions in the cell besides assisting protein folding, including maintaining stores of Ca^{2+} to be released as part of signal transduction, and synthesis of lipids for membrane formation. Although some proteins fold easily in solution, folding is a complex process in cells. Processes in the endoplasmic reticulum facilitate the folding of proteins, monitor the end result so that usually only correctly folded proteins are transported further along the secretory pathway, degrade many misfolded proteins, actively sort at least

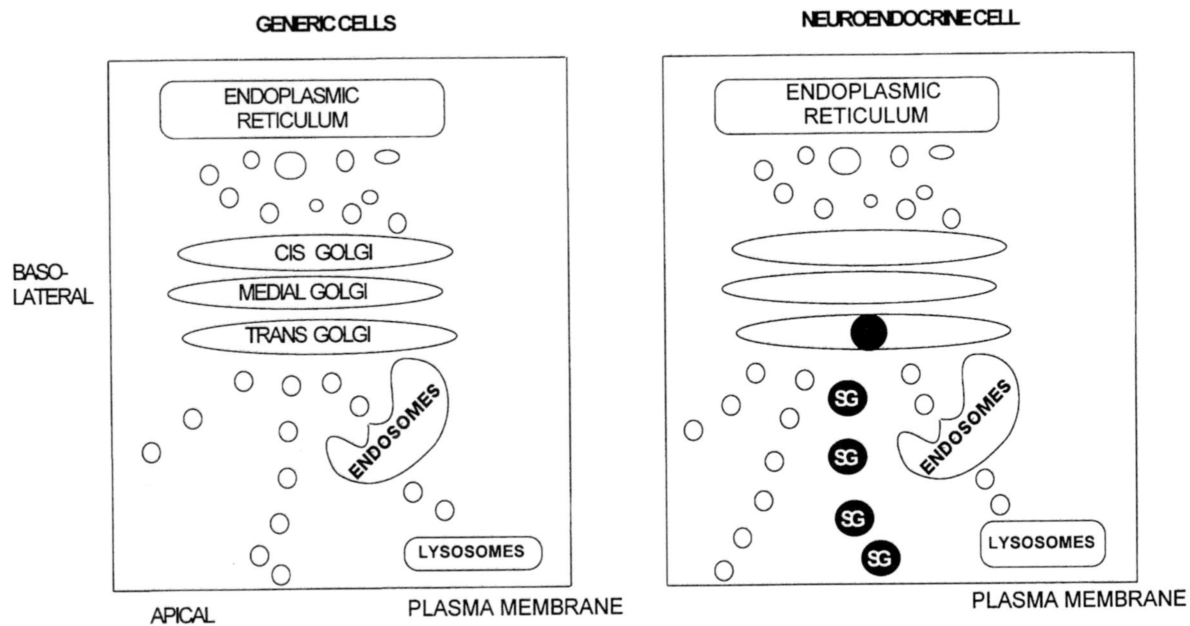

FIG. 1. Secretory pathway of cells. The left panel shows the secretory pathway that all cells have in which proteins are synthesized in the endoplasmic reticulum and transported through the Golgi complex to their final destination. The right panel shows the secretory pathway of neuroendocrine cells, in which large amounts of proteins are produced that are concentrated and stored in secretory granules.

some proteins for transport out of the endoplasmic reticulum, and communicate the state of folded proteins in the endoplasmic reticulum to the rest of the cell.

A. Protein Folding in the Endoplasmic Reticulum

Anfinsen and coworkers demonstrated that fully unfolded ribonuclease will refold to its native structure in solution without help (Anfinsen, 1973). This experiment indicated that the primary amino acid sequence of this protein contains the information for the tertiary structure of the protein. In the last decade, however, evidence has accumulated that proteins known as *chaperones* facilitate protein folding. There is agreement on the general principles by which chaperones assist protein folding. Chaperones do not supply information about the tertiary structure to other proteins, but increase the rate at which proteins achieve their native conformations. Protein folding is an equilibrium between conformations that are on the pathway to the folded state and other nonproductive conformations that will not lead to correct folding and that may lead to irreversible aggregation. Chaperones facilitate interconversion of conformations, so folding proteins do not get stuck in unproductive states (Ruddon and Bedows, 1997).

Chaperones in the endoplasmic reticulum that appear to be present in all cells include GRP78 (glucose-regulated protein 78, also called BiP), GRP94, ERP60, ERP72, calreticulin, calnexin, protein disulfide isomerase, and cyclophilin B, which is a peptidyl prolyl *cis-trans* isomerase. The ways in which these chaperones each specifically assists in folding are still being elucidated, and current knowledge has been summarized in a recent review (Ruddon and Bedows, 1997; Kim and Arvan, 1998), so this review will only emphasize two points.

The first point is that chaperones have some sequence specificity, even if they are binding unfolded proteins. One of the most intensively studied, GRP78 or BiP, binds to hydrophobic sequences, binding that is released by hydrolysis of ATP (Hendershot *et al.,* 1995). BiP appears to enhance folding by binding to hydrophobic patches of the folding protein, not normally exposed in the native state, that, because they are hydrophobic, have a tendency to form irreversible aggregates if left exposed. When the protein is released from BiP, it has a chance to achieve correct folding, and if not, is bound again. The peptide sequences to which BiP binds are 7-amino-acid stretches of hydrophobic residues alternating with other amino acids between each hydrophobic position in this stretch, which provides enough specificity so that a large protein like immunoglobulin G has only 24 predicted binding sites, 12 of which, when made as synthetic peptides, bind BiP (Knarr *et al.,* 1995). The presence of predicted BiP binding sites is alone not enough to ensure that BiP binds to the protein as it folds in cells; the rate of folding also plays a role (Hellman *et al.,* 1999). Therefore, although chaperones bind unfolded proteins, all chaperones are not involved in folding of specific proteins.

The second point is that chaperones may have more than one action. Protein disulfide isomerase facilitates interconversion of disulfide bonds so that incorrect disulfide bonds get a chance to reform correctly. When the disulfide isomerase activity is destroyed by mutating the necessary cysteines in the enzyme, however, the protein is still capable of helping proteins fold (Cai *et al.,* 1994; Hayano *et al.,* 1995; Lundstrom-Ljung *et al.,* 1995; Quan *et al.,* 1995). In addition, protein disulfide isomerase and BiP in solution under certain conditions cause proteins to aggregate, rather than preventing aggregation, a process referred to as *antichaperone activity* (Puig and Gilbert, 1994a,b; Puig *et al.,* 1994). Hsp47, a secretory pathway chaperone that specifically binds procollagen, and cyclophilin B have been proposed to serve both as chaperones and antichaperones for procollagen (Smith *et al.,* 1995).

Determining which chaperones are needed for which proteins in cells is difficult for many proteins because they fold relatively quickly, and the interactions with chaperones are transient. One method is to assess the effects of individual chaperones on folding of proteins in solution; the chaperones needed and the folding pattern of a completely synthesized protein may differ, however, from the requirements of the same protein that folds as it is synthesized (Frydman and Hartl, 1996). The most convincing way to demonstrate the involvement of a chaperone is to show its association with the newly synthesized protein, and that the association occurs only transiently, and does not occur with the mature protein. Methods commonly used to demonstrate interactions include indirect immunoprecipitation, that is, the ability to immunoprecipitate the protein in question with antibodies to chaperones, or vice versa; native gel electrophoresis; cosedimentation through sucrose; chemical or photo-cross-linking followed by immunoprecipitation (Gillikin *et al.,* 1995). Use of mutant proteins that do not fold as well as wild-type protein often makes detecting interactions with chaperones possible, but allows the possibility that the interaction may not occur in the wild-type protein. In all cases, specificity is demonstrated by showing that only newly synthesized protein, and not older forms, associates with the chaperone.

Such studies have shown that chaperones act sequentially on large proteins as they fold; some of the initial studies showed that BiP and

GRP94 interact in turn with immunogloblulin G in the endoplasmic reticulum (Melnick *et al.,* 1994), and calnexin and BiP interact with thyroglobulin (Kim and Arvan, 1995). Transmembrane proteins are expected to be likely to need help folding, and two hormone receptors have been shown to interact with chaperones. The insulin receptor undergoes disulfide bond formation while associated with calnexin and calreticulin (Bass *et al.,* 1998), and gonadotropin receptor precursors associate with calnexin (Rozell *et al.,* 1998). Smaller luminal proteins require chaperones as well. One of the better studied folding pathways for proteins in cells is human chorionic gonadotropin β-subunit; disulfide bond rearrangement occurs during folding, so that the original bonds formed are not present in the final molecule (Ruddon and Bedows, 1997). Perhaps because of processes occurring during this rearrangement, folding intermediates of the β-subunit are found linked through disulfide bonds to chaperones (Feng *et al.,* 1996). Even relatively small monomeric proteins may require chaperones; protein disulfide isomerase and cyclophilin interact with a fragment of prolactin transiently when it is synthesized in an isolated microsomal system (Klappa *et al.,* 1995), and, if prolactin is synthesized in a reconstituted microsomal system without chaperones, it aggregates (Haynes *et al.,* 1997).

B. Protein Degradation of Misfolded or Unfolded Proteins in the Endoplasmic Reticulum

It has been known for more than 10 years that proteins that do not fold properly frequently do not leave the endoplasmic reticulum (Pelham, 1989; Hammond and Helenius, 1994). Although some proteins accumulate as aggregates in the endoplasmic reticulum, others are degraded there. One means by which proteins that have not folded become degraded is that they are returned to the cytoplasm through the membrane of the endoplasmic reticulum, where they are degraded by the cytosolic system that uses ubiquitin addition followed by proteasome digestion (Werner *et al.,* 1996; Sommer and Wolf, 1997). Proteasomes are found associated with the endoplasmic reticulum, and those that are have different subunit compositions than those found free in the cytosol (Palmer *et al.,* 1996; Rivett, 1998). It is by no means clear how luminal proteins return through the membrane of the endoplasmic reticulum, but if they have developed any tertiary structure at all, they are likely to have to unfold again. A current vision of the process is that the unfolded protein returns through a channel very similar to, if not the same as, the one through which it proceeds into the lumen as it is synthesized. Related and unanswered questions include these: What are the initial steps to insert a soluble protein back into the channel? What is the driving force for pushing or pulling it through the channel?

At least two proteins in the lumen of the endoplasmic reticulum, ERP60 and ERP72, have calpain-like activity as well as chaperone-like attributes, and have been proposed to play a role in regulating unfolded protein accumulation (Urade *et al.,* 1993; Otsu *et al.,* 1995). Although it is clear that many proteins that do not fold properly do not leave the endoplasmic reticulum and are degraded, others that do not fold are secreted (Bedows *et al.,* 1994). The features that elicit degradation and what determines which system of degradation is used have not been determined.

C. Protein Sorting in the Endoplasmic Reticulum

Proteins that spend most of their time in the endoplasmic reticulum, such as chaperones, have signal sequences that retain them or cause them to be returned there if they leave. The best understood signal sequence for retention is the KDEL sequences (Nilsson and Warren, 1994; Pelham, 1995), but proteins without this sequence, such as cyclophilin B, are also located primarily in the endoplasmic reticulum and so other factors account for the location of these proteins as well (Arber *et al.,* 1992).

An early view was that proteins are passively carried from the endoplasmic reticulum unless they are actively retained or returned through recognition of the KDEL or other sequence. A more recent view is that proteins destined for transport from the endoplasmic reticulum are actively selected, at least in some cases. The first evidence that sorting occurs in the endoplasmic reticulum was that serum albumin and vesicular stomatitis virus glycoprotein are concentrated in discrete areas of the endoplasmic reticulum before transport (Mizuno and Singer, 1993; Balch *et al.,* 1994). Further evidence for sorting systems comes from genetic evidence in yeast (Gaynor and Emr, 1997). Vesicles that bud from the *cis*-Golgi area and return to the endoplasmic reticulum have a coat of protein components called COP I (coat protein complex I). Temperature-sensitive mutants in a COP I component, sec21, allow normal secretion of two proteins, invertase and HSP159, but two others, carboxypeptidase Y and proα-factor, remain in the endoplasmic reticulum. Carboxypeptidase Y–invertase hybrids leave the endoplasmic reticulum, indicating there is a signal for transport on invertase, rather than a retention signal on carboxypeptidase Y. The simplest explanation for why mutations in retrograde transport affect transport out of the endoplasmic reticulum is that COP I vesicles return components to the endoplasmic reticulum that are necessary for transport of certain secretory proteins (Gaynor and Emr, 1997). A candidate for these components is a member of the p24 family, because deletion of Emp24, a member of the family, affects transport of invertase and not carboxypeptidase Y (Schimmoller *et al.,* 1995).

In mammalian cells, transfer from the endoplasmic reticulum occurs through budding of irregularly shaped vesicles that fuse in the *cis*-Golgi region (Presley *et al.,* 1997), an area that has been called the endoplasmic reticulum–Golgi intermediate compartment, abbreviated ERGIC (Bannykh and Balch, 1997). ERGIC-53, a protein found in this area, may be involved in selective transport of certain proteins in mammalian cells. Combined deficiency of coagulation factors V and VII is an autosomal recessive bleeding disorder resulting from alterations to a single gene, ERGIC-53 (Nichols *et al.,* 1998), suggesting it is needed for their transport. ERGIC-53 associates with the COP II component, Sec23 (Kappeler *et al.,* 1997), and COP II is the coat protein complex of vesicles that transport proteins forward from the endoplasmic reticulum, providing further evidence that ERGIC-53 may be involved in sorting for transport from the endoplasmic reticulum. An unexpected result is that deficiencies in other secreted proteins have not been reported in people with these deficiencies.

Signals for transport have been identified on classes of proteins, such as the previously mentioned KDEL sequence, denoting localization in the endoplasmic reticulum. Another example is mannose-6-phosphate; proteins with this group attached are transported from the Golgi complex to lysosomes (Schmid, 1997). At this time, a signal for transport of a class of proteins out of the endoplasmic reticulum has not been identified. In addition to coagulation factors V and VII, there are other known cases where proteins cannot leave the endoplasmic reticulum without other proteins, and the interactions with the proteins necessary for transport are specific.

NinaA is a secretory pathway protein homolog of cyclophilin in the retinal cells of *Drosophila.* Cyclophilins have peptidyl-prolyl *cis, trans*-isomerase activity, but NinaA plays an additional specific role in transport of rhodopsin I. There are six forms of rhodopsin in the photoreceptor cells that after transport to the rhabdomeres, mediate phototransduction. Without NinaA, opsin I accumulates in the endoplasmic reticulum (Colley *et al.,* 1991; Baker *et al.,* 1994). It can only be transported to the rhabdomeres by forming a complex with NinaA. There is active sorting for transport out of the endoplasmic reticulum for opsin I, but the reason why it is necessary for one opsin and not others is not known.

Other examples of proteins necessary for transport have additional unusual properties. A family of proteins has been identified called RAMPs (receptor-activity modifying protein) (McLatchie *et al.,* 1998). RAMPs facilitate the expression of the calcitonin-receptor-like receptor on the surface of cells and therefore affect transport. In addition, RAMPs affect specificity. RAMP1 co-expressed with the calcitonin-receptor-like receptor causes the receptor to respond to calcitonin-gene related peptide; RAMP2 coexpressed with the same receptor reduces modification of glycosylation moieties on the receptor and changes the specificity to an adrenomedullin receptor (McLatchie *et al.,* 1998). The receptor that RAMPs affect is a 7-transmembrane G-protein requiring receptor, but not all receptors in this category require RAMPs.

These examples show stringent requirements for transport of certain proteins, and not others. Whether there are also general requirements for active transport from the endoplasmic reticulum in addition to correct folding is not yet clear.

D. Communication from the Endoplasmic Reticulum to the Rest of the Cell about Unfolded Proteins

A major means of communication is through a pathway called the *unfolded protein response* (Shamu *et al.,* 1994; Sidrauski *et al.,* 1998). When unfolded proteins accumulate in the endoplasmic reticulum, a transmembrane serine-threonine protein kinase resident in the endoplasmic reticulum is activated by dimerization followed by self-phosphorylation. Through an unusual mechanism, activation of this kinase results in increased transcription of secretory pathway chaperones, including BiP, GRP94, protein disulfide isomerase, and ERP72. The mechanism has been elucidated in yeast. A transcription factor, hac1, increases transcription of the secretory pathway chaperones. Hac1 mRNA is made continuously and found associated with polysomes, but, in the absence of induction by unfolded proteins, the mRNA is not completely spliced and not translated. Unfolded proteins cause activation of the kinase Ire1, probably by binding chaperones that normally prevent its dimerization (Sidrauski *et al.,* 1998). The activated kinase has a second activity as an endonuclease, which cleaves the mRNA for hac1 at splice sites that differ from normal mRNA splice sites. The cut mRNA is ligated by tRNA ligase. Only the correctly spliced mRNA is translated into active hac1; after synthesis, this transcription factor binds without further activation to an unfolded protein response element upstream of the genes for chaperones to increase their transcription. The result is that when the endoplasmic reticulum has trouble folding proteins, feedback through this pathway results in synthesis of more chaperones to help fold proteins.

Although the response has been characterized in yeast, it also exists in mammalian cells (Tirasophon *et al.,* 1998), and can be detected in cells that are expressing a protein that does not fold. It is frequently easier to assay the response by adding agents to cells that interfere with protein folding in the secretory pathway. β-Mercaptoethanol prevents disulfide bonds from forming. The endoplasmic reticulum is an oxidizing environment, and most sulfhydryl groups in secreted proteins are part of disulfide bonds, and must oxidize for proteins to fold. Tunicamycin prevents glycosylation, which occurs in the endoplasmic reticulum. Many proteins that are glycosylated do not fold well without that modification. Thapsigargin depletes the endoplasmic reticulum of Ca^{2+} stores by blocking the Ca^{2+} pump that takes Ca^{2+} into the lumen; folding of several proteins that has been examined depends on normal Ca^{2+} concentrations there. The stress these agents impose on cells may involve more than just unfolded proteins, but they all certainly cause abnormal protein folding in the endoplasmic reticulum and induce the unfolded protein response.

Several kinds of stress in cells result in a shutdown of protein synthesis, by phosphorylation of eukaryotic initiation factor 2α (eIF2α) (Harding *et al.,* 1999; Silverman and Williams, 1999). The kinase that is responsible for the stress-induced phosphorylation has recently been identified. The N-terminal portion resembles the luminal portion of Ire1, which is the part of Ire1 that senses unfolded proteins, most likely by detecting the absence of chaperones. The C-terminal portion resembles other kinases that shut off protein synthesis by phosphorylating eIF2α (Harding *et al.,* 1999). This newly identified kinase, called PERK (RNA-dependent protein kinase-like ER kinase) or PEK (pancreatic eIF2α kinase), therefore plays a role in the unfolded protein response by down-regulating all protein synthesis.

A different response to stress in the secretory pathway is called the *overload response* of the endoplasmic reticulum (Pahl and Baeuerle, 1997). In eukaryotic cells, when viral protein accumulates in the endoplasmic reticulum, Ca^{2+} begins to leak from the stores that are contained there, and the Ca^{2+} leak eventually results in activation of NF-κB (Pahl and Baeuerle, 1996). A continuous leak of Ca^{2+} stores could obviously be envisioned to have other profound effects on cells. This response is not the same as the unfolded protein response, because the agents that induce the responses are not all the same. Dithiothreitol, for example, induces the unfolded protein response, but not the overload response, and cycloheximide induces the overload response, but not the unfolded protein response (Pahl and Baeuerle, 1997). How the Ca^{2+} leak is mediated is not known.

III. Aspects of the Neuroendocrine Secretory Pathway

A. Production of Large Amounts of Secretory Proteins to Be Stored in Secretory Granules

All cells have secretory pathways, but only certain cell types store some proteins in a concentrated form in secretory granules; cells with this ability include neuroendocrine cells, peptidergic neurons, exocrine cells, mast cells, and some blood cells (Fig. 1). The mechanism by which these proteins are concentrated remains an active area of debate, without complete agreement on the basic principles involved (Arvan and Castle, 1998; Dannies, 1999). It is clear, however, that production of proteins to be stored in granules must produce special challenges for the cells that produce them for at least two reasons. One is that the cells produce large quantities of a few or even only one protein, so that the concentration of individual proteins will be extremely high. Second, many proteins that are cargo for secretory granules aggregate very easily when present at high concentrations, even at neutral pH, in the presence of divalent cations (Gorr *et al.,* 1989; Chanat and Huttner, 1991; Thompson *et al.,* 1992; Freedman and Scheele, 1993; Linard *et al.,* 1995; Colomer *et al.,* 1996). Although formation of large aggregates of cargo proteins is what forms the cores of secretory granules, it is usually assumed that these aggregates do not form until the late stages of the Golgi complex. Since divalent cation concentrations are relatively high in the endoplasmic reticulum (up to 0.4 μM free Ca^{2+}) (Miyawaki *et al.,* 1997), a question that could be asked is what keeps cargo proteins from aggregating prematurely, rather than why they aggregate when they reach the *trans*-Golgi complex. Proteins that bind Ca^{2+} and other divalent cations in the endoplasmic reticulum must be important, but all the factors that prevent and cause aggregation of luminal proteins in neuroendocrine cells may not be understood.

B. Proteins in the Secretory Pathway That Are Neuroendocrine Cell Specific

There are proteins that are known to have chaperone-like activities, or that are possible candidates for such roles, and that are specific for, or enriched in, neuroendocrine cells. These are possible candidates for helping with the problems associated with producing large amounts of easily aggregating proteins, although at least some of them have other roles as well.

1. *7B2*

The neuroendocrine protein 7B2 is composed of two domains. One is a 21-kDa N-terminal region, required for maturation of an enzyme that processes hormones, prohormone convertase 2 (PC2), and the second is the C-terminal region that inhibits PC2 at nanomolar concentrations (Zhu *et al.,* 1996). The protein 7B2 is required for proper transport and maturation of PC2 (Braks and Martens, 1994; Zhu and Lindberg, 1995; Muller *et al.,* 1997). The mRNA for the protein is found in many neuroendocrine cell types (Propato-Mussafiri *et al.,* 1992; Marcinkiewicz *et al.,* 1993). Whether the sole function of this protein is to ensure correct transport and activation of PC2 is not clear.

2. *Chromogranins*

The granin family (secretogranins, chromogranins) are found in neuroendocrine cells and are acidic sulfated proteins. The major proteins in this family are chromogranin A, secretogranin I (or chromogranin B), and secretogranin II (Iacangelo and Eiden, 1995). These proteins are processed to proteolytic fragments, which are what is usually secreted from the cells, rather than the intact proteins. Although these proteins have been known for over 10 years, their roles are still debated. Some of the proteolytic fragments have biological activities, and they have also been proposed to play a role in causing aggregation of secretory granule cargo proteins in the appropriate place in the cells (Iacangelo and Eiden, 1995; Kromer *et al.,* 1998). They do not appear to have a universal role in aggregation, since some cells that produce prolactin or insulin store these hormones without producing granins (Thompson *et al.,* 1992; Sher *et al.,* 1995).

3. *RESP18 (Regulated Endocrine-Specific Protein 18 kDa)*

RESP18 is located in the endoplasmic reticulum of neuroendocrine cells, and rapidly degrades with a half-life of 19 min (Schiller *et al.,* 1995). Preventing transport from the endoplasmic reticulum by either lowering the temperature or uncoupling energy production in the mitochondria greatly extends the half-life of RESP18 (Schiller *et al.,* 1995). This protein differs from most proteins, like chaperones, that reside in the endoplasmic reticulum, because it is not stable, and the main cause of its localization appears to be control of stability.

Concentrations of RESP18 are regulatable; in AtT20 cells, its production is induced by hydrocortisone, which prolongs its half-life. This protein is part of an endoplasmic reticulum to cytosol communication pathway, because when it is expressed in AtT20 cells on a tet-inducible vector, several mRNAs are induced when RESP18 expression is induced (Schiller *et al.,* 1997). Since RESP18 is located in neuroendocrine cells, this kind of communication will be found only in these cells; whether it activates general communication systems, such as the unfolded protein response or the overload response of the endoplasmic reticulum, or represents an entirely new pathway has yet to be determined.

IV. Isolated Dominant-Negative Hormone Deficiency: Problems in the Secretory Pathway

Mutations causing isolated autosomal-dominant hormone deficiency have been found in three hormones: vasopressin, parathyroid hormone, and growth hormone (Arnold *et al.,* 1990; Cogan *et al.,* 1994; Ito *et al.,* 1997). In these deficiencies, production of the mutant protein interferes with that of the wild-type protein; mechanisms for how the mutants interfere are not obvious, because these proteins are relatively simple. The two reasons that have been previously proposed are formation of toxic protein aggregations or formation of dysfunctional or toxic hetero-oligomers.

A. Vasopressin

Arginine vasopressin (AVP) is produced as a precursor polypeptide with a signal sequence for transport into the endoplasmic reticulum plus a sequence that is eventually processed by enzymatic proteolytic cleavage into AVP, the N-terminal portion, neurophysin, and a glycoprotein C-terminal fragment. Neurophysin has seven disulfide bonds when fully formed and has a chaperone-like activity because it is necessary for proper transport of AVP along the secretory pathway. Ito and coworkers expressed the wild type and four mutants of the precursor in neuro2A neuroblastoma cells (Ito *et al.,* 1997). Two were single amino acid changes in neurophysin, G57S-AVP, and desE47-AVP. One was a mutation causing a truncation in the middle of neurophysin, C67X-AVP, and one, a single amino acid substitution right before the signal sequence, A(-1T)-AVP, causes inefficient cleavage of the signal sequence. Neuro2A cells treated with valproic acid differentiate and cease dividing. Clones that stably expressed each mutant AVP were not different from that expressing wild type until they were treated with valproic

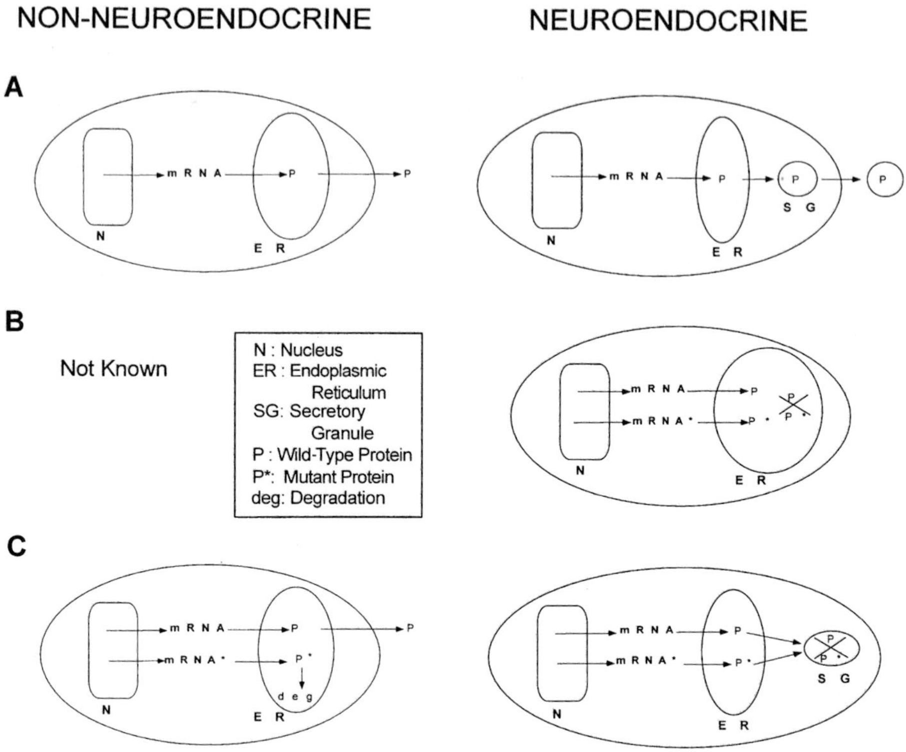

FIG. 2. Possible mechanisms for mutant hormones causing dominant-negative hormone deficiency. Mutant proteins may have more than one effect; it is likely from what is known so far about the different growth hormone mutants that effects will vary with different mutants. (A) Secretion in normal cells. Proteins are secreted directly from most cells, but select proteins are stored in secretory granules in neuroendocrine cells. (B) Mutant proteins accumulate in the endoplasmic reticulum of cells and suppress wild-type hormone production because the aggregates are toxic (Ito *et al.,* 1997); the mechanisms causing toxicity are not known. (C) Mutant proteins form dysfunctional aggregates in secretory granules, so the suppression is not manifest in non-neuroendocrine cells (Binder *et al.,* 1996). (D) Mutant proteins cause instability of wild-type hormone in neuroendocrine cells, either by inducing degradation systems more effectively in neuroendocrine cells, or competing for a chaperone important for folding that is limiting in neuroendocrine cells (Lee *et al.,* 1999). (E) Mutant proteins prevent wild-type hormone from being concentrated into secretory granules (Arrandale *et al.,* 1994). (F) Responses to the unfolded proteins in the secretory pathway reduce mRNA accumulation for secretory proteins (Sun *et al.,* 1997).

acid, in which case those expressing the mutants slowly died, while those expressing wild-type AVP lived. The mutants were secreted inefficiently and accumulated in the endoplasmic reticulum. These results led to the suggestion that accumulation of improperly folded proteins

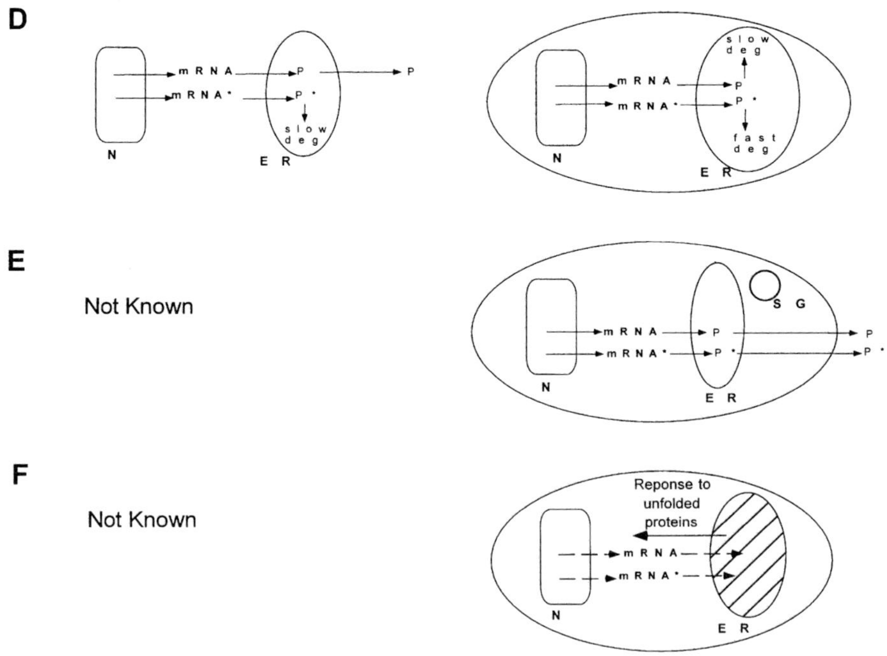

FIG. 2. (*continued*)

into aggregates must have toxic effects on the cells (Ito *et al.,* 1997), a suggestion that has also been made for other diseases of protein folding (Kim and Arvan, 1998) (Fig. 2). The mutants may not be able to be transported from the endoplasmic reticulum because they do not fold properly. It is also possible that if active transport is required then the mutants cannot be actively carried from the endoplasmic reticulum. Toxicity might occur preferentially in the differentiated cells, because they do not decrease concentrations of the precursor by dividing (Kim and Arvan, 1998), or the dividing cells may have more systems available, such as degradation systems, to handle the aggregates of unfolded protein that they lose when they differentiate. Deficiencies caused by retention of proteins in the endoplasmic reticulum, such as this one, have sometimes been referred to as *endoplasmic reticulum storage diseases,* although the term *storage* seems a misnomer in this context.

B. Growth Hormone

Human growth hormone is a monomeric protein with no covalent modifications, produced as a precursor of 217 amino acids. The only processing is enzymatic cleavage of the first 26 amino acids from the precursor, which removes the signal peptide necessary for entry into the endoplasmic reticulum (Baumann, 1991).

Familial isolated growth hormone deficiency type II (IGHDII) is inherited as autosomal dominant severe short stature. Because the structure of growth hormone is known (Ultsch *et al.,* 1994), it is possible to predict what mutations will affect structure. Several families with this disorder have mutations in the first or sixth base pair of the donor splice site of intron 3 of the growth hormone gene (*GH1*) (Phillips and Cogan, 1994; Cogan *et al.,* 1995). These mutations all result in missplicing of mRNA and loss of exon 3, so that growth hormone produced from this message lacks amino acids 32–71 (des32-71-GH) (Cogan *et al.,* 1995), which is the entire loop that connects helix 1 to helix 2 in the tertiary structure. Without this connection, des32-71-GH cannot fold normally.

Human growth hormone is a soluble protein that folds into its normal tertiary structure in solution (Brems *et al.,* 1990; Youngman *et al.,* 1995). The presence of a protein that cannot fold, such as des32-71-GH, would not necessarily be expected to interfere with the folding of wild-type hormone. Two examples of mutants that cannot fold, but do not interfere with wild-type growth hormone production, have been previously identified (Fig. 3). In one, the first 55 amino acids have the normal growth hormone sequences, but then a two-base-pair deletion results in a frameshift and therefore a different sequence (Igarashi *et al.,* 1993). In the second, an intron splice mutation results in an mRNA with a modified splice, which introduces a change in the reading frame after the first 102 amino acids (Cogan *et al.,* 1994, 1995). In each case, the changed amino acid sequence will alter the tertiary structure of the protein compared with wild-type hormone. Although these mutant proteins cannot fold normally, the mutations are recessive, and growth hormone is produced from the wild-type gene in heterozygotes in quantities sufficient to cause normal growth, indicating that the presence of a protein that does not fold is not sufficient to cause a dominant-negative effect. Although unfolded proteins may activate a kinase that down-regulates protein synthesis as a short-term effect in culture (Silverman and Williams, 1999), the recessive inheritance of these mutants suggests this effect is not lasting nor toxic to cells, if it occurs with all unfolded proteins.

Des32-71-GH is therefore unlikely to be dominant-negative simply because it cannot fold. Binder and coworkers (1996) used lymphocytes from heterozygotes with mutations causing des32-71-GH to investigate the cause further. Lymphocytes normally synthesize small amounts of growth hormone, but have no secretory granules in which to store the

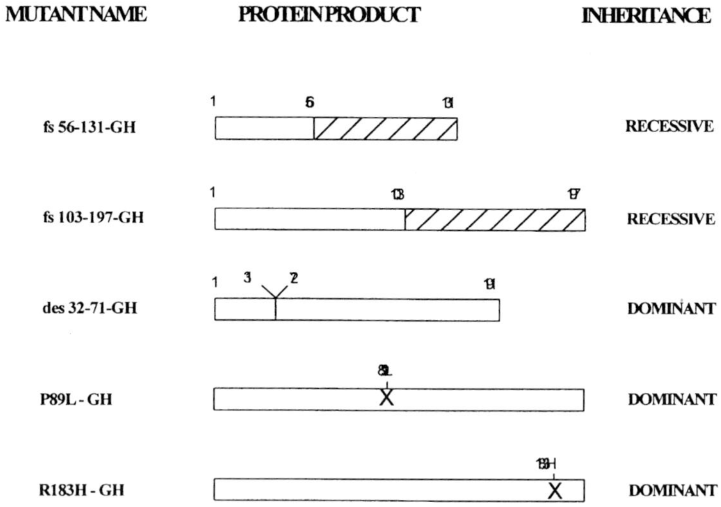

FIG. 3. The predicted protein products that result from characterized mutations in the human growth hormone gene, GH1. A deletion of two nucleotides results in fs56-131-GH, caused by a frameshift so that amino acids 56–131 are different than growth hormone, after which there is a termination codon (Igarashi *et al.,* 1993). A mutation in the donor splice site of intron IV results in fs103-197-GH, caused by missplicing, yielding a frameshift so that amino acids 103–197 are different than growth hormone, after which there is a termination codon (Cogan *et al.,* 1994). Several different splice site mutations in the donor splice site of intron III result in des32-71-GH, caused by missplicing to eliminate exon 3 (Cogan *et al.,* 1995). P89L-GH and R183-GH are single amino acid substitutions (Miyata *et al.,* 1997; Gertner *et al.,* 1998; Duquesnov *et al.,* 1998).

hormone; in heterozygotes with one *GH1* gene containing the mutation that results in des32-71-GH, there is no suppression of wild-type growth hormone (Binder *et al.,* 1996). These results are important because they indicate that there is tissue specificity for this effect. This finding, plus the presence of an unpaired sulfhydryl group in des32-71-GH, supported a model in which des32-71-GH is dominant negative because the mutant induces the formation of disulfide-bonded aggregates of mutant and wild-type hormone during storage on the secretory granules of somatotrophs (Fig. 2). In this model, there is no effect in lymphocytes because they do not store the hormone.

We transiently transfected pituitary and other cell lines with wild-type growth hormone, des32-71-GH, or half wild-type and half des32-71-GH, to resemble homozygotes and heterozygotes, and examined growth hormone production. The results were somewhat unexpected, because there was cell specificity in the response to this unfolded protein (Lee *et al.,* 1999). Des32-71-GH accumulated in COS and CHO cells and did not suppress production of wild-type growth hormone. Des32-71-GH did not accumulate in the neuroendocrine cell lines, GH_4C_1 and AtT20, but did suppress production of wild-type growth hormone. An important control is the cotransfection of a normally occurring deletion form of growth hormone, the 20-kDa growth hormone. The 20-kDa form results from use of an alternate splicing site in the wild-type gene and accounts for about one-fifth of the growth hormone secreted in normal subjects (Baumann, 1991). It is missing amino acids 32–46, which encompasses only part of the loop that connects helix 1 to helix 2, and is able to fold into almost the same conformation as wild type (Wada *et al.,* 1998). Unlike des32-71-GH, the 20-kDa form was easily detected after transfection into GH_4C_1 cells, and when coexpressed, did not decrease the stability of wild-type growth hormone (Lee *et al.,* 1999). We found no evidence for accumulation of disulfide-bonded aggregates in neuroendocrine cells because (1) there was no immunoreactivity in the loading gel and high-molecular-weight ranges of nonreducing SDS gels, (2) des32-71,C165A-GH, which lacks the unpaired cysteine that des32-71-GH has, also suppressed wild-type growth hormone secretion, and (3) we could not detect accumulation of des32-71-GH in neuroendocrine cells, but were able to detect it in COS and CHO cells. Des32-71-GH seems most likely to form disulfide-bonded aggregates when around in large amounts, but did not cause suppression in COS or CHO cells. We therefore conclude that there is a third possible mechanism for dominant-negative suppression of wild-type growth hormone in which the stability of wild-type growth hormone is affected (Fig. 2).

Possible mechanisms for decreasing instability of wild-type growth hormone include the following: (1) A direct interaction of wild-type growth hormone with des32–71-GH. The inability of des32-71-GH to suppress wild-type GH in COS cells, where the mutant accumulates, suggests that this mechanism does not operate. (2) Induction of proteasomes or some other degradation system to such an extent that proteins that fold as well as proteins that cannot fold are degraded. Such induction must occur to a much greater extent in neuroendocrine cells than in COS cells to explain the difference in the cell types. (3) Competition of des32-71-GH for a chaperone required for folding, and therefore stability of wild-type growth hormone. Such a chaperone must be present in limiting quantities in neuroendocrine cells.

Not all dominant-negative mutants have the same effect; two single amino acid mutations, P89L and R183H, are associated with autosomal dominant-negative isolated growth hormone deficiency (Miyata *et al.,*

1997; Duquesnov *et al.,* 1998; Gertner *et al.,* 1998), but have not been yet shown to have effects in culture. The mutation P89L changes a completely conserved proline in prolactin and growth hormone that forms a kink in the second α-helix (Ultsch *et al.,* 1994). Leucine does not form such a kink. The mutation R183H changes the highly conserved arginine to a more bulky histidine. Arginine 183 is adjacent to cysteine 182, which forms a disulfide bond, and the more bulky histidine may interfere with its formation. It is quite likely, therefore, that these two mutants cannot fold as easily as wild-type growth hormone, but how they exert their dominant effects is still unknown.

C. Prolactin

There are very few recognized cases of prolactin deficiency in people, and the cause of the deficiency is not known (Kauppila, 1994). Recognition may be limited because there are few signs of prolactin deficiency, unlike the phenotypes that make growth hormone and vasopressin deficiency easy to identify. We have, however, found effects of two forms of prolactin in culture that may be analogous to dominant-negative hormone deficiencies *in vivo,* in each case a suppression by human prolactin or its mutants of rat prolactin production. These suppressive effects may be a model for growth hormone mutants that do not directly suppress wild-type hormone production in culture.

We found that prolactin storage in rat pituitary GH_4C_1 cells was a specific inducible process; storage is the proportion of hormone retained in the cells relative to the total that they make, and in neuroendocrine cells, storage is in secretory granules. GH_4C_1 cells store little prolactin when grown in the absence of estradiol, but store more in its presence, especially with insulin and epidermal growth factor, and the number of secretory granules increases 50-fold (Scammell *et al.,* 1986). The intriguing fact about this induction is that it is specific for only one cargo protein, prolactin, because GH_4C_1 cells also make two others, growth hormone and secretogranin II, but their storage is not induced (Scammell *et al.,* 1986; Gorr, 1996). It was unexpected that only storage of prolactin increased when the cells contain so many more granules.

In clones transfected with human prolactin, hormone treatment induces neither rat nor human prolactin storage (Arrandale and Dannies, 1994). Genetic proof that the effect is caused by transfection of human prolactin is that two single amino acid mutations each ablate the ability of human prolactin to block rat prolactin; the mutants are N31T-prolactin and S34A-prolactin. Human prolactin therefore has a dominant-negative effect on rat prolactin storage (Fig. 2). If aggregation of prolactin into the dense cores of secretory granules is an actively controlled and specific process, then human prolactin could prevent rat prolactin storage by interfering with the process. Alternatively, human prolactin could interfere with the ability of membrane proteins essential for proper secretory granule retention and release to recognize the proper location as prolactin-containing secretory granules form, with the result that prolactin-containing granules are not retained (Dannies, 1999).

A separate kind of effect was found in clones transfected with human H27A-prolactin (Sun *et al.,* 1997). This mutant was not efficiently secreted, and 20–40% of the newly synthesized hormone was degraded. Rat prolactin production in the clones expressing human H27A-prolactin was also reduced, caused by a reduction in rat prolactin mRNA levels. Reduction of rat prolactin production is not mediated by the rat prolactin receptor, because mRNA sequences for the prolactin receptor were not detected in GH_4C_1 cells by RT-PCR. These results suggest that the instability of the mutant protein may trigger effects that suppress rat prolactin production directly, or that indirectly result in selection of clones with low rat prolactin production (Fig. 2). Either effect may be an as-yet-unrecognized response of the cells to specific unfolded proteins.

V. Conclusions

The growth hormone mutations causing growth hormone deficiency indicate that all unfolded proteins do not elicit the same response since some are dominant and some are recessive. Although the accumulation of toxic aggregates may account for some autosomal dominant hormone deficiencies, other causes are possible, including responses in the cells induced by some unfolded proteins. A long-term leak of Ca^{2+} from the endoplasmic reticulum because of an overload response (Pahl and Baeuerle, 1997), for example, would be predicted ultimately to damage mitochondria and even cause apoptosis (Berridge *et al.,* 1998). Further understanding of the responses that cells have to hormones that do not fold and what features trigger which responses will increase knowledge not only of hormone deficiencies, but also other diseases of protein folding, including amyloid and Alzheimer's diseases.

Acknowledgments

Supported in part by a grant from the American Diabetes Association and NIH grant DK-46807.

REFERENCES

Anfinsen, C. B. (1973). Principles that govern the folding of protein chains. *Science* **181,** 223–230.

Arber, S., Krause, K. H., and Caroni, P. (1992). s-cyclophilin is retained intracellularly via a unique COOH-terminal sequence and colocalizes with the calcium storage protein calreticulin. *J. Cell Biol.* **116,** 113–125.

Arnold, A., Horst, S. A., Gardella, T. J., Baba, H., Levine, M. A., and Kronenberg, H. M. (1990). Mutation of the signal peptide-encoding region of the preproparathyroid hormone gene in familial isolated hypoparathyroidism. *J. Clin. Invest.* **86,** 1084–1087.

Arrandale, J. M., and Dannies, P. S. (1994). Inhibition of rat prolactin (PRL) storage by coexpression of human PRL. *Mol. Endocrinol.* **8,** 1083–1090.

Arvan, P., and Castle, D. (1998). Sorting and storage during secretory granule biogenesis: Looking backward and looking forward. *Biochem. J.* **332,** 593–610.

Baker, E. K., Colley, N. J., and Zuker, C. S. (1994). The cyclophilin homolog NinaA functions as a chaperone, forming a stable complex in vivo with its protein target rhodopsin. *EMBO J.* **13,** 4886–4895.

Balch, W. E., McCaffery, J. M., Plutner, H., and Farquhar, M. G. (1994). Vesicular stomatitis virus glycoprotein is sorted and concentrated during export from the endoplasmic reticulum. *Cell (Cambridge, Mass.)* **76,** 841–852.

Bannykh, S. I., and Balch, W. E. (1997). Membrane dynamics at the endoplasmic reticulum-Golgi interface. *J. Cell Biol.* **138,** 1–4.

Bass, J., Chiu, G., Argon, Y., and Steiner, D. F. (1998). Folding of insulin receptor monomers is facilitated by the molecular chaperones calnexin and calreticulin and impaired by rapid dimerization. *J. Cell Biol.* **141,** 637–646.

Baumann, G. (1991). Growth hormone heterogeneity: Genes, isohormones, variants, and binding proteins. *Endocr. Rev.* **12,** 424–449.

Bedows, E., Norton, S. E., Huth, J. R., Suganuma, N., Boime, I., and Ruddon, R. W. (1994). Misfolded human chorionic gonadotropin beta subunits are secreted from transfected Chinese hamster ovary cells. *J. Biol. Chem.* **269,** 10574–10580.

Berridge, M. J., Bootman, M. D., and Lipp, P. (1998). Calcium—a life and death signal. *Nature (London)* **395,** 645–648.

Binder, G., Brown, M., and Parks, J. S. (1996). Mechanisms responsible for dominant expression of human growth hormone gene mutations. *J. Clin. Endocrinol. Metab.* **81,** 4047–4050.

Braks, J. A., and Martens, G. J. (1994). 7B2 is a neuroendocrine chaperone that transiently interacts with prohormone convertase PC2 in the secretory pathway. *Cell (Cambridge, Mass.)* **78,** 263–273.

Brems, D. N., Brown, P. L., and Becker, G. W. (1990). Equilibrium denaturation of human growth hormone and its cysteine-modified forms. *J. Biol. Chem.* **265,** 5504–5511.

Cai, H., Wang, C. C., and Tsou, C. L. (1994). Chaperone-like activity of protein disulfide isomerase in the refolding of a protein with no disulfide bonds. *J. Biol. Chem.* **269,** 24550–24552.

Chanat, E., and Huttner, W. B. (1991). Milieu-induced, selective aggregation of regulated secretory proteins in the trans-Golgi network. *J. Cell. Biol.* **115,** 1505–1519.

Cogan, J. D., Phillips, J. A., 3rd, Schenkman, S. S., Milner, R. D., and Sakati, N. (1994). Familial growth hormone deficiency: A model of dominant and recessive mutations affecting a monomeric protein. *J. Clin. Endocrinol. Metab.* **79,** 1261–1265.

Cogan, J. D., Ramel, B., Lehto, M., Phillips, J., 3rd, Prince, M., Blizzard, R. M., de Ravel, T. J., Brammert, M., and Groop, L. (1995). A recurring dominant negative mutation causes autosomal dominant growth hormone deficiency—a clinical research center study. *J. Clin. Endocrinol. Metab.* **80,** 3591–3595.

Colley, N. J., Baker, E. K., Stamnes, M. A., and Zuker, C. S. (1991). The cyclophilin homolog ninaA is required in the secretory pathway. *Cell (Cambridge, Mass.)* **67,** 255–263.

Colomer, V., Kicska, G. A., and Rindler, M. J. (1996). Secretory granule content proteins and the luminal domains of granule membrane proteins aggregate in vitro at mildly acidic pH. *J. Biol. Chem.* **271,** 48–55.

Dannies, P. S. (1999). Protein hormone storage in secretory granules: Mechanisms for concentration and sorting. *Endocr. Rev.* **20,** 3–21.

Duquesnov, P., Simon, D., Netchine, I., Dastot, F., Sobrier, M. L., Goossens, M., Czernichow, P., and Amselem, S. (1998). Familial isolated growth hormone deficiency with slight height reduction due to a heterozygote mutation in GH gene. *80th Annu. Meet., Endocr. Soc.,* Abstr. P2-202.

Feng, W., Bedows, E., Norton, S. E., and Ruddon, R. W. (1996). Novel covalent chaperone complexes associated with human chorionic gonadotropin beta subunit folding intermediates. *J. Biol. Chem.* **271,** 18543–18548.

Freedman, S. D., and Scheele, G. A. (1993). Regulated secretory proteins in the exocrine pancreas aggregate under conditions that mimic the trans-Golgi network. *Biochem. Biophys. Res. Commun.* **197,** 992–999.

Frydman, J., and Hartl, F. U. (1996). Principles of chaperone-assisted protein folding: Differences between in vitro and in vivo mechanisms. *Science* **272,** 1497–1502.

Gaynor, E. C., and Emr, S. D. (1997). COPI-independent anterograde transport: Cargo-selective ER to Golgi protein transport in yeast COPI mutants. *J. Cell Biol.* **136,** 789–802.

Gertner, J. M., Wajnrajch, M. P., and Leibel, R. L. (1998). Genetic defects in the control of growth hormone secretion. *Horm. Res.* **49** (Suppl. 1), 9–14.

Gillikin, J. W., Fontes, E. P., and Boston, R. S. (1995). Protein—protein interactions within the endoplasmic reticulum. *Methods Cell Biol.* **50,** 309–323.

Gorr, S. U. (1996). Differential storage of prolactin, granins (Chromogranin B and secretogranin II), and constitutive secretory markers in rat pituitary GH_4C_1 cells. *J. Biol. Chem.* **271,** 3575–3580.

Gorr, S. U., Shioi, J., and Cohn, D. V. (1989). Interaction of calcium with porcine adrenal chromogranin A (secretory protein-I) and chromogranin B (secretogranin I). *Am. J. Physiol.* **257**(2, Pt. 1), E247–E254.

Hammond, C., and Helenius, A. (1994). Quality control in the secretory pathway: Retention of a misfolded viral membrane glycoprotein involves cycling between the ER, intermediate compartment, and Golgi apparatus. *J. Cell Biol.* **126,** 41–52.

Harding, H. P., Zhang, Y., and Ron, D. (1999). Protein translation and folding are coupled by an endoplasmic-reticulum-resident kinase. *Nature (London)* **397,** 271–274.

Hayano, T., Hirose, M., and Kikuchi, M. (1995). Protein disulfide isomerase mutant lacking its isomerase activity accelerates protein folding in the cell. *FEBS Lett.* **377,** 505–511.

Haynes, R. L., Zheng, T., and Nicchitta, C. V. (1997). Structure and folding of nascent polypeptide chains during protein translocation in the endoplasmic reticulum. *J. Biol. Chem.* **272,** 17126–17133.

Hellman, R., Vanhove, M., Lejeune, A., Stevens, F. J., and Hendershot, L. M. (1999). The in vivo association of BiP with newly synthesized proteins is dependent on the rate and stability of folding and not simply on the presence of sequences that can bind to BiP. *J. Cell Biol.* **144,** 21–30.

Hendershot, L. M., Wei, J. Y., Gaut, J. R., Lawson, B., Freiden, P. J., and Murti, K. G. (1995). In vivo expression of mammalian BiP ATPase mutants causes disruption of the endoplasmic reticulum. *Mol. Biol. Cell* **6,** 283–296.

Iacangelo, A. L., and Eiden, L. E. (1995). Chromogranin A: Current status as a precursor for bioactive peptides and a granulogenic/sorting factor in the regulated secretory pathway. *Regul. Pept.* **58,** 65–88.

Igarashi, Y., Ogawa, M., Kamijo, T., Iwatani, N., Nishi, Y., Kohno, H., Masumura, T., and Koga, J. (1993). A new mutation causing inherited growth hormone deficiency: A compound heterozygote of a 6.7 kb deletion and a two base deletion in the third exon of the GH-1 gene. *Hum. Mol. Genet.* **2,** 1073–1074.

Ito, M., Jameson, J. L., and Ito, M. (1997). Molecular basis of autosomal dominant neurohypophyseal diabetes insipidus. Cellular toxicity caused by the accumulation of mutant vasopressin precursors within the endoplasmic reticulum. *J. Clin. Invest.* **99,** 1897–1905.

Kappeler, F., Klopfenstein, D. R., Foguet, M., Paccaud, J. P., and Hauri, H. P. (1997). The recycling of ERGIC-53 in the early secretory pathway. ERGIC-53 carries a cytosolic endoplasmic reticulum-exit determinant interacting with COPII. *J. Biol. Chem.* **272,** 31801–31808.

Kauppila, A. (1994). Isolated prolactin deficiency. *Curr. Ther. Endocrinol. Metab.* **5,** 29–31.

Kim, P. S., and Arvan, P. (1995). Calnexin and BiP act as sequential molecular chaperones during thyroglobulin folding in the endoplasmic reticulum. *J. Cell Biol.* **128,** 29–38.

Kim, P. S., and Arvan, P. (1998). Endocrinopathies in the family of endoplasmic reticulum (ER) storage diseases: Disorders of protein trafficking and the role of ER molecular chaperones. *Endocr. Rev.* **19,** 173–202.

Klappa, P., Freedman, R. B., and Zimmermann, R. (1995). Protein disulphide isomerase and a lumenal cyclophilin-type peptidyl prolyl cis-trans isomerase are in transient contact with secretory proteins during late stages of translocation. *Eur. J. Biochem.* **232,** 755–764.

Knarr, G., Gething, M. J., Modrow, S., and Buchner, J. (1995). BiP binding sequences in antibodies. *J. Biol. Chem.* **270,** 27589–27594.

Kromer, A., Glombik, M. M., Huttner, W. B., and Gerdes, H. H. (1998). Essential role of the disulfide-bonded loop of chromogranin B for sorting to secretory granules is revealed by expression of a deletion mutant in the absence of endogenous granin synthesis. *J. Cell Biol.* **140,** 1331–1346.

Kuznetsov, G., and Nigam, S. K. (1998). Folding of secretory and membrane proteins. *N. Engl. J. Med.* **339,** 1688–1695.

Lee, M. S., Wajnrajch, M. P., Kin, S. S., Plotnick, L. P., Gertner, J. M., Leibel, R. L., and Dannies, P. (1999). Autosomal-dominant growth hormone deficiency type II: The des32-71-growth hormone deletion mutant suppresses secretion of wild-type growth hormone. Manuscript in revision.

Linard, C. G., Tadros, H., Sirois, F., and Mbikay, M. (1995). Calcium-induced aggregation of neuroendocrine protein 7B2 in vitro and its modulation by ATP. *Mol. Cell. Biochem.* **151,** 39–47.

Lundstrom-Ljung, J., Birnbach, U., Rupp, K., Soling, H. D., and Holmgren, A. (1995). Two resident ER-proteins, CaBP1 and CaBP2, with thioredoxin domains, are substrates for thioredoxin reductase: Comparison with protein disulfide isomerase. *FEBS Lett.* **357,** 305–308.

Marcinkiewicz, M., Touraine, P., Mbikay, M., and Chrétien, M. (1993). Expression of neuroendocrine secretory protein 7B2 mRNA in the mouse and rat pituitary gland. *Neuroendocrinology* **58,** 86–93.

McLatchie, L. M., Fraser, N. J., Main, M. J., Wise, A., Brown, J., Thompson, N., Solari, R., Lee, M. G., and Foord, S. M. (1998). RAMPs regulate the transport and ligand specificity of the calcitonin-receptor-like receptor. *Nature (London)* **393,** 333–339.

Melnick, J., Dul, J. L., and Argon, Y. (1994). Sequential interaction of the chaperones BiP and GRP94 with immunoglobulin chains in the endoplasmic reticulum. *Nature (London)* **370,** 373–375.

Miyata, I., Cogan, J. D., Prince, M. A., Kamijo, T., Ogawa, M., and Phillips, J. A., 3rd (1997). Detection of growth hormone gene defects by dideoxy fingerprinting (ddF). *Endocrinol. Jpn.* **44,** 149–154.

Miyawaki, A., Llopis, J., Heim, R., McCaffery, J. M., Adams, J. A., Ikura, M., and Tsien, R. Y. (1997). Fluorescent indicators for Ca^{2+} based on green fluorescent proteins and calmodulin. *Nature (London)* **388,** 882–887.

Mizuno, M., and Singer, S. J. (1993). A soluble secretory protein is first concentrated in the endoplasmic reticulum before transfer to the Golgi apparatus. *Proc. Natl. Acad. Sci. U.S.A.* **90,** 5732–5736.

Muller, L., Zhu, X., and Lindberg, I. (1997). Mechanism of the facilitation of PC2 maturation by 7B2: Involvement in ProPC2 transport and activation but not folding. *J. Cell Biol.* **139,** 625–638.

Nichols, W. C., Seligsohn, U., Zivelin, A., Terry, V. H., Hertel, C. E., Wheatley, M. A., Moussalli, M. J., Hauri, H. P., Ciavarella, N., Kaufman, R. J., and Ginsburg, D. (1998). Mutations in the ER-Golgi intermediate compartment protein EFGIC-53 camuse combined deficiency of coagulation factors V and VIII. *Cell (Cambridge, Mass.)* **93,** 61–70.

Nilsson, T., and Warren, G. (1994). Retention and retrieval in the endoplasmic reticulum and the Golgi apparatus. *Curr. Opin. Cell Biol.* **6,** 517–521.

Otsu, M., Urade, R., Kito, M., Omura, F., and Kikuchi, M. (1995). A possible role of ER-60 protease in the degradation of misfolded proteins in the endoplasmic reticulum. *J. Biol. Chem.* **270,** 14958–14961.

Pahl, H. L., and Baeuerle, P. A. (1996). Activation of NF-kappa B by ER stress requires both Ca^{2+} and reactive oxygen intermediates as messengers. *FEBS Lett.* **392,** 129–136.

Pahl, H. L., and Baeuerle, P. A. (1997). The ER-overload response: activation of NF-kappa B. *Trends Biochem. Sci.* **22,** 63–67.

Palmer, A., Rivett, A. J., Thomson, S., Hendil, K. B., Butcher, G. W., Fuertes, G., and Knecht, E. (1996). Subpopulations of proteasomes in rat liver nuclei, microsomes and cytosol. *Biochem. J.* **316**(1, Pt. 2), 401–407.

Pelham, H. R. (1989). Control of protein exit from the endoplasmic reticulum. *Annu. Rev. Cell Biol.* **5,** 1–23.

Pelham, H. R. (1995). Sorting and retrieval between the endoplasmic reticulum and Golgi apparatus. *Curr. Opin. Cell Biol.* **7,** 530–535.

Phillips, J. A., 3rd, and Cogan, J. D. (1994). Genetic basis of endocrine disease. 6. Molecular basis of familial human growth hormone deficiency. *J. Clin. Endocrinol. Metab.* **78,** 11–16.

Presley, J. F., Cole, N. B., Schroer, T. A., Hirschberg, K., Zaal, K. J., and Lippincott-Schwartz, J. (1997). ER-to-Golgi transport visualized in living cells. *Nature (London)* **389,** 81–85.

Propato-Mussafiri, R., Kanse, S. M., Ghatei, M. A., and Bloom, S. R. (1992). Pituitary adenylate cyclase-activating polypeptide releases 7B2, adrenocorticotrophin, growth hormone and prolactin from the mouse and rat clonal pituitary cell lines AtT-20 and GH3. *J. Endocrinol.* **132,** 107–113.

Puig, A., and Gilbert, H. F. (1994a). Protein disulfide isomerase exhibits chaperone and anti-chaperone activity in the oxidative refolding of lysozyme. *J. Biol. Chem.* **269,** 7764–7771.

Puig, A., and Gilbert, H. F. (1994b). Anti-chaperone behavior of BiP during the protein disulfide isomerase-catalyzed refolding of reduced denatured lysozyme. *J. Biol. Chem.* **269,** 25889–25896.

Puig, A., Lyles, M. M., Noiva, R., and Gilbert, H. F. (1994). The role of the thiol/disulfide centers and peptide binding site in the chaperone and anti-chaperone activities of protein disulfide isomerase. *J. Biol. Chem.* **269,** 19128–19135.

Quan, H., Fan, G., and Wang, C. C. (1995). Independence of the chaperone activity of protein disulfide isomerase from its thioredoxin-like active site. *J. Biol. Chem.* **270,** 17078–17080.

Rivett, A. J. (1998). Intracellular distribution of proteasomes. *Curr. Opin. Immunol.* **10,** 110–114.

Rozell, T. G., Davis, D. P., Chai, Y., and Segaloff, D. L. (1998). Association of gonadotropin receptor precursors with the protein folding chaperone calnexin. *Endocrinology (Baltimore)* **139,** 1588–1593.

Ruddon, R. W., and Bedows, E. (1997). Assisted protein folding. *J. Biol. Chem.* **272,** 3125–3128.

Scammell, J. G., Burrage, T. G., and Dannies, P. S. (1986). Hormonal induction of secretory granules in a pituitary tumor cell line. *Endocrinology (Baltimore)* **119,** 1543–1548.

Schiller, M. R., Mains, R. E., and Eipper, B. A. (1995). A neuroendocrine-specific protein localized to the endoplasmic reticulum by distal degradation. *J. Biol. Chem.* **270,** 26129–26138.

Schiller, M. R., Mains, R. E., and Eipper, B. A. (1997). A novel neuroendocrine intracellular signaling pathway. *Mol. Endocrinol.* **11,** 1846–1857.

Schimmoller, F., Singer-Kruger, B., Schroder, S., Kruger, U., Barlowe, C., and Riezman, H. (1995). The absence of Emp24p, a component of ER-derived COPII-coated vesicles, causes a defect in transport of selected proteins to the Golgi. *EMBO J.* **14,** 1329–1339.

Schmid, S. L. (1997). Clathrin-coated vesicle formation and protein sorting: An integrated process. *Annu. Rev. Biochem.* **66,** 511–548.

Shamu, C. E., Cox, J. S., and Walter, P. (1994). The unfolded-protein-response pathway in yeast. *Trends Cell Biol.* **4,** 56–60.

Sher, E., Rosa, P., Bassetti, M., and Zanini, A. (1995). Immunolocalization of secretogranin II and insulin in a nerve growth factor-differentiated insulinoma cell line. *Eur. J. Cell Biol.* **67,** 15–22.

Sidrauski, C., Chapman, R., and Walter, P. (1998). The unfolded protein response: An intracellular signalling pathway with many surprising features. *Trends Cell Biol.* **8,** 245–249.

Silverman, R. H., and Williams, B. R. G. (1999). Translational control perks up. *Nature (London)* **397,** 209–211.

Smith, T., Ferreira, L. R., Hebert, C., Norris, K., and Sauk, J. J. (1995). Hsp47 and cyclophilin B traverse the endoplasmic reticulum with procollagen into pre-Golgi intermediate vesicles. A role for Hsp47 and cyclophilin B in the export of procollagen from the endoplasmic reticulum. *J. Biol. Chem.* **270,** 18323–18328.

Sommer, T., and Wolf, D. H. (1997). Endoplasmic reticulum degradation: Reverse protein flow of no return. *FASEB J.* **11,** 1227–1233.

Sun, Z., Lee, M. S., Rhee, H. K., Arrandale, J. M., and Dannies, P. S. (1997). Inefficient secretion of human H27A-prolactin, a mutant that does not bind Zn2+. *Mol. Endocrinol.* **11,** 1544–1551.

Thompson, M. E., Zimmer, W. E., Haynes, A. L., Valentine, D. L., Forss-Petter, S., and Scammell, J. G. (1992). Prolactin granulogenesis is associated with increased secretogranin expression and aggregation in the Golgi apparatus of GH_4C_1 cells. *Endocrinology (Baltimore)* **131,** 318–326.

Tirasophon, W., Welihinda, A. A., and Kaufman, R. J. (1998). A stress response pathway from the endoplasmic reticulum to the nucleus requires a novel bifunctional protein kinase/endoribonuclease (Ire1p) in mammalian cells. *Genes Dev.* **12,** 1812–1824.

Ultsch, M. H., Somers, W., Kossiakoff, A. A., and de Vos, A. M. (1994). The crystal structure of affinity-matured human growth hormone at 2 A resolution. *J. Mol. Biol.* **236,** 286–299.

Urade, R., Takenaka, Y., and Kito, M. (1993). Protein degradation by ERp72 from rat and mouse liver endoplasmic reticulum. *J. Biol. Chem.* **268,** 22004–22009.

Wada, M., Uchida, H., Ikeda, M., Tsunekawa, B., Naito, N., Banba, S., Tanaka, E., Hashimoto, Y., and Honjo, M. (1998). The 20-kilodalton (kDa) human growth hormone (hGH) differs from the 22-kDa hGH in the complex formation with cell surface hGH receptor and hGH-binding protein circulating in human plasma. *Mol. Endocrinol.* **12,** 146–156.

Werner, E. D., Brodsky, J. L., and McCracken, A. A. (1996). Proteasome-dependent endoplasmic reticulum-associated protein degradation: An unconventional route to a familiar fate. *Proc. Natl. Acad. Sci. U.S.A.* **93,** 13797–13801.

Youngman, K. M., Spencer, D. B., Brems, D. N., and DeFelippis, M. R. (1995). Kinetic analysis of the folding of human growth hormone. Influence of disulfide bonds. *J. Biol. Chem.* **270,** 19816–19822.

Zhu, X., and Lindberg, I. (1995). 7B2 facilitates the maturation of proPC2 in neuroendocrine cells and is required for the expression of enzymatic activity. *J. Cell Biol.* **129,** 1641–1650.

Zhu, X., Rouille, Y., Lamango, N. S., Steiner, D. F., and Lindberg, I. (1996). Internal cleavage of the inhibitory 7B2 carboxyl-terminal peptide by PC2: A potential mechanism for its inactivation. *Proc. Natl. Acad. Sci. U.S.A.* **93,** 4919–4924.

Structural Aspects of the G-Protein Receptor, Rhodopsin

ARLENE D. ALBERT AND PHILIP L. YEAGLE

Department of Molecular and Cell Biology, University of Connecticut, Storrs, Connecticut 06269

I. Introduction

The photoreceptor protein, rhodopsin, is a member of a large family of receptors that utilizes heterotrimeric GTP-binding proteins to transduce extracellular signals to intracellular events. Rhodopsin is one of the most extensively investigated members of this family because for many years it was the only G-protein receptor that could be isolated in quantities necessary for biochemical and biophysical studies. Rhodopsin is therefore of interest both because of its status as the archetype of G-protein receptors and because of the specific central role it plays in visual signal transduction.

Rhodopsin is imbedded in the disk membranes of rod photoreceptor cells. These disks are stacked along the length of the outer segment of the rod cell and are surrounded by the plasma membrane. The plasma membrane of the rod photoreceptor cell becomes hyperpolarized in response to exposure to light due to closure of Na^+ channels. To achieve this, the absorption of light by rhodopsin in the disks must be communicated to the plasma membrane.

The role of rhodopsin in visual signal transduction is to initiate the cascade of events that culminates in the hyperpolarization of the plasma membrane of the rod photoreceptor cell. On exposure to light, the rhodopsin chromophore, 11-*cis* retinal, undergoes a photoisomerization

0083-6729/00 $30.00

to the all-*trans* form, which triggers a conformational change in the rhodopsin structure. This conformational change activates the receptor and allows the binding and subsequent activation of the G protein, transducin. Transducin in turn activates a phosphodiesterase that hydrolyzes cGMP. Because it is cGMP that maintains Na^+ channels in the plasma membrane in the open position, these channels then close and the plasma membrane becomes hyperpolarized. This hyperpolarization triggers the neural response, which is interpreted as a visual signal in the brain.

At the beginning and end of the rhodopsin photocycle are two interconvertible conformations: the unbleached and bleached states. In the unbleached state 11-*cis* retinal is bound via a protonated Schiff base linkage to lysine 296 located on the seventh transmembrane helix. On exposure to light, the 11-*cis* retinal is converted to the all-*trans* form and eventually exits the binding pocket. The presence of the chromophore in the protein confers strong absorption maxima. Early studies on rhodopsin focused on these spectral properties. In the unbleached state rhodopsin absorbs at 498 nm. On exposure to light, several well-defined spectral intermediates have been identified using cryospectroscopy. Within picoseconds of the absorption of light the chromophore isomerizes to the all-*trans* form. Within milliseconds of the formation of these early intermediates, an equilibrium between Metarhodopsin I (λ_{max} = 480 nm) and Metarhodopsin II (λ_{max} = 380 nm) is established (Mathews *et al.,* 1963). The Schiff base becomes deprotonated on formation of Metarhodopsin II. Subsequently there is conversion to Metarhodopsin III (λ_{max} = 455) and then to the fully bleached state, opsin, which no longer has a bound chromophore. These spectral studies established that the protein undergoes a series of well-defined transitions on activation by light.

Although there are several spectrally identifiable earlier intermediates, it is the equilibrium between Metarhodopsin I and Metarhodopsin II that is of primary biological interest. This equilibrium occurs within milliseconds of light absorption (Mathews *et al.,* 1963). Metarhodopsin II has been identified as the activated state referred to as R* (Emeis *et al.,* 1982; Kibelbek *et al.,* 1991). This conformation of the receptor corresponds to the agonist bound form of other G-protein receptors. R* initiates the transduction cascade by binding the G protein, transducin. Upon activation, R* also becomes a substrate for rhodopsin kinase and for arrestin, both of which function to deactivate the receptor.

The spectral studies identified important spectroscopic and conformational intermediates. However the three-dimensional structure of these forms of rhodopsin and the conformational changes experienced by rhodopsin were unknown. Progress toward the three-dimensional structure occurred in the 1970s when circular dichroism (CD) studies of purified rhodopsin revealed that the structure of the protein was dominated by α-helix. These studies further suggested that the structure of the protein likely was built on a bundle of seven transmembrane α-helices (Albert and Litman, 1978).

The next major advance in understanding of rhodopsin structure was the reports from two laboratories that independently determined the amino acid sequence for bovine rhodopsin (Hargrave *et al.,* 1983; Ovchinnikov, 1982). This sequence contained indirect confirmation of the α-helical structure in this protein. Seven regions of hydrophobic amino acid sequences sufficiently long to traverse the membrane in the form of an α-helix were identified. Upon analysis of these helices it was observed that each of these hydrophobic helices was more hydrophobic on one side than the other. This was consistent with bundling of helices (Hargrave *et al.,* 1984). Although this analysis was not in itself sufficient to prove that the protein existed as a seven transmembrane helical protein, the seven transmembrane helical motif was widely identified as a fundamental architectural feature of rhodopsin (and other G-protein receptors) structure. Figure 1 shows a schematic representation of the primary structure of rhodopsin, organized in a format to show the hydrophobic stretches that putatively formed transmembrane helices.

That much of the mass of rhodopsin is buried in the hydrophobic core of the disk membrane bilayer was known early in the structural analysis of this protein through proteolytic studies. Rhodopsin is subject to only limited proteolysis by a variety of relatively nonspecific proteases (Saari, 1974; Trayburn *et al.,* 1974a,b; Pober, 1975; Sale *et al.,* 1977). Because the proteolytic fragments remained largely associated with the membrane, these studies suggested that the mass of the protein was largely buried in the membrane bilayer. The regions we now know to be transmembrane were strictly insensitive to proteolysis.

Proteolytic studies also demonstrated that the spectral properties of the rhodopsin were unaffected by cleavage (Albert and Litman, 1978; Trayburn *et al.,* 1974a). These spectral studies included both absorption and CD spectra in the UV as well as visible regions. Furthermore, the proteolyzed rhodopsin could be purified in detergents as a complex consisting of two to three fragments. This complex also retained the spectral properties of rhodopsin. Interestingly, these proteolytic fragments of rhodopsin could be separated on bleaching (Pober, 1975). However, physical separation of the fragments did not cause loss in secondary structure. This demonstrated that the protein existed in independently stable structural domains and suggested that domains

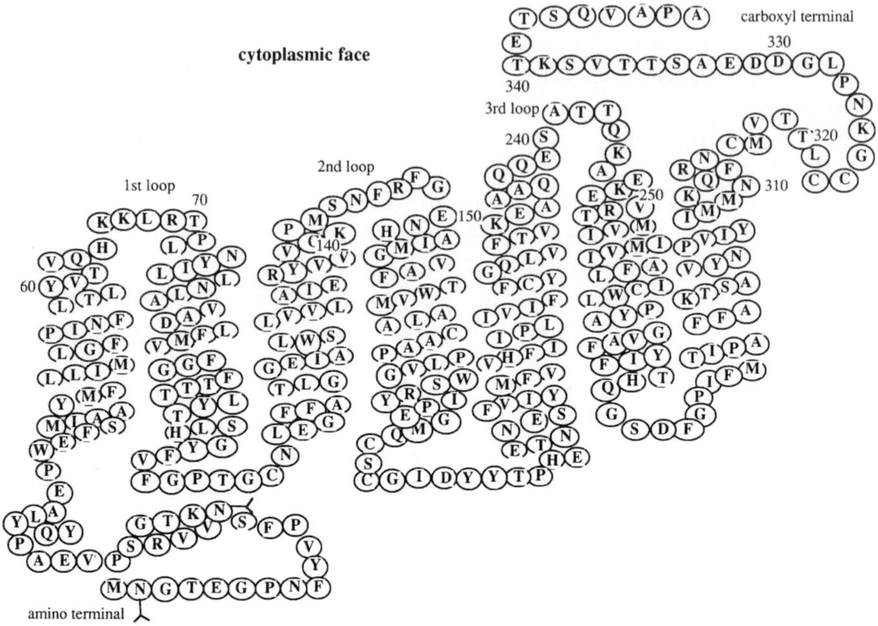

FIG. 1. Schematic representation of the primary structure of bovine rhodopsin. The helical portions are indicated and reflect the most recent data on the location of the ends of these helices.

of the protein moved relative to each other on bleaching (Albert and Litman, 1978). Because the chromophore was buried in the protein hydrophobic core it also suggested that the conformational change was transmitted to the transducin-binding surface via movement of the structural domains.

In addition to physical studies on rhodopsin, information on domains of the protein that are structurally and functionally important can be inferred by comparing opsins across species. Furthermore, many rhodopsin mutations are available that are either naturally occurring or have been experimentally created. Hargrave and coworkers have used this approach to investigate the structure of rhodopsin. Their analysis concluded that the intradiskal surface of rhodopsin (which corresponds to the extracellular surface of other G-protein receptors) plays an important role in the structural stability of the protein. Although this surface is not involved in any direct interactions with transducin, rhodopsin kinase, or arrestin, the structure of this region is important for stabilization of the receptor both in the activated and ground states (Hargrave *et al.*, 1998).

Rhodopsin does not exist in one simple biologically relevant structure. The unbleached (rhodopsin), activated (R*) and bleached (opsin) forms all represent important conformational states of the protein. Therefore, structural studies must take this into account. Pressure-dependent studies have implicated an increase in volume occupied by the protein as it is activated to R* (Meta II) state (Attwood and Gutfreund, 1980; Lamola *et al.,* 1974). Furthermore, FTIR measurements have suggested that R* represents the most expanded form of the protein (DeGrip *et al.,* 1988). These data imply that the protein expands on activation then collapses. It is likely that secondary structural elements remain intact, but that structural domains of the protein change their orientation relative to each other as rhodopsin undergoes its bleaching cycle.

Several studies have shown that the relationship of specifically labeled amino acid residues changes on activation of rhodopsin. By substituting histidines into the cytoplasmic ends of helices 3 and 6, metal binding sites were created. In the presence of metal ions, rhodopsin with these substitutions could not activate the G protein (Sheikh *et al.,* 1996). In other investigations it was shown that glutamine in helix 3 exhibits a change in hydrogen bonding strength (Fahmy *et al.,* 1993) and that the tryptophanes in helices 3 and 6 exhibit changes in their environments (Lin and Sakmar, 1996) on activation. In addition, distance measurements between a spin label at the cytoplasmic end of helix 3 and several specific locations at the cytoplasmic surface have indicated spatial changes on receptor activation (Farrens *et al.,* 1996; Yang *et al.,* 1996a,b). These data indicated that helices 3 and 6 move relative to each other, thus changing the receptor cytoplasmic surface. This change results in rhodopsin activation.

It is also important to recognize that as an integral membrane protein, rhodopsin inhabits a very special environment that is precisely engineered to accommodate the structural and functional requirements of the protein. The loops that connect the helices are exposed to the aqueous phase, while the helices are largely buried in the hydrophobic region of the membrane bilayer. The disk membrane bilayer is highly unsaturated with almost half the hydrocarbon chains being docosahexanoic acid (22:6). The sensitivity of rhodopsin to its lipid environment has been demonstrated by directly measuring the Meta I/MetaII equilibrium in different lipid environments and also by determining the effects of the lipid environment on the hydrolysis of cGMP. These studies have shown that rhodopsin is sensitive to the unsaturation of the hydrocarbon chains and to the level of cholesterol in the membrane.

A number of studies have focused on the effects of lipid composition on the functional properties of rhodopsin. The photolytic properties of rhodopsin are sensitive to the fatty acid composition of the surrounding bilayer. For example, the formation of Meta II is inhibited in bilayers containing fatty acid species less than 14 carbons in length (Baldwin and Hubbell, 1985a,b). In a set of elegant experiments, Litman and coworkers investigated the influence of lipid dynamics on rhodopsin photolysis. Changes in the cholesterol composition of recombinant phosphatidylcholine-rhodopsin vesicles were shown to have profound changes on the Metarhodopsin I/Metarhodopsin II equilibrium that corresponded to changes in the bilayer hydrocarbon dynamics (Mitchell *et al.,* 1990, 1991, 1992; Straume and Litman, 1987a,b, 1988; Straume *et al.,* 1990). High levels of cholesterol or saturated lipids favor Metarhodopsin I, whereas unsaturated lipids favor Metarhodopsin II. A reasonable explanation is that the conformational change required to form Metarhodopsin II is restricted in a high-cholesterol membrane or in a saturated membrane (Mitchell *et al.,* 1992). We have shown that cholesterol inhibits the activation of the cGMP cascade (Boesze-Battaglia and Albert, 1990). These studies showed that light activation of rhodopsin initiates minimal cascade activation in the rod out segment plasma membrane, whereas under the same conditions the cascade is fully activated in disks. When cholesterol was oxidized to cholestenone, activity approached that obtained with the disk membranes. These studies suggest that the lipids are responsible for modulating the dynamic environment of the membrane, which in turn can affect the properties of the protein.

Despite having achieved a substantial understanding of certain specific elements of rhodopsin structure and of the membrane properties that modulate its function, the complete structure of this receptor has remained elusive. Recently, however, considerable progress has been made toward obtaining a detailed three-dimensional structure for rhodopsin. This progress can be organized based on three aspects of rhodopsin structure: structural studies on the transmembrane helical bundle, studies on the structure of the retinal binding site, and the structure determination of the extramembranous portions. The latter includes the six loops that connect the helices as well as the N-terminal and C-terminal domains.

II. Seven Transmembrane Helical Hydrophobic Core of Rhodopsin

At this point it is clear that the membrane-spanning regions of rhodopsin assume an α-helical structure. The α-helical structure is an effective means of minimizing the energy of interaction between the nonpolar hydrocarbon region and the polar peptide backbone. The hydrogen bonding of the carbonyls and the N–H of the peptide bonds to each other effectively masks their polarity. If in the same sequence, hydrophobic side chains dominate, then an α-helical conformation for this sequence is thermodynamically compatible with the interior of the membrane lipid bilayer.

Baldwin has analyzed a large body of sequence data from the family of G-protein receptors to support a model of the transmembrane helices of rhodopsin and other related G-protein coupled receptors (Baldwin, 1993; Baldwin *et al.,* 1997). Several principles were used to extract the model from the sequence data. Among these principles were the following: (1) the amino acids facing the lipid hydrocarbon chains must be hydrophobic but not otherwise necessarily conserved; (2) conserved amino acids may be involved in helix–helix interactions; (3) charged amino acids within the (otherwise) hydrophobic helices are likely involved in helix–helix interactions.

Although, as mentioned earlier, the seven helices were predicted from the primary amino acid sequence and CD data provided early direct evidence for their presence, it was electron diffraction data that established this structure unequivocally. Henderson and colleagues had previously developed electron diffraction techniques on two-dimensional crystals of bacteriorhodopsin. They used these techniques to determine the three-dimensional structure of that protein (Grigorieff *et al.,* 1996; Leifer and Henderson, 1983). Their work showed that bacteriorhodopsin contains a bundle of seven transmembrane helices. The same approach was used on two-dimensional crystals of dark-adapted rhodopsin. A low-resolution structure of the transmembrane region was reported (Schertler *et al.,* 1993).

The seven transmembrane helices of bovine rhodopsin can be clearly seen in the structure generated by the electron density projection maps. These data, however do not identify which helix corresponds to each of the transmembrane helical domains. Therefore Schertler, Baldwin, and colleagues fitted a model of the seven transmembrane helix bundle to the electron density profile that represents the transmembrane helices (Baldwin *et al.,* 1997). The resulting structure shows helix 3 surrounded by the other helices and tilted. Helices 1 and 2 are also tilted, while helices 4, 5, 6, and 7 are nearly perpendicular to the membrane surface. Figure 2 shows a representation of this model. The arrangement of the helices is distinctly different from that observed in bacteriorhodopsin. The bundle of helices has an overall oval arrangement, viewed from the top. The helices appear to be packed in a closer arrangement at the cytoplasmic surface than at the intradiskal (extracellular) surface. Simi-

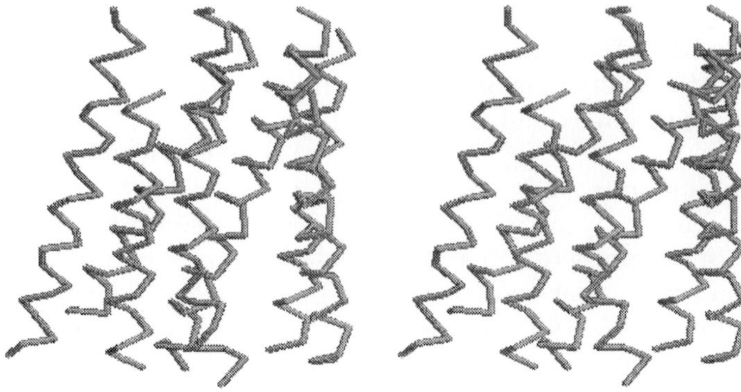

FIG. 2. Stereo view of the model derived for the transmembrane region of rhodopsin, as described in the text. (From Baldwin *et al.,* 1997.)

lar structures have now been reported for frog and octopus rhodopsin (Davies *et al.,* 1996; Unger *et al.,* 1997). The resolution of the structure from the two-dimensional crystals is not yet high enough to determine the accuracy of this model.

Herzyk and Hubbard (1998) have used modeling techniques to generate a structure for the seven helix bundle. Helices that are tilted with respect to the membrane are not well resolved in the electron density maps. Therefore the modeling studies concentrated on helices 1, 2, and 3, which are tilted with respect to the membrane, as well as helix 5, which is also not as well resolved. Structural constraints were derived from the experimental data generated from many different laboratories. These constraints were then used to direct a conformational search. The resultant conformation is consistent with the low-resolution three-dimensional map. Thus this approach may prove useful for future more detailed modeling of rhodopsin structure.

Another method to obtain information about the structure in the transmembrane domain of rhodopsin is the approach of Oprian, in which cysteines are introduced into a cys minus background and the ability of disulfide cross-links to form used to determine what residues are next to each other within the transmembrane helix bundle of rhodopsin. Using this method, evidence was obtained that residues 204 and 276 were close to each other in the three-dimensional structure of the transmembrane region of rhodopsin (Yu *et al.,* 1995). Through site-directed mutagenesis of rhodopsin, interactions between helices 3 and 5 were explored. The data suggested an interaction between residue 122 on helix 3 and residue 211 on helix 5 in the intact structure (Beck *et al.,* 1998).

As described later, it is also possible to obtain structural information about rhodopsin by determining the structures of peptides that encompass domains of this protein. One interesting example is found in the transmembrane domain of rhodopsin. Transmembrane helix 7 contains both the lysine that binds retinal and the conserved NPxxY sequence. Because this helix contains a proline, it is not expected to be a regular helix throughout its length. We have synthesized a peptide corresponding to this helix and determined its structure in DMSO. It forms an interesting structure in which the proline induces a kink in the helix and forms part of the binding pocket of the retinal. This structure is similar to other studies of hydrophobic helices containing proline (Berlose *et al.,* 1994; Snook *et al.,* 1998).

III. Structure of the Retinal and Its Binding Site in Rhodopsin

The 11-*cis* retinal chromophore is bound to lysine 296 in the seventh transmembrane helix via a protonated Schiff base linkage (Ovchinnikov, 1982; Wald, 1968). The long axis of the retinal is approximately 10 Å. It is buried within the protein structure and lies at an angle of approximately 16 degrees relative to the plane of the membrane (Liebman, 1962). Cross-linking studies using a photoactivatable retinal analog were used to investigate the relationship between retinal and the surrounding amino acids (Nakayama and Khorana, 1990). It was concluded from these studies that the β-ionone ring of the retinal is located between helices 3 and 6. Additional studies confirmed these conclusions and further indicated that glu 122 and trp 126 (helix 3) and trp 265 and trp 268 (helix 6) are close to retinal (Nakayama and Khorana, 1991). Glycine 121 (helix 3) has also been implicated in the retinal binding site (Han *et al.,* 1997). It was proposed that this residue interacts with the C-9 methyl group of the retinal.

The protonated Schiff base requires a counter ion. Normally, glutamic acid residue (glu 113) in helix 3 serves this function (Nathans, 1990; Sakmar *et al.,* 1989; Zhukovsky and Oprian, 1989). However, either a counterion positioned four residues away on helix 3 (Zhukovsky and Oprian, 1989) or an aspartic acid substituted for glycine (gly 90) on helix 2 (Rao *et al.,* 1994) can serve this function. Solid-state NMR studies have revealed some features about the binding site for retinal in rhodopsin and the conformation of retinal in its binding site. Studies using ^{13}C NMR of ^{13}C-labeled retinals reconstituted into rhodopsin and the sensitivity of the ^{13}C NMR chemical shifts to electron density distribution have suggested that one of the carboxylate oxygens from glu

113 is approximately 3 Å distant from the C_{12} carbon of the retinal (Han and Smith, 1995; Smith *et al.*, 1990).

^{2}H NMR studies on specifically deuterated retinals have provided information on the conformation of the retinal in its binding site on rhodopsin. The orientation of the methyl groups at C-19 and C-20 relative to the plane of the bilayer surface were determined for retinal in rhodopsin in oriented membranes. The C-19 group was found at an angle of 42 ± 5 degrees and the C-20 group was found at an angle of 30 ± 5 degrees in the dark-adapted state (Gröbner *et al.*, 1998). This information largely identifies the conformation and orientation of the retinal in the binding site. The suggestion arises from this work that the retinal is in a strained conformation when bound in the dark-adapted state.

The interaction of the chromophore with rhodopsin also has been shown to confer conformational stability on the protein. As stated earlier, the proteolytic fragments of rhodopsin solubilized in a detergent can only be separated after bleaching the rhodopsin (Albert and Litman, 1978). Furthermore, calorimetric studies have shown that the temperature at which the protein unfolds drops from approximately 72°C before bleaching to about 55°C after bleaching (Albert *et al.*, 1996; Khan *et al.*, 1991).

IV. Structure of Extramembranous Domains of Rhodopsin

Although the analysis of the two-dimensional crystals provides a low-resolution view of rhodopsin structure, much of the structural details remain to be defined. For example, the cytoplasmic face of rhodopsin that binds and activates the G-protein, transducin, was not defined by these studies and thus the critical mechanism for activation of transducin cannot be understood. More structural information is needed.

For many years considerable effort has been directed toward obtaining crystals of rhodopsin suitable for X-ray diffraction studies to obtain a complete, high-resolution three-dimensional structure of this receptor. However, as is the case for almost all membrane proteins, this has proved to be a formidable task. As is typical with membrane proteins, rhodopsin does not readily form crystals of the size and quality necessary for X-ray diffraction studies. This is largely due to the insolubility of rhodopsin in an isotropic solvent. One means that has proven effective for the crystallization of cytochrome oxidase was the co-crystallization of the protein with its antibody. The antibody allowed the formation of a three-dimensional crystal. An analogous approach to obtain crystals of rhodopsin is the co-crystallization of rhodopsin with the soluble G-protein, transducin. These approaches, however, have still not yielded a structure for rhodopsin.

Recent work has suggested that short-range interactions determine many elements of protein structure, in particular secondary structure. The structure of the β-turn has been shown to be stabilized by short-range interactions (Yang *et al.,* 1996). Many examples exist of small peptides that exhibit a β-turn in solution (Adler *et al.,* 1995; Blumenstein *et al.,* 1992; Fezoui *et al.,* 1997; Muñoz *et al.,* 1998; Yao *et al.,* 1994). If secondary structural elements of proteins are indeed determined by short-range interactions, one would expect that in some cases, peptides that correspond to protein sequences that contain a β-turn would also exhibit stable β-turn structures. Thus, peptides containing secondary structure such as β-turn and α-helix may act as small, structural domains of the protein that are stabilized by short-range interactions and reflect the structure of the intact protein.

In a recent study of myohemerythrin, the importance of short-range interactions was dramatically demonstrated (Dyson *et al.,* 1992). Myohemerythrin is a helical bundle, consisting of four helices and the turns that connect these helices. This protein was divided into several small peptides, and the structures of these peptides were determined by solution NMR. The structures of the individual peptides mimicked closely the structure of the intact protein. In particular, peptides that corresponded to the helical regions of the intact protein exhibited helical structure in solution.

In another example, peptides containing the sequences for turns formed structures in solution that mimicked the structures in the intact protein. Recent work on bacteriorhodopsin shows that a structural domain can be as small as a single loop, which connects two of the transmembrane helices (P. L. Yeagle and A. D. Albert, unpublished results). The structure of a 15-residue peptide which corresponds to loop DE of bacteriorhodopsin (amino acid residues 117–131) was determined. This structure is shown in Fig. 3 along with the corresponding structure from the electron diffraction structure of the intact bacteriorhodopsin. These data suggested that in proteins with substantial helix, the size of a domain that can independently fold correctly is surprisingly small.

Evidence that structural elements of proteins can be stabilized by relatively short-range interactions within the primary sequence suggested a new approach to the determination of the structure of integral membrane proteins such as the G-protein-coupled receptors. Suitable regions can be identified within the primary sequence that can be expected to form structural domains such as loops or helices. One can then obtain

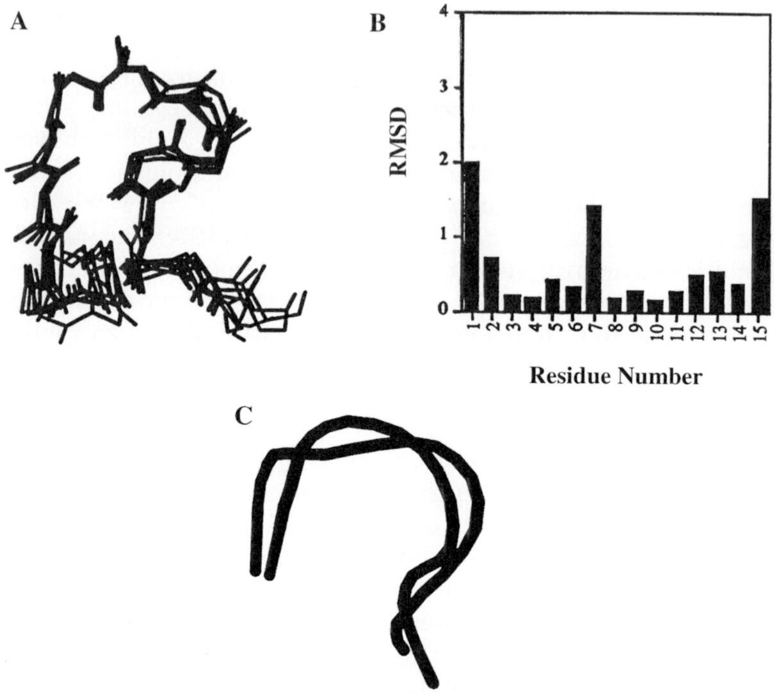

FIG. 3. Structure of the DE loop of bacteriorhodopsin, as an individual peptide. The structure was determined in solution by homonuclear two-dimensional ^{1}H NMR. (A) Family of DIANA structures for the peptide representing the DE loop of bacteriorhodopsin. (B) rmsd for the family of structures shown in part A, by residue. (C) Overlay of the average structure from the family of structures in part A and the same sequence from the structure of bacteriorhodopsin. (From Grigorieff *et al.,* 1996.)

the structure (and in some cases the basis of the biological activity) of these domains using established solution NMR methods of structural determination. The structures of these individual domains can then be assembled into the structure of the whole, like building blocks.

We have used this technique to study the structure of the G-protein-coupled receptor, rhodopsin. This protein is an appropriate choice for this approach. As shown by proteolytic cleavage experiments (Albert and Litman, 1978), rhodopsin has independent structural domains. Structural domains can be anticipated in the loop and helical regions as well. Furthermore, a number of physical parameters have been determined on the intact protein against which the validity of this approach can be tested. The initial results are encouraging. This building block approach is an exciting new approach that is applicable to other

G-protein-coupled receptors that are not readily available in quantity as is rhodopsin.

To exploit this concept to determine the structure of rhodopsin, peptides were synthesized that correspond to specific regions of rhodopsin. The regions initially chosen represented the cytoplasmic extramembranous portions of rhodopsin. These were chosen because it is the cytoplasmic surface that interacts with transducin. Furthermore, these extramembranous regions are not detected in the electron diffraction studies of the two-dimensional crystals. It was also anticipated that these peptides would be sufficiently soluble for NMR studies. This approach has produced a structure for both the cytoplasmic and intradiskal surfaces of rhodopsin.

V. Cytoplasmic Face of Rhodopsin

Upon activation by light to the R* state, rhodopsin is capable of activating the G protein, transducin. The activated rhodopsin also becomes a substrate for rhodopsin kinase. This phosphorylation then inhibits further transducin binding as well as enhancing arrestin binding. The cytoplasmic face of rhodopsin is the portion that directly interacts with the G protein, transducin. Several peptide studies have shown that regions of the carboxy terminus as well as loops 2 and 3 are involved in this interaction (Konig *et al.*, 1989; Phillips and Cerione, 1994; Takemoto *et al.*, 1985). Therefore the conformational properties of this region are key to activation of the G protein by rhodopsin.

A. Carboxyl-Terminal Domain of Rhodopsin

The carboxyl-terminal domain of bovine rhodopsin was synthesized, containing the last 43 amino acids of the protein sequence (residues 306–348). This sequence included the entire putative fourth cytoplasmic loop as well as a significant portion of helix 7, the transmembrane helix of this receptor to which the carboxyl terminal is attached. The solution structure was determined by multidimensional ^{1}H NMR. The peptide contained residues that were expected to be part of the α-helix at the cytoplasmic surface of transmembrane helix seven. Inclusion of these residues allowed unambiguous docking of the carboxyl-terminal domain to a model of the α-helix of the transmembrane domain.

The structure of the carboxyl-terminal domain (in addition to structures of other domains) is shown in Fig. 4. Several interesting observations can be made. Residues that were not predicted to be α-helical by

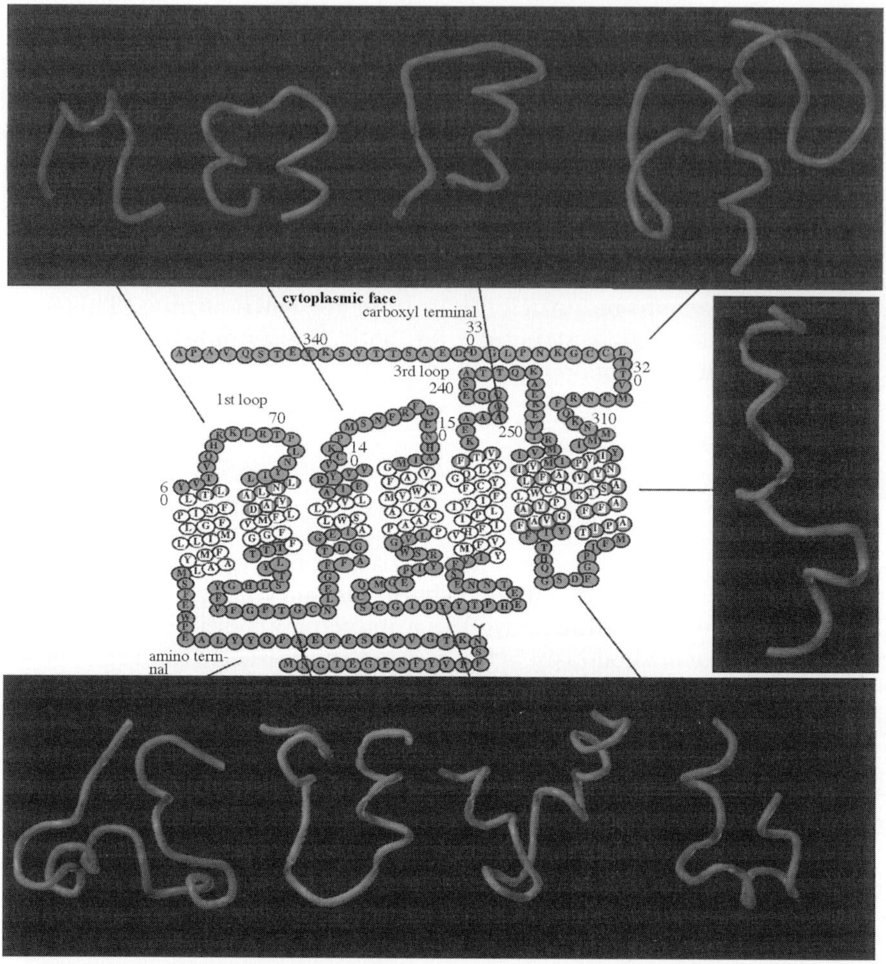

FIG. 4. Three-dimensional structures of peptides corresponding to loops, C terminus, N terminus, and one of the transmembrane helices of bovine rhodopsin as determined by solution NMR.

hydropathy analysis exhibit α-helical structure. Thus helix 7 is longer than was originally suggested and extends out of the membrane bilayer. The structure also revealed the fourth cytoplasmic loop. The palmitylation sites of rhodopsin are located at the bottom of this loop near the deduced membrane surface. Near the carboxy end of the structure is a region of β-sheet. This structural element was predicted by earlier FTIR studies (Pistorius and deGrip, 1994). The rhodopsin phosphorylation sites and mutations associated with certain autosomal dominant forms of retinitis pigmentosa can now be located in the three-dimensional structure of the carboxyl-terminal domain.

Phosphorylation of the receptor initiates desensitization. The sites of phosphorylation are in the region of β-sheet at the C terminus. This suggests that phosphorylation may initiate a conformational change by disruption of the β-sheet secondary structure. Although the molecular mechanism of desensitization by phosphorylation is unknown, at least two mechanisms are possible: Phosphorylation may change the conformation of the protein and inhibit transducin binding; phosphorylation will add negative charge, which may inhibit transducin binding. Recently NMR studies have been carried out on peptides that correspond to residues 330–348 of rhodopsin which were phosphorylated at one, three, and seven of the residues to mimic the sequence of phosphorylation of rhodopsin by rhodopsin kinase. These studies suggest that the addition of one phosphate by rhodopsin kinase introduces a conformational change in the carboxyl terminal, twisting the β-sheet. Additional phosphates do not introduce substantial further conformational changes. Therefore, the further diminution of the ability of the G protein to bind with increased phosphorylation of rhodopsin may be due to the increase in negative charge from the additional phosphates (Albert *et al.,* 1998).

It was shown that two cysteines (322 and 323) at the carboxy terminus are palmitolated (O'Brien and Zatz, 1984). This suggested that a region of the carboxy terminus forms an additional loop, the putative fourth cytoplasmic loop that was firmly established by the structural studies described earlier. The function of this acylation is not understood. These cysteines are at the base of the fourth cytoplasmic loop of rhodopsin. The structural studies on the carobxyl terminus have shown that the acyl groups are not required for the formation of the fourth cytoplasmic loop. This is consistent with mutagenesis studies, which have shown that acylation of these cysteines is not essential for transducin binding in experimental systems (Karnik *et al.,* 1993). However, the palmitylation could function in the native disks to enhance transducin binding. This could be through stabilization of the carboxyl-terminal region of rhodopsin or, since these fatty acyl groups are imbedded in the bilayer, they could affect the disk bilayer properties. It has also been suggested that the palmitylation plays a role in the regeneration of rhodopsin (Morrison *et al.,* 1991; Pepperberg *et al.,* 1995).

B. Cytoplasmic Loops of Rhodopsin

The structures of peptides containing the amino acid sequences of the first, second, and third cytoplasmic loops of rhodopsin were then determined. CD showed secondary structure in solution, indicating that the peptides were ordered. The structures were determined by two-dimensional NMR techniques and are shown in Fig. 4. All loops showed ordered structures in solution. In all loops the ends of the transmembrane helices unwind and form a β-turn. The structural motif is helix–turn–helix in which the helices are extensions of the transmembrane helix bundle. These data suggest a structural motif for short loops in transmembrane proteins. The well-ordered structures of these loops indicate that the primary sequences of these loops stabilize the β-turn. These data further suggest that the loops may contribute to the folding of such membrane proteins during their synthesis and insertion into membranes.

The third cytoplasmic loop contains an α-helix, which is an extension of transmembrane helix 5. That is in good agreement with spin label results which also suggested an extension of helix 5 (Farrens *et al.*, 1996). This structure is in remarkable agreement with the structure of the third cytoplasmic loop of the PTH/PTHrP receptor, another G-protein receptor (Mierke *et al.,* 1996).

C. Cytoplasmic Face of Rhodopsin

Having determined the individual structural domains of the cytoplasmic surface of rhodopsin, the next step is to assemble these units into a structure of the intact cytoplasmic face. To accomplish this, the structure and arrangement of the seven transmembrane helices determined by electron diffraction were used in conjunction with the experimentally determined structures of the four cytoplasmic domains of rhodopsin. These peptide structures were docked to the structure of the transmembrane domain to build the cytoplasmic face of this receptor. In this approach the helices are used as a scaffold which determined the placement of the peptide structures. The resultant structure is shown in Fig. 5A. To verify that the structure thus obtained was an accurate picture of the structure of intact rhodopsin, this structure was compared to the available information on the structure of the intact protein and found to be in good agreement with rhodopsin.

To further verify this structure, an alternate approach, utilizing a complex of the loop peptides and the carboxyl-terminal peptide in solution, was used. This approach required that the peptides form an ordered complex in solution. However, it had an advantage in that additional information can be obtained in the form of NMR structural constraints generated by close approach of residues in different peptides.

Previous data suggested that three of the domains may form a complex in solution. The second and third cytoplasmic loops as well as the carboxyl-terminal domain, which have biological activity individually in solution, are synergistic in their ability to inhibit the activation of the G protein by light-activated rhodopsin (Konig *et al.,* 1989). Additionally, CD data of these peptides combined in a 1:1:1:1 molar ratio confirmed the suggestion that a complex forms that is more ordered than the structures of the individual peptides. The structure of this complex (with the addition of the first cytoplasmic loop) was then determined using multidimensional NMR data. This structure is similar to the structure obtained by the docking procedure and is shown in Fig. 5B.

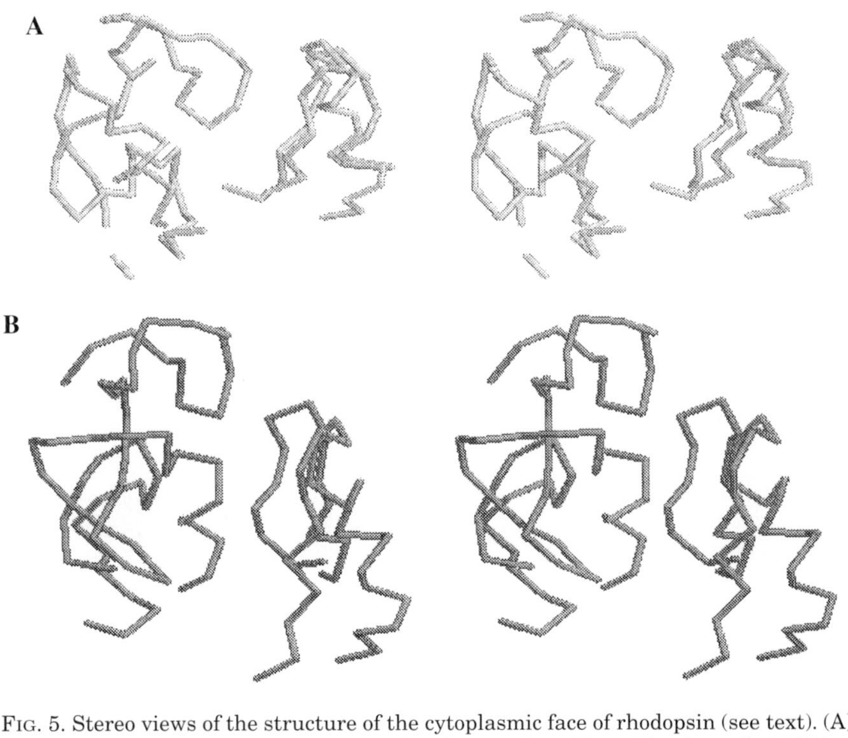

FIG. 5. Stereo views of the structure of the cytoplasmic face of rhodopsin (see text). (A) Docked structure. (B) Structure from simulated annealing. (From Yeagle *et al.,* 1997.)

D. Comparison of the Structure for the Cytoplasmic Face to Intact Rhodopsin

The solution of the high-resolution structure of the carboxyl-terminal domain and cytoplasmic loops of bovine rhodopsin together with the low-resolution structure of the transmembrane domains of this G-protein receptor has led to a partial model for the structure of the cytoplasmic face of rhodopsin. This structure can be validated by comparison to the whole protein. Distance measurements have been made between specific sites on intact rhodopsin that are in good agreement with the structure reported here. The laboratories of Khorana and Hubbell have used site-directed mutagenesis to introduce cysteines at specific positions in intact rhodopsin (Altenbach *et al.*, 1996; Farrens *et al.*, 1996). These sites were specifically spin labeled. Using the dipolar interactions between the spin labels, distances between specific sites on rhodopsin were measured in both the R and R* state. Table I compares these distances from the native protein with the same distances measured on the structure reported here. The structure reported here is in excellent agreement with the structure of the activated receptor, R*. This suggestion is in agreement with the observation of biological function associated with three of the four domains.

Furthermore, we have made a measurement (Albert *et al.*, 1997) between a spin labeled site (cys 140) and the predominant phosphorylation site on the carboxyl terminal. Rhodopsin was labeled with a nuclear spin label (^{31}P) by limited phosphorylation with rhodopsin kinase. Major phosphorylation occurs at serines 343 and 338 on the

TABLE I
Distances between Side Chains of Specific Amino Acid Residues in Rhodopsin, Measured on This Structure and the Same Distances Reported on Intact Rhodopsin[a]

	This structure (Å)	R*[b] (Å)	R (Å)
val 139 → lys 248	25	23–25	12–14
val 139 → glu 249	20	15–20	15–20
val 139 → val 250	13	12–14	15–20
val 139 → thr 251	20	23–25	12–14
val 139 → arg 252	11	23–25	15–20
his 65 → met 316	14	12–15	7–10

[a]Data from Farrens *et al.*, 1996; K. Yang *et al.*, 1996a,b.

[b]R* is the activated form of rhodopsin, and R is the dark-adapted state of rhodopsin.

carboxyl terminal. The phosphorylated rhodopsin was then specifically labeled on cysteine 140 with an electron spin label. Magic angle spinning ^{31}P NMR revealed the resonance arising from the phosphorylated protein. The enhancement of the transverse relaxation of this resonance by the paramagnetic spin label was observed. The strength of this perturbation was used to determine the through-space distance between the phosphorylation site(s) and the spin label position. A distance of 18 ± 3 Å was obtained. The distance calculated from the above-mentioned model was 15 Å, in good agreement with the measured distance.

Finally, the center-to-center distances of the ends of cytoplasmic loops 1, 2, and 3 are each about 10–12 Å, which corresponds to the center-to-center distance for the ends of the helices from the projection structure of rhodopsin. Each cytoplasmic loop is attached to two of these helices. This is also consistent with the electron diffraction data (Unger *et al.*, 1997) and with modeling data (Herzyk and Hubbard, 1998) that place the ends of the transmembrane helices at the cytoplasmic surface in close proximity.

The structure of the cytoplasmic face of rhodopsin exhibits several interesting features which are in agreement with suggestions from previously published data. This surface of rhodopsin has some β structure, both in turns in the loops and in a small antiparallel sheet in the carboxyl-terminal domain. These structural features are in good agreement with conclusions from FTIR studies on native rhodopsin (Pistorius and deGrip, 1994). Transmembrane helix 5 extends beyond the hydrophobic region of this membrane protein into the third cytoplasmic loop, consistent with those FTIR data and recent spin label experiments (Altenbach *et al.*, 1996). The cross-sectional area of the cytoplasmic face of rhodopsin determined in this work has dimensions of about 33 by 36 Å, in good agreement with previous low-resolution measurements that modeled rhodopsin with a cylinder of diameter approximately 36 Å (Osborne *et al.*, 1978; Sardet *et al.*, 1976). Previous work on boundary lipids with spin labeled lipids predicts a protein circumference consistent with the structure reported here (Watts *et al.*, 1979). The palmitylation sites on rhodopsin (cys 322 and cys 323) are found exposed at the putative membrane surface in this structure, where they can be readily acylated (Morrison *et al.*, 1991). In this structure, the fourth cytoplasmic loop forms without being acylated. That may explain why palmitylation was found to be nonessential to the activation of rhodopsin (Karnik *et al.*, 1993). Together, the body of data available on native rhodopsin is in excellent agreement with the structure of the cytoplasmic surface presented in Fig. 5.

VI. Intradiskal Face of Rhodopsin

The intradiskal face of rhosopsin is not directly involved in the visual signal transduction cascade. However, an analysis of various naturally occurring opsins and induced mutations of rhodopsin has suggested that this region is important for the stability of rhodopsin and its ability to function in transduction (Hargrave *et al.,* 1998). Furthermore, an understanding of the structure of this surface is important by analogy to the ligand binding surface of other G-protein receptors.

Rhodospin is glycosylated at asparagines 2 and 15. The oligosaccharides are largely hexasaccharides of mannose and *N*-acetylglucosamine that form branched chain structures (Fukuda *et al.,* 1979; Hargrave *et al.,* 1984; Liang *et al.,* 1979). Although rhodopsin will form a chromophore in the absence of glycosylation, it will not function properly in the transduction cascade.

A disulfide bridge forms between cysteine residues 110 and 187. This bridge links a cysteine near the bilayer surface with a cysteine in the central region of the loop connecting helices 4 and 5. The proper formation of this bridge is essential for correct folding of the protein. Mutations that cause packing defects in the transmembrane helical region also lead to incorrect disulfide bonds (Hwa *et al.,* 1997). The importance of the stability of this region has also been inferred by Hargrave based on studies of many vertebrate opsins as well as mutations (Hargrave *et al.,* 1998).

Three loops and the amino terminus form the intradiskal face of rhodopsin. The structures of each of these regions have been determined as were the structures for the cytoplasmic face of rhodopsin. The structure of the third intradiskal loop looks very much like the first and second cytoplasmic loops, following the helix–loop–helix motif. The structures of the remaining loops of the intradiskal face are somewhat larger and more complex, but still basically follow the same structural motif. These structures are shown in Fig. 4.

At this time experimental constraints have been obtained for about 75% of the residues of rhodopsin. These data are adequate for building a partial model of the overall structure of this protein. Efforts are currently under way to obtain structural information for the remainder of the α-helices of this protein to complete the structure.

VII. Future Directions

Having an overall structure for rhodopsin will enable examination of specific areas of the protein structure. This will allow the structure to be tested against experimental structural information obtained on the intact protein. Furthermore, changes in the protein as a result of conformational changes between rhodopsin, R*, and opsin can be investigated. Finally, the effect of mutations known to produce aberrant rhodopsins can be understood from a structural perspective.

REFERENCES

Adler, M., Sato, M. H., Nitecki, D. E., Lin, J.-H., Light, D. R., and Morser, J. (1995). The structure of a 19 residue fragment from the C loop of the fourth epidermal growth factor-like domain of thrombomodulin. *J. Biol. Chem.* **270,** 23366–23372.

Albert, A. D., and Litman, B. J. (1978). Independent structural domains in the membrane protein bovine rhodopsin. *Biochemistry* **17,** 3893–3900.

Albert, A. D., Boesze-Battaglia, K., Paw, Z., Watts, A., and Epand, R. M. (1996). Effect of cholesterol on rhodopsin stability in disk membranes. *Biochim. Biophys. Acta* **1297,** 77–82.

Albert, A. D., Watts, A., Spooner, P., Groebner, G., Young, J., and Yeagle, P. L. (1997). A distance measurement between specific sites on the cytoplasmic surface of bovine rhodopsin in rod outer segment disk membranes. *Biochim. Biophys. Acta* **1328,** 74–82.

Albert, A. D., Yeagle, P. L., Alderfer, J. L., Dorey, M., Vogt, T., Bhawasar, N., Hargrave, P. A., McDowell, J. H., and Arendt, A. (1999). Effects of phosphorylation on the structure of the G-protein receptor, rhodopsin: An NMR study of rhodopsin's phosphopeptides. *Biochim. Biophys. Acta* **1416,** 217–224.

Altenbach, C., Yang, K., Farrens, D. L., Farahbakhsh, Z. T., Khorana, H. G., and Hubbell, W. L. (1996). Structural features and light dependent changes in the cytoplasmic interhelical E-F loop regioni of rhodopsin: A site-directed spin-labeling study. *Biochemistry* **35,** 12470–12478.

Attwood, P. V., and Gutfreund, H. (1980). The application of pressure relaxation to the study of the equilibrium between metarhodopsin I and II from bovine retinas. *FEBS Lett.* **119,** 323–326.

Baldwin, J. M. (1993). The probable arrangement of the helices in G protein-coupled receptors. *EMBO J.* **12,** 1693–1703.

Baldwin, J. M., Schertler, G. F. X., and Unger, V. M. (1997). An alpha-carbon template for the transmembrane helices in the rhodopsin family of G-protein-coupled receptors. *J. Mol. Biol.* **272,** 144–164.

Baldwin, P. A., and Hubbell, W. L. (1985a). Effects of lipid environment on the light-induced conformational changes of rhodopsin. 1. Absence of metarhodopsin II production in dimyristoylphosphatidylcholine recombinant membranes. *Biochemistry* **24,** 2624–2632.

Baldwin, P. A., and Hubbell, W. L. (1985b). Effects of lipid environment on the light-induced conformational changes of rhodopsin. 2. Roles of lipid chain length, unsaturation, and phase state. *Biochemistry* **24,** 2633–2639.

Beck, M., Sakmar, T. P., and Siebert, F. (1998). Spectroscopic evidence for interaction between transmembrane helices 3 and 5 in rhodopsin. *Biochemistry* **37,** 7630–7639.

Berlose, J., Convert, O., Brunissen, A., Chassaing, G., and Lavielle, S. (1994). Three dimensional structure of the highly conserved seventh transmembrane domain of G-protein-coupled receptors. *FEBS Lett.* **225,** 827–843.

Blumenstein, M., Matsueda, G. R., Timmons, S., and Hawiger, J. (1992). A β turn is present in the 392-411 segment of the human fibrinogen γ chain. Effects of structural changes in this segment on affinity to antibody 4A5. *Biochemistry* **31,** 10692–10698.

Boesze-Battaglia, K., and Albert, A. (1990). Cholesterol modulation of photoreceptor function in bovine rod outer segments. *J. Biol. Chem.* **265,** 20727–20730.

Davies, A., Schertler, G. F., Gowen, B. E., and Saibil, H. R. (1996). Projection structure of an invertebrate rhodopsin. *J. Struct. Biol.* **117,** 36–44.

DeGrip, W. J., Gray, D., Gillespie, J., Bovee, P. H. M., Van den Berg, L. J., and Rothschild, K. J. (1988). Photoexcitation of rhodopsin: Conformation changes in the chromophore, protein and associated lipids as determined by FTIR difference spectroscopy. *Photochem. Photobiol.* **48,** 497–504.

Dyson, H. J., Merutka, G., Waltho, J. P., Lerner, R. A., and Wright, P. E. (1992). Folding of peptide fragments comprising the complete sequence of proteins. *J. Mol. Biol.* **226,** 795–817.

Emeis, D., Kuhn, H., Riechert, J., and Hofmann, K. P. (1982). Complex formation between metarhodopsin II and GTP-binding protein in bovine photoreceptor membranes leads to a shift of the photoproduct equilibrium. *FEBS Lett.* **43,** 29–34.

Fahmy, K., Jäger, F., Beck, M., Zvyaga, T. A., Sakmar, T. P., and Siebert, F. (1993). Protonation states of membrane-embedded carboxylic acid groups in rhodopsin and metarhodopsin II: A Fourier-transform infrared spectroscopy study of site-directed mutants. *Proc. Natl. Acad. Sci. U.S.A.* **90,** 10206–10210.

Farrens, D. L., Altenbach, C., Yang, K., Hubbell, W. L., and Khorana, H. G. (1996). Requirement of rigid-body motion of transmembrane helices for light activation of rhodopsin. *Science* **274,** 768–770.

Fezoui, Y., Connolly, P., and Osterhout, J. (1997). Solution structure of alpha t alpha, a helical hairpin protein of *de novo* design. *Protein Sci.* **6,** 1869–1877.

Fukuda, M. N., Papermaster, D. S., and Hargrave, P. A. (1979). Rhodopsin carbohydrate. Structure of small oligosaccharides attached at two sites near the NH_2 terminus. *J. Biol. Chem.* **254,** 8201–8207.

Grigorieff, N., Ceska, T. A., Downing, K. H., Baldwin, J. M., and Henderson, R. (1996). Electron-crystallographic refinement of the structure of bacteriorhodopsin. *J. Mol. Biol.* **259,** 393–421.

Gröbner, G., Choi, G., Burnett, I. J., Glaubitz, C., Verdegem, P. J. E., Lugtenburg, J., and Watts, A. (1998). Photoreceptor rhodopsin: Structural and conformational study of its chromophore, 11-cis retinal, in oiented membranes by deuterium solid state NMR. *FEBS Lett.* **422,** 201–204.

Han, M., and Smith, S. O. (1995). NMR constraints on the location of the retinal chromophore in rhodopsin and bathorhodopsin. *Biochemistry* **34,** 1425–1432.

Han, M., Groesbeek, M., Sakmar, T. P., and Smith, S. O. (1997). The C9 methyl group of retinal interacts with glycine-121 in rhodopsin. *Proc. Natl. Acad. Sci. U.S.A.* **94,** 13442–13447.

Hargrave, P. A., McDowell, J. H., Curtis, D. R., Wang, J. K., Juszczak, E., Fong, S. L., Rao, J. K. M., and Argos, P. (1983). The structure of bovine rhodopsin. *Biophys. Struct. Mech.* **9,** 235–244.

Hargrave, P. A., McDowell, J. H., Feldmann, R. J., Atkinson, P. H., Mohana Rao, J. K., and Argos, P. (1984). Rhodopsin's protein and carbohydrate structure: Selected aspects. *Vision Res.* **24,** 1487–1499.

Hargrave, P. A., Smith, W. C., and McDowell, J. H. (1998). Rhodopsin structure, function and topography; Selected aspects. *Rom. J. Biophys.* **8,** 35–47.

Herzyk, P., and Hubbard, R. E. (1998). Combined biophysical and biochemical information confirms arrangement of transmembrane helices visible from the three-dimensional map of frog rhodopsin. *J. Mol. Biol.* 741–745.

Hwa, J., Garriga, P., Liu, X., and Khorana, H. G. (1997). Structure and function in rhodopsin: Packing of the helices in the transmembrane domain and folding to a tertiary structure in the intradiscal domain are coupled. *Proc. Natl. Acad. Sci. U.S.A.* **94,** 10571–10576.

Karnik, S. S., Ridge, K. D., Bhattacharya, S., and Khorana, H. G. (1993). Palmitoylation of bovine opsin and its cysteine mutants in COS cells. *Proc. Natl. Acad. Sci. U.S.A.* **90,** 40–44.

Khan, S. M. A., Bolen, W., Hargrave, P. A., Santoro, M. M., and McDowell, J. H. (1991). Differential scanning calorimetry of bovine rhodopsin in rod-outer-segment membranes. *Eur. J. Biochem.* **200,** 53–59.

Kibelbek, J., Mitchell, D. C., Beach, J., and Litman, B. J. (1991). Functional equivalence of metarhodopsin II and the Gt-activating form of photolyzed bovine rhodpsin. *Biochemistry* **30,** 6761–6768.

Konig, B., Arendt, A., McDowell, J. H., Kahlert, M., Hargrave, P. A., and Hofmann, K. P. (1989). Three cytoplasmic loops of rhodopsin interact with transducin. *Proc. Natl. Acad. Sci. U.S.A.* **86,** 6878–6882.

Lamola, A. A., Yamane, T., and Zipp, A. (1974). Effects of detergents and high pressures upon the metarhodopsin I to metarhodopsin II equilibrium. *Biochemistry* **13,** 738–745.

Leifer, D., and Henderson, R. (1983). Three dimensional structure of orthorhombic purple membrane at 6.5 resolution. *J. Mol. Biol.* **163,** 451–466.

Liang, C.-J., Yamashita, K., Muellenberg, C. G., Shichi, H., and Kobata, A. (1979). Structure of the carbohydrate moieties of bovine rhodopsin. *J. Biol. Chem.* **254,** 6414–6418.

Liebman, P. A. (1962). In situ microspectrophotometric studies on the pigments of single retinal rods. *Biophys. J.* **2,** 161–178.

Lin, S. W., and Sakmar, T. P. (1996). Specific tryptophan UV-absorbance changes are probes of the transition of rhodopsin to its active state. *Biochemistry* **35,** 11149–11159.

Mathews, R. G., Hubbard, P. K., and Wald, G. (1963). Tautomeric forms of metarhodopsin. *J. Gen. Physiol.* **47,** 215–240.

Mierke, D. F., Royo, M., Pelligrini, M., Sun, H., and Chorev, M. (1996). Third cytoplasmic loop of the PTH/PTHrP receptor. *J. Am. Chem. Soc.* **118,** 8998–9004.

Mitchell, D. C., Straume, M., Miller, J., and Litman, B. J. (1990). Modulation of metarhodopsin formation by cholesterol-induced ordering of bilayers. *Biochemistry* **29,** 9143–9149.

Mitchell, D. C., Kibelbek, J., and Litman, B. J. (1991). Rhodopsin in dimyristoylphosphatidylcholine-reconstituted bilayers forms matarhodopsin II and activates Gt. *Biochemistry* **30,** 37–42.

Mitchell, D. C., Straume, M., and Litman, B. J. (1992). Role of sn-1-saturated, sn-2-polyunsaturated phospholipids in control of membrane receptor conformational equilibrium: Effects of cholesterol and acyl chain unsaturation on the matarhodopsin I-metarhodopsin II equilibrium. *Biochemistry* **31,** 662–670.

Morrison, D. F., O'Brien, P. J., and Pepperberg, D. R. (1991). Depalmitylation with hydroxylamine alters the functional properties of rhodopsin. *J. Biol. Chem.* **266,** 20118–20123.

Muñoz, V., Thompson, P. A., Henry, E. R., Hofrichter, J., and Eaton, W. A. (1998). Folding dynamics of a β-hairpin studies by laser temperature jump and kinetic modeling. *Biophys. J.* **74,** A175.

Nakayama, T. A., and Khorana, H. G. (1990). Orientation of retinal in bovine rhodopsin determined by cross-linking using a photoactivatable analog of 11-*cis* retinal *J. Biol. Chem.* **265,** 15762–15769.

Nakayama, T. A., and Khorana, H. G. (1991). Mapping of the amino acids in membrane-embedded helices that interact with the retinal chromophore in bovine rhodopsin. *J. Biol. Chem.* **266,** 4269–4275.

Nathans, J. (1990). Determinants of visual pigment absorbance: Identification of the retinylidene Schiff's base counterion in bovine rhodopsin. *Biochemistry* **29,** 9746–9752.

O'Brien, P. J., and Zatz, M. (1984). Acylation of bovine rhodopsin by [H3] palmitic acid. *J. Biol. Chem.* **259,** 5054–5057.

Osborne, H. B., Sardet, C., Michel-Villaz, M., and Charbre, M. (1978). Structural study of rhodopsin in detergent micelles by small-angle neutron scattering. *J. Mol. Biol.* **123,** 177–206.

Ovchinnikov, Y. A. (1982). Rhodopsin and bacteriorhodopsin: Structure-functional relationships. *FEBS Lett.* **148,** 179–191.

Pepperberg, D. R., Morrison, D. F., and O'Brien, P. J. (1995). Depalmitoylation of rhodopsin with hydroxylamine. *In* "Methods in Enzymology" (P. J. Casey and J. E. Buss, eds.), Vol. 250, pp. 348–361. Academic Press, San Diego, CA.

Phillips, W. J., and Cerione, R. A. (1994). A C-terminal peptide of bovine rhodopsin binds to the transducin alpha-subunit and facilitates its activation. *Biochem. J.* **299**(Pt. 2), 351–357.

Pistorius, A. M., and deGrip, W. J. (1994). Rhodopsin's secondary structure revisited: Assignment of structural elements. *Biochem. Biophys. Res. Commun.* **198,** 1040–1045.

Pober, J. S., and Stryer, L. (1975). Letter to the editor: Light dissociates enzymatically-cleaved rhodopsin into two different fragments. *J. Mol. Biol.* **95,** 477.

Rao, V. R., Cohen, G. G., and Oprian, D. D. (1994). Rhodopsin mutation G90D and a molecular mechanism for congenital night blindness. *Nature (London)* **367,** 639–642.

Saari, J. C. (1974). The accessibility of bovine rhodopsin in photoreceptor membranes. *J. Cell Biol.* **63,** 480–491.

Sakmar, T. P., Franke, R. R., and Khorana, H. G. (1989). Glutamic acid-113 serves as the retinylidene Schiff base counterion in bovine rhodopsin. *Proc. Natl. Acad. Sci. U.S.A.* **92,** 8309–8313.

Sale, G. J., Towner, P., and Akhtar, M. (1977). Functional rhodopsin complex consisting of three noncovalently linked fragments. *Biochemistry* **16,** 5641–5649.

Sardet, C., Tardieu, A., and Luzzati, V. (1976). Shape and size of bovine rhodopsin: A small-angle x-ray scattering study of a rhodopsin-detergent complex. *J. Mol. Biol.* **105,** 383–407.

Schertler, G. R. X., Villa, C., and Henderson, R. (1993). Projection structure of rhodopsin. *Nature (London)* **362,** 770–772.

Sheikh, S. P., Zvyaga, T. A., Lichtarge, O., Sakmar, T. P., and Bourne, H. R. (1996). Rhodopsin activation blocked by metal-ion binding sites linking transmembrane helices C and F. *Nature (London)* **383,** 347–350.

Smith, S. O., Palings, I., Miley, M. E., Courtin, J., de Groot, H., Lugtenburg, J., Mathies, R. A., and Griffin, R. G. (1990). Solid state NMR studies of the mechanism of the opsin shift in the visual pigment rhodopsin. *Biochemistry* **29,** 8158–8164.

Snook, C. F., Wolley, G. A., Oliva, G., Pattabhi, Wood, S. P., Blundell, T. L., and Wallace, B. A. (1998). The structure and function of antiamoebin I, a proline-rich membrane-active polypeptide. *Structure* **6,** 783–792.

Straume, M., and Litman, B. J. (1987a). Equilibrium and dynamic structure of large, unilamellar, unsaturated acyl chain phosphatidylcholine vesicles. Higher order analysis of 1,6-diphenyl-1,3,5-hexatriene and 1-[4-trimethylammonio]-6-phenyl-1,3,5-hexatriene anisotropy decay. *Biochemistry* **26,** 5113–5120.

Straume, M., and Litman, B. J. (1987b). Influence of cholesterol on equilibrium and dynamic bilayer structure of unsaturated acyl chain phosphatidylcholine vesicles as determined from higher order analysis of fluorescence anisotropy decay. *Biochemistry* **26,** 5121–5126.

Straume, M., and Litman, B. J. (1988). Equilibrium and dynamic bilayer structural properties of unsaturated acyl chain phosphatidylcholine-cholesterol-rhodopsin recombinant vesicles and rod outer segment disk membranes as determined from higher order analysis of fluorescence anisotropy decay. *Biochemistry* **27,** 7723–7733.

Straume, M., Mitchell, D., Miller, J., and Litman, B. J. (1990). Interconversions of Metarhodopsins I and II: A branched photointermediate decay model. *Biochemistry* **29,** 9135–9142.

Takemoto, D. J., Takemoto, L. J., Hansen, J., and Morrison, D. (1985). Regulation of retinal transducin by C-terminal peptides of rhodopsin. *Biochem. J.* **232,** 669–672.

Trayburn, P., Mandel, P., and Virmaux, N. (1974a). Removal of a large fragment of rhodopsin without changes in its spectral properties, by proteolysis of reginal rod outer segments. *FEBS Lett.* **38,** 351–353.

Trayburn, P., Mandel, P., and Virmaux, N. (1974b). Composition of the rhodopsin-core obtained by proteolysis of retinal rod outer segments with papain, and its regenerability after photobleaching. *Exp. Eye Res.* **19,** 259.

Unger, V. M., Hargrave, P. A., Baldwin, J. M., and Schertler, G. F. X. (1997). Arrangement of rhodopsin transmembrane α-helices. *Nature (London)* **389,** 203–206.

Wald, G. (1968). The molecular basis of visual excitation. *Nature (London)* **219,** 800–807.

Watts, A., Volovski, I. D., and Marsh, D. (1979). Rhodopsin-lipid associations in bovine rod outer segment membranes. Identification of immobilized lipid by spin-labels. *Biochemistry* **18,** 5006–5013.

Yang, A.-S., Hitz, B., and Honig, B. (1996). Free energy determinants of secondary structure formation: β turns and their role in protein folding. *J. Mol. Biol.* **259,** 873–882.

Yang, K., Farrens, D. L., Hubbell, W. L., and Khorana, H. G. (1996a). Structure and function in rhodopsin. Single cysteine substitution mutants in the cytoplasmic interhelical E-F loop region show position-specific effects in transducin activation. *Biochemistry* **35,** 12464–12469.

Yang, K., Farrens, D. L., Altenbach, C., Farahbakhsh, Z. T., Hubbell, W. L., and Khorana, H. G. (1996b). Structure and function in rhodopsin. Cysteines 65 and 316 are in proximity in a rhodopsin mutant as indicated by disulfide formation and interactions between attached spin labels. *Biochemistry* **35,** 14040–14046.

Yao, J., Dyson, H. J., and Wright, P. E. (1994). Three-dimensional structure of a type VI turn in a linear peptide in water solution. Evidence for stacking of aromatic rings as a major stabilizing factor. *J. Mol. Biol.* **243,** 736–753.

Yeagle, P. L., Alderfer, J. L., and Albert, A. D. (1997). Three-dimensional structure of the cytoplasmic face of the G protein receptor rhodopsin. *Biochemistry* **36,** 9649–9654.

Yu, H., Kono, M., McKee, T. D., and Oprian, D. D. (1995). A general method for mapping tertiary contacts between amino acid residues in membrane-embedded proteins. *Biochemistry* **34,** 14963–14969.

Zhukovsky, E. A., and Oprian, D. D. (1989). Effect of carboxylic acid side-chains on the absorption maximum of visual pigments. *Science* **246,** 928–930.

VITAMINS AND HORMONES, VOL. 58

Molecular Modeling of Mammalian Cytochromes *P*450: Application to Study Enzyme Function

GRAZYNA D. SZKLARZ,* SANDRA E. GRAHAM,† AND MARK D. PAULSEN‡

**Department of Basic Pharmaceutical Sciences, School of Pharmacy, West Virginia University, Morgantown, West Virginia 26506–9530, †Department of Biochemistry, University of Texas Southwestern Medical Center at Dallas, Dallas, Texas 75235–9038, and ‡Department of Chemistry, Northern Michigan University, Marquette, Michigan 49855*

Cytochromes *P*450 are important heme-containing enzymes that catalyze the oxidation of a vast array of endogenous and exogenous compounds, including drugs and carcinogens. One of the more successful approaches to study *P*450 function involves molecular modeling. Because none of the mammalian *P*450s have been crystallized, a number of homology models have been constructed based on the structures of known bacterial *P*450s. Molecular models, generated using molecular replacement or distance geometry methods, can be used to dock substrates and/or inhibitors in the active site to explain various aspects of enzyme function. The majority of modeling research has dealt with enzyme–substrate interactions in the active site. The analysis of these interactions has helped us to better understand the mechanism of *P*450 catalysis and provided the structural basis for the regio- and stereospecificity of substrate oxidation as well as susceptibility to inhibition or inactivation. The models have been utilized to identify and/or

0083-6729/00 $30.00

confirm key residues and to rationally interpret experimental data. The alteration in activity in a mutant *P*450 can be related to changes in enzyme–substrate/inhibitor interactions, such as the removal or appearance of van der Waals overlaps or changes in compound mobility. Homology models can also help to analyze *P*450–redox partner interactions and identify critical determinants of protein stability. We can expect further development of molecular modeling methods and their increasing contribution into research on *P*450 function as an integral part of a combined theoretical–experimental approach.

I. Introduction

Cytochromes *P*450 are a superfamily of hemoproteins that play a pivotal role in the metabolism of a wide variety of xenobiotic and endogenous compounds. There is considerable interest in the function of these enzymes due to their involvement in the detoxification of foreign compounds and bioactivation of drugs and carcinogens. Moreover, *P*450s catalyze key steps in steroidogenesis in animals, as well as the metabolism of such endogenous compounds as prostaglandins, biogenic amines, leukotrienes, bile acids, or vitamin D. Reactions catalyzed by the *P*450 monooxygenase system include hydroxylations, epoxidations, *O*-, *S*- and *N*-dealkylations, and *N*-oxidations (Guengerich, 1993). *P*450 enzymes are ubiquitous in living organisms, and more than 400 individual isoforms have been thus far identified and sequenced from plants, animals, bacteria, and yeast (Nelson *et al.,* 1996). Based on their sequence, cytochromes *P*450 are classified into gene families and subfamilies, with members of the same family exhibiting at least 40% amino acid sequence identity, while within a subfamily the identity is ≥55% (Nelson *et al.,* 1996). Furthermore, based on their requirements for the redox partner, *P*450 enzymes can be divided into two main groups: class I enzymes, found in mitochondria, and class II, more abundant and found in endoplasmic reticulum. Class I enzymes require both an iron-sulfur protein (ferredoxin) and an FAD-containing NAD(P)H-ferredoxin reductase for catalysis, whereas class II *P*450s require an FAD/FMN-containing NADPH-*P*450 reductase, although they may utilize an additional electron donor, cytochrome b_5. Finally, there are class III enzymes that do not require any exogenous source of electrons because they obtain them from their peroxide substrates (e.g., prostacycline synthase). In mammals, the main xenobiotic metabolizing enzymes belong to class II and comprise families 1–4 (Wrighton and

Stevens, 1992), while *P*450s involved in steroidogenesis are split between class I and class II with CYP11 and CYP27 being class I families and CYP17, 19, and 21 being class II families. Many forms of *P*450 display broad substrate specificities, yet they often exhibit strict regio- and stereospecificity toward a particular compound. The elucidation of the structural basis for such specificity is of great importance in understanding enzyme function and mechanism. This may help to predict the possible metabolic fate of drugs and carcinogens and also provide a foundation for the rational design of drugs and inhibitors.

The knowledge of the three-dimensional protein structure is essential for understanding enzyme catalysis. However, in the case of mammalian cytochromes *P*450, the systematic investigation of structural determinants of enzyme specificity has been hindered by the fact that none of the eukaryotic *P*450s has yet been crystallized. Nevertheless, structure–function relationships of these enzymes have been studied extensively, greatly assisted by the introduction of the concept of substrate recognition sites (SRSs). Six SRSs have been identified in *P*450 family 2 by Gotoh (1992), based on the comparative sequence alignment and the analogy with the bacterial enzyme, *P*450cam, for which substrate contact residues were known from the crystal structure (Poulos *et al.,* 1985, 1987). Numerous studies published to date confirm that amino acids critical for the substrate specificity of mammalian *P*450s fall within or near the putative SRSs (von Wachenfeldt and Johnson, 1995). Moreover, additional information on *P*450 specificity has been provided by the molecular modeling approach. Homology models of various mammalian enzymes have been constructed based on the known structures of bacterial enzymes, namely, *P*450cam (Poulos *et al.,* 1985, 1987), *P*450 BM-3 (Ravichandran *et al.,* 1993), *P*450terp (Hasemann *et al.,* 1994), *P*450eryF (Cupp-Vickery and Poulos, 1995), and *P*450nor (Park *et al.,* 1997). These models have been often applied to the study of *P*450 function.

The intent of this article is to describe molecular modeling methods and provide examples that illustrate how *P*450 models have been utilized in studies of structure–function relationships. The majority of modeling research has dealt with enzyme–substrate interactions in the active site, although some other aspects of *P*450 function have been successfully investigated, such as protein stability and *P*450–redox partner interactions. Molecular modeling techniques have been especially effective when combined with experimental approaches, such as site-directed mutagenesis. Additional perspective on modeling and mutagenesis can be found in a recent review (Szklarz and Halpert, 1997a)

II. Molecular Modeling Methods

A. Homology Modeling Methods

The basic assumptions underlying homology modeling are that there are a limited number of native folds which a protein might adopt and that the choice of which fold is adopted depends primarily on the sequence of the protein. The increasing number of proteins whose three-dimensional structures have been experimentally determined by either X-ray crystallography or NMR spectroscopy has improved the odds that an example of the fold characteristic of some new protein exists in the Protein Data Bank. If an appropriate template protein structure is available, it can be used as a basis for modeling an unknown structure. Choosing the appropriate template protein(s) is simplified if the function of the protein to be modeled is known. Often functionally related proteins will share a common fold. Thus in the case of modeling a *P*450 structure, it is reasonable to assume that the modeled protein should adopt a fold similar to other *P*450s.

The first step in a homology modeling exercise is dividing the template structure(s) into structurally conserved regions (SCRs) and variable regions. This process is facilitated if there are several known homologous structures as in the case of *P*450 where five members of the superfamily have had their structures determined by X-ray crystallography. A structure-based sequence alignment is generated for the known structures and is used as the basis for generating a sequence alignment between the new sequence and the known structures. In regions of low homology, secondary structure predictions, as well as additional information, such as site-directed mutagenesis data (Szklarz *et al.,* 1995), can be invaluable in determining an optimal alignment.

The initial modeled structure can be generated in different ways. One approach is molecular replacement. In this approach, the cartesian coordinates for backbone atoms and invariant side-chain atoms in SCRs are taken directly from the template protein(s). Coordinates for non-conserved side chains in these SCRs can be taken from rotamer libraries. If several template structures are available, the coordinates for different SCRs in the modeled protein can be taken from different templates based on the sequence similarity of the template and modeled protein in a particular SCR. The structurally conserved regions will be connected by a series of variable regions. In the case of distantly related proteins, these regions will often contain insertions or deletions. Such regions can be modeled by taking coordinates from loops of similar length and sequence in a set of known protein structures not limited to functionally related proteins.

A second approach for building an initial model is based on distance geometry. Distance geometry has been widely used in solving the three-dimensional structure of proteins starting with NMR data. In the distance geometry approach, the atom–atom distances from the SCRs of a set of known proteins are used to generate a set of constraints on the homologous distance in the modeled protein. In the variable regions, the constraints will be based primarily on basic structural considerations, typical bond length, and bond distance, for instance. In this modeling approach the only real distinction between SCRs and variable regions is the number of constraints used to fix atom–atom distances. Once the set of constraints has been determined, three-dimensional structures consistent with the constraints can be generated using appropriate algorithms. Because there will typically be insufficient constraints to determine a structure uniquely, a distance geometry approach will generate a family of structures that should be quite similar in the SCRs but may well have significant variability outside the SCRs.

With either approach, the initial structure(s) generated are often refined by either subjecting the structure to molecular mechanics energy minimization or running a molecular dynamics trajectory for the structure. The objective of these procedures is to relieve any close contacts generated in building the structure, particularly in the loop regions. Because the focus of these procedures is usually on improving the structure of the variable regions, the energy minimization and/or molecular dynamics calculations are often performed with the coordinates of atoms in the SCRs fixed.

B. Evaluation of the Model

Once a modeled three-dimensional structure has been constructed, the reasonableness of the model can be checked in various ways. These checks involve asking if the model structure resembles the known structures of properly folded proteins. Thus the structural characteristics of the modeled structure are compared to the characteristics of a set of experimentally determined structures. One characteristic that is often examined is the distribution of backbone dihedral angles, ϕ and ψ. An examination of known protein structures shows that the backbone dihedral angles for a properly folded protein generally fall into a few allowed regions based on steric and electrostatic constraints. If the modeled structure has a significant number of backbone dihedral angles for non-glycine residues falling in unallowed regions, the structure should be considered unreasonable and an improved model should be generated. Another property that is often examined is the solvent accessible surface area of the protein and of different classes of amino acid residues in the protein. The solvent accessible surface area of each amino acid residue in the modeled structure can be calculated (Lee and Richard, 1971) and the results can be compared with expectations based on similar calculations on known protein structure. A properly folded protein will have a relatively small contribution to its total solvent accessible surface area coming from nonpolar residues such as Leu, Ile, Val, and a much larger fraction coming from charged and polar residues. A related property is the overall packing of the buried residues (Gregoret and Cohen, 1990). A poorly packed protein will have too much volume on a per residue basis. One can also check whether the distribution of residues with respect to secondary structure assignment is reasonable. Different amino acid residues will have different statistical preferences for α-helices, β-sheets, and turns (Fasman, 1989). The modeled structure should be generally consistent with these preferences. In addition to examining these various properties individually, it is also possible to use a method such as Profiles 3-D (Eisenberg *et al.,* 1997), which scores protein structures based on a simultaneous evaluation of several of the aforementioned properties. All of these approaches focus on the question of whether the modeled structure is reasonable compared to a set of "typical" proteins. However, a far more important level of evaluation is determining whether the modeled structure is useful in understanding and explaining experimental data such as the regio- and/or stereoselectivity of monooxygenation of particular substrates, the effect on selectivity of particular point mutations, and the mode of action of different inhibitors.

C. Docking of Substrates/Inhibitors into the Active Site of *P*450 Models

To assess whether a model of a *P*450 is useful, it is common to try and dock substrates and/or inhibitors into the active site of the modeled protein. Docking of substrates or inhibitors into the active site of the homology model can help to explain enzyme–substrate interactions as well as the role of particular residues in catalysis. Docking can be done by either an automated approach or a manual approach. Automated approaches include such programs as AutoDock (Goodsell *et al.,* 1996), Dock (Kuntz *et al.,* 1982), and Ludi (Bohm, 1992). In these approaches, the interactions of the substrate and enzyme are evaluated for a number of different binding conformations and the various binding conformations are scored on the basis of favorable interaction energies. Such scoring may be based primarily on electrostatic and van der Waals energies or it may include specific considerations such as matching up H-bond donors and acceptors. One example of the use of automated docking with a *P*450 is a recent study in which Dock was used to examine possible novel substrates for *P*450cam (De Voss and Ortiz de Montellano, 1995).

In a manual docking approach, the researcher manipulates the coordinates of the substrate or inhibitor to give a set of energetically favorable binding geometries. The evaluation of a particular binding geometry may be based only on steric and electrostatic considerations or it may also include a consideration of known sites of metabolism. If the molecule being docked is a competitive inhibitor, then docking based only on sterics is appropriate. This approach when used for a substrate may reflect the preferred binding orientation of the substrate in the initial enzyme–substrate complex. If the docked molecule is a known substrate, then taking into account known sites of metabolism is appropriate. For known substrates, it is possible to use docking to create either a binary enzyme–substrate complex or a ternary complex with the substrate oriented toward the heme cofactor, which has a ferryl oxygen attached to it. If a ferryl oxygen is explicitly used in a docking calculation, then docking can be done in a way that ensures that the substrate is oriented in the active site to allow the initial hydrogen or electron abstraction. This approach has been successfully utilized to interpret changes in regiospecificity of substrate oxidation and susceptibility to inactivation upon residue replacement by site-directed mutagenesis (e.g., Szklarz *et al.*, 1995; Kent *et al.*, 1997; Kobayashi *et al.*, 1998). Using molecular dynamics calculations as a follow-up to an initial docking study may further increase our understanding of *P*450 catalysis and the importance of substrate mobility in the active site.

D. Limitations of Homology Modeling

The basic assumption in homology modeling is that a modeled protein resembles the structure(s) used as templates; therefore, the choice of the template(s) is of crucial importance. In the case of modeling a mammalian *P*450, the use of the bacterial P450s as templates, although appropriate, does neglect the fact that the mammalian variants are all membrane bound while the *P*450s of known structure are all cytosolic. However, the success of the various modeled structures in explaining a variety of experimental data suggests that the membrane-bound and cytosolic enzymes do share the same basic fold.

The final 3-D homology structure is highly dependent on the modeling procedure and is influenced by a number of factors. The most important factor is the initial sequence alignment. Alternate alignments may lead to differences in the location of some residues, and thus result in different models. For example, in *P*450 2B1 models, the identity of active site residues depended on the alignment (Szklarz *et al.,* 1994). Thus, the accuracy of the alignment is of crucial importance for the final structure. Another important factor is the choice of the modeling method. In molecular replacement methods, the coordinates of the SCRs are identical to those of a given reference protein, whereas in consensus models, they are averages of those of several templates. In the case of cytochromes *P*450, a model based on structures of several known enzymes should be more accurate than one based on the crystal structure of only a single protein, especially in view of the low sequence identity between mammalian and bacterial *P*450s. Moreover, the choice of coordinates for loops can alter the location of key residues, as shown in the case of *P*450 2B1 models (Szklarz *et al.,* 1994).

The geometry of the final model is also dependent on the force field used. A force field contains atomic parameters and energy terms utilized to calculate the energy of the molecule, and these may be different in different force fields. In particular, the choice of optimal binding orientations for a docked substrate or inhibitor may vary depending on the force field used. A variety of force fields have been developed to model proteins, and of these AMBER and CVFF have been used successfully in *P*450 modeling as well as molecular dynamics studies of crystallized *P*450s (Paulsen *et al.,* 1996). Additional parameters are required to describe the heme moiety, such as those to be used with the CVFF force field (Paulsen and Ornstein, 1991, 1992). Depending on the substrate or inhibitor being examined, additional new parameters may also be required. Finally, the choice of the refinement method, such as different minimization algorithms, or final minimization with or without water, may also influence the 3-D structure of the model.

E. Data Mining of Sequence Alignments

In the most recent version of "Entrez" from the National Center for Biotechnology Information on the Internet (http://www.ncbi.nlm.nih.gov), which contains data on proteins compiled from several databases (e.g., Genbank, EMBL, PIR), there are almost 2,000 sequences and fragments attributed to cytochromes *P*450. Some of those sequences were obtained from the genome project of *Caenorhabditis elegans* and *Arabidopsis thaliana.* By and large, the function of these sequences is completely unknown—which is not surprising because there are still many mammalian *P*450s whose functions and substrates are not known. In addition to understanding the function of these different *P*450s, there are questions about the role of different regions of the *P*450s: What are their functions in membrane association, substrate binding, redox partner binding, and catalysis? One way of answering these questions is through sequence alignments and analysis.

First, as always, note that the alignment of sequences is the most important since it is the basis of the analysis. So, how should the alignments best be done? Initially, an automated alignment can be done, followed by hand adjustments. Gapping should be kept at a minimum in α-helices and β-strands, and if possible placed into loops and turns. Locating turns is a key to aligning homologous proteins in that they signal the beginnings and ends of structural elements, and in proteins that are predominantly α-helical such as *P*450s, it is especially helpful. Also, if there is a clustering of potential-turn residues, it frequently indicates a coil or loop. What are these potential-turn residues? Of course, there are Pro and Gly—Pro causing a defined turn and Gly conferring the flexibility to make a turn in combination with bulkier and frequently charged residues such as Lys, Arg, and Gln. Also, Asn, Asp, Thr, and sometimes Ser can potentiate turns in combination with Pro or Gly or with other polar or charged residues. Several Asn are sometimes found clustered together in loops or forming turns.

With a reasonable sequence alignment in hand, one can start comparing regions of the *P*450s for insight into differences among their functions. An interesting region to look at is the redox-partner binding and charge-transfer region which appears to be found predominantly between the J′ helix through the L helix (see the section on Redox Partner Binding). In the alignment shown in Fig. 1, human *P*450 sequences are compared with those of rat CYP2B4. Here two residues in CYP2B4 that have been shown to be involved in reductase binding (Bridges *et al.*, 1998) are indicated with asterisks. Specifically in the L helix, where the alignment has a high probability of being correct, only half of the sequences have a basic residue in the same position. This may indicate an altered binding of the reductase with the *P*450s, which may affect the K_d or charge transfer.

On a more global level, regions in aligned sequences can be identified that have specific functions simply by the conservation of the character of that region. For example, the membrane-spanning region is defined by the hydrophobic clustering of residues at the N-terminal end, the heme-binding region by the conservation of residues around the cysteinyl ligand. So too can other regions be defined. Cohen and coworkers

```
Class II P450s:                         *      F  G R C G    *  ?
2b4 rat          PNTFNPGHFLD      ANGALKRNEGFMPFSLGKRICLGEGIARTELFLFFTTIL
1a1/p1           PSEFLPERFLT    PDGAIDKVLSEKVIIFGMGKRKCIGETIARWEVFLFLAILLQ
1a2/4            PSEFRPERFLTAD   GTAINKPLSEKMMLFGMGKRRCIGEVLAKWEIFLFLAILLQ
2c9/Mp4          PEMFDPHHFLDE      GGNFKKSKYFMPFSAGKRICVGEALAGMELFLFLTSILQ
2d6/db1          PFRFHPEHFLDA      QGHFVKPEAFLPFSAGRRACLGEPLARMELFLFFTSLLQ
2e1/J            PEKFKPEHFLNE      NGKFKYSDYFKPFSTGKRVCAGEGLARMELFLLLCAILQ
3A4              PEKFLPERFS      KKNKDNIDPYIYTPFGSGPRNCIGMRFALMNMKLALIRVLQ
17a              PDQFMPERFLN    PAGTQLISPSVSYLPFGAGPRSCIGEILARQELFLIMAWLLQ
21               PHEFWPDRFLE          PGKNSRALAFGCGARVCLGEPLARLELFVVLTRLLQ
19/arom          PNEFTLENFA          KNVPYRYFQPFGFGPRGCAGKYIAMVMMKAILVTLLR
4f2              PEVYDPFRFD      PENIKERSPLAFIPFSAGPRNCIGQTFAMAEMKVVLALTLL
4b1/ivb1         PEVFDSLRFS      TENASKRHPFAFMPFSAGPRNCIGQQFAMSEMKVVTAMCLL
4a11             PEVFDPFRFA        PGSAQHSHAFLPFSGGSRNCIGKQFAMNELKVATALTLL
102/P450BM3      DVEEFRPERFEN       PSAIPQHAFKPFGNGQRACIGQQFALHEATLVLGMMLK
Class I P450s:                                           ??
11a1 scc         DPENFDPTRWLS       KDKNITYRNLGFGWGVRQCLGRRIAELEMTIFLINMLEN
11b1              PERYNPQRWLDI      KGSGRNYHVPFGFGMRQCLGRRLAEVEMLLLLHHVLK
11b2              PERYNPQRWLDI      RGSGRNHHVPFGFGMRQCLGRRLAEVEMLLLLHHVLK
27 27oh           PESFQPHRWLRNSQPATPRIQHPFGSVPFGYGVRACLGRRIAELEMQLLLARLIQ
101/P450cam      ENACPMHVDFSR           QKVSHTTFGHGSHLCLGQHLARREIIVTLKEWLT
                  "Meander"                  HB Region     L Helix
```

FIG. 1. Sequence alignment of human *P*450s with rat CYP2B4 around the heme-binding region. Mutated residues in CYP2B4, which show reduced reductase binding (Bridges *et al.,* 1998), are double underlined and in bold, and the column of the alignment is indicated with an asterisk. The heme-binding region is indicated by "HB Region" and the conserved residues are at the top of the columns. Basic residues that may act in an analogous fashion to the CYP2B4 residues are indicated in bold with a single underline. CYP102 (*P*450BMP) and CYP101 (*P*450cam) are shown as structural references and their structural elements are indicated with underlining and labeled below the sequences. Lys344 and Arg364 in *P*450cam, shown in bold, are believed to be involved in putidaredoxin binding. Question marks above residues indicate possible involvement in eukaryotic *P*450 redox partner binding in the L helix.

(Lichtarge *et al.,* 1996a) have devised an "evolutionary trace method" that aids in identifying active sites and functional interfaces. They have used this method successfully to identify the G-protein–receptor complex interface (Lichtarge *et al.,* 1996b). This can be used for identification of redox-partner binding regions by locating clusters of conserved, charged residues; however, this is complicated by the fact that several different regions are probably involved in the binding. Peterson and Graham (1998) have used this method to identify microsomal versus mitochondrial *P*450 redox-partner binding surfaces, but it can be further refined to families of *P*450s. As shown in Fig. 1, potential redox-binding residues in the L helix are in different locations in the human microsomal CYP2 family as compared to the CYP4 family, and the mitochondrial basic residues fall into even another grouping, as might be expected. A similar analysis can be done at the binding site N terminal of the heme-binding region. Thus, with a good sequence alignment and some "data mining," one can use the sequence databases for a better understanding of *P*450 structure and function.

III. Homology Models in Studies of *P*450 Function

A number of homology models of mammalian *P*450s have been constructed based on one or more bacterial template(s). Early models were based on the structure of *P*450cam, a camphor hydroxylase from *Pseudomonas putida* (Poulos *et al.,* 1985; 1987), because that was the only crystal structure available at the time. These include models of *P*450scc (Vijayakumar and Salerno, 1992), aromatase (Laughton *et al.,* 1993), *P*450 1A1 (Zvelebil *et al.,* 1991), *P*450 2D6 (Koymans *et al.,* 1993), and *P*450 2B1 (Szklarz *et al.,* 1994). In some of these models, such as *P*450 2B1, docking of various substrates and/or inhibitors was used to explain enzyme specificity. When the hemoprotein domain of *P*450 BM-3, a fatty acid monooxygenase from *Bacillus megaterium* (Ravichandran *et al.,* 1993) was crystallized, it provided a second template for constructing homology models of mammalian enzymes. Because *P*450 BM-3 exhibits higher homology with microsomal *P*450s than *P*450cam, it is considered a better reference protein for modeling of these enzymes. Several *P*450 models based on the *P*450 BM-3 structure have been published, such as those of members of *P*450 1A subfamily (Lewis and Lake, 1996), including a new model of 1A2 (Lozano *et al.,* 1997), 2A subfamily (Lewis and Lake, 1995), as well as *P*450 2B1, 2B4, 2C3, and 2D6 (Lewis, 1995).

With the crystallization of additional bacterial *P*450s, *P*450terp, an α-terpineol monooxygenase from a *Pseudomonas* species (Hasemann *et al.,* 1994), and *P*450eryF, a 6-deoxyerythronolide B hydroxylase from *Saccharopolyspora erythrea* (Cupp-Vickery and Poulos, 1995), more information about *P*450 structure has become available. The comparison of known *P*450 enzymes has shown that their topology is quite similar, especially in the heme-binding core region, and a *P*450 structural fold can be easily defined (Cupp-Vickery and Poulos, 1995; Hasemann *et al.,* 1995; Graham-Lorence and Peterson, 1996). This prompted efforts to model mammalian *P*450s using several bacterial templates. A structure-based alignment of *P*450cam, *P*450 BM-3, and *P*450terp has been used to construct models of *P*450 2B1 (Szklarz *et al.,* 1995), aromatase (Graham-Lorence *et al.,* 1995), and *P*450 2D6 (de Groot *et al.,* 1996). More recent models of *P*450 3A4 (Szklarz and Halpert, 1997b) and *P*450 2B4 (Chang *et al.,* 1997) were based on four known structures. Certainly, with the availability of additional *P*450 structures, such as that of recently crystallized *P*450nor (Park *et al.,* 1997), more templates can be utilized to construct ever better *P*450 models. Moreover, further improvement in the accuracy of the models can be achieved by the incorporation of experimental data. In the case of 2B1 and 2D6 models, the sequence alignment was optimized using data from site-directed mutagenesis. Another model of *P*450 2D6, also utilizing a multiple alignment of the three bacterial proteins, *P*450cam, *P*450 BM-3, and *P*450terp, incorporated data from NMR studies of substrate binding (Modi *et al.,* 1996).

A. The Active Site: Enzyme–Substrate/Inhibitor Interactions and *P*450 Catalysis

1. *Active Site Architecture and Substrate Binding*

There are several reasons to study the active site structure: to determine the substrate specificity and the mechanism of substrate monooxygenation selectivity, to design inhibitors, and to understand the catalytic mechanism and possible intermediates. Generally, the steroidogenic *P*450s are substrate specific and monooxygenation selective; for example, *P*450scc catalyzes the conversion of cholesterol to pregnenolone, while the xenobiotic *P*450s can usually metabolize a range of substrates yet are relatively regio- and stereoselective.

Two places may allow substrate specificity: at the entrance to the access channel, and in the active site. A substrate has to be able to enter the access channel, as well as be accommodated in the active site. For example, long-chain fatty acid hydroxylases (e.g., *P*450 BM-3 or *P*450 4A11) generally have a basic residue at the entrance to the access channel, and that channel has to be long to accommodate the fatty acid chain length. Further, the active site above the heme iron has to be sufficiently narrow not to allow the ω-end of the fatty acid to fold back on itself too far, so that only the ω-1 or ω-2 carbons are presented for hydroxylation. These structural requirements are fulfilled in the crystal structure of *P*450 BM-P (the *P*450 domain of *P*450 BM-3). Finally, in the active site, hydrophobic residues sterically hold the substrate in a specific orientation so that hydroxylation is regio- and stereospecific. In *P*450 BM-P, when Phe-87, which is above pyrrole ring C of the heme, was mutated to Val, arachidonic acid metabolism changed from predominately 18*R*-hydroxylation to 14*S*,15*R*-epoxidation, but the initial rate was not substantially altered (Graham-Lorence *et al.,* 1997). On the other hand, when Arg-47, which is located at the entrance to the access channel, was changed to Ala, the initial rate was reduced, but the selectivity of substrate hydroxylation was not changed (Graham-Lorence *et al.,* 1997). Thus, as suggested in the previous example, the regions that confer stereo- and regioselectivity of oxidation are generally located directly above the heme in the active site, and are composed of five regions: the B-B′ and B′-C loop (SRS-1), the F helix (SRS-2), the center portion of the I helix (SRS-4), a small region between the K helix and β-strand 1–4 (SRS-5), and the residues in the turn of β-sheet 4 (SRS-6).

Docking of substrates or inhibitors into *P*450 models allows for the evaluation of the types of interactions that are responsible for binding the compound in the active site. As mentioned in the section on docking (Section II,D), a molecule can be placed in a productive binding orientation that allows for the initial oxidative event to occur, or in a nonproductive orientation based on steric considerations. In the case of *P*450 2B1 (Szklarz *et al.,* 1995) and 3A4 (Szklarz and Halpert, 1997b), the analysis of enzyme–substrate interactions indicated that hydrophobic interactions are mainly responsible for binding various substrates in the active site in a productive binding orientation. Likewise, the prevalence of hydrophobic interactions in binding of codeine in the active site of *P*450 2D6 has been proposed, although an ion pair formed between the basic nitrogen of the substrate and Asp-301 found in the I helix may also play a role (Modi *et al.,* 1996).

2. *Identification of Key Residues*

Homology models have been extensively utilized to locate and/or predict key amino acids that may interact with the substrate or inhibitor docked into the active site. The substitution of such a key residue with another amino acid results in a drastic alteration in activity of quantitative or qualitative nature. In early studies, key residues were first identified experimentally, and the models were utilized to confirm whether a given residue interacts with the substrate as expected. When little data on a given *P*450 are available, the identification of key residues from molecular models becomes of great importance to guide site-directed mutagenesis experiments that unequivocally verify those predictions.

A model of *P*450 2B1 based on the structure of *P*450cam (Szklarz *et al.,* 1994) correctly predicted the effect of substitutions at 9 out of 10 amino acid positions. Moreover, based on docking of steroid substrates in the active site of that model, we were able to identify additional residues likely to affect activity. Subsequently, the importance of key residues Leu-209, Ile-477, and Ile-480 was confirmed experimentally and led to the construction of an improved *P*450 2B1 model (Szklarz *et al.,* 1995). In that model, built using consensus methodology, key residues Ile-114 (SRS-1), Phe-206 (SRS-2), Ile-290 and Thr-302 (SRS-4), Val-363 and Val-367 (SRS-5), and Gly-478 (SRS-6), as well as newly identified residues 209, 477, and 480, constituted part of the active site of the enzyme, while other residues studied were farther from the active site, consistent with site-directed mutagenesis data for cytochromes *P*450 2B. In *P*450 2B1, one can distinguish the "upper part" of the pocket, with residues 114 and 290, and the "lower part," with residues 478 and 480. When a steroid substrate is docked in the active site in a productive binding orientation, it can fill either the upper or the lower pocket, where it interacts with amino acids located therein. As shown in Fig. 2, androstenedione docked into the upper part of the binding pocket can contact key residues 114, 206, 290, 302, 363, 367, and 477, but not 478 and 480. The latter two residues are able to interact with this substrate when it is bound in the lower part of the active site (Szklarz *et al.,* 1995).

The results from docking of substrates into the model of *P*450 3A4 made it possible to pinpoint residues that might be important for en-

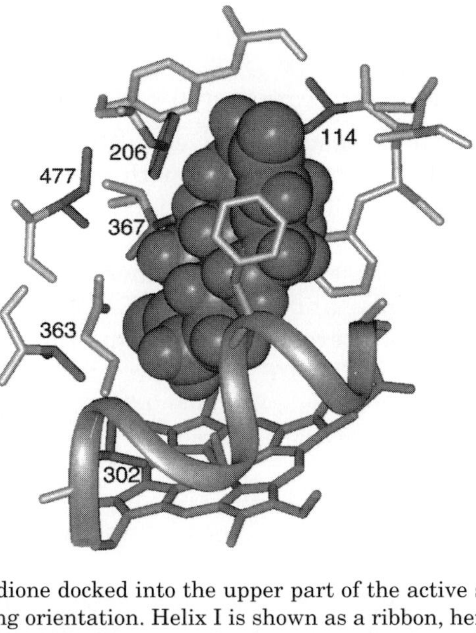

FIG. 2. Androstenedione docked into the upper part of the active site of the *P*450 2B1 model in a 16β-binding orientation. Helix I is shown as a ribbon, heme is dark gray, and key residues are in black. All of these residues are located within 5 Å from the substrate (Szklarz *et al.,* 1995).

zyme function and to target them for site-directed mutagenesis. Residues Ala-370 and Leu-373 from SRS-5 were indeed found to be of key importance in progesterone hydroxylation (He *et al.,* 1997), as were Ile-301, Phe-304, Ala-305, and Thr-309 located in helix I (SRS-4) (Domanski *et al.,* 1998). The most recent findings indicate that residue Ser-119 (SRS-1) is also essential for progesterone oxidation (Fabienne Roussel, private communication), as predicted by the model.

A model of *P*450 2D6, based on three bacterial structures (de Groot *et al.,* 1996), contained key amino acid residues Asp-301 (SRS-4) and Val-374 (SRS-5) in its active site. The model was able to accommodate three known enzyme substrates, debrisoquine, dextromethorphan, and GBR 12909. Another model of *P*450 2D6, which included NMR studies of substrate binding (Modi *et al.,* 1996), showed that the substrate codeine is able to interact with Asp-301, which forms an ion pair with the basic nitrogen of the substrate. A number of other residues interacting with codeine have also been identified.

It is important to remember that the identity of residues that contact the substrate or inhibitor depends on the structure of that compound and on the particular orientation it may assume in the active site. For example, in *P*450 2B1, mutation at position 209 (Ile → Ala) altered the regiospecificity progesterone hydroxylation (Fig. 3), while the metabolism of androstenedione was not affected (Szklarz *et al.,* 1995). In the case of *P*450 3A4, the number of residues that directly interact with the substrate docked in a productive binding orientation depends on the size of that substrate: 9 residues contact progesterone, but 18 contact much larger erythromycin (Szklarz and Halpert, 1997b).

3. *Elucidation of Enzyme Mechanisms*

In addition to the characterization of the active site, the mechanism of enzyme action can be analyzed using molecular models. An example of this is the conversion of androstenedione or testosterone by cytochrome *P*450 aromatase (*P*450arom) to estrone or β-estradiol, respectively. Three successive monooxygenation cycles take place consuming three equivalents of NADPH and three equivalents of molecular oxygen in which the 19-methyl group between the A and B steroid rings is sequentially monooxygenated first to an alcohol, then to an aldehyde. A hydrogen is then abstracted from the C_2 position followed by the release of the C_{19} carbon as formic acid and the aromatization of the A-ring forming estrogen.

This sequence of reactions raises several questions, such as these: What residues both sterically and electrostatically hold the substrate in the appropriate position to allow the three monooxygenations? Why does

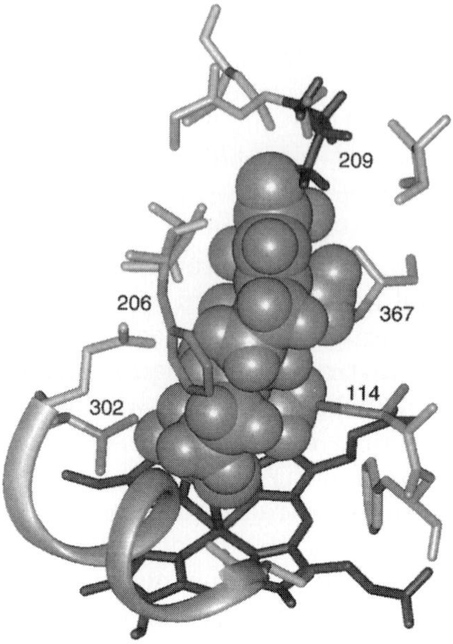

FIG. 3. Progesterone docked into the active site of the *P*450 2B1 L209A mutant in a 21-binding orientation. Helix I is shown as a ribbon, heme is black, as is the Leu-209 side chain. Key residues are labeled. The presence of Ala at position 209 (light gray) allows progesterone to assume a new 21 orientation, which is not possible in the wild-type enzyme due to van der Waals overlaps between the substrate and the side chain of Leu-209 (Szklarz *et al.*, 1995).

the third monooxygenation cause a carbon–carbon cleavage? A model of the active site of aromatase can help answer these questions. Thus, a homology model of *P*450arom was built by Graham-Lorence *et al.* (1995) helix by helix to optimize the interactions between helices yet generally following the *P*450 BM-P backbone, with β-sheet 1 and 2 modeled on the *P*450 BM-P sheets. When sequence homology was recognizable, for example, charged residues aligned with charged, and hydrophobic with hydrophobic, loops were designed to follow the backbone of one of the three structurally determined *P*450s known at that time; alternatively, the shortest path between two helices was taken except for the B′-C loop.

As seen in the crystallized *P*450s, the B′-C loop (SRS-1) is especially important in substrate orientation; for example, in *P*450cam it contains the active site residue Tyr-96, which hydrogen bonds to the 2-keto group of camphor, and in *P*450 BM-P, it contains Phe-87 as mentioned earlier. The secondary structure in this loop is determined by the residues that induce turns, and by the placement of the charged versus hydrophobic residues as demonstrated here in *P*450cam and *P*450 BM-P, and comparing them with the *P*450arom sequence:

```
P450cam....  CPFIPREAGEAYDFIPTSMDHPEQRQFRALANQVV...
P450BM-P ...SQALKFVRDFAGDGLFTSWTH.EKNWKKAHNILL...
P450arom.....  GLQCIGMHEKGIIF..NNNPELWKTTR...
                B′ Helix            C Helix
```

where the double-underlined residues are known/putative active site residues, and those in bold are potential-turn residues. Charged residues in the loop are most probably pointing out to the surface. Additionally, note that the B′ helix is not a conserved helix in that its orientation is different among the structurally determined *P*450s, and thus not readily aligned. In *P*450arom, Chen and coworkers mutated the "active site" residue His-128 to Gln and Ala, and found that the decrease in V_{max} was similar for both androstenedione and testosterone; however, the mutants exhibited a higher affinity for androstenedione than testosterone (Zhou *et al.,* 1992). These findings indicate that His-128 may orient testosterone in the active site by both a π-π stacking between the five-member steroid D ring and the His-128 pyrrole ring, and hydrogen–bond interactions between the histidine pyrrole ring and the C_{17}-hydroxy group. On the other hand, androstenedione may be oriented only by the π-π stacking. If this is the case, as we believe it is, then the C_{19} methyl group is oriented close to Asp-309 and Thr-310 (SRS-4) in the center of the I helix above pyrrole ring B. Moreover, in the case of androstenedione the β-face of the steroid is presented, whereas in testosterone both the α- and β-faces are presented to Asp-309 and Thr-310.

In the center of the I helix, there is a highly conserved motif, AGh(D/E)T, where "h" is usually a hydrophobic residue, and where a mild bend is encoded in the helix and where the conserved acidic residue and the Thr have been shown to be involved in *P*450 monooxygenase reactions. In *P*450arom, the actual sequence AAPDT appears to encode a stronger bend/kink due to the Pro in the I helix than is seen in most other *P*450s. With the substrates aligned with His-128, androstenedione or testosterone will fit into the cleft with the methyl group close to the Asp and Thr residues. From this configuration, it was postulated that the first two monooxygenations would occur as usual forming 19-hydroxy and 19-oxo product, but with the presence of the 19-oxo group proximal to Thr-310, Thr-310 would hydrogen bond to it, polarizing the carbonyl double bond and promoting a peroxidative anion attack by the Fe-O-O$^-$, breaking the C10–C19 bond. Thus, the *P*450 model was able to suggest a plausible catalytic mechanism for the aromatase reaction.

4. *Understanding P450 Inhibition*

Docking of inhibitors coupled with the evaluation of enzyme–inhibitor interaction energy may help to explain the inhibitory effect of the compound. An important consideration in docking is the mechanism of action of the inhibitor. In the case of aromatase, the enzyme inhibitor, *S*-vorozole, was docked in such a fashion that its imidazole ring points toward the heme, which allowed one of the nitrogens to bind to the heme iron, in agreement with the known inhibition mechanism (Graham-Lorence *et al.,* 1995). Similarly, de Groot *et al.* (1996) docked a 2D6 inhibitor, ajmalicine, into the model of *P*450 2D6, and the most energetically favorable protein–inhibitor complex was formed when the A-ring of ajmalicine was placed above pyrrole ring B of the heme moiety.

In addition to competitive inhibitors, mechanism-based inactivators constitute an important group of compounds that affect *P*450 function. The inactivation is a multistep process that involves the initial binding of the inactivator followed by its oxidation to a reactive intermediate that irreversibly inactivates the enzyme. In the case of *P*450s, this intermediate may react with one of the amino acid side chains or with the heme moiety. For example, secobarbital (SB), a mechanism-based inactivator of *P*450 2B1, can be oxidized by the enzyme to form a reactive intermediate that alkylates the heme or the protein. Docking of SB into the active site of *P*450 2B1 helped to identify Thr-302 as a target residue for protein alkylation thus providing additional insight into the mechanism of inactivation (He *et al.,* 1996).

When a compound that acts as a mechanism-based inactivator is metabolized at different sites of the molecule, different products may be formed, some of them unable to inactivate the enzyme. In the case of SB, the oxidation of this compound leads to the formation of the epoxide product or to the inactivation of *P*450 2B1 by heme-adduct formation or protein modification. Thus, a thorough analysis of 2B1 inactivation by SB requires the examination of enzyme-SB interactions when the compound is docked in distinct binding orientations leading either to the epoxide product or to the inactivating reactive intermediate (He *et al.,* 1996). Another *P*450 2B1 inactivator, *N*-benzyl-1-aminobenzotriazole (BBT), can be oxidized either at the 1-amino nitrogen, which results in the generation of products that inactivate the enzyme, or at the 7-benzyl carbon, which leads to the formation of a stable metabolite. Therefore, this compound has been docked in two different binding orientations to account for the known inactivation mechanism and to correctly interpret experimental results (Kent *et al.,* 1997). Further details concerning the effect of mutations on *P*450 2B1 inactivation by SB and BBT are given in the subsequent section on the interpretation of experimental results.

B. The Active Site: Interpretation of Experimental Data

Homology models have been found very useful in the interpretation of alterations in activity in mutant *P*450 enzymes. They have been successfully used to explain changes in regio- and stereospecificity of substrate oxidation as well as alterations in inhibition and activation on residue substitution by site-directed mutagenesis (Szklarz and Halpert, 1997a). The usual procedure involves the replacement in the model of a given amino acid to mimic the mutant, and docking of the substrate or inactivator in an orientation leading to the expected product. The binding of the substrate in the active site of the mutant is then compared to that of the wild-type enzyme with respect to both enzyme–substrate interaction energies and the possibility of van der Waals overlaps.

1. *Van der Waals Overlaps*

The appearance or removal of van der Waals overlaps can be related to quantitative or qualitative changes in activity. For example, the elimination of the overlaps between the substrate and a side chain of a key residue in the mutant often enables the substrate to assume a new orientation in the active site leading to a novel activity compared with the wild-type enzyme. As shown in Fig. 3, the replacement of Leu-209 (SRS-2) with Ala in *P*450 2B1 confers a novel 21-progesterone hydroxylase activity due to an enlarged active site of the mutant, which allows progesterone to bind in a new orientation. The larger Ile present in the wild-type enzyme hinders the substrate from assuming this orientation due to van der Waals overlaps (Szklarz *et al.,* 1995). A similar mechanism can explain the appearance of a novel product, 6β-OH androstenedione, in the *P*450 2B1 V367A mutant (Szklarz *et al.,* 1995). Likewise, the replacement of Val-363 (SRS-5) with a smaller Ala in *P*450 2B1 allows for hydroxylation of androstenedione in a 15α position (Szklarz *et al.,* 1994). This mutation also permits *O*-dealkylation of 7-butoxycoumarin, since the Ala side chain is small enough to allow bending of the butoxy chain not possible in the wild-type enzyme due to van der Waals overlaps between Val-363 and the substrate (Kobayashi *et al.,* 1998).

The appearance of van der Waals overlaps is often linked to the decrease or loss of activity. In the mutant enzyme, the overlaps that ensue between a key residue and a substrate hinder or prevent substrate binding. In *P*450 2B4, the mutation of Ile-114 to Phe (SRS-1) hinders binding of benzyloxyresorufin as well as androstenedione in a 16β-binding orientation, resulting in the loss of activity (Szklarz *et al.,* 1996). Van der Waals overlaps also prevent *O*-deethylation of 7-ethoxycoumarin in the 2B1 G478S mutant, as well as *O*-dealkylation of 7-propoxy- or 7-butoxycoumarin upon the replacement of Val-363 with Leu in *P*450 2B1 (Kobayashi *et al.,* 1998).

Similar phenomena can be observed when the mutation alters the susceptibility of the enzyme to inactivation. One of the compounds studied was secobarbital (SB), which is both a substrate and an inactivator of *P*450 2B1, that undergoes epoxidation or hydroxylation to a reactive intermediate that can alkylate heme (He *et al.,* 1996). Modeling of the SB-heme adduct demonstrated that residues 302 and 363, which are important for the heme alkylation, are close to the (OH)-SB product. Consequently, the decrease in adduct formation with V363L mutant appears to be due to the presence of van der Waals overlaps between Leu-363 and the oxidized SB product as a result of the longer side chain of Leu (Fig. 4). In the case of another 2B1-specific inactivator, BBT, a single substitution of Gly-478 to Ala prevents enzyme inactivation due to steric hindrance, which suppresses oxidation of BBT at the 1-amino nitrogen (Kent *et al.,* 1997). Thus, no inactivating products can be generated.

2. *Analysis of Compound Mobility*

The mutation of a key residue may alter compound mobility in the active site leading to the decrease or increase in activity. In *P*450 2B1, the replacement of Ile at position 477 or 480 with Ala leads to a drastic reduction in androgen 16α- and 16β-hydroxylase activity (Szklarz *et al.,* 1995). As shown in Fig. 5, these two residues appear to play a role in holding the substrate in a 16β-binding orientation through multiple hydrophobic interactions. On mutation of either residue to Ala, the side chains are too far from the substrate for the interactions to be maintained, which in turn may increase the mobility of the substrate in the active site and result in a consequent loss of activity (Szklarz *et al.,* 1995).

In some cases, the mobility of the compound is reduced on mutation. For example, in the *P*450 3A4 A370V mutant, Val stabilizes a 16α-binding orientation of progesterone through the increase in van der Waals contacts with the substrate, leading to the increase in nonbond interaction energy (He *et al.,* 1997). This phenomenon is likely to be responsible for the increased activity of the mutant. The decrease in substrate mobility may also lead to oxidation of the compound in alternate ori-

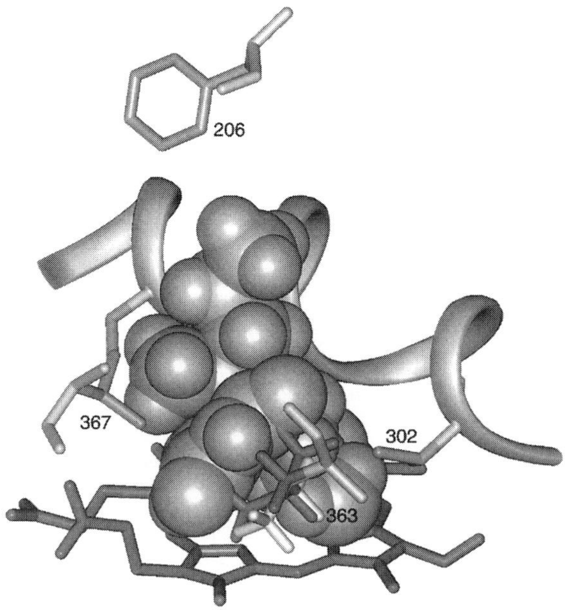

FIG. 4. Secobarbital-heme adduct of *P*450 2B1. The (S)OH-SB product is bonded to N_A of pyrrole ring A and shown in a space-filled representation. Residues 302, 363, and 367 that affect heme adduct formation are displayed. When Val-363 (gray) is replaced by Leu (black), van der Waals overlaps between the side chain of this residue and the oxidized SB product hinder heme adduct formation (He *et al.,* 1996).

entations, yielding novel products. In the *P*450 2B1 V363L mutant, the presence of a large Leu evidently limits the mobility of 7-butoxycoumarin and enables its binding in new orientations resulting in ω and (ω-1) hydroxylation of the alkoxy chain (Kobayashi *et al.,* 1998). Figure 6 shows 7-butoxycoumarin docked into the active site of this mutant in an orientation allowing for its (ω-1) hydroxylation.

A detailed analysis of compound mobility can be performed using molecular dynamics. In very recent studies, we utilized this technique to explain differences in susceptibility to inactivation by 2-ethynylnaphthalene (2EN) between two highly related enzymes, *P*450 2B4 and 2B5 (Strobel *et al.,* 1999). In the case of 2B4, which is highly susceptible to 2EN, the inactivator remained in a productive binding orientation close to heme and Thr-302, the proposed site of enzyme modification, during 5 ps of simulation. On the other hand, in *P*450 2B5, the 2EN molecule moves out of the productive binding orientation within the first 60 fs and assumes a different orientation far from heme and residue 302.

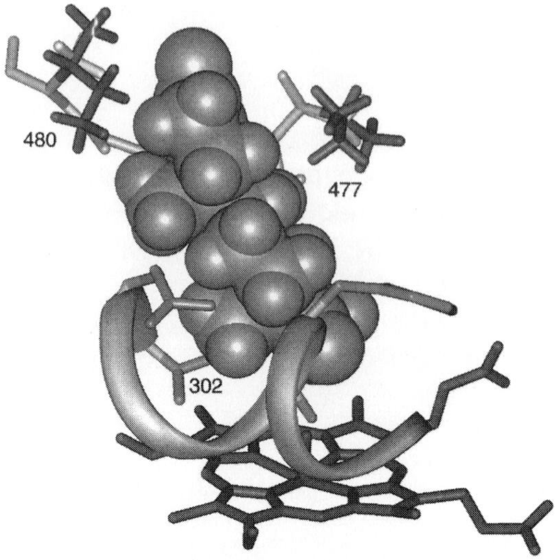

FIG. 5. Androstenedione docked into the active site of the *P*450 2B1 model in a 16β-binding orientation, in the lower part of the binding pocket. Helix I is shown as a ribbon and heme is black. Ile-477 and Ile-480 (the side chains are black) may play a role in holding the substrate in a proper binding orientation. When either of these residues is substituted with Ala (light gray), the enzyme–substrate interactions are no longer maintained, leading to the increased mobility of androstenedione in the active site of the mutant (Szklarz *et al.,* 1995).

Moreover, this new orientation remains quite stable throughout the 5-ps simulation. Because both *P*450 2B4 and 2B5 catalyzed the oxidation of 2EN to the stable product, 2-naphthylacetic acid, the susceptibility to inactivation must be related to the metabolic fate of the reactive ketene intermediate. Based on the results of molecular dynamics, it is likely that in *P*450 2B5, this intermediate preferentially hydrolyzes with water instead of reacting with active site residues. Thus, the resistance of 2B5 toward inactivation by 2EN can be related to high mobility of this compound in the active site, which prevents enzyme modification.

3. *Docking of Several Molecules in the Active Site*

In *P*450 3A4, the active site of the enzyme must be big enough to allow for binding of large substrates such as erythromycin. The recent model of this enzyme fulfills this requirement well (Szklarz and Halpert, 1997b). The large size of the active site also makes it possible to dock several smaller molecules simultaneously. This phenomenon can explain both homotropic and heterotropic enzyme stimulation. Thus, two molecules of progesterone can easily fit into the active site of *P*450 3A4 leading to enzyme autostimulation, and the simultaneous presence of a steroid molecule and that of α-naphthoflavone (α-NF) may be responsible for enzyme activation by flavonoids (He *et al.,* 1997). Figure 7 shows progesterone and α-NF docked into the active site of *P*450 3A4. More than one molecule of α-NF can be docked simultaneously with progesterone. The plausibility of that hypothesis was confirmed by recent studies, in which residues 210, 211, and 363 were found to affect flavonoid stimulation in *P*450 3A4 (He *et al.,* 1997; Harlow and Halpert, 1997). These residues constitute part of the active site of the enzyme and are located relatively close to the bound effector (He *et al.,* 1997). Mutations at these positions may alter the volume of the active site and thus affect binding of α-NF. This hypothesis is further supported by most recent findings that the replacement of Leu-211 and Asp-214 with the larger Phe and Glu, respectively, led to the decrease in flavonoid stimulation of steroid hydroxylation (Harlow and Halpert, 1998). In

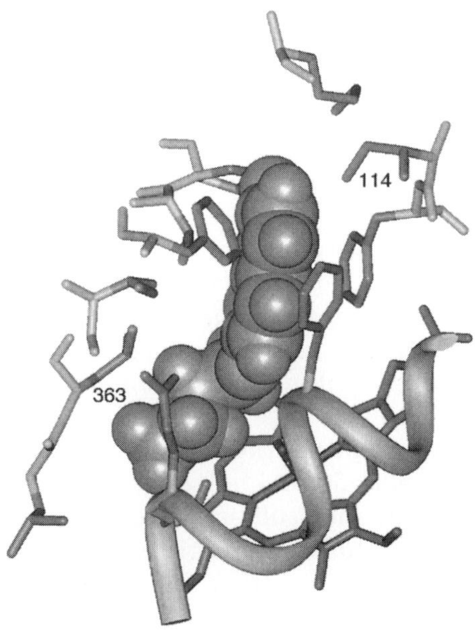

FIG. 6. 7-Butoxycoumarin docked into the active site of the *P*450 V363L mutant in an orientation that allows for its (ω-1) hydroxylation. Helix I is shown as a ribbon and heme is dark gray. The substitution of Val-363 with Leu may limit the mobility of the substrate and enable its binding in a new orientation.

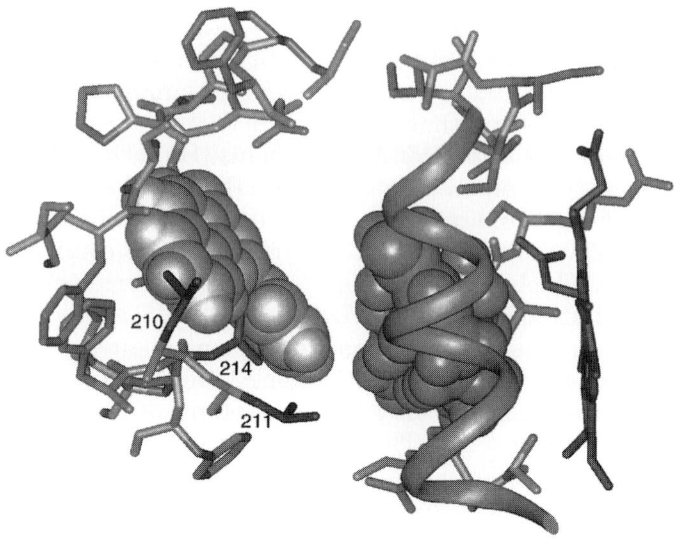

FIG. 7. Progesterone and α-NF docked into the active site of the *P*450 3A4 model. Progesterone in a 6β-binding orientation is in dark gray close to heme (black), and α-NF is in lighter gray. Helix I is shown as a ribbon. Residues 210, 211, and 214, shown to affect 3A4 stimulation by α-NF (Harlow and Halpert, 1997, 1998), are within 5 Å from the effector molecule.

this case, the residue substitution mimicked the action of the effector by reducing the size of the active site. Thus, the enzyme model suggested a plausible mechanism of *P*450 3A4 activation by flavonoids and pinpointed residues that play a key role in this process.

Positive cooperativity is also observed in the metabolism of aflatoxin B1 (AFB1) by *P*450 3A4, and may be due to simultaneous binding of two AFB1 molecules in the active site of the enzyme. As shown in our recent studies (Wang *et al.,* 1998), several amino acid substitutions affected the rate of AFB1 oxidation as well as the partition ratio between 3α-hydroxylation and 8,9-epoxidation. Docking of substrate molecules in the homology model of *P*450 3A4 helped to explain experimental results. When one substrate molecule is docked in a productive binding orientation, molecular dynamics simulations indicated that the second molecule may move freely in the active site cavity and be able to interact with a number of residues found to affect AFB1 metabolism. This molecule would thus act as an "activator." Major effects observed for residue replacements at positions 206, 210, and 369 can be linked to the alteration in the size of the amino acid side chain and the subsequent changes in packing and mobility of substrate molecules in the active site. It is also of interest that residue 210 was earlier discovered to affect 3A4 stimulation by α-NF (see earlier comments). Thus, in the case of *P*450 3A4, residues that constitute the active site may interact with one or more substrate molecules in a complex manner that determines the regio- and stereospecificity of substrate oxidation.

C. *P*450 Redox Binding Partners

In the reaction cycle of *P*450s, first the substrate binds followed by the binding of the redox partner, which donates an electron. Molecular oxygen then binds followed by the donation of a second electron, product formation, and product release. Cytochromes *P*450s can be divided into several classes depending on their redox partner requirement. As mentioned earlier, class I *P*450s, found in mitochondria and prokaryotes, require an iron-sulfur protein and an FAD-containing reductase. Class II *P*450s, all found in the endoplasmic reticulum of eukaryotes except for the prokaryote *P*450 BM-3 (Narhi and Fulco, 1986), require an FAD/FMN-containing NAD(P)H-reductase, whereas class III *P*450s do not require an exogenous source of electrons but obtain them from their substrates, which are peroxides. Furthermore, some *P*450s interact with cytochrome b_5 through a mechanism that is not fully understood. Two different types of interactions are postulated: one in which cytochrome b_5 donates the second-required electron to the *P*450 rather than the reductase (Vergeres and Waskell, 1995), and the other interaction affects the *P*450 presumably in an allosteric fashion facilitating the transfer of electrons from the reductase to the *P*450 (Yamazaki *et al.*, 1996; Auchus *et al.*, 1998); however, this latter interaction is still in question.

It has been shown by Wolf and coworkers with the FAD/FMN-containing reductase (CPR) of eukaryotes (Smith *et al.*, 1994), and in similar experiments by Peterson and coworkers with the FAD/FMN-containing reductase domain (BMR) of *P*450 BM-3 (Sevrioukova *et al.*, 1996) that the class II reductase can be molecularly engineered into two functional domains—an FAD domain and an FMN domain. These domains have been separately purified and then reconstituted in combination with their *P*450 in a productive fashion. Thus, these two domains can function separately. Recently, two crystal structures for class II redox partners have been determined: the eukaryotic *P*450 FAD/FMN-containing reductase (Wang *et al.*, 1997), as well as the complex between the FMN domain and the *P*450 domain of *P*450 BM-3 (Sevrioukova *et al.*, 1999). The structure of adrenodoxin (Adx), the class I iron-sulfur protein for eukaryotes (Muller *et al.*, 1998), has also been determined, but the coordinates are not yet available; therefore, we focus on class II interactions.

In the crystal structure of CPR, it can be seen that the FMN domain is attached by a linker region to the FAD domain, and the globular FMN domain is rotated such that its FMN is pointed toward and in proximity to the FAD of the crescent-shaped FAD domain (Wang *et al.,* 1997). On the other hand, in the cocrystallized structure of the FMN and *P*450 domains of *P*450 BM-3, the FMN of the FMN domain is oriented proximal to the heme of the *P*450 domain (Sevrioukova *et al.,* 1998). Thus, it would appear that the FMN domain might rotate from FAD to heme in transferring electrons while still embraced between the FAD and *P*450 domains. (Note that in both structures, the proteins are in their fully oxidized state, and thus the FMN domains of CPR and BMR may not be in their optimum orientation in the crystal structures.) Therefore, the FMN domain has to be capable of orienting and coupling to the FMN domain and then to the heme domain/protein depending on the charge and potential of the prosthetic groups.

The correct protein–protein alignment of the CPR or BMR with their respective *P*450(s) is most likely a combination of electrostatic and hydrophobic interactions. Waskell and coworkers (Bridges *et al.,* 1998) have mutated more than 20 residues in CYP2B4 and shown that an Arg just upstream of the heme-binding region and another in the L helix are involved in CPR binding. Surprisingly, none of three Arg residues that were mutated in the C helix affected CPR binding, but rather they reduced the binding cytochrome b_5. Thus, it would appear that on the proximal face of *P*450s appropriate docking/alignment of charged residues in the heme-binding region and the L helix are important for CPR binding, while residues in the C helix are more important for cytochrome b_5 binding. Finally, although a residue was mutated by Waskell and coworkers between the J and J′ helices, there was no residue mutated in the J′-K helix region, which is on the proximal face of *P*450s ringing the putative docking region and may be involved in reductase binding. On the other hand, in the complexed crystal structure of the *P*450 BM-3 heme domain and FMN domain, the C helix also does not associate with residues in the FMN domain, but more importantly, the interactions between the FMN and *P*450 domain are entirely hydrophobic (Sevrioukova *et al.,* 1998). That is, those basic residues found in the CYP2 family to be involved in CPR binding are not present in BMP. The comparison of the region around the FMN group on the FMN domain of CPR to that of BMR shows that there are three acidic residues and one basic residue on CPR that are not present on BMR. When comparing the FMN-binding region of other *P*450s with that of the CYP2 family, the CYP1 family has fewer basic residues; the CYP3 family has more; and the CYP17 has none. Therefore, it appears that hydrophobic interactions may be the primary force for the protein–protein complex, with electrostatic interactions playing a secondary role in the eukaryotic *P*450s, possibly indicating the preferences of CPR for one family of *P*450s over another.

Homology models may help to identify residues that might interact with the redox partner in a specific *P*450. For example, Graham-Lorence *et al.* (1995) indicated the region of *P*450 aromatase that might be responsible for redox partner binding and suggested that *P*450–reductase interactions may include both hydrophobic and ionic interactions. Both basic and acidic residues found in regions of helices B, J, K, and L are postulated to interact with the reductase. Hydrophobic and aromatic residues appear to be in abundance on the proximal face of aromatase mostly in the meander and in the region between the meander and the heme-binding region, creating a hydrophobic center over the heme ring similar to that of *P*450 BMP.

Homology models can be also utilized to study *P*450–cytochrome b_5 interactions. A model of *P*450 2B5 predicts that residue 120 is located in the N-terminal part of helix C, far from the active site of the enzyme, and thus cannot directly interact with the substrate progesterone (He *et al.*, 1998). Nevertheless, the substitution of His at this position with Arg decreased enzymatic activity. Based on the model and earlier data, it has been suggested that residue 120 plays a role in binding of cytochrome b_5. The subsequent experiments have shown that 2B5 H120R and 2B5 wild type were sensitive to the presence of cytochrome b_5 as predicted. Thus, the analysis of *P*450 models can help in prediction of specific residues that interact with one or another redox partner and to plan suitable experiments to verify those predictions.

D. Protein Stability

*P*450 models can be effectively utilized to analyze interactions between various side chains within the 3-D protein structure. Some of those interactions may be essential for protein stability or the ability of the enzyme to bind the heme prosthetic group. Consequently, mutations at critical positions would have profound effects on *P*450 stability and activity.

In *P*450 3A4, several Arg mutants at positions 365, 372, and 375 exhibited very low expression levels or were not expressed at all (He *et al.*, 1997). The effect of mutations on protein stability was analyzed using molecular modeling. In the case of Arg-372, the replacement of this residue with Lys did not affect the enzyme, but the mutant containing

Ser was unable to bind CO, based on its CO difference spectrum. Based on the enzyme model, an ion pair may be formed between Arg-372 and Asp-61. Thus, it is likely that a substitution of Arg-61 with Ser leads to misfolding or unfolding of the protein accompanied by the loss of heme. In contrast, a positively charged Lys-372 can maintain charge–charge interactions with Asp and thus successfully replace Arg.

Another important residue is Arg-375 of *P*450 3A4, which seems to play a role in heme binding. This residue aligns to other Arg residues in *P*450cam, *P*450eryF, and *P*450terp (Szklarz and Halpert, 1997b), which, in crystal structures, interact with the heme propionate group. A similar situation is found in the *P*450 3A4 model. Thus, a conservative, it would seem, mutation, such as the substitution of Arg-375 with Lys, may disrupt the formation of hydrogen bonds with the heme propionate group since the side chain of Lys is shorter than that of Arg. As a result, the mutant would be unable to bind heme, as confirmed experimentally by the analysis of its *P*450 reduced CO-difference spectrum (He *et al.,* 1997).

However, the most drastic effect was observed on the replacement of Arg-365. The mutant containing a Lys in this position was expressed only transiently at very low levels. The model of *P*450 3A4 indicates that Arg-365 participates in a complex network of hydrogen bonding, which may stabilize the region of helices K and the meander proximal to heme. As shown in Fig. 8, the guanidino group of Arg-365 is hydrogen bonded to the backbone carbonyl oxygen atoms of residues 402 and 411, the side-chain oxygen atom of Glu-362, and the side chain of Trp-408. Glu and Arg at positions equivalent to 362 and 365 in 3A4, respectively, are absolutely conserved in known bacterial enzymes, as are hydrogen bonds equivalent to those found in 3A4, except the bond between Arg-365 and Trp-408. Upon mutation of Arg-365 to Lys in 3A4, these hydrogen bonds are lost, which may decrease the stability of the mutant in the region of helices K and the meander, and lead to the subsequent unfolding of the whole structure. Thus, Arg-365 and its hydrogen bond network may contribute to the maintenance of the tertiary structure of cytochrome *P*450 3A4. The analysis of *P*450 models can provide important information about enzyme stability based on the interactions between various residues within the 3-D protein structures.

IV. Perspectives and Conclusions

During the last decade, there has been great progress in homology modeling methods, and their impact on studies of *P*450 function increased tremendously. Early models, laboriously constructed, were of-

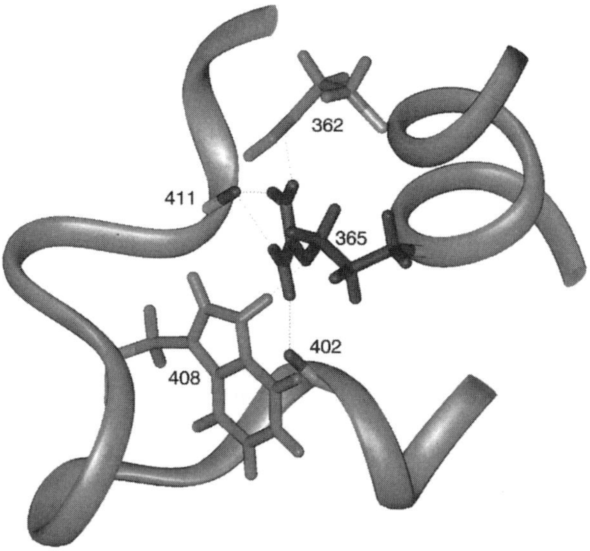

FIG. 8. Topology of the *P*450 3A4 model in the region of helices K and the meander. Hydrogen bonds between Arg-365 and residues 362, 402, 408, and 411 are shown in dotted gray lines. This hydrogen bond network stabilizes the tertiary structure of the enzyme (He *et al.*, 1997).

ten not too accurate, but nevertheless opened a new vista in *P*450 research. In recent years, molecular modeling, especially in conjunction with experimental methods, such as site-directed mutagenesis, has provided an important insight into enzyme structure–function relationships. The complexity of scientific issues involved has been increasing continually, from a relatively simple and straightforward identification of key residues that determine substrate specificity and inhibitor susceptibility, to analyses of compound mobility in the active site.

The majority of modeling research has focused on enzyme–substrate interactions in the active site, although other aspects of *P*450 function have been successfully investigated, such as protein stability and *P*450–redox partner interactions. Homology models make it possible to visualize binding of substrates and/or inhibitors in the active site and to determine predominant interactions involved, which helps to better understand enzyme mechanism. Furthermore, key residues, able to contact the substrate/inhibitor molecule, can now be easily identified and/or confirmed. Docking of the substrate into the active site of the enzyme model can indicate likely candidates for mutagenesis and thus guide experimental research.

Molecular models have been successfully utilized to explain experimental data, in particular alterations in regio- and stereospecificity of substrate oxidation, and susceptibility toward inactivation upon site-directed mutagenesis. These changes can be related to the removal or appearance of van der Waals overlaps in the mutant protein, which may lead to the formation of novel metabolites or a decrease in activity, respectively. The mutation may also cause alterations in substrate/inhibitor mobility in the active site, which can be investigated with molecular dynamics. Recently, molecular dynamics methods have been successfully utilized to analyze differential inactivation of related enzymes, *P*450 2B4 and 2B5. Docking simulations may also suggest plausible mechanisms for such aspects of *P*450 catalysis as heterotropic and homotropic enzyme activation. With a homology model of *P*450 3A4, it has been demonstrated that the simultaneous presence of two or more substrate and/or activator molecules may account for the observed enzyme stimulation. Modeling predictions were verified experimentally and found to be consistent with this hypothesis.

In addition to analyzing enzyme–substrate interactions in the active site, *P*450 models can be used to identify residues that may play a role in redox partner binding. Until very recently, not much has been known about structures and mechanisms of action of *P*450 reductase or ferredoxin (adrenodoxin). With *P*450 models, the regions of a molecule able to bind redox partners, reductase and cytochrome b_5, can be easily located, and residues likely to participate in binding unique to a given *P*450 can be pinpointed. Furthermore, the analysis of homology models can also provide an explanation for changes in enzyme stability, based on residue–residue interactions within the tertiary structure, as demonstrated with *P*450 3A4.

In summary, homology modeling allows for a mechanistic interpretation of various aspects of *P*450 function. These include enzyme catalysis and inhibition, interactions with redox partners, and protein stability. We can expect further development of homology modeling methods, as well as methods for structural verification of the models. Modeling methods are becoming more automated and faster, and homology modeling is moving slowly from the realm of "art" to that of "science." With available software, it is now relatively easy to construct a reasonable protein model and structural verification methods will ensure that the model is within allowed parameters. In the case of *P*450s, an important factor is the type of force field used and, in particular, parameters for heme and iron-oxo intermediate. The improvement of force fields for protein modeling and generation of better parameters for heme will allow for the more extensive utilization of molecular dynamics methods. This, in turn, will enable us to analyze substrate or inhibitor motion in the active site and to calculate binding free energies, as has been done for the crystal structure of *P*450cam (Paulsen and Ornstein, 1992, 1996; Paulsen *et al.,* 1993). The trend toward a more rigorous computational approach to *P*450 catalysis can thus be expected to grow. We can also anticipate that new types of functional studies will be undertaken, such as those concerning substrate access to the active site and electron transfer from the reductase and/or cytochrome b_5 to a *P*450. With the improvement of homology modeling methods, as well as methods to analyze various aspects of enzymatic catalysis, it will be more important than ever to verify modeling predictions experimentally. A "structurally" correct model need not be "functionally" correct, thus the experiment will remain the ultimate test of the model. Consequently, modeling research is likely to become an integral part of a new approach combining theoretical and experimental studies, highly advantageous in the field of structure–function research.

At present, a structure of a eukaryotic *P*450 has not been solved, although there is much effort in various laboratories directed toward that aim. Thus, it is likely that one of the mammalian enzymes will be crystallized in the very near future. When the structure of a eukaryotic enzyme is solved, it will provide a better template to construct homology models of mammalian *P*450s. Moreover, the methodology established and verified in structure–function studies utilizing homology models can be easily adapted and refined for use with the "real" structures.

Acknowledgments

This work was supported in part (GDS) by NIH grants ES03619 and GM54995 (Dr. James R. Halpert), Center Grant ES06694 (University of Arizona, Tucson, Arizona), and by the West Virginia University Computational Chemistry and Molecular Modeling Laboratory, and also in part by NIH grant GM43479 (for SEG).

REFERENCES

Auchus, R. J., Lee, T. C., and Miller, W. L. (1998). Cytochrome b_5 augments the 17,20-lyase activity of human P450c17 without direct electron transfer. *J. Biol. Chem.* **273,** 3158–3165.

Bohm, H.-J. (1992). LUDI: Rule-based automatic design of new substituents for enzyme-inhibitor leads. *J. Comput.-Aided Mol. Des.* **6,** 593–606.

Bridges, A., Gruenke, L., Chang, Y.-T., Vakser, I. A., Loew, G., and Waskell, L. (1998). Identification of the binding sites on cytochrome P450 2B4 for cytochrome b5 and cytochrome P450 reductase. *J. Biol. Chem.* **273,** 17036–17049.

Chang, Y.-T., Stiffelman, O. B., Vakser, I. A., Loew, G. H., Bridges, A., and Waskell, L. (1997). Construction of a 3D model of cytochrome P450 2B4. *Protein Eng.* **10,** 119–129.

Cupp-Vickery, J. R., and Poulos, T. L. (1995). Structure of cytochrome P450eryF involved in erythromycin biosynthesis. *Struct. Biol.* **2,** 144–153.

de Groot, M. J., Vermeulen, N. P. E., Kramer, J. D., van Acker, F. A. A., and Donne-Op den

Kelder, G. M. (1996). A three-dimensional protein model for human cytochrome P450 2D6 based on the crystal structures of P450 101, P450 102, and P450 108. *Chem. Res. Toxicol.* **9,** 1079–1091.

De Voss, J. J., and Ortiz de Montellano, P. R. (1995). Computer-assisted, structure-based prediction of substrates for cytochrome P450cam. *J. Am. Chem. Soc.* **117,** 4185–4186.

Domanski, T. L., Liu, J., Harlow, G. R., and Halpert, J. R. (1998). Analysis of four residues within substrate recognition site 4 of human cytochrome P450 3A4: Role in steroid hydroxylase activity and α-naphthoflavone stimulation. *Arch. Biochem. Biophys.* **350,** 223–232.

Eisenberg, D., Luthy, R., and Bowie, J.(1997). VERIFY3D: Assessment of protein models with three-dimensional profiles. *In* "Methods in Enzymology" (C. W. Carter, Jr. and R. M. Sweet, eds.), Vol. 277, pp. 396–404. Academic Press, San Diego, CA.

Fasman, G. D. (1989). The development of the prediction of protein structure. *In* "Prediction of Protein Structure and Principles of Protein Conformation" (G. D. Fasman, ed.), pp. 193–301. Plenum, New York.

Goodsell, D. S., Morris, G. M., and Olson, A. J. (1996). Automated docking of flexible ligands: applications of AutoDock. *J. Mol. Recognition* **9,** 1–5.

Gotoh, O. (1992). Substrate recognition sites in cytochrome P450 family 2 (CYP2) proteins inferred from comparative analyses of amino acid and coding nucleotide sequences. *J. Biol. Chem.* **267,** 83–90.

Graham-Lorence, S., and Peterson, J. A., (1996). P450s: Structural similarities and functional differences. *FASEB J.* **10,** 206–214.

Graham-Lorence, S., Amarneh, B., White, R. E., Peterson, J. A., and Simpson, E. R. (1995). A three-dimensional model of aromatase cytochrome P450. *Protein Sci.* **4,** 1065–1080.

Graham-Lorence, S., Truan, G., Peterson, J. A., Falck, J. R., Wei, S., Helvig, C., and Capdevila, J. H (1997). An active site substitution, F87V, converts cytochrome P450 BM-3 into a regio- and stereoselective (14S,15R)-arachidonic acid epoxygenase. *J. Biol. Chem.* **272,** 1127–1135.

Gregoret, L. M., and Cohen, F. E. (1990). A novel method for the rapid evaluation of packing in protein structures. *J. Mol. Biol.* **211,** 959–974.

Guengerich, F. P. (1993). Metabolic reactions: Types of reactions of cytochrome P450 enzymes. *Handb. Exp. Pharmacol.* **105,** 89–103.

Harlow, G. R., and Halpert, J. R. (1997). Alanine-scanning mutagenesis of a putative substrate recognition site in human cytochrome P450 3A4. Role of residues 210 and 211 in flavonoid activation and substrate specificity. *J. Biol. Chem.* **272,** 5396–5402.

Harlow, G. R., and Halpert, J. R. (1998). Analysis of human cytochrome P450 3A4 cooperativity: Construction and characterization of a site-directed mutant that displays hyperbolic steroid hydroxylation kinetics. *Proc. Natl. Acad. Sci. U.S.A.* **95,** 6636–6641.

Hasemann, C. A., Ravichandran, K. G., Peterson, J. A., and Deisenhofer, J. (1994). Crystal structure and refinement of cytochrome $P450_{terp}$ at 2.3 Å resolution. *J. Mol. Biol.* **236,** 1169–1185.

Hasemann, C. A., Kurumbail, R. G., Boddupalli, S. S., Peterson, J. A., and Deisenhofer, J. (1995). Structure and function of cytochromes P450: A comparative analysis of three crystal structures. *Structure* **3,** 41–62.

He, K., He, Y. A., Szklarz, G. D., Halpert, J. R., and Correia, M. A. (1996). Secobarbital-mediated inactivation of cytochrome P450 2B1 and its active site mutants. Partitioning between heme and protein alkylation and epoxidation. *J. Biol. Chem.* **271,** 25864–25872.

He, Y. A., He, Y. Q., Szklarz, G. D., and Halpert, J. R. (1997). Identification of three key residues in substrate recognition site 5 of human cytochrome P450 3A4 by cassette and site-directed mutagenesis. *Biochemistry* **36,** 8831–8839.

He, Y. Q., Harlow, G. R., Szklarz, G. D., and Halpert, J. R. (1998). Structural determinants of progesterone hydroxylation by cytochrome P450 2B5: The role of nonsubstrate recognition site residues. *Arch. Biochem. Biophys.* **350,** 333–339.

Kent, U. M., Hanna, I. H., Szklarz, G. D., Vaz, A. N. D., Halpert, J. R., Bend, J. R., and Hollenberg, P. F. (1997). Significance of glycine 478 in the metabolism of *N*-benzyl-1-aminobenzotriazole to reactive intermediates by cytochrome P450 2B1. *Biochemistry* **36,** 11707–11716.

Kobayashi, Y., Fang, X., Szklarz, G. D., and Halpert, J. R. (1998). Probing the active site of cytochrome P450 2B1: Metabolism of 7-alkoxycoumarins by the wild type and five site-directed mutants. *Biochemistry* **37,** 6679–6688.

Koymans, L. M. H., Vermeulen, N. P. E., Baarslag, A., and Donne-Op den Kelder, G. M. (1993). A preliminary 3D model for cytochrome P450 2D6 constructed by homology model building. *J. Comput.-Aided Mol. Des.* **7,** 281–289.

Kuntz, I. D., Blaney, J. M., Oatley, S. J., Langridge, R., and Ferrin, T. E. (1982). A geometric approach to macromolecule-ligand interactions *J. Mol. Biol.* **161,** 269–288.

Laughton, C. A., Zvelebil, M. J. J. M., and Neidle, S. J. (1993). A detailed molecular model for human aromatase. *J. Steroid Biochem. Mol. Biol.* **44,** 399–407.

Lee, B., and Richard, F. M. (1971). The interpretation of protein structures: Estimation of static accessibility. *J. Mol. Biol.* **55,** 379–400.

Lewis, D. F. V. (1995). Three-dimensional models of human and other mammalian cytochromes P450s constructed from an alignment with P450102 ($P450_{bm3}$). *Xenobiotica* **25,** 333–366.

Lewis, D. F. V., and Lake, B. G. (1995). Molecular modeling of members of the P4502A subfamily: Application to studies of enzyme specificity. *Xenobiotica* **25,** 585–598.

Lewis, D. F. V., and Lake, B. G. (1996). Molecular modeling of CYP1A subfamily members based on an alignment with CYP102: Rationalization of CYP1A substrate specificity in terms of active site amino acid residues. *Xenobiotica* **26,** 723–753.

Lichtarge, O., Bourne, H. R., and Cohen, F. E. (1996a). An evolutionary trace method defines binding surfaces common to protein families. *J. Mol. Biol.* **257,** 342–358.

Lichtarge, O., Bourne, H. R., and Cohen, F. E. (1996b). Evolutionarily conserved $G_{\alpha\beta\gamma}$ binding surfaces support a model of the G protein-receptor complex. *Proc. Natl. Acad. Sci. U.S.A.* **93,** 7507–7511.

Lozano, J. J., Lopez-de-Brinas, E., Centeno, N. B., Guigo, R., and Sanz, F. (1997). Three-dimensional modeling of human cytochrome P450 1A2 and its interaction with caffeine and MeIQ. *J. Comput.-Aided Mol. Des.* **11,** 395–408.

Modi, S., Paine, M. J., Sutcliffe, M. J., Lian, L.-Y., Primrose, W. U., Wolf, C. R., and Roberts, G. C. K. (1996). A model for human cytochrome P_{450} 2D6 based on homology modeling and NMR studies of substrate binding. *Biochemistry* **35,** 4540–4550.

Muller, A., Muller, J. J., Muller, Y. A., Uhlmann, H., Bernhardt, R., and Heinemann, U. (1998). New aspects of electron transfer revealed by the crystal structure of a truncated bovine adrenodoxin, Adx(4–108). *Structure* **6,** 269–280.

Narhi, L. O., and Fulco, A. J. (1986). Characterization of a catalytically self-sufficient 119,000-dalton cytochrome P-450 monooxygenase induced by barbiturates in *Bacillus megaterium*. *J. Biol. Chem.* **261,** 7160–7169.

Nelson, D. R., Koymans, L., Kamataki, T., Stegeman, J. J., Feyereisen, R., Waxman, D. J., Waterman, M. R., Gotoh, O., Coon, M. J., Estabrook, R. W., Gunsalus, I. C., and Nebert, D. W. (1996). P450 superfamily: Update on new sequences, gene mapping, accession numbers and nomenclature. *Pharmacogenetics* **6,** 1–42.

Park, S. Y., Shimizu, H., Adachi, S., Nakagawa, A., Tanaka, I., Nakahara, K., Shoun, H., Obayashi, E., Nakamura, H., Iizuka, T., and Shiro, Y. (1997). Crystal structure of nitric oxide reductase from denitrifying fungus Fusarium oxysporum. *Nat. Struct. Biol.* **4**, 827–832.

Paulsen, M. D., and Ornstein, R. L. (1991). A 175-psec molecular dynamics simulation of camphor-bound cytochrome $P450_{cam}$. *Proteins* **11,** 184–204.

Paulsen, M. D., and Ornstein, R. L. (1992). Predicting the product specificity and coupling of cytochrome P450cam. *J. Comput.-Aided Mol. Des.* **6,** 449–460.

Paulsen, M. D., and Ornstein, R. L. (1996). Binding free energy calculations for P450cam-substrate complexes. *Protein Eng.* **9,** 567–571.

Paulsen, M. D., Filipovic, D., Sligar, S. G., and Ornstein, R. L. (1993). Controlling the regiospecificity and coupling of cytochrome P450cam: T185F mutant increases coupling and abolishes 3-hydroxynorcamphor product. *Protein Sci.* **2,** 357–365.

Paulsen, M. D., Manchester, J. I., and Ornstein, R. L. (1996). Using molecular modeling and molecular dynamics simulations to predict P450 oxidation products. *In* "Methods in Enzymology" (E. F. Johnson and M. R. Waterman, eds.), Vol. 272, pp. 347–357. Academic Press, San Diego, CA.

Peterson, J. A., and Graham, S. E. (1998). A close family resemblance: The importance of structure in understanding cytochromes P450. *Structure* **6,** 1079–1085.

Poulos, T. L., Finzel, B. C., Gunsalus, I. C., Wagner, G. C., and Kraut, J. (1985). The 2.6-Å crystal structure of *Pseudomonas putida* cytochrome P450. *J. Biol. Chem.* **260,** 16122–16130.

Poulos, T. L., Finzel, B. C., and Howard, A. J. (1987). High-resolution crystal structure of cytochrome P450cam. *J. Mol. Biol.* **195,** 687–700.

Ravichandran, K. G., Boddupalli, S. S., Hasemann, C. A., Peterson, J. A., and Deisenhofer, J. (1993). Crystal structure of hemoprotein domain of P450BM-3, a prototype for microsomal P450's. *Science* **261,** 731–736.

Sevrioukova, I., Truan, G., and Peterson, J. A. (1996). Reconstitution of the fatty acid hydroxylase activity of cytochrome P450BM-3 utilizing its functional domains. *Biochemistry* **35,** 7528–7535.

Sevrioukova, I., Li, H., Zhang, H., Peterson, J. A., and Poulos, T. (1999). Structure of a cytochrome P450-redox partner electron transfer complex. *Proc. Natl. Acad. Sci. U.S.A.* **96,** 1863–1868.

Smith, G. C., Tew, D. G., and Wolf, C. R. (1994). Dissection of NADPH-cytochrome P450 oxidoreductase into distinct functional domains. *Proc. Natl. Acad. Sci. U. S. A.* **91,** 8710–8714.

Strobel, S. M., Szklarz, G. D., He, Y., Foroozesh, M., Alworth, W. L., Roberts, E. S., Hollenberg, P. F., and Halpert, J. R. (1999). Identification of mechanism-based inactivators of cytochromes P450 2B4 and 2B5, and determination of the molecular basis for differential susceptibility. *J. Pharm. Exp. Ther.* **290,** 445–451.

Szklarz, G. D., and Halpert, J. R. (1997a). Use of homology modeling in conjunction with site-directed mutagenesis for analysis of structure-function relationships of mammalian cytochromes P450. *Life Sci.* **61,** 2507–2520.

Szklarz, G. D., and Halpert, J. R. (1997b). Molecular modeling of cytochrome P450 3A4. *J. Comput.-Aided Mol. Des.* **11,** 265–272.

Szklarz, G. D., Ornstein, R. L., and Halpert, J. R. (1994). Application of 3-dimensional homology modeling of cytochrome P450 2B1 for interpretation of site-directed mutagenesis results. *J. Biomol. Struct. Dyn.* **12,** 61–78.

Szklarz, G. D., He, Y. A., and Halpert, J. R. (1995). Site-directed mutagenesis as a tool for molecular modeling of cytochrome P450 2B1. *Biochemistry* **34,** 14312–14322.

Szklarz, G. D., He, Y. Q., Kedzie, K. M., Halpert, J. R., and Burnett, V. L. (1996). Elucidation of amino acid residues critical for unique activities of rabbit cytochrome P450 2B5 using hybrid enzymes and reciprocal site-directed mutagenesis with rabbit cytochrome P450 2B4. *Arch. Biochem. Biophys.* **327,** 308–318.

Vergeres, G., and Waskell, L. (1995). Cytochrome b_5, its functions, structure and membrane topology. *Biochimie* **77,** 604–620.

Vijayakumar S., and Salerno, J. C. (1992). Molecular modeling of the 3-D structure of cytochrome $P450_{scc}$. *Biochim. Biophys. Acta* **1160,** 281–286.

von Wachenfeldt, C., and Johnson, E. F. (1995). Structures of eukaryotic cytochrome P450 enzymes. *In* "Cytochrome P450: Structure, Mechanism and Biochemistry" (P. R. Ortiz de Montellano, ed.), 2nd ed., pp. 183–223. Plenum, New York.

Wang, H., Dick, R., Yin, H., Licad-Coles, E., Kroetz, D. L., Szklarz, G., Harlow, G., Halpert, J. R., and Correia, M. A. (1998). Structure-function relationships of human liver cytochromes P450 3A: Aflatoxin B1 metabolism as a probe. *Biochemistry* **37,** 12536–12545.

Wang, M., Roberts, D. L., Paschke, R., Shea, T. M., Masters, B. S. S., and Kim, J.-J. P. (1997). Three-dimensional structure of NADPH-cytochrome P450 reductase: Prototype for FMN- and FAD-containing enzymes. *Proc. Natl. Acad. Sci. U.S.A.* **94,** 8411–8416.

Wrighton, S. A., and Stevens, J. C. (1992). The human hepatic cytochromes P450 involved in drug metabolism. *Crit. Rev. Toxicol.* **22,** 1–21.

Yamazaki, H., Johnson, W. W., Ueng, Y-F., Shimada, T., and Gueugerich, P. F. (1996). Lack of electron transfer from cytochrome b5 in stimulation of catalytic activities of cytochrome P450 3A4. Characterization of a reconstituted cytochrome P450 3A4/NADPH-cytochrome P450 reductase system and studies with apo-cytochrome b5. *J. Biol. Chem.* **271,** 27438–27444.

Zhou, D., Korekwa, K. R., Poulos, T., and Chen, S. (1992). A site-directed mutagenesis study of human placental aromatase. *J. Biol. Chem.* **267,** 762–768.

Zvelebil, M. J. J. M., Wolf, C. R., and Sternberg, M. J. E. (1991). A predicted three-dimensional structure of human cytochrome P450: Implications for substrate specificity. *Protein Eng.* **4,** 271–282.

VITAMINS AND HORMONES, VOL. 58

Prostaglandin D Synthase: Structure and Function

YOSHIHIRO URADE AND OSAMU HAYAISHI

Core Research for Evolutional Science and Technology, Japan Science and Technology Corporation, and Department of Molecular Behavioral Biology, Osaka Bioscience Institute, 6-2-4 Furuedai, Suita, Osaka 565-0874, Japan

---

Prostaglandin (PG) D synthase catalyzes the isomerization of PGH_2, a common precursor of various prostanoids, to produce PGD_2 in the presence of sulfhydryl compounds. PGD_2 induces sleep, regulates nociception, inhibits platelet aggregation, acts as an allergic mediator, and is further converted to $9\alpha,11\beta$-PGF_2 or the J series of prostanoids, such as PGJ_2, Δ^{12}-PGJ_2, and 15-deoxy-$\Delta^{12,14}$-PGJ_2. We have purified two distinct types of PGD synthase; one is the lipocalin-type enzyme and the other is the hematopoietic enzyme. We isolated the cDNA and the gene for each enzyme and determined the tissue distribution profile and the cellular localization in several animal species. Lipocalin-type PGD synthase is localized in the central nervous system and male genital organs of various mammals and the human heart and is secreted into cerebrospinal fluid, sem-

0083-6729/00 $30.00

inal plasma, and plasma, respectively. The human enzyme was identified as β-trace, which is a major protein in human cerebrospinal fluid. This enzyme is considered to be a dual-function protein; it acts as a PGD_2-producing enzyme and also as a lipophilic ligand-binding protein, because the enzyme binds retinoids, thyroids, and bile pigments, with high affinities. Hematopoietic PGD synthase is widely distributed in the peripheral tissues and localized in the antigen-presenting cells, mast cells, and megakaryocytes. The hematopoietic enzyme is the first recognized vertebrate homolog of the sigma class of glutathione *S*-transferase. X-ray crystallographic analyses and generation of gene-knockout and transgenic mice for each enzyme have been performed.

I. Introduction

Prostaglandin (PG) D synthase (PGDS, EC 5.3.99.2) catalyzes the isomerization of the 9,11-endoperoxide group of PGH_2, a common precursor of various prostanoids, to produce PGD_2 with 9-hydroxy and 11-keto groups, in the presence of sulfhydryl compounds (Fig. 1). There are two distinct types of PGDS; one is the lipocalin-type PGDS (L-PGDS), which was previously known as the brain-type enzyme or glutathione (GSH)-independent enzyme, and the other is hematopoietic PGDS (H-PGDS), the spleen-type enzyme or GSH-requiring enzyme. We have purified these two types of PGDS, isolated their cDNAs and genes, and produced the recombinant proteins for use in structural analyses and screening for inhibitors. The X-ray crystallographic analyses (Kanaoka *et al.,* 1997) and generation of gene-knockout (KO) mice (Eguchi *et al.,* 1999) and transgenic (TG) mice for each PGDS have been performed by our group. We recently reviewed the earlier studies on PGDS (Urade *et al.,* 1995b, 1996). In this paper, several recent findings on the structures and functions of two distinct types of PGDS are summarized. A part of these findings is also reviewed elsewhere (Urade and Hayaishi, 1999).

II. Pharmacologic Actions of Prostaglandin D_2 and Its Metabolites

PGDS is considered to be involved in a variety of physiologic and pathologic functions as described next by forming PGD_2 and its further metabolites.

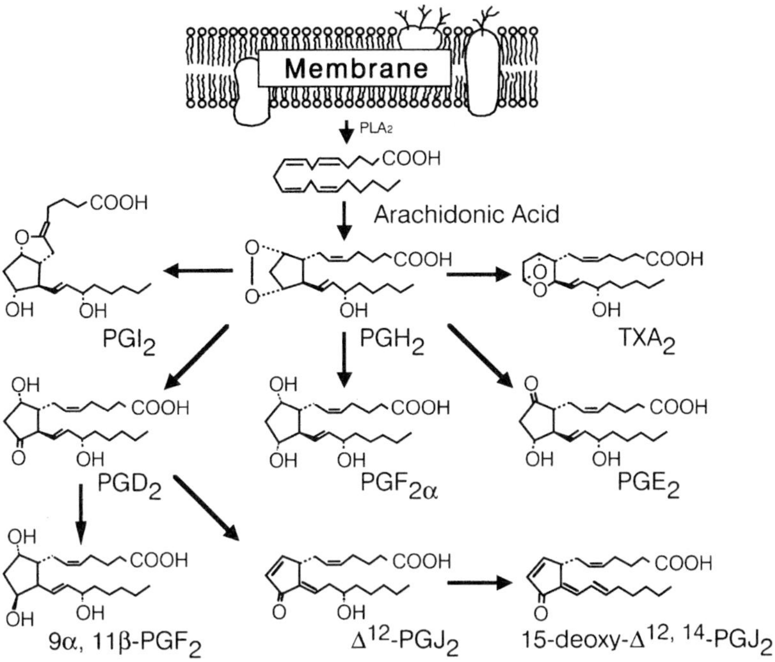

FIG. 1. Metabolic pathway of prostanoid cascade. PLA_2, phospholipase A_2; TXA_2, thromboxane A_2.

A. Pharmacologic Actions of Prostaglandin D_2

PGD_2 is actively produced in a variety of tissues as a major prostanoid (Ujihara *et al.*, 1988b). In the central nervous system (CNS), it induces sleep (Ueno *et al.*, 1983; Onoe *et al.*, 1988; Hayaishi, 1988, 1991; Hayaishi and Matsumura, 1995; Urade *et al.*, 1996; Urade and Hayaishi, 1999) and regulates body temperature (Ueno *et al.*, 1982; Moritomo *et al.*, 1988; Sri Kantha *et al.*, 1994), hormone release (Kinoshita *et al.*, 1982; Nishi *et al.*, 1984; Terao *et al.*, 1995), and nociception (Horiguchi *et al.*, 1986; Uda *et al.*, 1990; Minami *et al.*, 1996, 1997; Eguchi *et al.*, 1999). In peripheral tissues, PGD_2 inhibits platelet aggregation (Whittle *et al.*, 1978) and nitric oxide release (Nagoshi *et al.*, 1998), and induces vasodilation and bronchoconstriction (Narumiya and Toda, 1985; Giles and Leff, 1988; Ito *et al.*, 1989). PGD_2 is also released from mast cells upon stimulation with various immunologic stimuli and functions as an allergic and inflammatory lipid mediator (Lewis *et al.*, 1982).

B. Prostaglandin D_2 Receptor (DP Receptor)

The actions of PGD_2 just mentioned are mediated by a prostanoid receptor specific for PGD_2, that is, the D type of prostanoid (DP) receptor (Coleman *et al.*, 1994; Negishi *et al.*, 1993, 1995). The cDNA for the DP receptor was cloned from mice (Hirata *et al.*, 1994), humans (Boie *et al.*, 1995), and rats (Gerashchenko *et al.*, 1998a). This receptor is a member of the G-protein-coupled, rhodopsin-type receptor family, which contains seven hydrophobic transmembrane domains. The activation of the DP receptor induces an increase in intracellular cAMP and Ca^{2+} (Hirata *et al.*, 1994; Boie *et al.*, 1995).

The tissue distribution profile of the mRNA for the DP receptor varied significantly among mice, rats, and humans, which variation is consistent with the highly species-specific pharmacologic activities of PGD_2 (Narumiya and Toda, 1985). The tissue distribution profile of the mRNA for the DP receptor overlaps those profiles of L-PGDS and H-PGDS (Gerashchenko *et al.*, 1998a). In mouse (Oida *et al.*, 1997) and rat (Gerashchenko *et al.*, 1998a) brain, the mRNA for the DP receptor was dominantly expressed in the leptomeninges, similar to the case of L-PGDS, as described later. In rat retina, iris, and ciliary processes, the distribution of DP receptor is very similar to that of L-PGDS (Gerashchenko *et al.*, 1998a,b). These results are in agreement with the classical idea that PGD_2 acts as a local mediator in an autocrine or paracrine fashion.

C. $9\alpha,11\beta$-Prostaglandin F_2 and J Series of Prostaglandins

PGD_2 is converted by PGF synthase, which catalyzes the reduction of PGH_2 to $PGF_{2\alpha}$ (Watanabe *et al.*, 1985, 1988), to produce $9\alpha,11\beta$-PGF_2 (Liston and Roberts, 1985; Watanabe *et al.*, 1986, 1991; Urade *et al.*, 1990b; Chen *et al.*, 1992; Suzuki *et al.*, 1999), which is a stereoisomer of $PGF_{2\alpha}$ and exerts various pharmacologic actions different from those induced by $PGF_{2\alpha}$ (Smith *et al.*, 1991).

PGD_2 is also converted to the J series of PGs, such as PGJ_2, Δ^{12}-PGJ_2, and 15-deoxy-$\Delta^{12,14}$-PGJ_2, in human plasma (Kikawa *et al.*, 1984) urine (Hirata *et al.*, 1988), and seminal plasma (Tokugawa *et al.*, 1998). Δ^{12}-PGJ_2 has an antiproliferative effect against tumor cells (Fukushima, 1992). A PGJ_2 metabolite, 15-deoxy-$\Delta^{12,14}$-PGJ_2 is considered to promote adipocyte differentiation by binding to a nuclear receptor, the peroxisome proliferator-activated receptor γ (PPARγ) (Forman *et al.*, 1995; Kliewer *et al.*, 1995). Recently, ligand activation of PPARγ was found to regulate macrophage and monocyte functions (Ricote *et al.,* 1998a,b; Jiang *et al.,* 1998; Nagy *et al.,* 1998; Tontonoz *et al.,* 1998).

The contribution of PGDS to the formation of the above PGs is still speculative at present. However, note that those PGs might play important roles in certain circumstances under physiologic and pathologic conditions.

III. Lipocalin-Type Prostaglandin D Synthase (β-Trace)

A. Enzymatic Properties

L-PGDS was originally purified from rat brain as a soluble, GSH-independent enzyme with M_r of 80,000 (Shimizu *et al.,* 1979). The M_r value of L-PGDS was then reevaluated to be 26,000 (Urade *et al.,* 1985). From studies of cDNA cloning (Urade *et al.,* 1989a), the protein sequence, and chemical analysis (Hoffmann *et al.,* 1994) described later, L-PGDS is now known to be an *N*-glycosylated monomeric protein identical to β-trace, which was originally discovered as a major protein of human cerebrospinal fluid (CSF) (Clausen, 1961; Hochwald and Thorbecke, 1962). Therefore, L-PGDS has recently been purified from various body fluids of several mammals, such as CSF (Kuruvilla *et al.,* 1991; Zahn *et al.,* 1993; Hoffmann *et al.,* 1993, 1994; Watanabe *et al.,* 1994; Giacomelli *et al.,* 1996), seminal plasma (Tokugawa *et al.,* 1998; Gerena *et al.,* 1998; Fouchécourt *et al.,* 1999), and amniotic fluid (Melegos *et al.,* 1996a).

L-PGDS requires free sulfhydryl compounds for the reaction (Urade *et al.,* 1985). However, L-PGDS does not absolutely require GSH, because it is also active with other thiol compounds such as β-mercaptoethanol, dithiothreitol, cysteine, and cysteamine. Inorganic quadrivalent selenium (Se^{4+}) compounds (Islam *et al.,* 1991) and several nonsubstrate hydrophobic ligands, such as fatty acids (Osama *et al.,* 1983), retinoids (Tanaka *et al.,* 1997), bile pigments, and thyroid hormones (Beuckman *et al.,* 1999) are noncompetitive and reversible inhibitors of L-PGDS. Thiol modifiers, such as iodoacetoamide and *N*-ethylmaleimide, are irreversible inhibitors of L-PGDS (Urade *et al.,* 1995a).

L-PGDS is a very stable enzyme and highly resistant against heat treatment (Urade *et al.,* 1985) or protease digestion (Urade *et al.,* 1987a). For example, more than 50% of the activity was maintained after heating the enzyme at 100°C for 5 min at alkaline pH. However, it is easily inactivated during the enzymic reaction (Urade *et al.,* 1985).

B. cDNA Cloning and Mutational Study

The cDNA for L-PGDS was first cloned from a rat brain cDNA library (Urade *et al.,* 1989a). The cDNA was subsequently isolated from many species, including humans (Nagata *et al.,* 1991), mice (Hoffmann *et al.,* 1996b), and two amphibians (Achen *et al.,* 1992; Lepperdinger *et al.,* 1996). The cDNA for L-PGDS encodes a protein composed of 180–190 amino acid residues (Fig. 2). A signal peptide of about 20 amino acid residues is commonly observed at the N terminus of L-PGDS and is removed to form the mature enzyme. Two *N*-glycosylation sites are con-

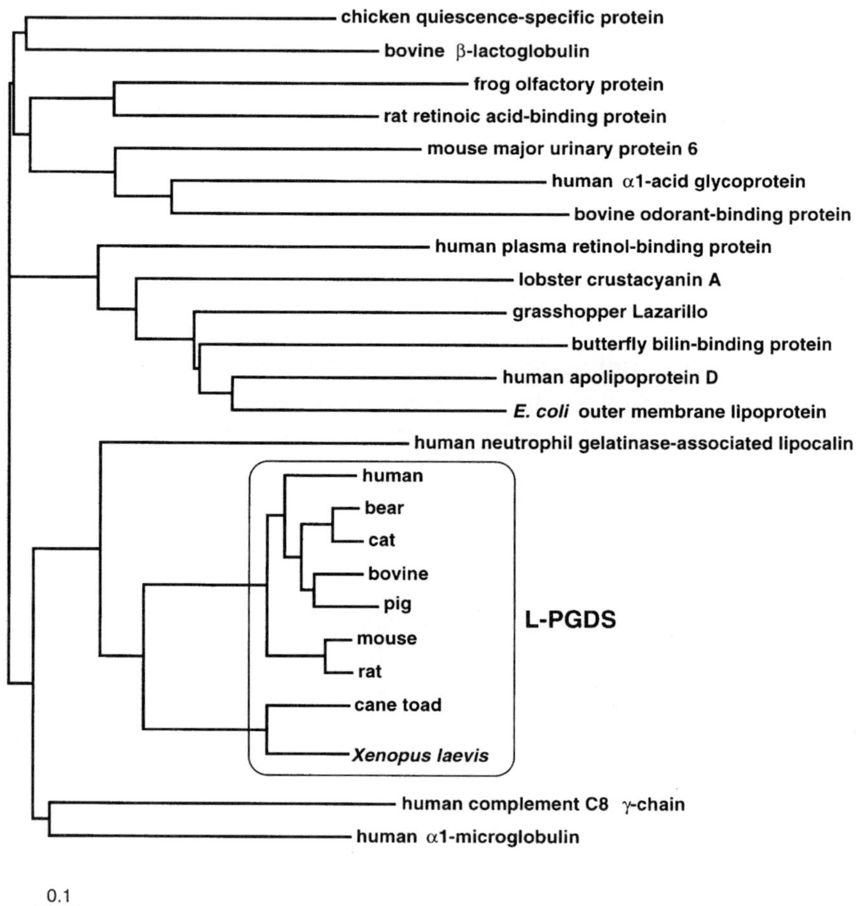

FIG. 2. Phylogenetic tree of L-PGDS and members of the lipocalin superfamily. Bar represents 0.1 amino acid substitution/site.

served in the enzyme of all mammalian species but are not found in the amphibian homologs. Three Cys residues, corresponding to Cys^{65}, Cys^{89}, and Cys^{186} in the rat and human L-PGDS, are completely conserved among all species; and two of these residues, Cys^{89} and Cys^{186}, form a disulfide bridge. Cys^{65} was identified to be the key residue for the catalytic activity of L-PGDS by site-directed mutagenesis and chemical modification studies with the recombinant rat enzyme (Urade *et al.,* 1995a). When Cys^{65} of L-PGDS was replaced with Ser/Ala or chemically modified, the enzyme activity disappeared completely. Quadrivalent selenium (Se^{4+}) compounds are predicted to interact with this free sulfhydryl group and thus inhibit L-PGDS. We predict that the rapid inactivation of L-PGDS during catalysis is also due to modification of the active thiol of Cys^{65}.

C. Evolutional Analysis (Lipocalin)

A homology search in databases of protein primary structure and comparison of the gene structure described later revealed the L-PGDS is a member of the lipocalin (lipophilic ligand-carrier protein) superfamily (Nagata *et al.,* 1991; Igarashi *et al.,* 1992; White *et al.,* 1992) consisting of a variety of secretory proteins involved in the binding and transport of small hydrophobic molecules (Pervaiz and Brew, 1987; Toh *et al.,* 1996; Flower, 1995, 1996) (Fig. 2). β-Lactoglobulin, plasma retinol-binding protein, major urinary protein, and epididymal retinoic acid-binding protein are well-known members of the lipocalin family.

Among almost 100 members of the lipocalin family thus far identified (Toh *et al.,* 1996; Flower, 1996), L-PGDS is the only lipocalin that is associated with an enzymatic activity (Nagata *et al.,* 1991; Peitsch and Boguski, 1991). The two Cys residues constructing a disulfide bridge in L-PGDS are also highly conserved among most, but not all, lipocalins; whereas the Cys^{65} residue, the key residue for the enzymic reaction, is completely conserved in L-PGDS but never found in other lipocalins. Therefore, L-PGDS is considered to have evolved from a lipophilic ligand-binding protein to the enzyme by acquiring an active residue, Cys^{65}. L-PGDS is an interesting protein for the study of molecular evolution of an enzyme from a nonenzymatic protein and is a useful model to design a new functional protein from other lipocalins (Toh *et al.,* 1996).

D. Binding of Nonsubstrate Lipophilic Ligands

Although L-PGDS is considered to have evolved from some lipophilic-ligand carrier proteins as described earlier, it retains the ancestral characteristic of binding lipophilic ligands. By using recombinant rat L-PGDS, we found that L-PGDS binds all-*trans*- or 9-*cis*-retinoic acid and all-*trans*- or 13-*cis*-retinaldehyde, but not all-*trans*-retinol, at a molar ratio of 1:1 (Tanaka *et al.,* 1997). The binding affinity of L-PGDS for those retinoids (70–80 n*M*) is higher than that of other previously characterized extracellular retinoid transporters, such as β-lactoglobulin, plasma retinol-binding protein, and plasma retinoic acid-binding protein. By monitoring the quenching of the intrinsic tryptophan fluorescence of the recombinant protein, the *Xenopus* homolog of L-PGDS was recently identified to be a secretory retinaldehyde-binding protein that also binds thyroid hormone; although the affinity constants remain to be determined (Lepperdinger *et al.,* 1996). We also recently found that the recombinant L-PGDS binds thyroid hormone with a K_d of about 0.7–2 μ*M* and also biliverdin and bilirubin with a very high affinity of K_d of about 30–40 n*M* (Beuckmann *et al.,* 1999).

Therefore, we propose that L-PGDS is a dual-function protein; it acts as a PGD_2-producing enzyme and may also function as a lipophilic ligand-binding protein.

E. Tertiary Structure

Although the homology of the amino acid sequences of members of the lipocalin superfamily is rather weak because of the high evolutionary divergence, the tertiary structure is well conserved to form a remarkably similar β-barrel structure (Flower, 1995, 1996; Toh *et al.,* 1996). The X-ray crystallographic structure has been identified for β-lactoglobulin (Papiz *et al.,* 1986; Monaco *et al.,* 1987; Brownlow *et al.,* 1997; Wu *et al.,* 1999), plasma retinol-binding protein (Newcomer *et al.,* 1984; Cowan *et al.,* 1990; Zanotti *et al.,* 1993), major urinary protein (Böcskel *et al.,* 1992), epididymal retinoic acid-binding protein (Newcomer *et al.,* 1993), insecticyanin (Holden *et al.,* 1987), bilin-binding protein (Huber *et al.,* 1987), odorant-binding protein (Monaco and Zunotti, 1992; Tegoni *et al.,* 1996; Bianchet *et al.,* 1996; Spinelli *et al.,* 1998), nitrophorin 1 (Weichsel *et al.,* 1998; Andersen *et al.,* 1998), and lipocalin allergen (Rouvinen *et al.,* 1999). We constructed the tertiary structure of rat L-PGDS by homology modeling based on the crystal structure of the lipocalins mentioned (Tanaka *et al.,* 1997; Toh *et al.,* 1996). In the model structure, the Cys^{65} residue was located in the hydrophobic pocket with a size sufficient to bind retinoids, biliverdin, bilirubin, and thyroid hormones (Fig. 3A).

Most recently we succeeded in the crystallization of recombinant L-PGDS (Fig. 4A) and obtained X-ray diffraction images sufficient for cal-

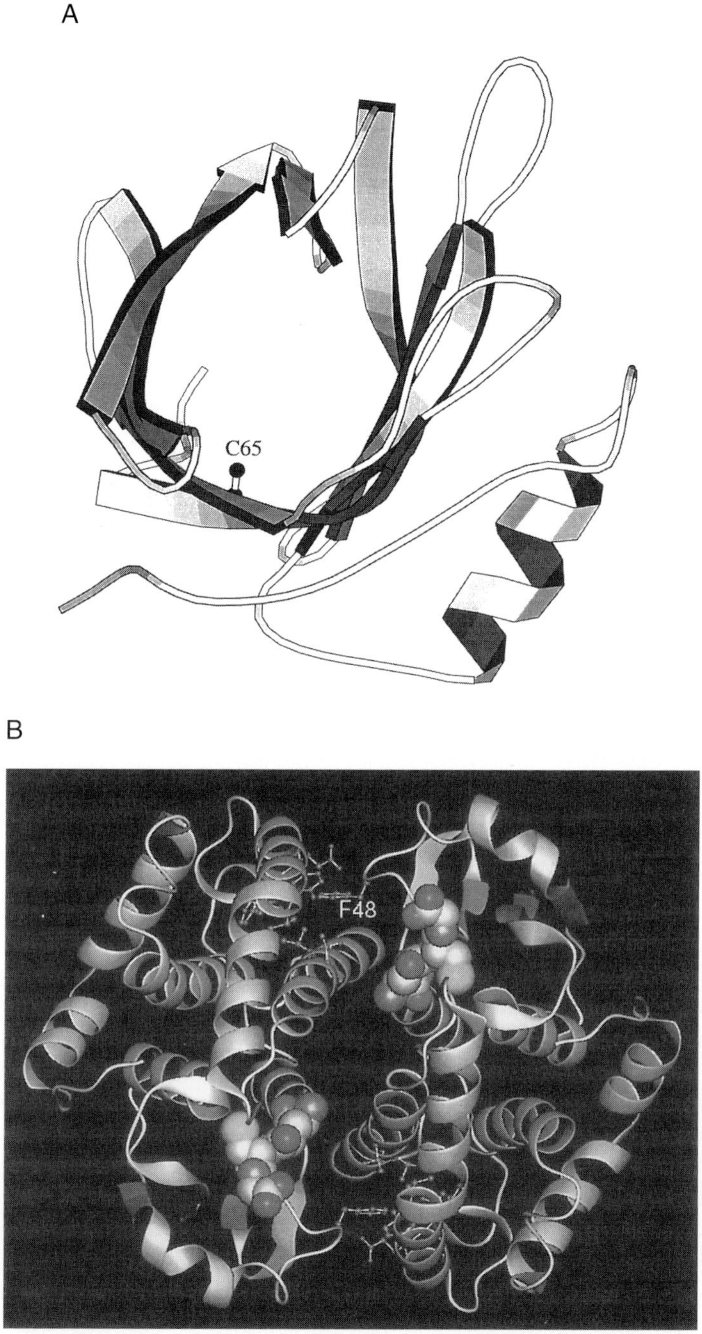

FIG. 3. Ribbon modeling of L-PGDS and H-PGDS. (A) Active site (C65) of L-PGDS is indicated. (B) GSH molecules bound to the active site of H-PGDS dimer are shown in CPK model. The dimer interface is composed of a hydrophobic side chain (F48) from one monomer and a hydrophobic cavity on the other.

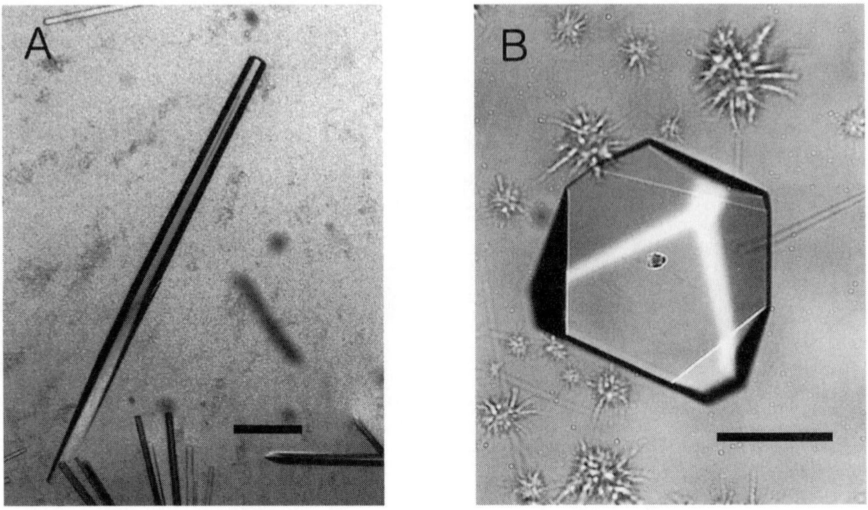

FIG. 4. Light micrographs of crystals of (A) L-PGDS and (B) H-PGDS. Bar = 0.1 mm.

culation of its tertiary structure. Therefore, the tertiary structure of L-PGDS will be determined in the near future.

F. IDENTIFICATION AS β-TRACE

β-Trace was discovered in the early 1960s as a protein specific to the human CSF (Clausen, 1961; Hochwald and Thorbecke, 1962, 1963; Laterre *et al.,* 1964; Link, 1965). Although extensive studies had been conducted in the 1960s and 1970s to examine the physicochemical properties and the tissue-specific synthesis of β-trace (Hochwald *et al.,* 1967; Pepe and Hochwald, 1967; Olsson *et al.,* 1973; Olsson and Link, 1973; Olsson and Nord, 1973; Olsson and Sandberg, 1975), its structure, cellular localization, sites of synthesis, and function of β-trace remained unclear until recently.

From 1991 to 1993, several research groups independently and almost simultaneously succeeded in the purification of β-trace from human CSF (Kuruvilla *et al.,* 1991; Zahn *et al.,* 1993; Hoffmann *et al.,* 1993) and found that the N-terminal partial amino acid sequences were highly homologous to that of the rat and human L-PGDS that we deposited in the DNA database (Urade *et al.,* 1989a; Nagata *et al.,* 1991). Finally, the full amino acid sequence of β-trace was determined to be essentially identical to that of human L-PGDS except for the absence of the signal peptide in β-trace (Hoffmann *et al.,* 1993). We also confirmed that L-PGDS purified from human CSF was structurally, enzymatically, and immunologically identical to β-trace (Watanabe *et al.,* 1994). Due to the identification of β-trace as L-PGDS, the structure, cellular localization of the synthesis sites, and function of β-trace have been revealed as described here.

G. Genomic Cloning and Generation of Gene-Manipulated Mice

The genes have been cloned from rats (Igarashi *et al.,* 1992), humans (White *et al.,* 1992), and mice (Eguchi *et al.,* 1999). The gene organization of L-PGDS was shown to be remarkably analogous to that of other lipocalins, in terms of number and size of exons and phase of splicing of introns (Igarashi *et al.,* 1992; White *et al.,* 1992). The genes were mapped to mouse chromosome 2 B-C1 (Chan *et al.,* 1994) and human chromosome 9 q34.2-34.3 (White *et al.,* 1992), both of which were localized within the lipocalin gene cluster (Chan *et al.,* 1994).

The mRNA for L-PGDS was found to be down-regulated in hypothyroid rats (Muñoz *et al.,* 1991; García-Fernández *et al.,* 1993, 1997). The thyroid hormone response element was identified in the promoter region of the rat and human genes (White *et al.,* 1997; García-Fernández *et al.,* 1998). The L-PGDS concentration in rat CSF was recently reported to decrease by subcutaneous injection of all-*trans*-retinoic acid, although the regulation mechanism remains unidentified (Kabeya *et al.,* 1998).

L-PGDS-KO mice were generated and used to determine the function of PGD_2 *in vivo* as described later (Eguchi *et al.,* 1999). Transgenic (TG) mice overexpressing the human L-PGDS have also already been generated by our group.

H. Tissue Distribution, Cellular Localization, and Functional Relevance

L-PGDS is localized in the CNS and male genital organs of various mammals (Ujihara *et al.,* 1988b; Urade *et al.,* 1989a; Blödorn *et al.,* 1996; Hoffmann *et al.,* 1996b; Gerena *et al.,* 1998a; Tokugawa *et al.,* 1998) and is also distributed in the heart (Eguchi *et al.,* 1997) of several species. Information has been accumulated about the functional relevance of L-PGDS to the regulation of sleep and nociception, and the enzyme may be a possible clinical marker for several diseases. We describe below the cellular localization and functional relevance of L-PGDS in each tissue.

1. *Central Nervous System*

a. Cellular Localization. By immunoperoxidase staining with specific polyclonal or monoclonal antibodies and by *in situ* hybridization with the antisense RNA, L-PGDS was shown to be mainly produced in the leptomeninges (pia-arachnoid membrane) and choroid plexus of the brain and spinal cord of rats (Urade *et al.*, 1993), mice (Hoffmann *et al.*, 1996a; Eguchi *et al.*, 1999), and humans (Blödorn *et al.*, 1996; Yamashima *et al.*, 1997) and secreted into the CSF as β-trace, as described earlier. The production and secretion of L-PGDS/β-trace were also demonstrated in cultured rat leptomeningeal cells (Ohe *et al.*, 1996) and porcine choroid plexus epithelial cells (Hoffmann *et al.*, 1996b). By immunoelectron microscopy, L-PGDS was shown to be localized in the rough endoplasmic reticulum of human arachnoid and meningioma cells (Yamashima *et al.*, 1997).

In the parenchyma of the brain of adult rats (Urade *et al.*, 1987a; García-Fernández *et al.*, 1997), mice (Hoffmann *et al.*, 1996a; Eguchi *et al.*, 1999), and humans (Blödorn *et al.*, 1996), L-PGDS was detected in oligodendrocytes but not in Schwann cells, another type of myelin-producing cells, in the peripheral nervous system. The immunoreactivity of L-PGDS in oligodendrocytes was electronmicroscopically detected in the rough endoplasmic reticulum and outer nuclear membrane (Urade *et al.*, 1987a), similar to cyclooxygenase, which catalyzes the production of PGH_2, that is, the substrate for PGDS. The up-regulation of L-PGDS was recently found in oligodendrocytes in a murine model of a demyelinating disease, suggesting that the cytoprotective role of L-PGDS against oligodendroglial dysfunction associated with a demyelinating lesion (M. Taniike *et al.*, unpublished results).

Although most astrocytes in the rat brain were negative for immunostaining with anti-L-PGDS antibody or for *in situ* hybridization with the antisense RNA for L-PGDS, the synthesis and secretion of PGD_2 (Keller *et al.*, 1985; Seregi *et al.*, 1987, 1992; Yamamoto *et al.*, 1988) and L-PGDS (Giacomelli *et al.*, 1996) were reportedly detected in cultured rat astrocytes. However, the production of PGD_2 by cultured astrocytes was recently claimed to be due to microglial contamination (Matsuo *et al.*, 1995).

In the infant rat brain, the immunoreactivity for L-PGDS was also abundantly detected in immature neurons (Urade *et al.*, 1987a; Garciá-Fernández *et al.*, 1997). The immunoreactivity was detected in small neurons of the B_2 subclass in chick dorsal root ganglia (Vesin *et al.*,

1995a,b,c), in small neurons of laminae II and III, and in motor neurons in the ventral horn of the chick spinal cord (Vesin *et al.,* 1995a). In those neurons of chick spinal cord, the immunoreactivity was localized in the rough endoplasmic reticulum and limited subsurface areas of their perikarya and dendrites (Vesin *et al.,* 1995a).

b. Relation to Sleep Regulation. In the 1980s, PGD_2 was demonstrated to induce sleep in rats (Ueno *et al.,* 1983) and monkeys (Onoe *et al.,* 1988) after cerebroventricular infusion of the prostanoid. Interestingly and most importantly, the PGD_2-induced sleep was indistinguishable from physiologic sleep, as judged by the electroencephalogram, electromyogram, brain temperature, heart rate, and general behavior of animals injected with it. Among more than 30 so-called "endogenous sleep substances" (Borbely and Tobler, 1989), PGD_2 is recognized as the most potent endogenous sleep-promoting substance whose action mechanism is the best characterized among the various sleep substances thus far reported (Hayaishi and Matsumura, 1995; Urade *et al.,* 1996; Urade and Hayaishi, 1999).

Because administration of SeCl4, an inhibitor of L-PGDS (Islam *et al.,* 1991), into the third ventricle of rats inhibited the sleep of the animals in a time- and dose-dependent manner (Matsumura *et al.,* 1991), L-PGDS is considered to be the key enzyme in the regulation of physiologic sleep (Hayaishi *et al.,* 1993). The PGD_2 concentration in rat CSF shows a circadian change coupled to the sleep–wake cycle (Pandey *et al.,* 1995) and elevates with an increase in sleep propensity during sleep deprivation (Ram *et al.,* 1997). L-PGDS is dominantly produced in the arachnoid membrane and choroid plexus of the brain, and is secreted into the CSF to become β-trace, as described earlier. L-PGDS/β-trace as well as the PGD_2 thus produced circulates in the ventricular system, subarachnoidal space, and extracellular space in the brain system. PGD_2 is, therefore, not a typical neurotransmitter but rather a neurohormone that circulates through the CSF and transmits chemical messages to promote sleep (Urade *et al.,* 1996; Urade and Hayaishi, 1999). The mode of communication through the CSF in the ventricular system and the extracellular space has advantages for global regulation of the brain to induce sleep or to increase the propensity for sleep. PGD_2 then interacts with DP receptors in the chemosensory region of the ventromedial surface of the rostral basal forebrain (Matsumura *et al.,* 1994) to initiate the signal to promote sleep probably via the activation of adenosine A_{2A} receptive neurons (Satoh *et al.,* 1996, 1998). The activation of DP receptors in the PGD_2-sensitive chemosensory region results in activation of a cluster of neurons within the ventrolateral preoptic area, a recently identified sleep center (Sherin *et al.,* 1996), and in suppression of activities in the tuberomammillary nucleus, a classically identified histaminergic arousal system (Scammell *et al.,* 1998).

Profound lethargy in patients with mastocytosis (Roberts *et al.,* 1980) and African sleeping sickness (Pentreath *et al.,* 1990) was proposed to be primarily due to the overproduction of endogenous PGD_2. Recently we also observed a transient increase in the amount of slow-wave sleep in TG mice that overexpressed the human L-PGDS after thier tails were clipped for a genetic analysis (E. Pinzar *et al.,* unpublished results). Therefore, our TG mice are considered to be useful for development of a novel animal model to study profound lethargy or excess pathologic sleep mediated by PGD_2.

c. Nociception. Intrathecal administration of PGD_2 into mice induced hyperalgesia, which was blocked by simultaneous injection of a substance P antagonist (Uda *et al.,* 1990), and inhibited allodynia (tactile pain) induced by PGE_2 (Minami *et al.,* 1996) or nociceptin (Minami *et al.,* 1997). L-PGDS was localized in the arachnoid membrane and oligodendrocytes of rat and mouse spinal cord (Urade *et al.,* 1993; Eguchi *et al.,* 1999), in small neurons of laminae II and III, and in motor neurons in the ventral horn of the chick spinal cord (Vesin *et al.,* 1995a). Therefore, PGD_2 was also considered to be involved in pain responses at the spinal cord level.

More recently, PGD_2 produced by L-PGDS was demonstrated to be essential for allodynia induced by PGE_2 or bicuculline, the latter being a γ-aminobutyric acid $(GABA)_A$ receptor antagonist (Eguchi *et al.,* 1999). Allodynic responses induced by PGE_2 or bicuculline were selectively abolished in the L-PGDS-KO mice, although the mutant mice showed allodynia induced by $PGF_{2\alpha}$, strychnine, α-amino-3-hydroxy-5-methyl-4-isoxalole-4-propionic acid (AMPA), *N*-methyl-D-aspartate (NMDA), arginine, and sodium nitroprusside to the same extent as wild-type mice. The PGE_2- or bicuculline-induced allodynia was recovered in L-PGDS-KO mice by supplementation with PGD_2 or a DP receptor agonist, BW 245C, in a femtogram amount to the same extent as in wild-type mice. Moreover, the PGE_2- or bicuculline-induced allodynia in wild-type and in PGD_2-supplemented L-PGDS-KO mice was blocked by a DP receptor antagonist, BW A868C, given in a femtogram amount.

2. *Auditory and Visual Systems*

L-PGDS is also distributed in the cochlea (Tachibana *et al.,* 1984; Hoffmann *et al.,* 1996a) and ocular tissues (Goh *et al.,* 1987; Beuckmann *et al.,* 1996; Hoffmann *et al.,* 1996a; Gerashchenko *et al.,* 1998b) of rats and mice. In the rat cochlea, the immunoreactivity was localized in the inner and outer hair cells, Claudius' cells, Deiters' cells, marginal cells, basal cells, and cells of Reissner's membrane (Tachibana *et al.*, 1984). In the rat ocular system, L-PGDS is produced in pigmented epithelial cells of the retina and nonpigmented epithelial cells of the iris and the ciliary body, and accumulates within the interphotoreceptor matrix (Beuckmann *et al.*, 1996) and vitreous and aqueous humors (Gerashchenko *et al.*, 1998b). In the eye, L-PGDS was also abundant in extracellular compartments and less abundant within the tissues. The intracellular ocular enzyme was smaller than the extracellular form because of the difference in the *N*-glycosylation moiety. In the rat eye, the DP receptor was localized in photoreceptor cells of the retina and nonpigmented epithelial cells of the iris and ciliary body, suggesting the involvement of PGD_2 produced by L-PGDS in the physiologic regulation of the vision process and intraocular pressure (Gerashchenko *et al.*, 1998a). Alternatively, extracellular L-PGDS (β-trace) may also act as a transporter of lipophilic ligands, such as retinoids and/or docosahexanoic acid, in those compartments.

3. *Male Genital Organs*

In situ hybridization and immunohistochemistry revealed that L-PGDS was localized in Leydig cells and Sertoli cells of the testis and in epithelial cells of the ductus epididymis of mice (Hoffmann *et al.*, 1996a) and humans (Blödorn *et al.*, 1996; Tokugawa *et al.*, 1998). Secretion of L-PGDS/β-trace into the seminal plasma and epididymal fluid was reported in various mammals including humans (Olsson, 1975; Tokugawa *et al.*, 1998), bulls (Gerena *et al.*, 1998), rats (Sorrentino *et al.*, 1998), rams, and stallions (Fouchécourt *et al.*, 1999).

The physiologic role of L-PGDS in the male genital organs remains unclear. However, it is worthwhile to note the reports indicating that the human seminal plasma concentration of L-PGDS/β-trace was significantly lower in oligozoospermic patients than in the normal group (Olsson, 1975; Tokugawa *et al.*, 1998). Alternatively, a high fertility-associated protein in bovine seminal plasma (Killian *et al.*, 1993) was recently identified to be L-PGDS (Gerena *et al.*, 1998). These results, taken together, suggest that L-PGDS may play an important role in both the development and the maturation of sperm.

4. *Cardiovascular System*

We recently found that mRNA for human L-PGDS was expressed most intensely in the heart among various tissues examined including the brain (Eguchi *et al.*, 1997). The mRNA for L-PGDS was also found to be expressed in the heart of monkeys and mice but not in that of rats.

Immunostaining of human autopsy specimens revealed that L-PGDS was localized in myocardial cells and atrial endocardial cells. Interestingly, the immunoreactivity was also localized in a synthetic phenotype of smooth muscle cells in the arteriosclerotic intima, and was found to be accumulated in the atherosclerotic plaque of coronary arteries with severe stenosis. Therefore, L-PGDS is considered to be a useful marker for identification of the functional or differentiation stage of myocardial, endocardial, and smooth muscle cells.

We also showed that human L-PGDS/β-trace was secreted into and accumulated in the plasma of the coronary circulation of angina patients (Eguchi *et al.,* 1997). The plasma level of L-PGDS/β-trace in patients with stable angina was significantly higher in the great cardiac vein than in the coronary artery. However, in normal subjects without stenosis, the venoarterial difference in the plasma L-PGDS concentration was not observed. After percutaneous transluminal coronary angioplasty (PTCA) of the patients to compress the stenotic atherosclerotic plaques, the L-PGDS concentration in the cardiac vein decreased and reached the arterial level within 1 h.

Clinically, the acute occulusion by thrombosis often happens within several hours after PTCA but hardly occurs 7 days after PTCA when the plasma L-PGDS concentration has increased as compared with that at 1 and 2 days after PTCA. These results suggest that reproliferation of smooth muscle cells in a cracked atherosclerotic plaque may contribute to an increase in the L-PGDS level in the peripheral vein and that the disappearance of the homeostatic synthesis and secretion of L-PGDS in the atherosclerotic plaque after PTCA may result in stimulation of platelet aggregation mediated by loss of the antiaggregatory function of PGD_2. Because PGD_2 is classically known as an antiaggregatory prostanoid against human platelets (Whittle *et al.,* 1978), L-PGDS may be up-regulated to protect against platelet aggregation in atherosclerotic blood vessels.

5. *Diagnostic Marker*

In the 1970s, due to the relatively specific occurrence of β-trace in the CNS and genital organs, the β-trace concentration was determined in several body fluids and demonstrated to be useful as a potential diagnostic marker of several neurologic disorders (Link and Olsson, 1972) and dysfunction of sperm formation (Olsson, 1975). However, the previously used methods were semiquantitative and insensitive to measure a low level of the protein.

Because β-trace was identified as L-PGDS, highly sensitive immunoassay methods for L-PGDS/β-trace were developed with monoclonal antibodies against recombinant human L-PGDS/β-trace and used to determine its content in several human body fluids (Oda *et al.,* 1996; Melegos *et al.,* 1996b). The L-PGDS/β-trace in human CSF was also detemined by capillary electrophoresis (Hiraoka *et al.,* 1997, 1998) and several other methods (Harrington *et al.,* 1993; Giacomelli *et al.,* 1996).

Although somewhat contraversial, the relationship between the CSF concentration of L-PGDS/β-trace and neurologic diseases has been suggested. For example, the concentration was found to be decreased in patients with multiple sclerosis, schizophrenia (Harrington *et al.,* 1993), brain tumor (Saso *et al.,* 1998), and bacterial meningitis (Tunami *et al.,* 1998a,b) and to be increased in patients with aneurysmal subarachnoid hemorrhage (Mase *et al.,* 1999). On the other hand, the CSF concentration was reported to be unchanged in patients with multiple sclerosis, HIV/AIDS related neuropathies, viral meningitis, and fibromyalgia (Melegos *et al.,* 1997). The amniotic fluid concentration was proposed to be associated with fetal abnormality (Melegos *et al.,* 1996a). The plasma and urinary concentrations of L-PGDS are also considered to be a potential diagnostic marker for renal diseases (Hoffmann *et al.,* 1997).

IV. Hematopoietic Prostaglandin D Synthase

A. Enzymatic Properties

H-PGDS was originally purified from rat spleen as a cytosolic, GSH-requiring enzyme with M_r of 26,000 (Christ-Hazelhof and Nugteren, 1979). Because its molecular weight is identical to that of L-PGDS (Urade *et al.,* 1985), we reexamined the properties of H-PGDS, compared them with those of L-PGDS, and showed that H-PGDS and L-PGDS were biochemically, enzymatically, and immunologically different from each other (Urade *et al.,* 1987b).

H-PGDS absolutely requires GSH for the reaction and is inactive with other thiol compounds. H-PGDS is associated with the activity of GSH *S*-transferase (GST; EC 2.5.1.18) and catalyzes the conjugation of GSH to 1-chloro-2,4-dinitro-benzene, a common substrate for various isozymes of GST, with about 10% of the turnover rate of PGDS activity (Urade *et al.,* 1987b). H-PGDS is also inactivated during the PGDS reaction similar to L-PGDS, whereas the inactivation was not observed during the GST reaction (Urade *et al.,* 1987b).

By partial amino acid sequence analysis (Urade *et al.,* 1987b, 1989b; Meyer and Thomas, 1995) and finally by cDNA cloning (Kanaoka *et al.,* 1997; Thomson *et al.,* 1998), H-PGDS was identified to be an isozyme of GST of the sigma class, as described next. The catalytic properties of

H-PGDS as GST were recently investigated in detail by using the recombinant enzyme of the chicken homolog by Thomson *et al.* (1998).

B. cDNA Cloning and Evolutionary Analysis (A Novel Vertebrate Sigma Class Type of Glutathione *S*-Transferase)

The cDNA for H-PGDS was first cloned by RT-PCR with poly-A RNA from rat spleen (Kanaoka *et al.,* 1997). The cDNA was then isolated from chicken (Thomson *et al.,* 1998), mice, and humans (Y. Kanaoka and Y. Urade, unpublished). The cDNA for H-PGDS encodes a protein composed of 199 amino acid residues without signal peptide or *N*-glycosylation sites (Fig. 5).

A homology search in databases of protein primary structure revealed that H-PGDS is a member of the GST family, as previously predicted. However, H-PGDS showed a weak homology against mammalian GST isozymes of the previously known four classes (alpha, mu, pi, and theta) and a high homology with invertebrate GST isozymes of the sigma class. Since the sigma class GST had been observed only in invertebrates, H-PGDS was demonstrated to be the first vertebrate homolog of the sigma class GST (Kanaoka *et al.,* 1997; Thomson *et al.,* 1998).

C. X-Ray Crystallographic Structure

We then crystallized the recombinant rat H-PGDS (Fig. 4B) and determined the tertiary structure of the enzyme complexed with GSH

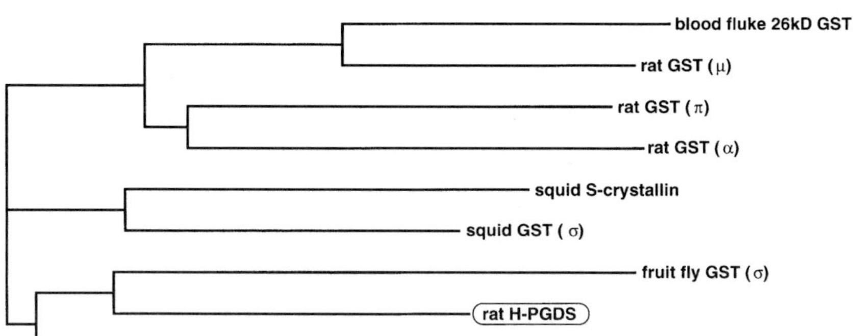

FIG. 5. Phylogenetic tree of H-PGDS and members of the GST family. Bar represents 0.1 amino acid substitution/site.

with a resolution of 2.3 Å by X-ray diffraction analysis (Kanaoka *et al.*, 1997) (Fig. 3B). This is the first report of the tertiary structure of an enzyme that utilizes PGH_2 as a substrate. The X-ray crystallographic analysis revealed that the overall structure of H-PGDS was similar to that of other GSTs, yet it possessed a unique catalytic site within a wide cleft near the GSH-binding site. Such a prominent cleft identified as the putative active site of H-PGDS has never been seen among other members of the GST family. This is consistent with the fact that H-PGDS catalyzes the highly selective conversion of PGH_2 to produce PGD_2, whereas other GST isozymes catalyze the conversion of PGH_2 to produce a mixture of PGD_2, PGE_2, and/or $PGF_{2\alpha}$ (Chang *et al.*, 1987; Ujihara *et al.*, 1988c).

D. Genomic Cloning and Generation of Gene-Manipulated Mice

The human and mouse genes for H-PGDS have already been cloned and their chromosomal localizations were mapped to 4q21-22 and 3D-E, respectively (Y. Kanaoka and Y. Urade, unpublished). These locations are different from those of any of human and mouse genes for the alpha, mu, pi, and theta classes of the GST family. When the gene structure was compared with that of other members of the GST family, the H-PGDS gene was found to be the largest in size and the most similar to the genes for members of the sigma class. Furthermore, the positions of introns were highly conserved upon multiple alignment of the amino acid sequences between H-PGDS and other members of the GST family. Therefore, the H-PGDS gene is considered to have evolved from a common ancestor for invertebrate sigma GSTs.

KO mice for the H-PGDS gene and TG mice overexpressing human H-PGDS have already been generated in our group and used to examine functional abnormalities in inflammatory and allergic reactions.

E. Tissue and Cellular Distribution and Possible Function

The tissue distribution profile and cellular localization of H-PGDS have been most extensively investigated in rats. As examined by measuring the enzyme activity, immunoabsorption tests (Ujihara *et al.*, 1988b), and Northern blot analysis (Kanaoka *et al.*, 1997), rat H-PGDS was expressed most abundantly in spleen and moderately in oviduct, thymus, bone marrow, Peyer's patch, and ileum. However, the chicken H-PGDS was detected abundantly in liver, kidney, small intestine, and colon; moderately in lung, pancreas, ovary, and bone marrow: and weakly in spleen, thymus, heart, and brain (Thomson *et al.*, 1998). Recently we found that the tissue distribution profiles of the mRNA for H-PGDS in humans and mice were remarkably different from those in rats and chickens (Y. Kanaoka *et al.,* unpublished). The human H-PGDS was expressed strongly in placenta; moderately in lung, fetal liver, lymph nodes, and appendix; and weakly in brain, heart, thymus, bone marrow, and spleen. But it was undetectable in adult liver and pancreas. Alternatively, the mouse transcripts were found in the oviduct and skin, but were negligible in content in other tissues including spleen and thymus.

Rat H-PGDS was immunohistochemically localized in antigen-presenting cells, such as Langerhans cells in the skin (Ujihara *et al.,* 1988a), Kuppfer cells in the liver, dendritic cells in the thymus and Peyer's patch, and histiocytes in the spleen and lamina propria of the intestine and colon (Urade *et al.,* 1989b). H-PGDS is also localized in mast cells (Urade *et al.,* 1990a). The induction of H-PGDS is reportedly involved in mast cell maturation (Murakami *et al.,* 1995a,b,c) and also in human megakaryocytic differentiation (Mahmud *et al.,* 1997; Suzuki *et al.,* 1997). H-PGDS is considered to be involved in the deep sleep of mastocytotic patients (Roberts *et al.,* 1980). Profound lethargy in patients with this disease was considered to be primarily due to the remarkable increase in endogenous production of PGD_2. The immunoreactivity of H-PGDS was localized in satellite and Schwann cells of chick dorsal root ganglia (Vesin *et al.,* 1995b). However, the existence of H-PGDS in the CNS and its cellular localization there, if any, remain to be elucidated.

V. Future Studies

We recently crystallized recombinant mouse L-PGDS and human H-PGDS and have started the X-ray diffraction analyses. The three-dimensional coordinates of these enzymes will be useful for designing selective and nonselective inhibitors for each enzyme. KO and TG mice for L-PGDS or H-PGDS have also already been generated by our group. Further investigation to examine the functional abnormality in these genetically engineered mutant mice should provide us new insight into the physiologic relevance of PGD_2 and its metabolites. Recently, we found that the human malaria parasite, *Plasmodium falciparum,* possesses a PG-producing system that is different from the mammalian system (Kubata *et al.,* 1998a,b). Therefore, other types of PGDS may exist in the malaria parasite and also in other species.

ACKNOWLEDGMENTS

We are grateful to Drs. N. Eguchi, S. Sri Kantha, C. T. Beuckmann, Y. Kanaoka, D. Y. Gerashchenko, E. Pinzar, K. B. Kubata, T. Inui, Y. Fujitani, K. Fujimori, T. Mochizuki, M. Lazarus, and S. Satoh of Osaka Bioscience Institute; Dr. H. Toh of Biomolecular Engineering Research Institute; and Dr. M. Miyano of Central Pharmaceutical Research Institute, Japan Tabacco Inc., for valuable discussions. We also thank D. Irikura, Y. Kuwahata, S. Matsumoto, N. Uodome, and R. Matsumoto for technical assistance and S. Matsumoto-Sakae, S. Ueda, M. Yamaguchi, and T. Nishimoto for secretarial assistance.

This work was supported in part by grants from the program Grants-in-Aid for Scientific Research of the Ministry of Education, Science, Sports, and Culture of Japan (07558108, 07457033 and 09044352 to YU and 06508003 to OH), the program for Core Research for Evolutional Science and Technology from Japan Science and Technology Corporation (to YU), and by grants from the Ministry of Health and Welfare of Japan (100107 to OH), the Ground Research for Space Utilization promoted by NASDA and Japan Space Forum (to YU), the Suntory Institute for Bioorganic Research (to YU), and the Japan Foundation for Applied Enzymology (to YU).

REFERENCES

Achen, M. G., Harms, P. J., Thomas, T., Richardson, S. J., Wettenhall, R. E. H., and Schreiber, G. (1992). Protein synthesis at the blood-brain barrier. The major protein secreted by amphibian choroid plexus is a lipocalin. *J. Biol. Chem.* **267,** 23170–23174.

Andersen, J. F., Weichsel, A., Balfour, C. A., Champagne, D. E., and Montfort, W. R. (1998). The crystal structure of nitrophorin 4 at 1.5 A resolution: Transport of nitric oxide by a lipocalin-based heme protein. *Structure* **15,** 1315–1327.

Beuckmann, C. T., Gordon, W. C., Kanaoka, Y., Eguchi, N., Marcheselli, V. L., Gerashchenko, D. Y., Urade, Y., Hayaishi, O., and Bazan, N. G. (1996). Lipocalin-type prostaglandin D synthase (β-trace) is located in pigmented epithelial cells of rat retina and accumulates within interphotoreceptor matrix. *J. Neurosci.* **16,** 6119–6124.

Beuckmann, C. T., Aoyagi, M., Okazaki, I., Hiroike, T., Toh, H., Hayaishi, O., and Urade, Y. (1999). Binding of biliverdin, bilirubin, and thryoid hormones to lipocalin-type prostaglandin D synthase. *Biochemistry* **38,** 8006–8013.

Bianchet, M. A., Bains, G., Pelosi, P., Pevsner, J., Snyder, S. H., Monaco, H. L., and Amzel, L. M. (1996). The three-dimensional structure of bovine odorant binding protein and its mechanism of odor recognition. *Nat. Struct. Biol.* **11,** 934–939.

Blödorn, B., Mäder, M., Urade, Y., Hayaishi, O., Felgenhauer, K., and Bruck, W. (1996). Choroid plexus: The major site of mRNA expression for the β-trace protein (prostaglandin D synthase) in human brain. *Neurosci. Lett.* **209,** 117–120.

Böcskel, Z., Groom, C. R., Flower, D. R., Wright, C. E., Philips, S. E. V., Cavaggioni, A., Findlay, J. B. C., and North, A. C. T. (1992). Pheromone binding to two rodent urinary proteins revealed by X-ray crystallography. *Nature (London)* **360,** 186–188.

Boie, Y., Sawyer, N., Slipetz, D. M., Metters, K. M., and Abramovitz, M. (1995). Molecular cloning and characterization of the human prostanoid DP receptor. *J. Biol. Chem.* **270,** 18910–18916.

Borbely, A. A., and Tobler, I. (1989). Endogenous sleep-promoting substances and sleep regulation. *Physiol. Rev.* **69,** 605–670.

Brownlow, S., Morais Cabral, J. H., Cooper, R., Flower, D. R., Yewdall, S. J., Polikarpov, I., North, A. C., and Sawyer, L. (1997). Bovine beta-lactoglobulin at 1.8 A resolution—still an enigmatic lipocalin. *Structure* **5,** 481–495.

Chan, P., Simon-Chazottes, D., Mattei, M. G., Guenet, J. L., and Salier, J. P. (1994). Comparative mapping of lipocalin genes in human and mouse: The four genes for complement C8 γ chain, prostaglandin-D-synthase, oncogene-24p3 and progestagen-associated endometrial protein map to HSA9 and MMU2. *Genomics* **23,** 145–150.

Chang, M., Hong, Y., Burgess, J. R., Tu, C.-P. D., and Reddy, C. C. (1987). Isozyme specificity of rat liver glutathione S-transferases in the formation of $PGF_{2\alpha}$ and PGE_2 from PGH_2, *Arch. Biochem. Biophys.* **259,** 548–557.

Chen, L.-Y., Watanabe, K., and Hayaishi, O. (1992). Purification and characterization of prostaglandin F synthase from bovine liver. *Arch. Biochem. Biophys.* **296,** 17–26.

Christ-Hazelhof, E., and Nugteren, D. H. (1979). Purification and characterization of prostaglandin endoperoxide D-isomerase, a cytoplasmic, glutathione-requiring enzyme. *Biochim. Biophys. Acta* **572,** 43–51.

Clausen, J. (1961). Proteins in normal cerebrospinal fluid not found in serum. *Proc. Soc. Exp. Biol. Med.* **107,** 170–172.

Coleman, R. A., Smith, W. L., and Narumiya, S. (1994). International Union of Pharmacology classification of prostanoid receptors: Properties, distribution, and structure of the receptors and their subtypes. *Pharmacol. Rev.* **46,** 205–229.

Cowan, S. W., Newcomer, N. E., and Jones, T. A. (1990). Crystallographic refinement of human serum retinal binding protein at 2 A resolution. *Proteins* **8,** 44–61.

Eguchi, N., Minami, T., Shirafuji, N., Kanaoka, Y., Tanaka, T., Nagata, A., Yoshida, N., Urade, Y., Ito, S., and Hayaishi, O. (1999). Lack of tactile pain (allodynia) in lipocalin-type prostaglandin D synthase-deficient mice. *Proc. Natl. Acad. Sci. U.S.A.* **96,** 726–730.

Eguchi, Y., Eguchi, N., Oda, H., Seiki, K., Kijima, Y., Matsu-ura, Y., Urade, Y., and Hayaishi, O. (1997). Expression of lipocalin-type prostaglandin D synthase (β-trace) in human heart and its accumulation in the coronary circulation of angina patients. *Proc. Natl. Acad. Sci. U.S.A.* **94,** 14689–14694.

Flower, D. R. (1995). Multiple molecular recognition properties of the lipocalin family. *J. Mol. Recognition* **8,** 185–195.

Flower, D. R. (1996). The lipocalin protein family: Structure and function. *Biochem. J.* **318,** 1–14.

Forman, B. M., Tontonoz, P., Chen, J., Brun, R. P., Spiegelman, B. M., and Evans, R. M. (1995). 15-Deoxy-delta 12, 14-prostaglandin J_2 is a ligand for the adipocyte determination factor PPAR gamma. *Cell (Cambridge, Mass.)* **83,** 803–812.

Fouchécourt, S., Dacheux, F., and Dacheux, J.-L. (1999). Glutathione-independent prostaglandin D_2 synthase in ram and stallion epididymal fluid: Origin and regulation. *Biol. Reprod.* **60,** 558–566.

Fukushima, M. (1992). Biological activities and mechanisms of action of PGJ_2 and related compounds: An update. *Prostaglandins Leukot. Essent. Fatty Acids* **47,** 1–12.

García-Fernández, L. F., Iniguez, M. A., Rodríguez-Peña, A., Muñoz, A., and Bernal, J. (1993). Brain-specific prostaglandin D_2 synthetase mRNA is dependent on thyroid hormone during rat brain development. *Biochem. Biophys. Res. Commun.* **196,** 396–401.

García-Fernández, L. F., Rausell, E., Urade, Y., Hayaishi, O., Bernal, J., and Muñoz, A. (1997). Hypothyroidism alters the expression of prostaglandin D_2 synthase/β-trace in specific areas of developing rat brain. *Eur. J. Neuroci.* **9,** 1566–1573.

García-Fernández, L. F., Urade, Y., Hayaishi, O., Bernal, J., and Muñoz, A. (1998). Identification of a thyroid hormone response element in the promoter region of the rat lipocalin-type prostaglandin D synthase (β-trace) gene. *Mol. Brain Res.* **55,** 321–330.

Gerashchenko, D., Beuckmann, C. T., Kanaoka, Y., Eguchi, N., Gordon, W. C., Urade, Y., Bazan, N. G., and Hayaishi, O. (1998a). Dominant expression of rat prostanoid DP receptor mRNA in leptomeninges, inner segments of photoreceptor cells, iris epithelium, and ciliary processes. *J. Neurochem.* **71,** 937–945.

Gerashchenko, D. Y., Beuckmann, C. T., Marcheselli, V. L., Gordon, W. C., Kanaoka, Y., Eguchi, N., Urade, Y., Hayaishi, O., and Bazan, N. G. (1998b). Localization of lipocalin-type prostaglandin D synthase (β-trace) in iris, ciliary body, and eye fluids. *Invest. Ophthalmol. Visual Sci.* **39,** 198–203.

Gerena, R. L., Irikura, D., Urade, Y., Eguchi, N., Chapman, D. A., and Killian, G. J. (1998). Identification of a fertility-associated protein in bull seminal plasma as lipocalin-type prostaglandin D synthase. *Biol. Reprod.* **58,** 826–833.

Giacomelli, S., Leone, M.-G., Grima, J., Silvestrini, B., and Cheng, C. Y. (1996). Astrocytes synthesize and secrete prostaglandin D synthase in vitro. *Biochim. Biophys. Acta* **1310,** 269–276.

Giles, H., and Leff, P. (1988). The biology and pharmacology of PGD_2. *Prostaglandins* **35,** 277–300.

Goh, H., Urade, Y., Fujimoto, N., and Hayaishi, O. (1987). Content and formation of prostaglandins and distribution of prostaglandin-related enzyme activities in the rat ocular system. *Biochim. Biophys. Acta* **921,** 302–311.

Harrington, M. G., Aebersold, R., Martin, B. M., Merril, C. R., and Hood, L. (1993). Identification of a brain-specific human cerebrospinal fluid glycoprotein, beta-trace protein. *Appl. Theor. Electrophor.* **3,** 229–234.

Hayaishi, O. (1988). Sleep-wake regulation by prostaglandins D_2 and E_2. *J. Biol. Chem.* **263,** 14593–14596.

Hayaishi, O. (1991). Molecular mechanisms of sleep-wake regulation: Roles of prostaglandins D_2 and E_2. *FASEB J.* **5,** 2572–2581.

Hayaishi, O., and Matsumura, H. (1995). Prostaglandins and sleep. *Adv. Neuroimmunol.* **5,** 211–216.

Hayaishi, O., Matsumura, H., and Urade, Y. (1993). Prostaglandin D synthase is the key enzyme in the promotion of physiological sleep. *J. Lipid Mediators* **6,** 429–431.

Hiraoka, A., Arato, T., Tominaga, I., Eguchi, N., Oda, H., and Urade, Y. (1997). Analysis of low-molecular-mass proteins in cerebrospinal fluid by sodium dodecyl sulfate capillary gel electrophoresis. *J. Chromatogr.* **697,** 141–147.

Hiraoka, A., Arato, T., Tominaga, I., Eguchi, N., Oda, H., and Urade, Y. (1998). Sodium dodecyl sulfate-capillary gel electrophoretic analysis of molecular mass microheterogeneity of β-trace protein in cerebrospinal fluid from patients with central nervous system diseases. *J. Chromatogr.* **802,** 143–148.

Hirata, M., Kakizuka, A., Aizawa, M., Ushikubi, F., and Narumiya, N. (1994). Molecular characterization of a mouse prostaglandin D receptor and functional expression of the cloned gene. *Proc. Natl. Acad. Sci. U.S.A.* **91,** 11192–11196.

Hirata, Y., Hayashi, H., Ito, S., Kikawa, Y., Ishibashi, M., Sudo, M., Miyazaki, H., Fukushima, M., Narumiya, S., and Hayaishi, O. (1988). Occurrence of 9-deoxy-Δ^9, Δ^{12}-13,14-dihydroprostaglandin D_2 in human urine. *J. Biol. Chem.* **263,** 16619–16625.

Hochwald, G. M., and Thorbecke, G. J. (1962). Use of an antiserum against cerebrospinal fluid in demonstration of trace proteins in biological fluids. *Proc. Soc. Exp. Biol. Med.* **109,** 91–95.

Hochwald, G. M., and Thorbecke, G. J. (1963). Trace proteins in cerebrospinal fluid and other biological fluids. I. Effect of various fractionation procedures on beta-trace and gamma-trace proteins and methods for isolation of both proteins. *Arch. Biochem. Biophys.* **101,** 325–334.

Hochwald, G. M., Pepe, A. J., and Thorbecke, G. J. (1967). Trace proteins in biological fluids. IV. Physicochemical properties and sites of formation of gamma-trace and beta-trace proteins. *Proc. Soc. Exp. Biol. Med.* **124,** 961–966.

Hoffmann, A., Conradt, H. S., Gross, G., Nimtz, M., Lottspeich, F., and Wurster, U. (1993). Purification and chemical characterization of β-trace protein from human cerebrospinal fluid: Its identification as prostaglandin D synthase. *J. Neurochem.* **61,** 451–456.

Hoffmann, A., Nimtz, M., Wurster, U., and Conradt, H. S. (1994). Carbohydrate structures of β-trace proteins from human cerebrospinal fluid: Evidence for "brain-type" N-glycosylation. *J. Neurochem.* **63,** 2185–2196.

Hoffmann, A., Bächner, D., Betar, N., Lauber, J., and Gross, G. (1996a). Developmental expression of murine β-trace in embryos and adult animals suggests a function in maturation and maintenance of blood-tissue barriers. *Dev. Dyn.* **207,** 332–343.

Hoffmann, A., Gath, U., Gross, G., Lauber, J., Getzlaff, R., Hellwig, S., Galla, H. J., and Conradt, H. S. (1996b). Constitutive secretion of beta-trace protein by cultivated porcine choroid plexus epithelial cells: Elucidation of its complete amino acid and cDNA sequences. *J. Cell. Physiol.* **169,** 235–241.

Hoffmann, A., Nimtz, M., and Conradt, H. S. (1997). Molecular characterization of β-trace protein in human serum and urine: A potential diagnostic marker for renal diseases. *Glycobiology* **7,** 499–506.

Holden, H. M., Rypniewski, W. R., Law, J. H., and Rayment, I. (1987). The molecular structure of insecticyanin from the tobacco hornworm Manduca sexta L. at 2.6 A resolution. *EMBO J.* **6,** 1565–1570.

Horiguchi, S., Ueno, R., Hyodo, M., and Hayaishi, O. (1986). Alterations in nociception after intracisternal administration of prostaglandin D_2, E_2 or $F_{2\alpha}$ to conscious mice. *Eur. J. Pharmacol.* **122,** 173–179.

Huber, R., Schneider, M., Epp, O., Mayr, I., Messerschmidt, A., Pflugrath, J., and Kayser, H. (1986). Crystallization, crystal structure analysis and preliminary molecular model of the bilin binding protein from the insect *Pieris brassicae. J. Mol. Biol.* **195,** 423–434.

Igarashi, M., Nagata, A., Toh, H., Urade, Y., and Hayaishi, O. (1992). Structural organization of the gene for prostaglandin D synthase in the rat brain. *Proc. Natl. Acad. Sci. U.S.A.* **89,** 5376–5380.

Islam, F., Watanabe, Y., Morii, H., and Hayaishi, O. (1991). Inhibition of rat brain prostaglandin D synthase by inorganic selenocompounds. *Arch. Biochem. Biophys.* **289,** 161–166.

Ito, S., Narumiya, S., and Hayaishi, O. (1989). Prostaglandin D_2: A biochemical perspective. *Prostaglandins Leukotrienes Essent. Fatty Acids* **37,** 219–234.

Jiang, C., Ting, A. T., and Seed, B. (1998). PPAR-gamma agonists inhibit production of monocyte inflammatory cytokines. *Nature (London)* **391,** 82–86.

Kabeya, K., Ishikawa, K., Katakai, K., Watanabe, M., Ohe, Y., Wakabayashi, K., and Mori, M. (1998). Prostaglandin-D-synthase (β-trace protein) levels in rat cerebrospinal fluid. *NeuroReport* **9,** 915–919.

Kanaoka, Y., Ago, H., Inagaki, E., Nanayama, T., Miyano, M., Kikuno, R., Fujii, Y., Eguchi, N., Toh, H., Urade, Y., and Hayaishi, O. (1997). Cloning and crystal structure of hematopoietic prostaglandin D synthase. *Cell (Cambridge, Mass.)* **90,** 1085–1095.

Keller, M., Jackisch, R., Seregi, A., and Hertting, G. (1985). Comparison of prostanoid-forming capacity of neuronal and astroglial cells in primary cultures. *Neurochem. Int.* **7,** 655–665.

Kikawa, Y., Narumiya, S., Fukushima, M., Wakatsuka, H., and Hayaishi, O. (1984). 9-Deoxy-Δ^9, Δ^{12}-13,14-dihydro-prostaglandin D_2, a novel metabolite of prostaglandin D_2 formed in human plasma. *Proc. Natl. Acad. Sci. U.S.A.* **81,** 1317–1321.

Killian, G. L., Chapman, D. A., and Rogowski, L. A. (1993). Fertility-associated proteins in Holstein bull seminal plasma. *Biol. Reprod.* **49,** 1202–1207.

Kinoshita, F., Nakai, Y., Katakami, H., Imura, H., Shimizu, T., and Hayaishi, O. (1982). Suppressive effect of prostaglandin (PG)D_2 on pulsatile leuteinizing hormone release in conscious castrated rats. *Endocrinology (Baltimore)* **110,** 2207–2209.

Kliewer, S. A., Lenhard, J. M., Willson, T. M., Patel, I., Morris, D. C., and Lehmann, J. M. (1995). A prostaglandin J_2 metabolite binds peroxisome proliferator-activated receptor gamma and promotes adipocyte differentiation. *Cell (Cambridge, Mass.)* **83,** 813–819.

Kubata, B. K., Eguchi, N., Urade, Y., Yamashita, K., Horii, T., and Hayaishi, O. (1998a). Evidence of prostaglandin production by the human malaria parasite, *Plasmodium falciparum. S. Afr. J. Sci.* **94,** 285–286.

Kubata, B. K., Eguchi, N., Urade, Y., Yamashita, K., Mitamura, T., Tai, K., Hayaishi, O., and Horii, T. (1998b). *Plasmodium falciparum* produces prostaglandins that are pyrogenic, somnogenic, and immunosuppressive substances in humans. *J. Exp. Med.* **188,** 1197–1202.

Kuruvilla, A. P., Hochwald, G. M., Ghiso, J., Castano, E. M., Pizzolato, M., and Frangione, B. (1991). Isolation and amino terminal sequence of β-trace, a novel protein from human cerebrospinal fluid. *Brain Res.* **565,** 337–340.

Laterre, E. C., Heremans, J. F., and Carbonara, A. (1964). Immunological comparison of some proteins found in cerebrospinal fluid, urine and extracts from brain and kidney. *Clin. Chim. Acta* **10,** 197–201.

Lepperdinger, G., Strobl, B., Jilek, A., Weber, A., Thalhamer, J., Flöckner, H., and Mollay, C. (1996). The lipocalin Xlcp11 expressed in the neural plate of *Xenopus laevis* embryos is a secreted retinaldehyde binding protein. *Protein Sci.* **5,** 1250–1260.

Lewis, R. A., Soter, N. A., Diamond, P. T., Austen, K. F., Oates, J. A., and Roberts, L. J. (1982). Prostaglandin D_2 generation after activation of rat and human mast cells with anti-IgE. *J. Immunol.* **129,** 1627–1631.

Link, H. (1965). Isolation and partial characterization of "trace" proteins and immunoglobulin G from cerebrospinal fluid. *J. Neurosurg. Psychiatry* **28,** 552–559.

Link, H., and Olsson, J. E. (1972). Beta-trace protein concentration in CSF in neurological disorders. *Acta Neurol. Scand.* **48,** 57–68.

Liston, T. E., and Roberts, L. J., II (1985). Transformation of prostaglandin D_2 to 9α,11β-(15*S*)-trihydroxyprosta-(5Z,13E)-dien-1-oic acid (9α,11β-prostaglandin F_2): A unique biologically active prostaglandin produced enzymatically *in vivo* in humans. *Proc. Natl. Acad. Sci. U.S.A.* **82,** 6030–6034.

Mahmud, I., Ueda, N., Yamaguchi, H., Yamashita, R., Yamamoto, S., Kanaoka, Y., Urade, Y., and Hayaishi, O. (1997). Prostaglandin D synthase in human megakaryoblastic cells. *J. Biol. Chem.* **272,** 28263–28266.

Mase, M., Yamada, K., Iwata, A., Matsumoto, T., Seiki, K., Oda, H., and Urade, Y. (1999). Acute and transient increase of lipocalin-type prostaglandin D synthase (*β*-trace) level in cerebrospinal fluid of patients with aneurysmal subarachnoid hemorrhage. *Neurosci. Lett.* **270,** 188–190.

Matsumura, H., Takahata, R., and Hayaishi, O. (1991). Inhibition of sleep in rats by inorganic selenium compounds, inhibitors or prostaglandin D synthase. *Proc. Natl. Acad. Sci. U.S.A.* **88,** 9046–9050.

Matsumura, H., Nakajima, T., Osaka, T., Satoh, S., Kawase, K., Kubo, E., Kantha, S. S., Kasahara, K., and Hayaishi, O. (1994). Prostaglandin D_2-sensitive, sleep-promoting zone defind in the ventral surface of the rostral basal forebrain. *Proc. Natl. Acad. Sci. U.S.A.* **91,** 11998–12002.

Matsuo, M., Hamasaki, Y., Fujiyama, F., and Miyazaki, S. (1995). Eicosanoids are produced by microglia, not by astrocytes, in rat glial cell cultures. *Brain Res.* **685,** 201–204.

Melegos, D. N., Yu, H., and Diamandis, E. P. (1996a). Prostaglandin D_2 synthase—a component of human amniotic fluid and its association with fetal abnormalities. *Clin. Chem. (Winston-Salem, N.C.)* **42,** 1042–1050.

Melegos, D. N., Diamandis, E. P., Oda, H., Urade, Y., and Hayaishi, O. (1996b). Immunofluorometric assay of prostaglandin D synthase in human tissue extracts and fluids. *Clin. Chem. (Winston-Salem, N.C.)* **42,** 1984–1991.

Melegos, D. N., Freedmann, M. S., and Diamandis, E. P. (1997). Prostaglandin D synthase concentration in cerebrospinal fluid and serum of patients with neurological disorders. *Prostaglandins* **54,** 463–474.

Meyer, D. J., and Thomas, M. (1995). Characterization of rat spleen prostaglandin H D-isomerase as a sigma-class GSH transferase. *Biochem. J.* **311,** 739–742.

Minami, T., Okuda-Ashitaka, E., Mori, H., Ito, S., and Hayaishi, O. (1996). Prostaglandin D_2 inhibits prostaglandin E_2-induced allodynia in conscious mice. *J. Pharmacol. Exp. Ther.* **278,** 1146–1152.

Minami, T., Okuda-Ashitaka, E., Nishizawa, M., Mori, H., and Ito, S. (1997). Inhibition of nociceptin-induced allodynia in conscious mice by prostaglandin D_2. *Br. J. Pharmacol.* **122,** 605–610.

Monaco, H. L., and Zanotti, G. (1992). Three dimensional structure and active site of three hydrophobic molecule-binding proteins with significant amino acid sequence similarity. *Biopolymers* **32,** 457–465.

Monaco, H. L., Zanotti, G., Spadon, P., Bolognesi, M., Sawyer, L., and Eliopoulos, E. E. (1987). Crystal structure of the trigonal form of bovine beta-lactoglobulin and its complex with retinol at 2.5 A resolution. *J. Mol. Biol.* **197,** 695–706.

Moritomo, A., Murakami, N., Nakamori, T., and Watanabe, T. (1988). Multiple control of fever production in the central nervous system of rabbits. *J. Physiol. (London)* **397,** 269–280.

Muñoz, A., Rodríguez-Peña, L. F. A., Pérez-Castillo, A., Ferreiro, B., Sutcliffe, J. G., and Bernal, J. (1991). Effects of neonatal hypothyroidism on rat bain gene expression. *Mol. Endocrinol.* **5,** 273–280.

Murakami, M., Bingham, C. D., III, Matsumoto, R., Austen, K. F., and Arm, J. P. (1995a). IgE-dependent activation of cytokine-primed mouse cultured mast cells induces a delayed phase of prostaglandin D_2 generation via prostaglandin endoperoxide synthase-2. *J. Immunol.* **155,** 4445–4453.

Murakami, M., Matsumoto, R., Urade, Y., Austen, K. F., and Arm, J. P. (1995b). c-Kit ligand mediates increased expression of cytosolic phospholipase A_2, prostaglandin endoperoxide synthase-1, and hematopoietic prostaglandin D_2 synthase and increased IgE-dependent prostaglandin D_2 generation in immature mouse mast cells. *J. Biol. Chem.* **270,** 3239–3246.

Murakami, M., Penrose, J. F., Urade, Y., Austen, K. F., and Arm, J. P. (1995c). Interleukin 4 suppresses c-*kit* ligand-induced expression of cytosolic phospholipase A_2 and prostaglandin endoperoxide synthase 2 and their roles in separate pathways of eicosanoid synthesis in mouse bone marrow-derived mast cells. *Proc. Natl. Acad. Sci. U.S.A.* **92,** 6107–6111.

Nagata, A., Suzuki, Y., Igarashi, M., Eguchi, N., Toh, H., Urade, Y., and Hayaishi, O. (1991). Human brain prostaglandin D synthase has been evolutionarily differentiated from lipophilic-ligand carrier proteins. *Proc. Natl. Acad. Sci. U.S.A.* **88,** 4020–4024.

Nagoshi, H., Uehara, Y., Kanai, F., Maeda, S., Ogura, T., Goto, A., Toyo-oka, T., Esumi, H., Shimizu, T., Omata, M. (1998). Prostaglandin D_2 inhibits inducible nitric oxide synthase expression in rat vascular smooth muscle cells. *Circ. Res.* **82,** 204–209.

Nagy, L., Tontonoz, P., Alvarez, J. G. A., Chen, H., and Evans, R. M. (1998). Oxidized LDL regulates macrophage gene expression through ligand activation of PPARγ. *Cell (Cambridge, Mass.)* **93,** 229–240.

Narumiya, S., and Toda, N. (1985). Different responsiveness of prostaglandin D_2-sensitive systems to prostaglandin D_2 and its analogues. *Br. J. Pharmacol.* **85,** 367–375.

Negishi, M., Sugimoto, Y., and Ichikawa, A. (1993). Prostanoid receptors and their biological actions. *Prog. Lipid Res.* **32,** 417–434.

Negishi, M., Sugimoto, Y., and Ichikawa, A. (1995). Molecular mechanisms of diverse actions of prostanoid receptors. *Biochim. Biophys. Acta* **1259,** 109–120.

Newcomer, M. E., Jones, T. A., Åqvist, J., Sundelin, J., Eriksson, U., Rask, L., and Peterson, P. A. (1984). The three-dimensional structure of retinol-binding protein. *EMBO J.* **3,** 1451–1454.

Newcomer, M. E., Pappas, R. S., and Ong, D. E. (1993). X-ray crystallographic identification of a protein-binding site for both all-trans- and 9-cis-rertinoic acid. *Proc. Natl. Acad. Sci. U.S.A.* **90,** 9223–9227.

Nishi, S., Seino, Y., Seino, S., Tsuda, K., Takemura, J., Shimizu, T., Hayaishi, O., and Imura, H. (1984). Different effects of prostaglandin E_1, E_2 and D_2 on pancreatic somatostatin release. *Horm. Metab. Res.* **16** (Suppl. 1), 114–118.

Oda, H., Eguchi, N., Urade, Y., and Hayaishi, O. (1996). Quantitative sandwich enzyme-linked immunosorbent assay for human secretory prostaglandin D synthase (β-trace). *Proc. Jpn. Acad.* **72,** 108–111.

Ohe, Y., Ishikawa, K., Itoh, Z., and Tatemoto, K. (1996). Cultured leptomeningeal cells secrete cerebrospinal fluid proteins. *J. Neurochem.* **67,** 964–971.

Oida, H., Hirata, M., Sugimoto, Y., Ushikubi, F., Ohishi, H., Mizuno, N., Ichikawa, A., and Narumiya, S. (1997). Expression of messenger RNA for the prostaglandin D receptor in the leptomeninges of the mouse brain. *FEBS Lett.* **417,** 53–56.

Olsson, J. E. (1975). Correlation between the concentration of beta-trace protein and the number of spermatozoa in human semen. *J. Reprod. Fertil.* **42,** 149–151.

Olsson, J. E., and Link, H. (1973). Distribution of serum proteins and beta-trace protein within the nervous system. *J. Neurochem.* **20,** 837–846.

Olsson, J. E., and Nord, L. (1973). Immunochemical and immunofluorescence studies of beta-trace protein in different species and organs, with special reference to the central nervous system. *J. Neurochem.* **21,** 625–633.

Olsson, J. E., and Sandberg, M. (1975). Demonstration of synthesis of beta-trace protein in different tissues of squirrel monkey. *Neurobiology* **5,** 270–276.

Olsson, J. E., Link, H., and Nosslin, B. (1973). Metabolic studies on ^{125}I-labelled beta-trace protein, with special reference to synthesis within the central nervous system. *J. Neurochem.* **21,** 1153–1159.

Onoe, H., Ueno, R., Fujita, I., Nishino, H., Oomura, Y., and Hayaishi, O. (1988). Prostaglandin D_2, a cerebral sleep-inducing substance in monkeys. *Proc. Natl. Acad. Sci. U.S.A.* **85,** 4082–4086.

Osama, H., Narumiya, S., Hayaishi, O., Iinuma, H., Takeuchi, T., and Umezawa, H. (1983). Inhibition of brain prostaglandin D synthetase and prostaglandin D_2 dehydrogenase by some saturated and unsaturated fatty acids. *Biochim. Biophys. Acta* **752,** 251–258.

Pandey, H. P., Ram, A., Matsumura, H., Satoh, S., and Hayaishi, O. (1995). Circadian variations of prostaglandin D_2, E_2, and $F_{2\alpha}$ in the cerebrospinal fluid of anesthetized rats. *Biochem. Biophys. Res. Commun.* **213,** 625–629.

Papiz, M. Z., Sawyer, L., Eliopoulos, E. E., North, A. T., Findlay, J. B. C., Sivaprasadarao, R., Jones, T. A., Mewcomer, M. E., and Kraulis, P. J. (1986). The structure of beta-lactoglobulin and its similarity to plasma retinol-binding protein. *Nature (London)* **324,** 383–385.

Peitsch, M., and Boguski, M. (1991). The first lipocalin with enzymatic activity. *Trends Biochem. Sci.* **16,** 363.

Pentreath, V. W., Rees, K., Owolabi, O. A., Philip, K. A., and Doua, F. (1990). The somnogenic T lymphocyte suppressor prostaglandin D_2 is selectively elevated in cerebrospinal fluid of advanced sleeping sickness patients. *Trans. R. Soc. Trop. Med. Hyg.* **84,** 795–799.

Pepe, A. J., and Hochwald, G. M. (1967). Trace proteins in biological fluids, III. Quantitation of gamma-trace and major spinal fluid proteins including beta-trace. *Proc. Soc. Exp. Biol. Med.* **126,** 630–635.

Pervaiz, S., and Brew, K. (1987). Homology and structure-function correlations between α1-acid glycoprotein and serum retinol-binding protein and its relatives. *FASEB J.* **1,** 209–214.

Ram, A., Pandey, H. P., Matsumura, H., Kasahara-Orita, K., Nakajima, T., Takahata, R., Satoh, S., Terao, A., and Hayaishi, O. (1997). CSF levels of prostaglandins, especially the level of prostaglandin D_2, are correlated with increasing propensity towards sleep in rats. *Brain Res.* **751,** 81–89.

Ricote, M., Huang, J., Fajas, L., Li, A., Welch, J., Najib, J., Witztum, J. L., Auwerx, J., Palinski, W., and Glass, C. K. (1998a). Expression of the peroxisome proliferator-activated receptor γ (PPARγ) in human atherosclerosis and regulation in macrophages by colony stimulating factors and oxidized low density lipoprotein. *Proc. Natl. Acad. Sci. U.S.A.* **95,** 7614–7619.

Ricote, M., Li, A. C., Willson, T. M., Kelly, C. J., and Glass, C. K. (1998b). The peroxisome proliferator-activated receptor-gamma is a negative regulator of macrophage activation. *Nature (London)* **391,** 79–82.

Roberts, J. L., II, Sweetman, B. J., Lewis, R. A., Austen, K. F., and Oates, J. A. (1980). Increased productin of prostaglandin D_2 in patients with systemic mastocytosis. *N. Engl. J. Med.* **303,** 1400–1404.

Rouvinen, J., Rautiainen, J., Virtanen, T., Zeiler, T., Kauppinen, J., Taivainen, A., and Mantyjarvi, R. (1999). Probing the molecular basis of allergy. three-dimensional structure of the bovine lipocalin allergen Bos d2. *J. Biol. Chem.* **274,** 2337–2343.

Saso, L., Leone, M. G., Sorrentio, C., Giacomelli, S., Silvestrini, B., Grima, J., Li, J. C. H., Samy, E., Mruk, D., and Cheng, C. Y. (1998). Quantification of prostaglandin D synthase in cerebrospinal fluid: A potential marker for brain tumor. *Biochem. Mol. Biol. Int.* **46,** 643–656.

Satoh, S., Matsumura, H., Suzuki, F., and Hayaishi, O. (1996). Promotion of sleep mediated by the A_{2a}-adenosine receptor and possible involvement of this receptor in the sleep induced by prostaglandin D_2 in rats. *Proc. Natl. Acad. Sci. U.S.A.* **93,** 5980–5984.

Satoh, S., Matsumura, H., and Hayaishi, O. (1998). Involvement of adenosine A_{2a} receptor in sleep promotion. *Eur. J. Pharmacol.* **351,** 155–162.

Scammell, T., Gerashchenko, D., Urade, Y., Onoe, H., Saper, C., and Hayaishi, O. (1998). Activation of ventrolateral preoptic neurons by the somnogen prostaglandin D_2. *Proc. Natl. Acad. Sci. U.S.A.* **95,** 7754–7759.

Seregi, A., Keller, M., and Hertting, G. (1987). Are cerebral prostanoids of astroglial origin? Studies on the prostanoid forming system in developing rat brain and primary cultures of rat astrocytes. *Brain Res.* **404,** 113–120.

Seregi, A., Doll, S., Schobert, A., and Hertting, G. (1992). Functionally diverse purinergic

P2Y-receptors mediate prostanoid synthesis in cultured rat astrocytes: The role of ATP-induced phosphatidyl-inositol breakdown. *Eicosanoids* **5,** S19–S22.

Sherin, J. E., Shiromani, P. J., McCarley, R. W., and Saper, C. B. (1996). Activation of ventrolateral preoptic neurons during sleep. *Science* **271,** 216–219.

Shimizu, T., Yamamoto, S., and Hayaishi, O. (1979). Purification and properties of prostaglandin D synthetase from rat brain. *J. Biol. Chem.* **254,** 5222–5228.

Smith, W. L., Marnett, L. J., and DeWitt, D. L. (1991). Prostaglandin and thromboxane biosynthesis. *Pharmacol. Ther.* **49,** 153–179.

Sorrentino, C., Siverstrini, B., Braghiroli, L., Chung, S. S. W., Giacomelli, S., Leone, M.-G., Xie, Y., Sui, Y., Mo, M., and Cheng, C. Y. (1998). Rat prostaglandin D_2 synthase: Its tissue distribution, changes during maturation, and regulation in the testis and epididymis. *Biol. Reprod.* **59,** 843–853.

Spinelli, S., Ramoni, R., Grolli, S., Bonicel, J., Cambillau, C., and Tegoni, M. (1998). The structure of the monomeric procine odorant binding protein sheds light on the domain swapping mechanism. *Biochemistry* **37,** 7913–7918.

Sri Kantha, S., Matsumura, H., Kubo, E., Kawase, K., Takahata, R., Serhan, C. N., and Hayaishi, O. (1994). Effects of prostaglandin D_2, lipoxins and leukotrienes on sleep and brain temperature of rats. *Prostaglandins Leukotrienes Essent. Fatty Acids* **51,** 87–93.

Suzuki, T., Watanabe, K., Kanaoka, Y., Sato, T., and Hayaishi, O. (1997). Induction of hematopoietic prostaglandin D synthase in human megakaryocytic cells by phorbol ester. *Biochem. Biophys. Res. Commun.* **241,** 288–293.

Suzuki, T., Fujii, Y., Miyano, M., Chen, L.-Y., Takahashi, T., and Watanabe, K. (1999). cDNA cloning, expression, and mutagenesis study of liver-type prostaglandin F synthase. *J. Biol. Chem.* **274,** 241–248.

Tachibana, M., Fex, J., Urade, Y., and Hayaishi, O. (1984). Brain-type prostaglandin D synthetase occurs in the rat cochlea. *Proc. Natl. Acad. Sci. U.S.A.* **84,** 7677–7680.

Tanaka, T., Urade, Y., Kimura, H., Eguchi, N., Nishikawa, A., and Hayaishi, O. (1997). Lipocalin-type prostaglandin D synthase (β-trace) is a newly recognized type of retinoid transporter. *J. Biol. Chem.* **272,** 15789–15795.

Tegoni, M., Ramoni, R., Bignetti, E., Spinelli, S., and Cambillau, C. (1996). Domain swapping creates a third putative combining site in bovine odorant binding protein dimer. *Nat. Struct. Biol.* **10,** 863–867.

Terao, A., Kitamura, H., Asano, A., Kobayashi, M., and Saito, M. (1995). Roles of prostaglandins D_2 and E_2 in interleukin-1-induced activation of norepinephrine turnover in the brain and peripheral organs of rats. *J. Neurochem.* **65,** 2742–2747.

Thomson, A. M., Meyer, D. J., and Hayes, J. D. (1998). Sequence, catalytic properties and expression of chicken glutathione-dependent prostaglandin D_2 synthase, a novel class Sigma glutathione S-transferase. *Biochem. J.* **333,** 317–325.

Toh, H., Kubodera, H., Nakajima, N., Sekiya, T., Eguchi, N., Tanaka, T., Urade, Y., and Hayaishi, O. (1996). Glutathione-independent prostaglandin D synthase as a lead molecule for designing new functional proteins. *Protein Eng.* **9,** 1067–1082.

Tokugawa, Y., Kunishige, I., Kubota, Y., Shimoya, K., Nobunaga, T., Kimura, T., Saji, F., Murata, Y., Eguchi, N., Oda, H., Urade, Y., and Hayaishi, O. (1998). Lipocalin-type prostaglandin D synthase in human male reproductive organs and seminal plasma. *Biol. Reprod.* **58,** 600–607.

Tontonoz, P., Nagy, L., Alvarez, J. G. A., Thomazy, V. A., and Evans, R. M. (1998). PPARγ promotes monocyte/macrophage differentiation and uptake of oxidized LDL. *Cell (Cambridge, Mass.)* **93,** 241–252.

Tunami, H., Nau, R., and Felgenhauer, K. (1998a). β-Trace protein in cerebrospinal fluid: A blood-CSF barrier-related evaluation in neurological diseases. *Ann. Neurol.* **44,** 882–889.

Tunami, H., Reiber, H., Nau, R., Prange, H. W., Kauffmann, K., Mäder, M., and Felgenhauer, K. (1998b). Beta-trace protein concentration in cerebrospinal fluid is decreased in patients with bacterial meningitis. *Neurosci. Lett.* **242,** 5–8.

Uda, R., Horiguchi, S., Ito, S., Hyodo, M., and Hayaishi, O. (1990). Nociceptive effects induced by intrathecal administration of prostaglandin D_2, E_2, and $F_{2\alpha}$ to conscious mice. *Brain Res.* **510,** 26–32.

Ueno, R., Narumiya, S., Ogorochi, T., Nakayama, T., Ishikawa, Y., and Hayaishi, O. (1982). Role of prostaglandin D_2 in the hypothermia of rats caused by bacterial lipopolysaccharide. *Proc. Natl. Acad. Sci. U.S.A.* **79,** 6093–6097.

Ueno, R., Honda, K., Inoue, S., and Hayaishi, O. (1983). Prostaglandin D_2, a cerebral sleep inducing substance in rats. *Proc. Natl. Acad. Sci. U.S.A.* **80,** 1735–1737.

Ujihara, M., Horiguchi, Y., Ikai, K., and Urade, Y. (1988a). Characterization and distribution of prostaglandin D synthetase in rat skin. *J. Invest. Dermatol.* **90,** 448–451.

Ujihara, M., Urade, Y., Eguchi, N., Hayashi, H., Ikai, K., and Hayaishi, O. (1988b). Prostaglandin D_2 formation and characterization of its synthetases in various tissues of adult rats. *Arch. Biochem. Biophys.* **260,** 521–531.

Ujihara, M., Tsuchida, S., Satoh, K., Sato, K., and Urade, Y. (1988c). Biochemical and immunological demonstration of prostaglandin D_2, E_2, and $F_{2\alpha}$ formation from prostaglandin H_2 by various rat glutathione S-transferase isozymes. *Arch. Biochem. Biophys.* **264,** 428–437.

Urade, Y., and Hayaishi, O. (1999). Prostaglandin D_2 and sleep regulation. *Bioichim. Biophys. Acta* **1436,** 606–615.

Urade, Y., Fujimoto, N., and Hayaishi, O. (1985). Purification and characterization of rat brain prostaglandin D synthetase. *J. Biol. Chem.* **260,** 12410–12415.

Urade, Y., Fujimoto, N., Ujihara, M., and Hayaishi, O. (1987a). Biochemical and immunological characterization of rat spleen prostaglandin D synthetase. *J. Biol. Chem.* **262,** 3820–3825.

Urade, Y., Fujimoto, N., Kaneko, T., Konishi, A., Mizuno, N., and Hayaishi, O. (1987b). Postnatal changes in the localization of prostaglandin D synthetase from neurons to oligodendrocytes in the rat brain. *J. Biol. Chem.* **262,** 15132–15136.

Urade, Y., Nagata, A., Suzuki, Y., Fujii, Y., and Hayaishi, O. (1989a). Primary structure of rat brain prostaglandin D synthetase deduced from cDNA sequence. *J. Biol. Chem.* **264,** 1041–1045.

Urade, Y., Ujihara, M., Horiguchi, Y., Ikai, K., and Hayaishi, O. (1989b). The major source of endogenous prostaglandin D_2 production is likely antigen presenting cells: localization of glutathione-requiring prostaglandin D synthetase in histiocytes, dendritic and Kupffer cells in various tissues. *J. Immunol.* **143,** 2982–2989.

Urade, Y., Ujihara, M., Moriguchi, Y., Igarashi, M., Nagata, A., Ikai, K., and Hayaishi, O. (1990a). Mast cells contain spleen-type prostaglandin D synthetase. *J. Biol. Chem.* **265,** 371–375.

Urade, Y., Watanabe, K., Eguchi, N., Fujii, Y., and Hayaishi, O. (1990b). 9α,11β-Prostaglandin F_2 formation in various bovine tissues. Different isozymes of prostaglandin D_2 11-ketoreductase, contribution of prostaglandin F synthetase and its cellular localization. *J. Biol. Chem.* **265,** 12029–12035.

Urade, Y., Kitahama, K., Ohishi, H., Kaneko, T., Mizuno, N., and Hayaishi, O. (1993). Dominant expression of mRNA for prostaglandin D synthase in leptomeninges, choroid plexus and oligodendrocytes of the adult rat brain. *Proc. Natl. Acad. Sci. U.S.A.* **90,** 9070–9074.

Urade, Y., Tanaka, T., Eguchi, N., Kikuchi, M., Kimura, H., Toh, H., and Hayaishi, O. (1995a). Structural and functional significance of cysteine residues of glutathione-independent prostaglandin D synthase. Identification of Cys^{65} as an essential thiol. *J. Biol. Chem.* **270,** 1422–1428.

Urade, Y., Watanabe, K., and Hayaishi, O. (1995b). Prostaglandin D, E, and F synthases. *J. Lipid Mediators Cell Signall.* **12,** 257–273.

Urade, Y., Hayaishi, O., Matsumura, H., and Watanabe, K. (1996). Molecular mechanism of sleep regulation by prostaglandin D_2. *J. Lipid Mediators Cell Signall.* **14,** 71–82.

Vesin, M.-F., and Droz, B. (1995). Immunodetection of prostaglandin D synthase: Conditions of localization in a defined subclass of primary sensory neurons. *J. Neurochem.* **43,** 681–687.

Vesin, M.-F., Billotte, C., and Droz, B. (1995a). Biosynthesis of prostaglandin D_2 by motorneurons and dorsal horn microneurons: A biochemical and high resolution immunocytochemical study in chick spinal cord. *Neuroscience* **69,** 967–975.

Vesin, M.-F., Urade, Y., Hayaishi, O., and Droz, B. (1995b). Neuronal and glial prostaglandin D synthase isozymes in chick dorsal root ganglia: A light and electron microscopic immunocytochemical study. *J. Neurosci.* **15,** 470–476.

Watanabe, K., Yoshida, R., Shimizu, T., and Hayaishi, O. (1985). Enzymatic formation of prostaglandin $F_{2\alpha}$ from prostaglandin H_2 to D_2. Purification and properties of prostaglandin F synthase from bovine lung. *J. Biol. Chem.* **260,** 7035–7041.

Watanabe, K., Iguchi, Y., Iguchi, S., Arai, Y., Hayaishi, O., and Roberts, L. J., II (1986). Stereospecific converstion of prostaglandin D_2 to (5Z, 13E)-(15S)-9α,-11β,15-trihydroxyprosta-5,13-dien-1-oic acid (9α,11β-prostaglandin F_2) and of prostaglandin H_2 to prostaglandin $F_{2\alpha}$ by bovine lung prostaglandin F synthase. *Proc. Natl. Acad. Sci. U.S.A.* **83,** 1583–1587.

Watanabe, K., Fujii, Y., Nakayama, K., Ohkubo, H., Kuramitsu, S., Kagamiyama, H., Nakanishi, S., and Hayaishi, O. (1988). Structural similarity of bovine lung prostaglandin F synthase to lens ϵ-crystallin of the European common frog. *Proc. Natl. Acad. Sci. U.S.A.* **85,** 11–15.

Watanabe, K., Fujii, Y., Ohkubo, H., Kuramitsu, S., Kagamiyama, H., Nakanishi, S., and Hayaishi, O. (1991). Expression of bovine lung prostaglandin F synthase in *Escherichia coli. Biochem. Biophys. Res. Commun.* **181,** 272–278.

Watanabe, K., Urade, Y., Mäder, M., Murphy, C., and Hayaishi, O. (1994). Identification of β-trace as prostaglandin D synthase. *Biochem. Biophys. Res. Commun.* **203,** 1110–1116.

Weichsel, A., Andersen, J. F., Champagne, D. E., Walker, F. A., and Montfort, W. R. (1998). Crystal structures of a nitric oxide transport protein from a blood-sucking insect. *Nat. Struct. Biol.* **5,** 304–309.

White, D. M., Mikol, D. D., Espinosa, R., Weimer, B., Le Beau, M. M., and Stefansson, K. (1992). Structural and chromosomal localization of the human gene for a brain form of prostaglandin D_2 synthase. *J. Biol. Chem.* **267,** 23202–23208.

White, D. M., Takeda, T., de Groot, L. J., Stefansson, K., and Arnason, B. G. W. (1997). β-trace gene expression is regulated by a core promoter and a distal thyroid hormone response element. *J. Biol. Chem.* **272,** 14387–14393.

Whittle, B. J. R., Moncada, S., and Vane, J. R. (1978). Comparison of the effects of prostacyclin (PGI_2), prostaglandin E_1 and D_2 on platelet aggregation in different species. *Prostaglandins* **16,** 373–388.

Wu, S. Y., Perez, M. D., Puyol, P., and Sawyer, L. (1999). Beta-lactoglobulin binds palmitate within its central cavity. *J. Biol. Chem.* **274,** 170–174.

Yamamoto, K., Miwa, T., Ueno, R., and Hayaishi, O. (1988). Muramyl dipeptide-elicited production of PGD_2 from astrocytes in culture. *Biochem. Biophys. Res. Commun.* **156,** 882–888.

Yamashima, T., Sakuda, K., Tohma, Y., Yamashita, J., Oda, H., Irikura, D. Eguchi, N., Beuckmann, C. T., Kanaoka, Y., Urade, Y., and Hayaishi, O. (1997). Prostaglandin D synthase (β-trace) in human arachnoid and meningioma cells: Roles as a cell marker or in CSF absorption, tumorigenesis, and calcification process. *J. Neurosci.* **17,** 2376–2382.

Zahn, M., Mäder, M., Schmidt, B., Bollensen, E., and Felgenhauer, K. (1993). Purification and N-terminal sequence of β-trace, a protein abundant in human cerebrospinal fluid. *Neurosci. Lett.* **154,** 93–95.

Zanotti, G., Berni, R., and Monaco, H. L. (1993). Crystal structure of liganded and unliganded forms of bovine plasma retinol-binding protein. *J. Biol. Chem.* **268,** 10728–10738.

VITAMINS AND HORMONES, VOL. 58

Steroid Dehydrogenase Structures, Mechanism of Action, and Disease

WILLIAM L. DUAX,* DEBASHIS GHOSH,*,† AND VLADIMIR PLETNEV‡

Hauptman-Woodward Medical Research Institute, Inc., Buffalo, New York 14203, †Roswell Park Cancer Institute, Buffalo, New York 14263, and ‡Russian Academy of Science, Shemyakin Institute, Moscow, Russia

---

Steroid dehydrogenase enzymes influence mammalian reproduction, hypertension, neoplasia, and digestion. The three-dimensional structures of steroid dehydrogenase enzymes reveal the position of the catalytic triad, a possible mechanism of keto-hydroxyl interconversion, a molecular mechanism of inhibition, and the basis for selectivity. Glycyrrhizic acid, the active ingredient in licorice, and its metabolite carbenoxolone are potent inhibitors of human 11β-hydroxysteroid dehydrogenase and bacterial 3α, 20β-hydroxysteroid dehydrogenase (3α,20β-HSD). The three-dimensional structure of the 3α,20β-HSD carbenoxolone complex unequivocally verifies the postulated active site of the enzyme, shows that inhibition is a result of direct competition with the substrate for binding, and provides a plausible model for the mechanism of inhibition of 11β-hydroxysteroid dehydrogenase by carbenoxolone. The structure of the ternary complex of human 17β-hydroxysteroid dehydrogenase type 1 (17β-HSD) with the cofactor $NADP^+$ and the antiestrogen equilin reveals the details of binding of an inhibitor in the active site of the enzyme and the possible roles of various amino acids in the catalytic cleft. The short-chain dehydrogenase reductase (SDR) family includes these steroid dehydrogenase enzymes and more than 60 other proteins from human, mammalian, insect, and bacterial sources. Most members of the family contain the tyrosine and lysine of the catalytic triad in a YxxxK sequence. X-ray crystal structures of 13 members of the family have been completed. When the α-carbon backbone of the

0083-6729/00 $30.00

cofactor binding domains of the structures are superimposed, the conserved residues are at the core of the structure and in the cofactor binding domain, but not in the substrate binding pocket.

I. Introduction

Most steroids bind to specific receptors in the cytosolic and nuclear compartments of target tissues and induce or stabilize conformations appropriate for selective interaction with DNA-recognition elements leading to hormone function (Duax and Griffin, 1989; Evans, 1988). Specific enzymes play a critical role in controlling levels of hormonally active steroids and work in concert with the receptors. Because the enzymes have much greater tissue-dependent substrate specificity than the receptors, they have emerged as attractive targets for the design of potent and selective drugs to combat steroid-related disorders. To design molecules that will inhibit or alter the action of these enzymes, structural details of how they work and how they differ from one another are needed. The most reliable way to gain this information is by X-ray crystallographic analysis of the enzymes with and without cofactors, substrates, inhibitors, and other effectors.

The short-chain dehydrogenase reductase (SDR) family includes more than 60 enzymes from humans, mammals, insects, and bacteria that have steroids, sugars, prostaglandins, alcohols, dyes, and other small molecules as substrates. The SDR family includes dozens of mammalian enzymes that work in concert with steroid receptors to control hormone actions, including fertility, growth, and hypertension, as well as neoplastic processes. These enzymes typically exhibit only 15–30% amino acid sequence identity (Jõrnvall *et al.,* 1995). A small number of partially conserved residues (using amino acid single-letter code, where x represents any amino acid), including an YxxxK sequence, characterize members of the family. The most conserved residue in all members of family of 60 enzymes is a tyrosine that is a member of the catalytic triad (Jõrnvall *et al.,* 1995). There is no significant degree of sequence homology with members of the long/medium-chain dehydrogenase family that includes the lactate (LDH), malate (MDH), and D-glyceraldehyde-3-phosphate dehydrogenases (GDPH) (reviewed in Birktoft and Banaszak, 1984) that were among the earliest studied by crystallographic techniques (Watson and Banaszak, 1964; Adams *et al.,* 1970; Hill *et al.,* 1972). These were found to have a sequence of alternating α-helices and β-strands. These β-strands form a four- or five-stranded parallel β-sheet with two or three α-helices residing on either side. This tertiary structure has become known as the *Rossmann fold* and is characteristic of many NADH-binding and NADPH-binding domains. Conserved sequences of GxGxxG and GxGxxA found in the long/medium-chain dehydrogenases were identified as being associated with NADH-cofactor and NADPH-cofactor preferences, respectively (Scrutton *et al.,* 1990). The SDR enzymes have a similar conserved sequence including three glycine residues (GxxxGxG) at a comparable point in their sequences, which forms a turn between a β-strand and α-helix that borders on the cofactor-binding site. The first SDR structure determination reported was that of *Streptomyces hydrogenans* (3α,20β-HSD) (Ghosh *et al.,* 1994a). Since then the structures of 12 additional members of the SDR family have been reported and their atomic coordinates deposited in the Protein Data Bank (Bernstein *et al.,* 1977). The enzymes discussed in this review, for which crystal structures have been reported and atomic coordinates are available, are listed in Table I. The sequences of a subset of the SDR enzymes, including the three

TABLE I
SDR Enzyme Discussed in This Review[a]

Enzyme	Abbreviation	Reference
3α,20β(-Hydroxysteroid dehydrogenase (*Streptomyces hydrogenans*)	3α,20β-HSD	Ghosh *et al.* (1994a)
Dihydropteridine reductase (rat liver)	DHPR	Varughese *et al.* (1992)
Carbonyl reductase (mouse lung)	MLCR	Tanaka *et al.* (1996a)
7α(-Hydroxysteroid dehydrogenase (*Escherichia coli*)	7α-HSD	Tanaka *et al.* (1996b)
17β(-Hydroxysteroid dehydrogenase (human)	17β-HSD	Ghosh *et al.* (1995)
Enoyl acyl carrier protein reducase (*E. coli*)	ENRE	Baldock *et al.* (1996)
Enoyl acyl carrier protein reducase (*Brassico napus*)	ENRB	Rafferty *et al.* (1995)
Enoyl acyl carrier protein reducase (*Mycobacterium tuberculosis*)	ENRM	Rozwarski *et al.* (1998) Dessen *et al.* (1995)
UDP-galactase 4 epimerase (*E. coli*)	UDPE	Bauer *et al.* (1992)
Alcohol dehydrogenase (*Drosophila lebanonensis*)	ADH	Benach *et al.* (1998)
1,3,8-Trihydroxynapthalene reductase (*Magnaporthe grisea*)	THNR	Andersson *et al.* (1997)
Sepiapterin reductase (mouse)	SEP	Auerbach *et al.* (1997)
Cis biphenyl-2-3-dihydrodiol-2,3-dehydrogenase (*Pseudomonas* sp.)	BDB	Huelsmeyer *et al.*, (unpublished)

[a] Enzymes for which crystal structures are reported and coordinates are in the Protein Data Bank.

```
                                                                     ↓    ↓ 18
1  17β-HSD_hum ..........  ..........  ..........  ..ARTVVLIT  GCSSGIGLHL
2  MLCR        ..........  ..........  .......MKL  NFSGLRALVT  GAGKGIGRDT
3  7α-HSD      ..........  ..........  ...MFNSDNL  RLDGKCAIIT  GAGAGIGKEI
4  3α,20β-HSD  ..........  ..........  ........MN  DLSGKTVIIT  GGARGLGAEA
5  DHPR_hum    ..........  ..........  ....MAAAAA  AGEARRVLVY  GGRGALGSRC
6  HCR         ..........  ..........  ..........  SSGIHVALVT  GGNKGIGLAI
7  11β-HSD_hum MAFMKKYLLP  ILGLFMAYYY  YSANEEFRPE  MLQGKKVIVT  GASKGIGREM
                                                                 ↑    ↑

   19          30 ↓        40              50                    60         68
1  AVRLASDPSQ  SFKVYATLRD  LK..TQ.GRL  WEA.ARALAC  PPG.SL....  .ETLQLDVR.....D
2  VKALHASGA.  ..KVVAVTRT  NS..DL.VSL  AKE.C.....  ..  PGI....  .EPVCVDLG.....D
3  AITFATAGA.  ..SVVVSDI.  ....NA.DAA  NHV.VDEIQQ  LGG QA....  .FACRCDIT.....S
4  ARQAVAAGA.  ..RVVLADVL  DE..EGAATA  RE.......L  .GD.AA....  .RYQHLDVT.....I
5  VQAFRARNWW  ...VASVDVV  E....N.E..  E..A......  .....S....  .ATIIVKMTDSFTEQ
6  VRDLC.RLF.  SGDVVLTARD  ..VTRGQAAV  .........Q  .QL.QAEGLS  PRFHQLDID.....D
7  AYHLAKMGA.  ..HVVVTARS  KETLQKVVSH  CL.......E  .LG.AA....  .SAHYIAGT.....M
                 ↑

   69                    80   ↓↓      90↓         100         110     118
1  S....K...S  VAAARER.V.  T..EGRVDVL  VCNAGLGLLG  P.LEALGEDA  VASVLDVNVVGT
2  WDATEK...A  LGGIG.....  .....PVDLL  VNNAALVIMQ  P.FLEVTKEA  FDRSFSVNLRSV
3  E....Q...E  LSALADF.AI  SKL.GKVDIL  VNNAGGGGPK  P.F.DMPMAD  FRRAYELNVFSF
4  E....E...D  WQRVVAYA..  REEFGSVDGL  VNNAGISTGM  F.LETESVER  FRKVVDINLTGV
5  A....D...Q  VTAEVGKLL.  .G.EEKVDAI  LCVAGGWAGG  NAKSKSLFKN  CDLMWKQSIWTS
6  L....Q...S  IRALRDFL..  RKEYGGLDVL  VNNAGIAFKV  A.DPTPFHIQ  AEVTMKTNFFGT
7  E....DMTFA  EQFV.AQA..  GKLMGGLDML  ILNHITNTSL  N.LFHDDIHH  VRKSMEVNFLS.
                                   ↑

  119          130    ↓   140           151
1  VRMLQAF.LPD MKRR.GSGRV  LVTGSVGGLM  GLPF......  ..........
2  FQVSQMV.ARD MINRGVPGSI  VNVSSMVAHV  TFPN......  ..........
3  FHLSQLV.APE M.EKNGGGVI  LTITSMAAEN  KNIN......  ..........
4  FIGMKTV.IPA MKD.AGGGSI  VNISSAAGLM  GLAL......  ..........
5  .TISSHLATKH LKE...GGLL  TLAGAKAALD  GTPG......  ..........
6  RDVCTEL.LPL IK...PQGRV  VNVSSIMSVR  ALKSCSPELQ  QKFRSETITE
7  YVVLTVAA.LP MLK.QSNGSI  VVVSSLAGKV  AYPM......  ..........
                     ↑

                                      152  ↓      ↓       ↓167  ↓  175
1  ..........  ..........  .....NDVYC  ASKFALEGLC  ESLAVLLLP
2  ..........  ..........  .....LITYS  STKGAMTMLT  KAMAMELGP
3  ..........  ..........  .....MTSYA  SSKAAASHLV  RNMAFDLGE
4  ..........  ..........  .....TSSYG  ASKWGVRGLS  KLAAVELGT
5  ..........  ..........  .....MIGYG  MAKGAVHQLC  QSLAGK...
6  EELVGLMNKF  VEDTKKGVHQ  KEGWPSSAYG  VTKIGVTVLS  RIHARKLSE
7  ..........  ..........  .....VAAYS  ASKFALDGFF  SSIRKEYSV
                                   ↑      ↑

      176 180   185    ↓         200         210           219      227
1  .....PGVHLSL IECGPVHTAF  MEKVLGSPEE  VLDRTDIHTF  HRF...YQYL  AHSKQVFR
2  .....HKIRVNS VNPTVVLTDM  GKKV......  ..........  ..........  ........
3  .....KNIRVNG IAPGAILTDA  LKSV......  ..........  ..........  ........
4  .....DRIRVNS VHPGMTYTPM  TAET......  ..........  ..........  ........
5  NSGMPPGAAAIA VLPVTLDTPM  NRKS......  ..........  ..........  ........
6  .QRKGDKILLNA CCPGWVRTDM  AGPK......  ..........  ..........  ........
7  .S....RVNVSI TLCVLGLIDT  ETAM......  ..........  ..........  ........

                                 228           240                    250         259
1  ..........  ..........  .....EAAQ.  NPEEVAEVFL  TALR..A..P  KPT.LRYFTT.ERF
2  SA.DP.....  EFARKLK..E  RHPL.RKFA.  EVEDYVNSIL  FLLSDRS..A  STSGGGILVD.AGY
3  I..TP.....  EIEQKML..Q  HTPI.RRLG.  QPQDIANAAL  FLCSPAA..S  WVSGQILTVS.GGG
4  G..IRQGEGN  ......Y..P  NTPMG RVGN  EPGEIAGAVV  KLLS..DTSS  YVTGAELAVDGGWT
5  MPEADFS...  ..........  ......SWT.  PLEFLVETFH  DWIT..GKNR  PSSGSLIQVVT...
6  A..TKSPEEG  ......AETP  VYLAL.PVYL  ALL PP.DAE  GPHGQ FVSE  K..R..VEQW....
7  K..AVSGIVH  ......M..Q  AAPKE ECAL  EIIKGGALRQ  EEV YY.DSS  LWTTL LIRNP..C

1  - 67 more residues
2  -  3 more residues
3  -  5 more residues
4  - 11 more residues
5  - 12 more residues
6
7  - 20 more residues
```

steroid dehydrogenase enzymes for which X-ray coordinates are available, are compared in Fig. 1.

II. 3α,20β-Hydroxysteroid Dehydrogenase

3α,20β-Hydroxysteroid dehydrogenase (3α,20β-SD) from *S. hydrogenans,* is an NADH-linked enzyme involved in the reversible oxidation of the 3α group of androstane derivatives and the 20β group of pregnane derivatives (Sweet and Samant, 1980). It is interesting that the *S. hydrogenans* gene encodes a steroid dehydrogenase enzyme because there is no known steroid hormone function in the bacteria. The enzyme has unusual stereochemical specificites in that it oxidizes 3α- and 20β-hydroxyl groups but not 3β- and 20α-hydroxyl groups. Because the active form of the enzyme is a tetramer, two models had been proposed for the mechanism of action. In one model the substrate binding site lies between two cofactor binding sites so that one cofactor can oxidize the 3α-hydrogen group and a second cofactor is in a position to oxidize the 20β-hydrogen group. In this model the steroid fits in the active site in one orientation only. In an alternative model the substrate binding site is flexible enough to allow the steroid to take up two possible orientations. The 3α-hydrogen group in one orientation and 20β-hydrogen group in the second orientation approach a common cofactor binding site.

Structure determination of the holoenzyme revealed that the four identical monomers in the active tetramer are related by noncrystallographic symmetry, three mutually perpendicular twofold rotation axes. Each monomer contains a single domain incorporating the Rossmann fold cofactor binding motif and a solvent-filled pocket (Fig. 2). Because the fourfold symmetry is not crystallographically imposed, there are subtle differences in the conformations of individual monomers and in the occupancy of the cofactor in the four binding sites. These subtle differences could play a role in enzyme cooperativity. The position of the

Fig. 1. Sequence comparison of short-chain dehydrogenase reductases: human type 1 17β-hydroxysteroid dehydrogenase (17β-HSD hum); mouse lung carbonyl reductase (MLCR); *S. hydrogenans* 3α,20β-hydroxysteroid dehydrogenase (3α,20β-HSD); human dihydropteridine reductase (DHPR hum); human carbonyl reductase (HCR); and human 11β-hydroxysteroid dehydrogenase (11β-HSD hum). Twelve residues that are fully conserved in structures 1 to 5 are indicated by arrows on the top of the column. Eight of these residues are conserved in six of the structures and seven (denoted by arrows at the bottom of the column) are conserved in all seven structures. (From Duax *et al.,* 1996.)

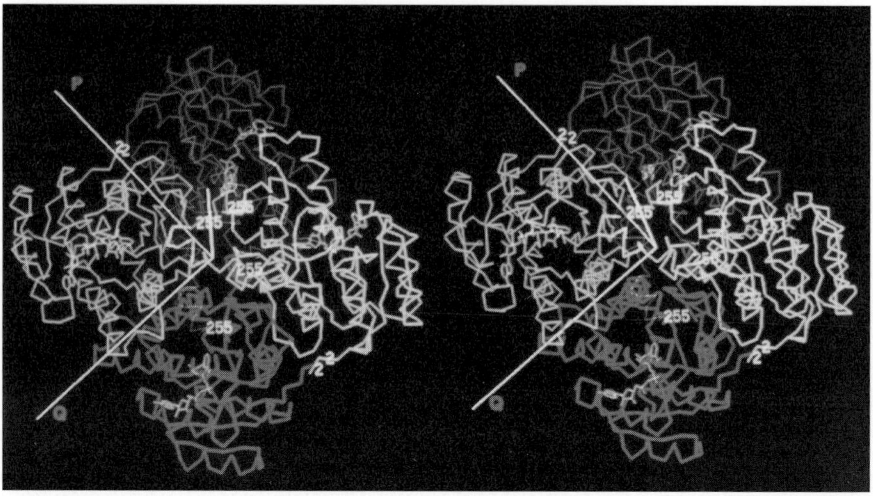

FIG. 2. A stereographic diagram of tetrameric 3α,20β-HSD. The monomers are drawn as α-carbon traces of the polypeptide chains. The noncrystallographic 222 symmetry elements are shown as P, Q, and R axes. Four NADH molecules are also shown. (From Ghosh *et al.*, 1994a.)

NAD cofactor relative to the nucleotide-binding fold is identical to that observed in the long/medium-chain dehydrogenase family except for the *syn* conformation of the nicotinamide ring. This conformation is consistent with the Pro S hydride transfer in the SDR enzymes, in contrast with the Pro R hydride transfer in most long/medium-chain dehydrogenases, and in enzymes of the aldehyde and aldose reductase family. The locations of the YxxxK sequence and a nearly conserved serine that lies between the cofactor and the solvent-filled pocket led to the proposal that Tyr-152, Lys-156, and Ser-139 were involved in the mechanism of action (Fig. 3). Without altering the structure of the enzyme, a cortisone molecule (Declerq *et al.*, 1972) was modeled in the substrate-binding cleft, placing C20 of the corticoid side chain within 2 Å of the C4 position of the nicotinamide ring, mimicking the transition state (Fig. 4, see color plate). A mechanism by which the conserved Tyr-152, Lys-156, and Ser-135 residues and solvent molecules in the catalytic cavity could catalyze the reaction was proposed by Ghosh *et al.* (1994a). In the proposed mechanism, the Tyr-152 hydroxyl proton initiates the electrophilic attack on the 20-keto oxygen of the steroid (Fig. 5). The presence of the positively charged side chain of Lys-156, in proximity to the Tyr-152 hydroxyl oxygen, facilitates proton transfer and Ser-

139 also participates in catalysis, either by stabilizing the reaction intermediates or as part of a proton-relay network. Once a proton is transferred from Tyr-152 to the 20-keto oxygen of the steroid, it can be replenished by the solvent network surrounding these residues in the catalytic cavity.

The ternary complex of 7α-hydroxysteroid dehydrogenase (7α-HSD) from *Escherichia coli,* NADH and the reaction product 7-oxo glycochenodeoxycholic acid provided further evidence that the tyrosine residue is directly involved in hydride transfer (Tanaka *et al.,* 1996b).

A

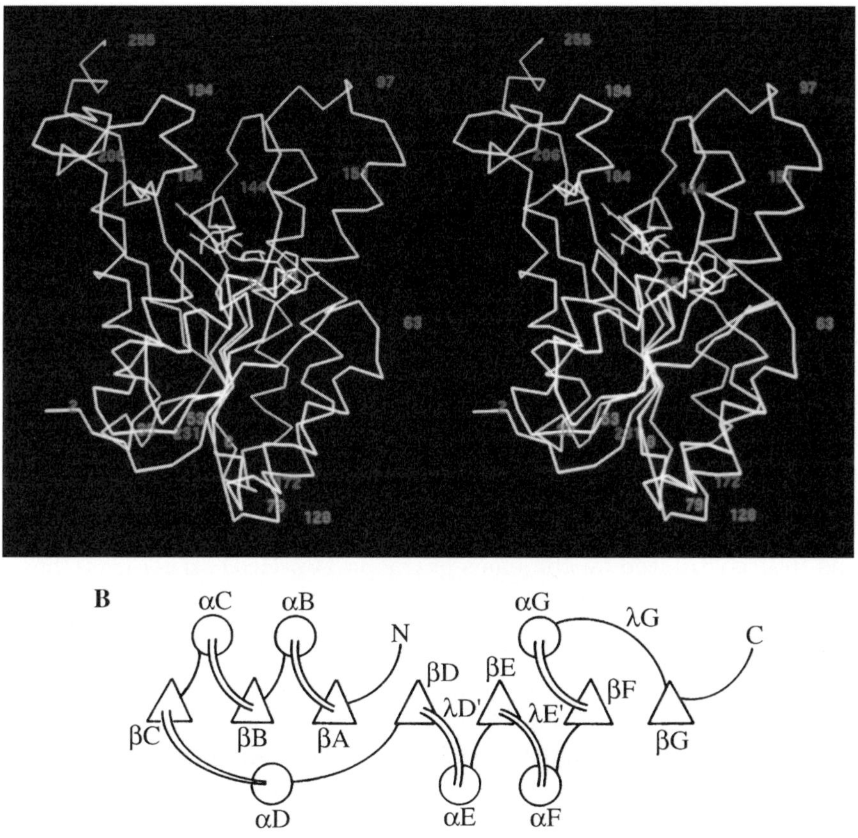

FIG. 3. (A) Stereoview of one subunit of the tetramer of 3α,20β-HSD shown as an α-carbon atom trace. The view is nearly along the edge of the central β-sheet. The NAD molecule is shown in light gray. (B) Folding topology of 3α,20β-HSD. Helices are represented by circles and strands by triangles. (From Ghosh *et al.,* 1994a.)

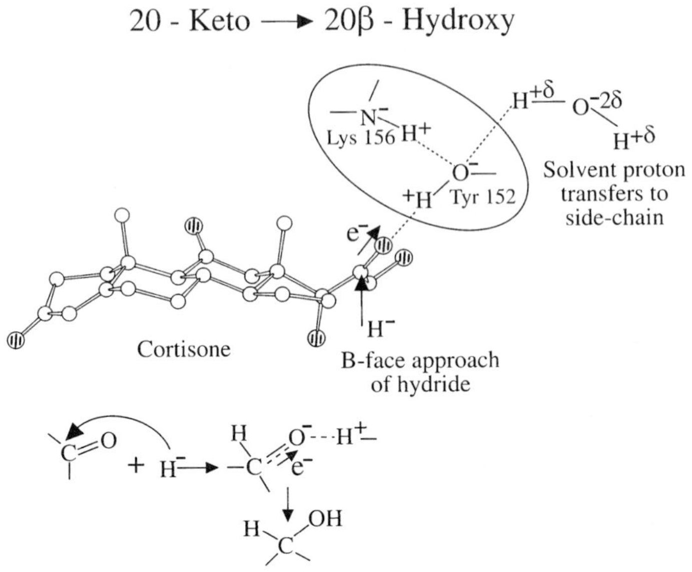

FIG. 5. A possible mechanism for stereospecific hydride transfer and proton relay during 20-keto to 20β-hydroxyl conversion. (From Ghosh *et al.,* 1994a.)

The comparison of the tertiary structure with the binary complex of the enzyme with NAD indicates that a significant conformational change accompanies substrate binding in 7α-HSD. A C-terminal "binding loop" closes over the substrate. A similar role involving an interaction with the cofactor was postulated for residue 192 in α-helix F of *Drosophila* alcohol dehydrogenase, another member of the SDR family (Ribos-de-Pouplona and Fothergall-Gilmore, 1994). A dynamic role for this "flap" was first postulated on the basis of the 3α,20β-HSD structure (Ghosh *et al.,* 1994a).

III. Aldosterone, Cortisol, Licorice, and Hypertension

Although a correlation between licorice consumption and elevation of blood pressure is well established, the physiologic basis for the effect is not fully understood (Girerd *et al.,* 1958; Ulick *et al.,* 1979). When glycyrrhizic acid, the active ingredient responsible for the hypertensive response, was isolated, its ability to compete for the mineralocorticoid receptor was tested, but no evidence of binding could be detected. On the basis of small-molecule crystal structure analysis and its comparison with data on receptor binding and activity, it was proposed that the A-ring plays a dominant role in initiating receptor binding and the D-ring modulates hormone potency (Duax and Griffin, 1989). The determination of the crystal structure of the steroid binding domain of the estrogen receptor with estradiol and relaxofin bound to it (Brzozowski, *et al.,* 1997) reveals that such a model correctly predicted close association of the phenolic rings of the estrogen agonist and antagonist in nearly identical orientations in the binding site. Differences in stereochemistry of the D-ring and associated difference in the tertiary structure of the receptor are responsible for variation in hormonal response. Subtle differences in the A/B-ring structure appear to be responsible for the selective binding of estrogens, progestins, and corticoids to their respective receptors. Hydrogen bonds involving the C(3) carbonyl oxygen and the C(11) hydroxyl may account for the competition of aldosterone and cortisol for the mineralocorticoid receptor-binding site, and the failure of cortisone to bind. Cortisone has a C(11) carbonyl instead of a hydroxyl, as in cortisol. The fact that cortisol is present in the body at 100-fold higher concentration relative to aldosterone raises a question of how the aldosterone signal is detected by the mineralocorticoid receptor in the presence of a 100-fold excess of cortisol.

In the 1970s some newborn infants showed clinical signs of hyperaldosteronism, no detectable circulating aldosterone, and an inability to oxidize cortisol to cortisone. Urinary 11β-hydroxy cortisol metabolites in these infants were greatly increased, and the corresponding 11-oxo metabolites were diminished. The condition was named apparent mineralocorticoid excess (AME) (Ulick *et al.,* 1979; New *et al.,* 1982). The altered steroid metabolism in AME was traced to a defective 11β-hydroxysteroid dehydrogenase (11β-HSD). One consequence of the defect was a persistently active mineralocorticoid receptor saturated by excess accumulated cortisol. The critical role that 11β-HSD plays in mineralocorticoid receptor action, metabolizing cortisol to cortisone in the kidney, led investigators to question whether glycyrrhizic acid might inhibit 11β-HSD (Monder *et al.,* 1989), producing an increased concentration of cortisol and associated hypertension. As predicted, inhibitor studies revealed that glycyrrhizic acid is a potent competitor for binding to the active site of 11β-HSD. The role of 11β-HSD in controlling cortisol/cortisone equilibrium, the relationship between that balance and corticoid binding to the mineralocorticoid receptor, and the inhibiting effect of glycyrrhizic acid are illustrated schematically in Fig. 6.

Because 11β-HSD is a member of the SDR family, we tested glycyrrhizic acid and its metabolite carbenoxolone and found them to be very potent competitive inhibitors of 3α,20β-HSD (Ghosh *et al.,* 1992).

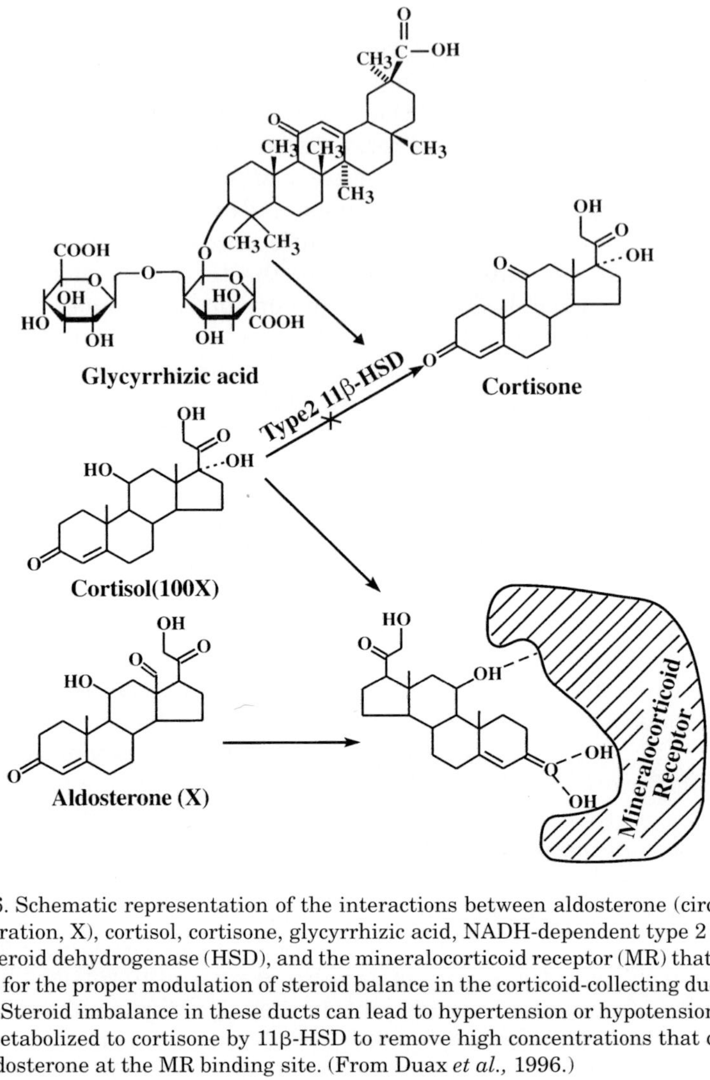

FIG. 6. Schematic representation of the interactions between aldosterone (circulating concentration, X), cortisol, cortisone, glycyrrhizic acid, NADH-dependent type 2 11β-hydroxysteroid dehydrogenase (HSD), and the mineralocorticoid receptor (MR) that are important for the proper modulation of steroid balance in the corticoid-collecting duct of the kidney. Steroid imbalance in these ducts can lead to hypertension or hypotension. Cortisol is metabolized to cortisone by 11β-HSD to remove high concentrations that compete with aldosterone at the MR binding site. (From Duax *et al.,* 1996.)

Lineweaver-Burk plots illustrate the competitive inhibition of 3α,20β-HSD by these compounds (Fig. 7). Crystal structure analysis of the complex of carbenoxolone and 3α,20β-HSD (Ghosh *et al.,* 1994b) revealed the presence of the inhibitor in the binding site (Fig. 8). The hemisuccinate terminal oxygen of carbenoxolone forms a hydrogen bond to Tyr-152, one of the conserved residues of the catalytic triad. No significant change was detected in the quaternary association of the enzyme when the complexed and uncomplexed forms were compared. In addition to hydrogen bonding to the catalytic tyrosine, the hemisuccinate side chain occupies the binding site of the nicotinamide ring of the cofactor. The cofactor is either prevented from binding to or is displaced by the inhibitor molecule. The fact that carbenoxolone occupies the catalytic cleft, displaces the cofactor, and hydrogen bonds directly to a member of the catalytic triad makes it a very potent high-affinity inhibitor and a good model on which to base the design of selective inhibitors for other enzymes, particularly other members of the SDR family.

IV. 17β-Hydroxysteroid Dehydrogenase, Inhibitors, and Cancer

The principal human estrogen, 17β-estradiol, is a potent stimulator of certain endocrine-dependent forms of breast cancer (Dickson and

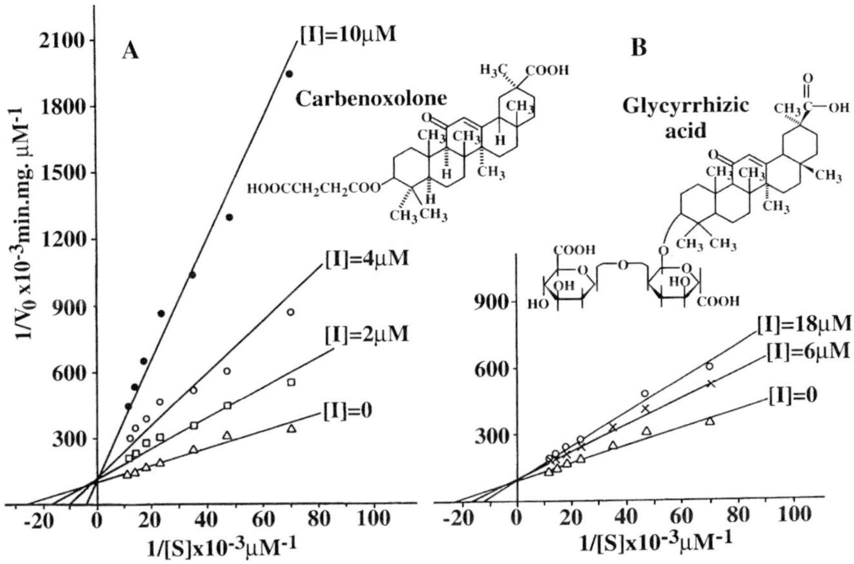

FIG. 7. Double reciprocal (Lineweaver-Burk) plots relating the substrate concentrations with the initial velocities at various concentrations of the following inhibitors: (A) carbenoxolone with cortisone as substrate and (B) glycyrrhizic acid with cortisone as substrate. (Reprinted from *J. Steroid Biochem. Mol. Biol.*, **42,** Ghosh, D., Erman, M., Pangborn, W., Duax, W. L., and Baker, M. E., Inhibition of *Streptomyces hydrogenans,* 3α,20β-hydroxysteroid dehydrogenase by licorice-derived compounds and crystallization of an enzyme-cofactor-inhibitor complex, pp. 849–853, © 1992, with permission from Elsevier Science.)

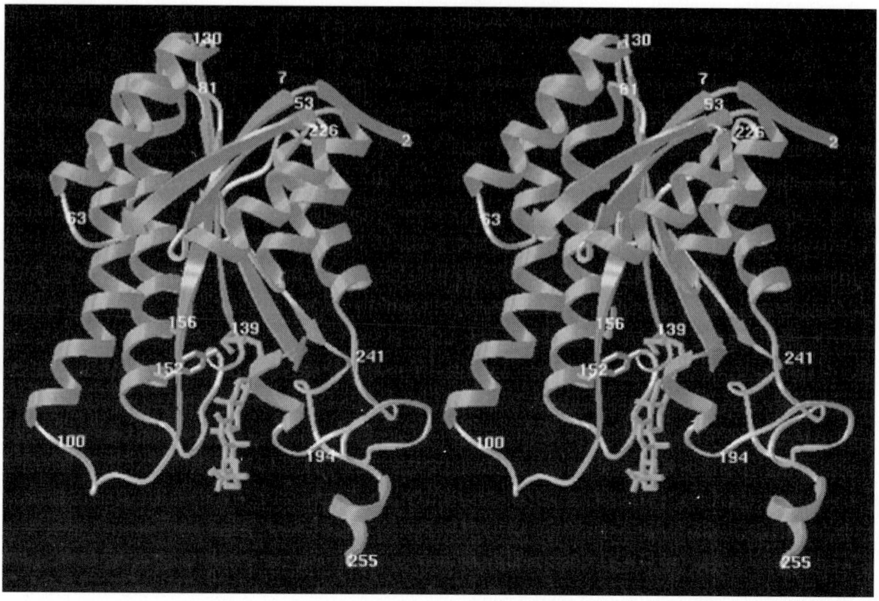

FIG. 8. Stereo illustration of the monomer of 3α,20β-HSD in complex with carbenoxolone. (From Ghosh *et al.,* 1994b.)

Lippman, 1987; Horwitz and McGuire, 1987). Because human estrogenic 17β-hydroxysteroid dehydrogenase (type 1 17β-HSD) catalyzes the last step in the biosynthesis of 17β-estradiol from the less potent estrogen, estrone, it is an attractive target for the design of inhibitors of estrogen production and tumor growth (Labrie *et al.,* 1992). Six distinct 17β-HSD isozymes, numbered 1 through 6, have been identified and cloned (Peltoketo *et al.,* 1988; Luu-The *et al.,* 1989; Geissler *et al.,* 1994, Adamski *et al.,* 1995; Deyashiki *et al.,* 1995). These isozymes differ in substrate and tissue specificities and in the preferred direction of the reaction. In human breast tissues, the most active estrogen, 17β-estradiol (E2), is formed by reduction of the inactive estrogen, estrone (E1), catalyzed by 17β-HSD type 1 (17β-HSD1). The estrogenic specificity of 17β-HSD1 as well as its preference for the reduction reaction have been well established (Poutanen *et al.,* 1993; Puranen *et al.,* 1997a,b).

17β-HSD1 is expressed in steroidogenic tissues including estrogen target tissues, such as normal and malignant endometrium and breast tissues (Fournet-Dulguerov *et al.,* 1987; Mäentausta *et al.,* 1991, 1992; Dupont *et al.,* 1991; Ghersevich *et al.,* 1994; Poutanen *et al.,* 1992). Because of its estrogenic specificity and preference for the E1 to E2 reduction reaction, the enzyme is considered to be primarily responsible for E2 biosynthesis in gonads and in peripheral tissues. This enzyme has been proposed to be involved in maintaining high E2 levels found in breast tumors of postmenopausal women (Poutanen *et al.,* 1995). A direct correlation between higher concentrations of E2 and onset of breast cancer, especially in postmenopausal cases, is well established (Poutanen *et al.,* 1995). There are reports of elevated E2/E1 concentration ratios in breast tumors in comparison with the E2/E1 ratio in circulating blood (James *et al.,* 1986). Furthermore, a number of studies appear to indicate higher levels of 17β-HSD1 activities in the outer quadrant of the breast where tumors most commonly occur (James *et al.,* 1986, 1990; Adams *et al.,* 1988). 17β-HSD1, therefore, poses an attractive target for structure-based rational drug design for the prevention and control of breast tumor growth. This human enzyme shares less than 15% sequence identity with bacterial 3α,20β-HSD. Despite limited sequence homology and insertions, crystal structure determination of type 1 17β-HSD revealed a fold characteristic of the short-chain dehydrogenases (Ghosh *et al.*, 1995). The active site contains the YxxxK sequence and a serine residue, in nearly identical locations relative to the positions of these residues in 3α,20β-HSD. The structure contains three α-helices and a helix–turn–helix motif, not observed in short-chain dehydrogenase structures reported previously. The inserted helices, which are located at one end of the substrate-binding cleft away from the catalytic triad (Fig. 9), restrict access to the active site and appear to influence substrate specificity. No cofactor density could be located.

When an estradiol molecule was docked in the substrate-binding site with the 17-hydroxyl oriented toward the catalytic triad, the steroid fit well in the pocket (Fig. 10). On the basis of this modeling it was proposed that a histidine on the helical insert of residues 209 to 229 could form a hydrogen bond to oxygen O(3) on the steroid A-ring, thus introducing specificity of binding of estranes and androstanes. A site-directed mutagenesis study (Puranen *et al.,* 1994) showed that His-221 as well as Try-155 and essential for enzyme activity.

A superposition drawing of the protein backbone of 17β-HSD and 3α,20β-HSD (Fig. 11) reveals the similarity in the Rossmann fold and cofactor-binding regions of the structures. The insertions in 17β-HSD are seen to be at the end of the binding pocket distal from the cofactor. The insertions reduce the openness of the binding pocket and introduce specificity for estranes and androstanes as described earlier. The constriction of the binding pocket also prevents carbenoxolone from

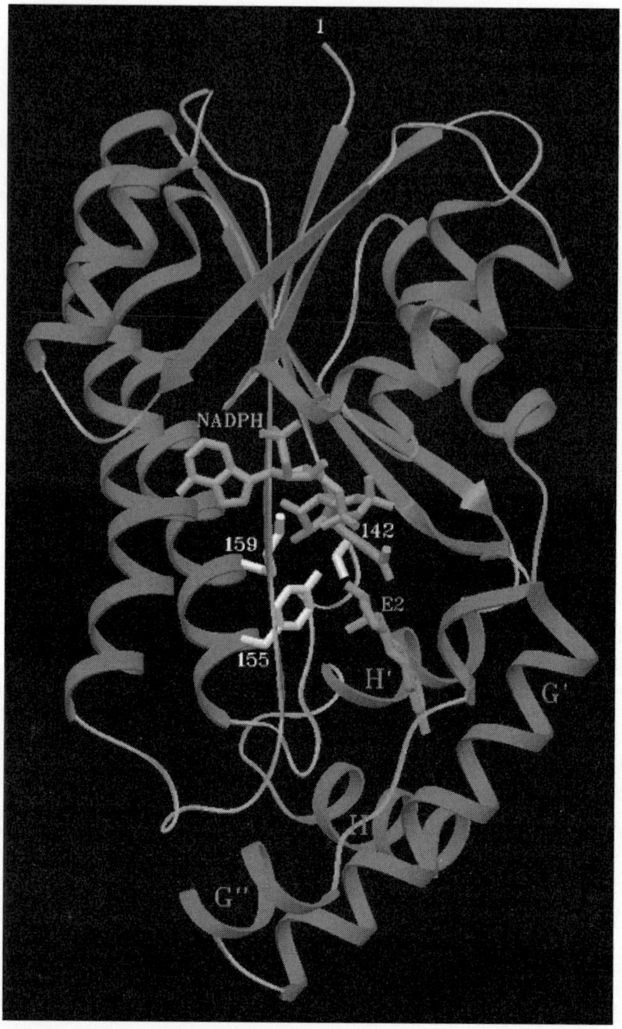

FIG. 9. A ribbon diagram of the 17β-HSD structure with the substrate-binding domain, in which helical insertion αG″, αG′, αH′, and αH make up the end distal to the catalytic triad and cofactor location. The 17β-estradiol molecule is modeled in the site.

binding in the site, as verified by the failure of carbenoxolone to inhibit 17β-HSD or to have any effect on circulating levels of estranes and androstanes. The structure of the active site provides a rational basis for designing more specific inhibitors of this breast cancer-associated enzyme.

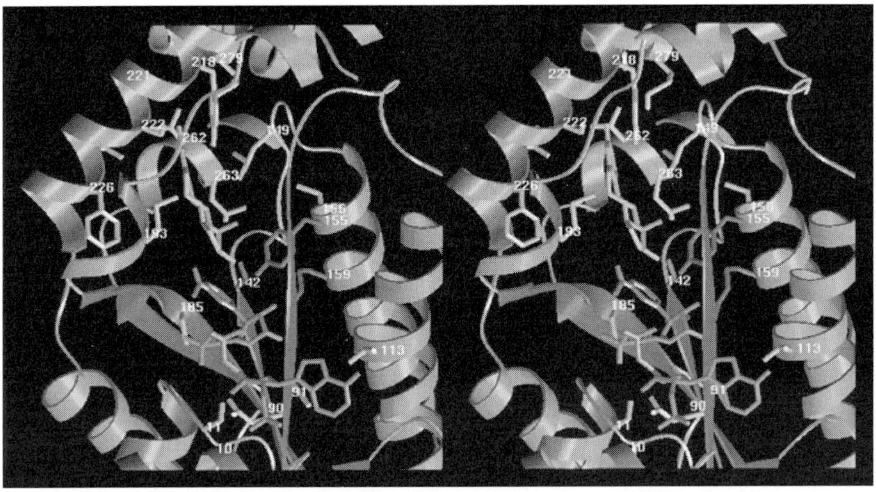

FIG. 10. Stereo view of the active site of human estrogenic 17β-HSD, showing the cofactor and the steroid-binding clefts. The steroid molecule is positioned relative to Tyr-155 and the nicotinamide ring of the cofactor in accordance with the proposed transition state of estrone to estradiol interconversion. (From Ghosh *et al.,* 1995.)

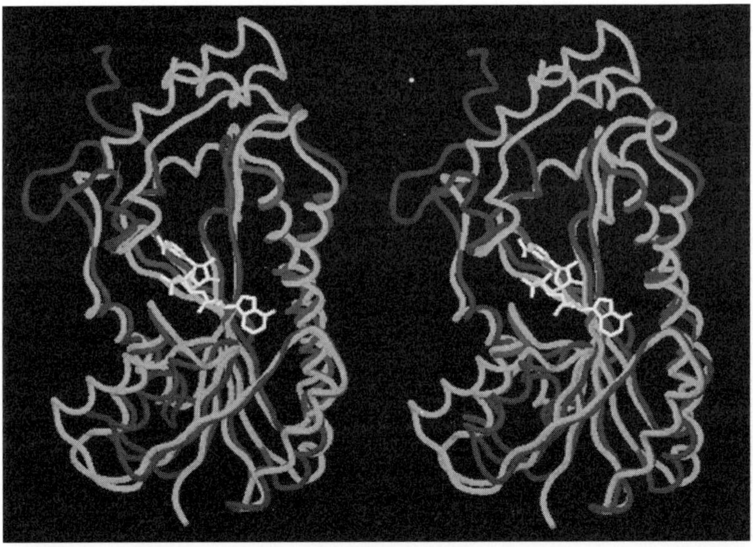

FIG. 11. A stereodiagram of superimposed Cα chains of bacterial 3α,20β-HSD (dark) and human estrogenic 17β-HSD (light). The sequence insertions and C-terminal additions on 17β-HSD are observed to enclose the substrate binding site. The superposition was achieved by least-squares fitting 20 Cα atoms that are at similar positions in both structures. (From Ghosh *et al.,* 1995.)

Our X-ray crystal structure determination of the ternary complex of 17β-HSD1 with the cofactor $NADP^+$ and equilin unequivocally reveals the nature of antiestrogen binding in the active site (Fig. 12) (Sawicki *et al.,* 1999a). Equilin (3-hydroxyestra-1,3,5,7-tetraen-17-one) is used in estrogen replacement therapy (ERT), in conjunction with estrone and 17α-dihydroequilin. These conjugated estrogens are administered as salts of their sulfate esters, which are subsequently hydrolyzed to free estrogens. Kinetic data indicate that equilin inhibits the *in vitro* reduction of E1 to E2 by 17β-HSD. This structure determination confirms

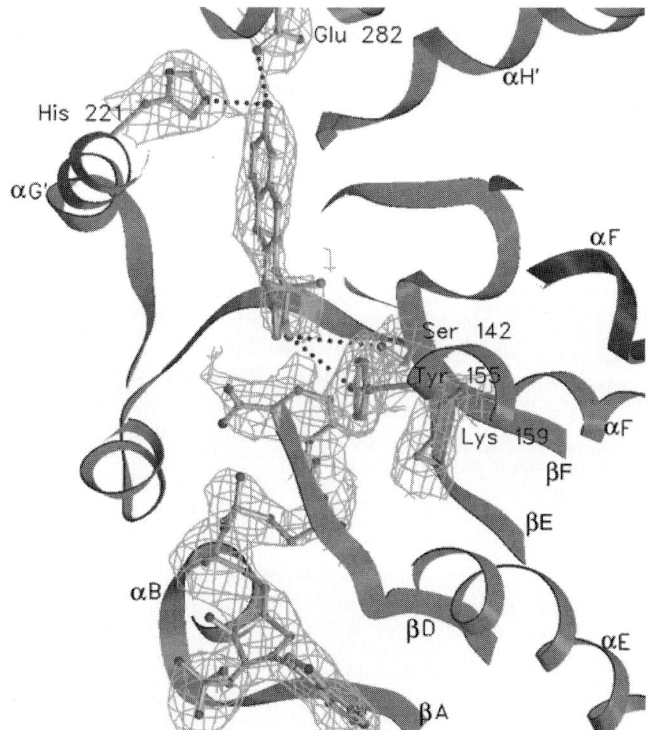

FIG. 12. A view of the active site within the A-subunit of 17β-HSD. $NADP^+$, equilin, and side chains of amino acid residues Ser-142, Tyr-155, and Lys-159 at the catalytic end and His-221 and Glu-282 at the recognition end of the active site cleft are shown in their electron densities. The equilin molecule was not included in the protein model. The visible secondary structure elements are labeled. This figure was generated using SETOR (Evans, 1993). (From Sawicki *et al.,* 1999a.)

our postulated mechanism of action and suggests ways to design more potent and selective inhibitors. The 17β-HSD1–equilin complex crystallized with a dimer in the asymmetric (crystallographic repeat) unit. Consequently, there are subtle differences between the monomers that are related by noncrystallographic pseudo symmetry. Such differences reflect the variation possible in the active dimeric form of the enzyme *in vitro* and *in vivo*. This is in contrast to the observed crystal structures of apo- and wild-type 17β-HSD1–E2 complexes (Ghosh *et al.*, 1995; Azzi *et al.*, 1996; Breton *et al.*, 1996) in which the monomers in the dimer are identical because they are related by a crystallographic symmetry. A homodimer is known to be the functional unit of the enzyme (Lin *et al.*, 1992). The holo-form observed in the crystal represents a true ternary complex of the wild-type enzyme, with the cofactor and a steroidal ligand. Both equilin and $NADP^+$ have well-defined electron density in the A-subunit of the dimeric enzyme. However, the ligand density in the B-subunit is poorly defined. The active site structure of the 17β-HSD1–equilin complex for the A-subunit is shown in Fig. 13. The equilin molecule makes four hydrogen bond contacts with protein atoms. The 17-keto oxygen accepts protons from catalytic residues Tyr-155 and Ser-142 (2.7 and 2.8 Å, respectively) at the catalytic end of the steroid binding cleft. The 3-hydroxyl group of the ligand simultaneously forms hydrogen bonds to His-221 and Glu-282 (2.9 and 2.9 Å, respectively) at the recognition end of the cleft. With the exception of these two hydrophilic ends, the rest of the steroid-binding cleft is almost exclusively hydrophobic. In the apo-enzyme structure, the substrate-entry loop adopts an *open* conformation providing unrestricted access to the active site cleft (Ghosh *et al.*, 1995). In contrast, the substrate-entry loop forms a *closed* conformation in the 17β-HSD1–equilin complex, where the polypeptide chain with residues 186–201 moves toward the catalytic cleft, restricting the access to the active site, as depicted in Fig. 13. The three-dimensional structures of the substrate, estrone, and the inhibitor equilin are strikingly different at the C-D-ring systems, because of the presence of the C7=C8 double-bond in equilin. The difference in the torsion angle C7–C8–C9–C11 of −179° for estrone and 121° for equilin (Sawicki *et al.*, 1999b) is caused by the C7=C8 double bond and results in a 0.9-Å displacement between the C17 carbon atoms. In Fig. 14 the structures of both the equilin and estradiol (Breton *et al.*, 1996) complexes are superimposed using the main-chain Cα carbons of the catalytic triad (Ser-142, Tyr-155, and Lys-159) and His-221. The distance between the C4 atom of the nicotinamide ring of the cofactor and C17 of equilin is 3.4 Å. However, to evaluate the structural origin of inhibition by equilin, the active site structure of the E1–17β-HSD1 complex should ideally be compared with that of the 17β-HSD1–equilin complex, since E2 is not a substrate, but the product of the catalysis. The origin of the inhibitory property of equilin, from the structural perspective, is, therefore, a manifestation of its altered C-D-ring structure, and location and orientation of its C17 keto group, with respect to the catalytic machinery at the catalytic end of the active site.

The *open* and *closed* conformations of the substrate-entry loop are illustrated in Fig. 15 (see color plate), where the dimeric Cα-backbones of the apo-enzyme (Ghosh *et al.,* 1995) the H221L–mutant–NAD complex (Mazza *et al.* 1998) and the 17β-HSD1–equilin complex are superimposed. The overall tertiary structures are nearly identical, except for the substrate-entry loop (shown as thicker cross-section of the backbone) between the strand βF and the helix αG". The location and overall conformation of the cofactor $NADP^+$ are similar to those in crystal structures of other SDRs (Sawicki *et al.,* 1999a).

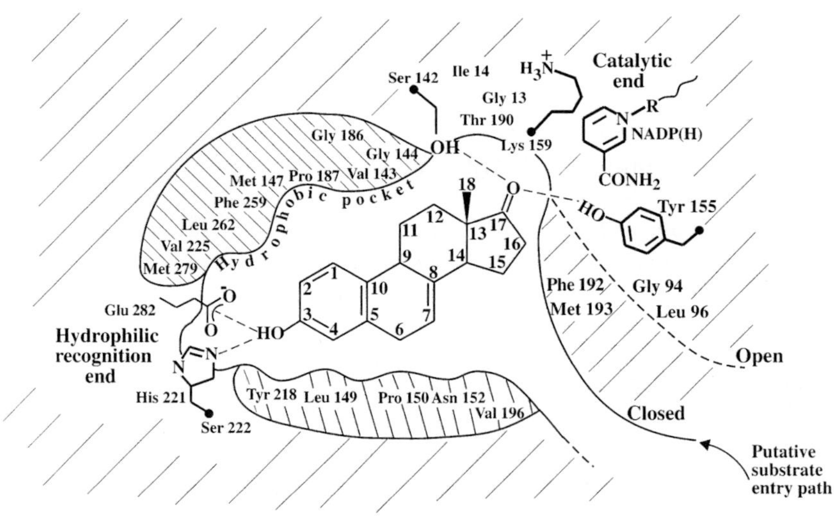

FIG. 13. A schematic diagram of the ligand-binding pocket, viewed roughly perpendicular to the β-face of the equilin molecule, derived from the description of the active site provided in the text. The hydrophobic environment depicted above and below the steroid molecule is, in fact, comprised of residues that surround the ligand in the pocket. The *open* conformation of the substrate-entry path is shown by a dashed line. In the ternary complex representing the *closed* conformation of the entry path, shown by a solid line, residues Phe-192 and Met-193 form hydrophobic contacts with the ligand. (From Sawicki *et al.,* 1999a.)

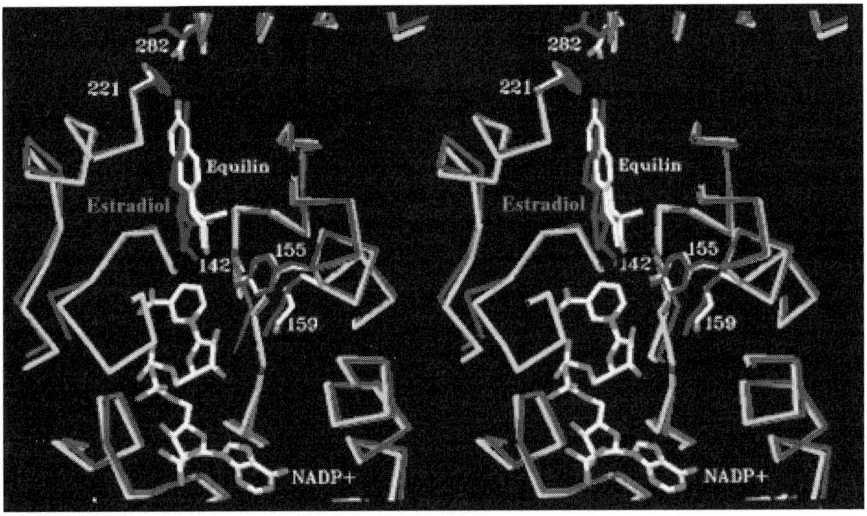

FIG. 14. Least-squares superposition of the structures of ternary complexes with E2 (dark) and equilin (light) fitting the catalytic triad (Ser-142, Tyr-155, and Lys-159) and His-221 Cα carbons. E2 and equilin are labeled, and the $NADP^+$ molecule shown is from the equilin structure. The difference in orientation of the two ligands, especially of their B-, C-, and D-rings is evident. (From Sawicki *et al.,* 1999a.)

V. PROTEIN SEQUENCE AND THREE-DIMENSIONAL CONFORMATION

Comparative analyses of protein sequences and unequivocal identification of residue having structural and functional significance are most productive when full three-dimensional structures of several members of a family of proteins are known. The published coordinates from 13 members of the SDR family have been the basis for several publications describing the role of conserved residues in enzyme structures and function.

A comparison of the conformations of five SDR structures [bacterial 3α,20β-HSD (Ghosh *et al.,* 1994a); human 17β-HSD, (Ghosh *et al.,* 1995); bacterial 7α-HSD (Tanaka *et al.,* 1996b); mouse DHPR (Varughese *et al.,* 1992); and mouse MLCR (Tanaka *et al.,* 1996a), revealed that although there are only 11 fully conserved residues common to the five structures, the three-dimensional conformation is highly conserved (Duax *et al.,* 1996). Two members of the catalytic triad proposed for the mechanism of action of 17β-HSD and 3α,20β-HSD (Tyr-155 and Lys-159) are included in all five, and Ser-142 is present in four of the five structures (replaced by Ala-136 in DHPR). The superposition of the five structures including the three enzymes that have steroids as their substrate is illustrated in Fig. 16A (see color plate). The α carbon atoms of the residues in the β-sheet were used in the fitting process. The cofactor NAD(P)(H) was present in four of the five crystals. When the cofactor positions are compared (Fig. 17A, see color plate), they are found to be as well conserved as the catalytic triad (Tanaka *et al.,* 1996a). Of the 11 fully conserved residues in the five structures, only Tyr-155 and Thr-190 either make contact with the cofactor or reside in its vicinity.

The recent publication of the high resolution three-dimensional structure of alcohol dehydrogenase from *Drosophila lebanonensis* (Benach *et al.,* 1998) included a structure-based sequence analysis of 10 members of the SDR family (3α,20β-HSD, 17β-HSD, 7α-HSD, DHPR, MLCR, ADH, THNR, BDB, UDPE, and ENRM, Table I). These 10 structures, which include the 5 illustrated in Fig. 16A, have only two fully conserved residues, those corresponding to Asp-84 and the catalytic Lys-159 in 17β-HSD. Nine of the 10 structures have a tyrosine in the sequence in a position analogous to the catalytic Tyr-155 of 17β-HSD and 8 of the 10 have a serine in a position analogous to the catalytic Ser-142 of 17β-HSD. The sequence difference in one of these structures, enoyl-acyl carrier protein from *Mycobacterium tuberculosis* (ENCM), is particularly interesting. This is the only structure of the 10 that does not have a catalytic tyrosine in common with the others. When the three-dimensional structures of the enoyl-acyl carrier protein for *B. napua* (ENRB) and 3α,20β-HSD were compared by Rafferty *et al.* (1995) the catalytic lysine residues superimposed nearly perfectly but the catalytic Tyr (152) in 3α,20β-HSD is Met 202 in ENRB. However, a nearby tyrosine residue Tyr-198 in ENRB (corresponding to residue 149 in 3α,20β-HSD) is oriented in the structure in such a way as to occupy the position and perform a function analogous to the generally conserved catalytic tyrosine. The superposition of the structures of three different enoyl-acyl carrier proteins (from *E. coli,* ENRE; *B. napus,* ENRB; and *M. tuberculosis,* ENRM) and 17β-HSD reveals that a tyrosine corresponding to Y190 in ENRB is conserved in ENRE and ENRM (Fig. 17B). In two structures (ENRB and ENRE) the tyrosine hydroxyl oxygen is within 1.5 Å of the tyrosine hydroxyl oxygen in 17β-HSD, and rotation around the α-β bond of the tyrosine of ENRM would bring its hydroxyl oxygen into alignment with the others. The fact that the enoyl-acyl carrier proteins have a conserved tyrosine that can occupy a comparable position relative to the cofactor in this subgroup of the SDR family suggests that the mechanism of action involving three residues (usually Tyr, Lys, and Ser) that was first proposed on the basis of the structure of 3α,20β-HSD applies to all members of the family for which X-ray analysis has been reported thus far.

We have now compared the three-dimensional structures of a total of 13 members of the SDS family using 17β-HSD as the reference structure and overlapping the α carbon backbone of the NAD/NADP binding domains (Fig. 16). Figures 16A and 16B are viewed from the same perspective and Figs. 16C and 16D are rotated approximately 90° about the vertical axis of the view in Figs. 16A and 16B to better illustrate similarities and differences of the structures. The seven-stranded parallel β-sheet is the feature that retains the greatest stereochemical similarity in the 13 structures. There are three prominent α-helices on either side of this sheet. Those at the back as viewed in Figs. 16A and 16B are at the right in Figs. 17C and 17D and include the two longest and most well-conserved backbone α-helices, αE and αF (Fig. 3B). The central short helix, αB in the foreground of Figs. 16A and 16B and to the left in Figs. 16D and 16C, retains a very similar backbone conformation in all 13 structures.

It is clear that the greatest similarity is seen in the cofactor binding, Rossmann fold, especially β-strands and α-helices that are in the most direct contact with the NAD(P) cofactor. The short α-helices that are farthest from the cofactor binding site (αC, and αG, Fig. 3B) exhibit greater variation in the 13 structures. The substrate binding cleft is the most variable part of the structure. It has already been noted that the insertion of an additional long helix in 17β-HSD at the top of all four views in Fig. 16 distinguishes it from other steroid dehydrogenases. We can see in Figs. 16B, 16C, and 16D that several other members of the family have loops and helices enclosing the substrate binding site in an analogous fashion but there is little direct overlap of the α carbon chain backbones. This is especially true of ADH and SEP. The UDPE substrate binding domain differs most extensively, and yet its cofactor binding domain is essentially unchanged from all other structures despite little or no sequence identity.

The variability of the conformation of the protein backbone surrounding the substrate binding pocket contrasts sharply with the consistency of the Rossmann fold associated with cofactor binding. The NADP binding domains have sequence variations but architectural integrity, and the substrate binding domain of different SDR enzymes has neither sequence nor structural similarity. The variability in the substrate binding pocket is consistent with the wide variety of structures used as substrates by different members of the family. It is less readily understood how a protein conformation as consistent as the Rossmann fold has no structurally conserved residues. Clearly certain individual residues are interchangeable without disrupting the stability of the fold. As more data on sequences and three-dimensional structures of families of enzymes such as the SDR become available, a clearer understanding of the relationship between protein sequence and conformation will be achieved.

A number of steroid dehydrogenase enzymes, including human 3α-hydroxysteroid/dihydrodiol dehydrogenase (3α-HSD), belong to the aldo-keto reductase (AKR) superfamily (Bohren *et al.,* 1989). The X-ray structure determination of 3α-HSD and its NADP+ complex (Hoog *et al.,* 1994, Bennett *et al.,* 1996) shows that the protein folds into an α/β barrel and does not have a Rossmann fold. Residues Tyr-55, Lys-84, and His-117, which are implicated in catalysis, reside at one end of the barrel cavity. A stereochemical fit of the tyrosine, lysine, serine/histidine triads and enzyme cofactors in the SDR and AKR families has been demonstrated by Penning *et al.* (1996). This is especially remarkable because of the total absence of sequence homology between the families, and because the residues are located on different strands of the folded proteins.

VI. Conclusions

It is possible to draw the following conclusions from the structural studies of steroid-dehydrogenase enzymes that have been carried out thus far. The short-chain steroid dehydrogenase enzymes studied crystallographically (Table I) contain a common cofactor binding domain (Rossmann fold) made up of a seven-strand parallel β-sheet flanked by three α-helices on either face. The enzymes share a common catalytic triad (Tyr, Lys, Ser) and mechanism of action. The mechanism of action and possible roles of the highly conserved catalytic triad were first proposed on the basis of the structure of bacterial 3α,20β-HSD. That determination also revealed that the bifunctional nature of the bacterial enzyme arose because the substrate binding site could accommodate binding of steroids in two orientations. Twelve other short-chain dehydrogenase enzymes for which X-ray studies have been reported have similar conformations of the cofactor binding domain of the enzyme and identical or analogous catalytic residues suggesting a similar mechanism of action. The similarity in overall three-dimensional structure of the cofactor binding domains in 13 members of the SDR family is remarkable in that there are only two fully conserved residues in this group of 13 (Table I) and no residues that are fully conserved in the whole family.

The observation of inhibition of activity of 3α,20β-HSD by glycyrrhizic acid and carbenoxolone, the determination of the crystal structure of the complex of 3α,20β-HSD carbenoxolone and the fact that 11β-hydroxysteroid dehydrogenase is a member of the short-chain steroid dehydrogenase family supports the postulate that licorice ingestion influences blood pressure by inhibition of 11β-HSD in the kidney. The determination of the structure of the complex between 3α,20β-HSD and carbenoxolone revealed the precise basis for inhibition and provided a model for the design of potent inhibitors of SDR enzymes. The most potent enzyme inhibitors would be transition state analogs that function as carbenoxolone does, to displace substrate and cofactor, and form strong hydrogen bonds to one or more amino acid residues directly involved in catalysis.

It is noteworthy that the only conserved residues in the substrate binding cleft of the SDR family members of Table I are residues in the catalytic triad. Most of the partially conserved residues are at the core of the Rossmann fold and are primarily hydrophobic. It is not surprising that there are no conserved residues in the substrate binding site, since the individual members of the family of enzymes have selectivity for a variety of substrates, including steroids, prostaglandins, sugars, and alcohols.

The critically important tyrosine, lysine, and serine appear to maintain a fixed position relative to the scaffolding of the Rossmann fold and the cofactor position, while the substrate-binding pocket alters to accommodate binding of various substrates in such a way that the point of dehydrogenation or reduction is brought into bonding distance to the tyrosine hydroxyl. The conservation of the Rossmann fold in the SDR family, despite little sequence conservation, suggests that some mutations at every site in the Rossmann fold are tolerated without loss of function. Once substrate specificity evolved, changes in the active site could lead to loss of function or change in specificity. It appears that in this family, function evolved early, while specificity evolved later. All of the SDR family members studied thus far are active as dimers or tetramers. The dimer interface of which 17β-OH provides a typical example (Fig. 16) is also present in tetramers (3α,20β-HSD, Fig. 2). The interface is stabilized by interactive association of symmetry related α-helices E and F. It has been shown that while conservative variation in amino acid (i.e., substitution of one hydrophobic residue for another) is tolerated nonconservative mutations can disrupt dimer formation and cause loss of function. As the crystal structures of more members of the family are determined, it may be possible to determine exactly which amino acids at each site in the Rossmann fold are compatible with retaining effective cofactor binding dimer formation and catalytic action. Well-designed mutation experiments could also determine which amino acid substitutions disrupt the integrity of the Rossmann fold and enzyme aggregation.

The enoyl-acyl carrier protein subgroup of the SDR family exhibits an interesting sequence variation in which an alternative catalytic tyrosine residue substitutes for the one common to most members of the family. This underscores the potency of the Tyr, Lys, Ser triad-based catalysis. The similarity in the arrangement of the Tyr, Lys, and His triad in the active site of 3α-HSD (a member of the aldo-keto reductase family) and of the Tyr, Lys, Ser catalytic triad in the SDR family further supports that conclusion.

ACKNOWLEDGMENT

This research was funded by grant DK26546 from the National Institutes of Health.

REFERENCES

Adams, E. F., Newton, C. J., Tait, G. J., Braumsberg, H., Reed, M. J., and James, V. H. T. (1988). Paracrine influence of human breast stromal fibroblasts on breast epithelial cells: Secretion of a polypeptide which stimulates reductive 17β-oestradiol dehydrogenase activity. *Int. J. Cancer* **42,** 119–122.

Adams, M. J., Ford, G. C., Koekock, R., Lentz, P. J., Jr., McPherson, A., Jr., Rossmann, M. G., and Smiley, I. E. (1970). Structure of lactate dehydrogenase at 2.8 Å resolution. *Nature (London)* **227,** 1098–1103.

Adamski , J., Normand, T., Leenders, F., Monte, D., Begue, A., Stehelin, D., Jungblut, P. W., and Launoit, Y. (1995). Molecular cloning of a novel widely expressed human 80 kDa 17β-hydroxysteroid dehydrogenase IV. *Biochem. J.* **311,** 437–443.

Andersson, A., Jordan, D., Schneider, G., and Lindqvist, Y. (1997). Crystal structure of the ternary complex of 1,3,8-trihydroxysteroid naphtalene reductase from *Magnaporthe grisea* with NADPH and an active-site inhibitor. *Structure* **4,** 1161–1170.

Auerbach, G., Herrmann, A., Gütlich, M., Fischer, M., Jacob, U., Bacher, A., and Huber, R. (1997). The 1.25Å crystal structure of sepiapterin reductase reveals its binding mode to pterins and brain neurotransmitters. *EMBO J.* **16,** 7219–7230.

Azzi, A., Rehse, P. H., Zhu, D. W., Campbell, R. L., Labrie, F., and Lin, S. C. (1996). Crystal structure of human estrogenic 17β-hydroxysteroid dehydrogenase complexed with 17β-estradiol. *Nat. Struct. Biol.* **8,** 665–668.

Baldock, C., Rafferty, J. B., Sedelnikova, S. E., Baker, P. J., Stuitje, A. R., Slabas, A. R., Hawkes, T. R., and Rice, D. W. (1996). A mechanism of drug action revealed by structural studies of enoyl reductase. *Science* **274,** 2107–2110.

Bauer, A. J., Rayment, I., Frey, P. A., and Holden, H. M. (1992). The molecular structure of UDP-galactose 4-epimerase from *Escherichia coli* determined at 2.5Å resolution. *Proteins: Struct., Funct., Genet.* **12,** 372–381.

Benach, J., Atrian, S., Gonzalez-Duarte, R., and Ladenstein, R. (1998). The refined crystal structure of *Drosophila lebanonensis* alcohol dehydrogenase at 1.9 Å resolution. *J. Mol. Biol.* **282,** 383–399.

Bennett, M. J., Schlegel, B. P., Jex, J. M., Penning, T. M., and Lewis, M. (1996). Structure of 3α-hydroxysteroid/dihytdrodiol dehydrogenase complexed with NADP+. *Biochemistry* **35,** 10702–10711.

Bernstein, F. C., Koetzle, T. F., Williams, G. J. B., Meyer, E. R., Brice, M. D., Rodgers, J. R., Kennard, O., Shimanouchi, T., and Tasumi, M. (1977). The Protein Data Bank: A computer-based archival file for macromolecular structures. *J. Mol Biol.* **112,** 535–542.

Birktoft, J. J., and Banaszak, L. J. (1984). Structure-function relationships among nicotinamide-adenine dinucleotide dependent oxidoreductases. *In* "Peptide and Protein Review" (T. Milkton and W. Hearn, eds.), Vol. 4, pp. 1–44. Dekker, New York.

Bohren, K. M., Bullock, B., Wermuth, B., and Gabbay, K. H. (1989). The aldoketo reductase superfamily cDNAs and deduced amino acid sequences of human aldehyde and aldose reductases. *J. Biol. Chem.* **264,** 9547–9551.

Breton, R., Housset, D., Mazza, C., and Fontecilla-Camps, J. C. (1996). The structure of a complex of human 17β-hydroxysteroid with estradiol and NADP+ identifies two principal targets for the design of inhibitors. *Structure* **4,** 905–915.

Brzozowski, A. M., Pike, A. C. W., Dauter, Z., Hubbard, R. W., Bonn, T., Engström, O, Ohman, L., Greene, G. L., Gustafsson, J.-A., and Carlquist, M. (1997). Molecular basis of agonism and antagonism in the oestrogen receptor. *Nature (London)* **389,** 753–758.

Declercq, J. P., Germain, G., and Van Meerssche, M. (1972). 17α,20β-dihydroxy-4-pregnene-3,11,20-trione $C_{21}H_{28}O_5$. *Cryst. Struct. Commun.* **1,** 13–15.

Dessen, A., Quemard, A., Blanchard, J. S., Jacobs, W. R., Jr., and Sacchettini, J. C. (1995). Crystal structure and function of the isoniazid target of mycobacterium tuberculosis. *Science* **267,** 1638–1641.

Deyashiki, Y., Ohshima, K., Nakanishi, M., Sato, K., and Hara, A. (1995). Molecular cloning of a novel widely expressed human 80 kDA 17β-dehydrogenase (A-specific), a member of the aldo-keto reductase family. *J. Biol. Chem.* **270,** 10461–10467.

Dickson, R. B., and Lippman, M. E. (1987). Estrogenic regulation of growth and polypeptide growth factor secretion in human breast carcinoma. *Endocrinol. Rev.* **8,** 29–43.

Duax, W. L., and Griffin, J. F. (1989). The structure and receptor binding of steroid hormones. In "Advances in Drug Research" (B. Testa, ed), pp. 115–138. Academic Press, New York.

Duax, W. L., Griffin, J. F., and Ghosh, D. (1996). The fascinating complexities of steroid-binding enzymes. *Curr. Opin. Struct. Biol.* **6,** 813–823.

Dupont, E., Labrie, F., Luu-Thé, V., and Pelletier, G. (1991). Localization of 17β-hydroxysteroid dehydrogenase throughout gestation in human placenta. *J. Histochem. Cytochem.* **39,** 1403–1407.

Evans, R. M. (1988). The steroid and thyroid hormone receptor superfamily. *Science* **240,** 889–895.

Evans, S. V. (1993). SETOR: Hardware lighted three-dimensional solid model representation of macromolecules. *J. Mol. Graphics* **11,** 134–138.

Fournet-Dulguerov, N., MacLusty, N. J., Leranth, C. Z., Todd, R., Mendelson, C. R., Simpson, E. R., and Naftolin, F. (1987). Immunohistochemical localization of aromatase cytochrome P-450 and estradiol dehydrogenase in the syncytiotrophoblast of the human placenta. *J. Clin. Endocrinol. Metab.* **65,** 757–764.

Geissler, W. M., Davis, D. L., Wu, L., Bradshaw, K. D., Patel, S., Mendonca, B. B., Elliston, K. O., Wilson, J. D., Russell, D. W., and Andersson, S. (1994). Male pseudohermaphroditism caused by mutations of testicular 17β-hydroxysteroid dehydrogenase 3. *Nat. Genet.* **7,** 34–39.

Ghersevich, S. A., Poutanen, M. H., Marttikainen, H. K., and Vihko, K. (1994). Expression of 17β-hydroxysteroid dehydrogenase in human granulosa cells: Correlation with follicular size, cytochrome P450 aromatase activity and oestradiol production. *J. Endocrinol.* **143,** 139–150.

Ghosh, D., Erman, M., Pangborn, W., Duax, W. L., and Baker, M. E. (1992). Inhibition of Streptomyces hydrogenans 3α,20β-hydroxysteroid dehydrogenase by licorice-derived compounds and crystallization of an enzyme-cofactor-inhibitor complex. *J. Steroid Biochem. Mol. Biol.* **42,** 849–853.

Ghosh, D., Wawrzak, Z., Weeks, C. M., Duax, W. L., and Erman, M. (1994a). The refined three-dimensional structure of 3α,20β-hydroxysteroid dehydrogenase and possible roles of the residues conserved in "short-chain" dehydrogenases. *Structure* **2,** 629–640.

Ghosh, D., Erman, M., Wawrzak, Z., Duax, W. L., and Pangborn, W. (1994b). Mechanism of Inhibition of 3α,20β-hydroxysteroid dehydrogenase by a licorice-derived steroidal inhibitor. *Structure* **2,** 973–980.

Ghosh, D., Pletnev, V. Z., Zhu, D.-W., Wawrzak, Z., Duax, W. L., Pangborn, W., Labrie, F., and Lin, S.-X. (1995). Structure of human estrogenic 17β-hydroxysteroid dehydrogenase at 2.20Å resolution. *Structure* **3,** 503–513.

Girerd, R. J., Rassaert, C. L., DiPasquale, G., and Kroc, R. L. (1960). Endocrine involvement in licorice hypertension. *Am. J. Physiol.* **198,** 718–720.

Hill, E., Tsernoglau, D., Webb, L., and Banaszak, L. J. (1972). Polypeptide conformation of cytoplasmic malate dehydrogenase from an electron density map at 3.0Å resolution. *J. Mol. Biol.* **72,** 577–589.

Hoog, S. S., Pawlowski, J. E., Alzari, P. M., Penning, T. M., and Lewis, M (1994). Three-dimensional structures of rat liver 3α-hydroxysteroid/dihydrodiol dehydrogenase: A member of the aldo-keto reductase superfamily. *Proc. Natl. Acad. Sci. U.S.A.* **91,** 2517–2521.

Horwitz, K. B., and McGuire, W. L. (1987). Nuclear mechanisms of estrogen action. Effects of estradiol and anti-estrogens on estrogen receptors and nuclear receptor processing. *J. Biol. Chem.* **253,** 8185–8191.

James, V. H. T., McNeil, J. M., Beranek, P. A., Bonney, R. C., and Reed, M. J. (1986). The role of tissue steroids in regulating aromatase and oestradiol 17β-hydroxysteroid dehydrogenase activities in breast and endometrial cancer. *J. Steroid Biochem.* **25,** 787–790.

James, V. H. T., Reed, M. J., and Coldham, N. G. (1990). Regulation of estrogen concentrations in human breast tissues. *Ann. N.Y. Acad. Sci.* **595,** 227–235.

Jörnvall, H., Persson, B., Krook, M., Atrian, S., Gonzàles-Duarte, J. J., and Ghosh, D. (1995). Short-chain dehydrogenase/reductases (SDR). *Biochemistry* **34,** 6003–6013.

Labrie, C., Martel, C., Dufour, J.-M., Levesque, C., Merand, Y., and Labrie, F. (1992). Novel compounds inhibit estrogen formation and action. *Cancer Res.* **52,** 610–615.

Lin, S. X., Yang, F., Jin, J. Z., Breton, R., Zhu, D. W., Luu-Thé, V., and Labrie, F. (1992). Subunit identity of the dimeric 17β-hydroxysteroid dehydrogenase from human placenta. *J. Biol Chem.* **267,** 16182–16187.

Luu-Thé, V., Lachance, Y., Labrie, C., Leblanc, G., Thomas, J. L., Strickler, R. C., and Labrie, F. (1989). Characterization of cDNAs for human estradiol 17β-dehydrogenase and assignment of the gene to chromosome 17: Evidence of two mRNA species with distinct 5' termini in human placenta. *Mol. Endocrinol.* **3,** 1301–1309.

Mäentausta, O., Sormunen, R., Isomaa, V., Lehto, V. P. Jouppila, P., and Vihko, R. (1991). Immunohistochemical localization of 17β-hydroxysteroid dehydrogenase in the human endometrium during the menstrual cycle. *Lab. Invest.* **65,** 582–587.

Mäentausta, O., Boman, K., Isomaa, V., Stendahl, U., Backstrom, T., and Vihko, R. (1992). Immunohistochemical study of the human 17β-hydroxysteroid dehydrogenase and steroid receptors in endometrial adenocarcinoma. *Cancer (Philadelphia)* **70,** 1551–1555.

Mazza, C., Breton, R., Housset, D., and Fontecilla-Camps, J. C. (1998). Unusual charge stabilization of NADP+ in 17β-hydroxysteroid dehydrogenase. *J. Biol. Chem.* **273,** 8145–8152.

Monder, C., Stewart, P. M., Lakshmi, V., Valentino, R., Burt, D., and Edwards, C. R. W. (1989). Licorice inhibits corticosteroid dehydrogenase of rat kidney and liver: *In vivo* and *in vitro* studies. *Endocrinology (Baltimore)* **125,** 1046–1053.

New, M. I., Oberfield, S. E., Carey, R., Greig, F., Ulick, S., and Levind, L. S. (1982). A genetic defect in cortisol metabolism as the basis for the syndrome of apparent mineralocorticoid excess. *Proc. Serono Sympo.* **50,** 85–101.

Peltoketo, H., Isomaa, V., Mäentausta, O. I., and Vihko, R. K. (1988). Complete amino acid sequence of human placental 17β-hydroxysteroid dehydrogenase deduced from cDNA. *FEBS Lett.* **239,** 73–77.

Penning, T. M., Pawlowski, J. E., Schlegel, B. P., Jez, J. M., Lin, H.-K., Hoog, S. S., Bennett, M. J., and Lewis, M. (1996). Mammalian 3α-hydroxy steroid dehydrogenase. *Steroids* **61,** 508–523.

Poutanen, M. H., Isomaa, V., Lehto, V. P., and Vihko, R. K. (1992). Immunological analysis of 17β-hydroxysteroid dehydrogenase in benign and malignant human breast tissue. *Int. J. Cancer* **50,** 386–390.

Poutanen, M. H., Miettinen, M., and Vihko, R. K. (1993). Differential estrogen substrate specificity for transiently expressed human placental 17β-hydroxysteroid dehydrogenase and an endogenous enzymne expressed in cultured COS-m6 cells. *Endocrinology (Baltimore)* **133,** 2639–2644.

Poutanen, M., Isomaa, V., Peltoketo, H., and Vihko, R. K. (1995). Role of 17β-hydroxysteroid dehydrogenase type 1 in endocrine and intracrine estradiol biosynthesis. *Steroid Biochem. Mol. Biol.* **55** 525–532.

Puranen, T. J., Poutanen, M. H., Peltoketo, H. W., Vihko, P. T., and Vihko, R. K. (1994). Site-directed mutagenesis of the putative active site of human 17β-hydroxysteroid dehydrogenase type 1. *Biochem J.* **304,** 289–293.

Puranen, T. J., Poutanen, M. H., Ghosh, D., Vihko, P. T., and Vihko, R. K. (1997a). Characterization of structural and functional properties of human 17β-hydroxysteroid dehydrogenase type 1 using recombinant enzymes and site directed mutagenesis. *Mol. Endocrinol.* **11,** 77–86.

Puranen, T. J., Poutanen, M. H., Ghosh, D., Vihko, P. T., and Vihko, R. K. (1997b). Origin of substrate specificity of human and rat 17β-hydroxysteroid dehydroigenase type 1, using chimeric enzymes and site directed mutagenesis. *Endocrinology (Baltimore)* **138,** 3532–3539.

Rafferty, J. B., Simon, J. W., Baldock, C., Artymiuk, P. J., Baker, P. J., Stuije, A. R., Slabas, A. R., and Rice, D. W. (1995). Common themes in redox chemistry emerge from the X-ray structure of oilseed rape (*Brassica napus*) enoyl acyl carrier protein reductase. *Structure* **3,** 927–938.

Ribos-de-Pouplona, L., and Fothergall-Gilmore, L. A. (1994). The active site architecture of a short chain dehydrogenase defined by site-directed mutagenesis and structure-modeling. *Biochemistry* **33,** 7047–7055.

Rozwarski, D. A., Grant, G. A., Barton, D. H. R., Jacobs, W. R., Jr., and Sacchettini, J. C. (1998). Modification of the NADH of the isoniazid target (INHA) from Mycobacterium tuberculosis. *Science* **279,** 98–102.

Sawicki, M. W., Erman, M., Puranen, T. J., Pirkko, V., and Ghosh, D. (1999a). Structure of the ternary complex of human 17β-hydroxysteroid dehydrogenase type 1 with

3-hydroxyestra-1,3,5,7-tetraen-17-one (equilin) and NADP+. *Proc. Natl. Acad. Sci. U.S.A.* **96,** 840–845.

Sawicki, M. W., Li, N., and Ghosh, D. (1999b). Crystal structure of equilin. *Acta Crystallogr. C* (in press).

Scrutton, N. S., Berry, A., and Perham, R. N. (1990). Redesign of the coenzyme specificity of a dehydrogenase by protein engineering. *Nature (London)* **343,** 38–43.

Sweet, F., and Samant, B. R. (1980). Bifunctional enzyme activity at the same active site: Study of 3α,20β-hydroxysteroid dehydrogenase with 17-(bromoacetoxy)steroids. *Biochemistry* **19,** 978–986.

Tanaka, N., Nonaka, T., Nakanishi, M., Deyashiki, Y., Hara, A., and Mitsui, Y. (1996a). Crystal structure of the ternary complex of mouse lung carbonyl reductase at 1.8Å resolution: The structural origin of coenzyme specificity in the short-chain dehydrogenase/reductase family. *Structure* **4,** 33–45.

Tanaka, N., Nonaka, T., Tanabe, T., Yashimoto, T., Tsuru, D., and Mitsui, Y. (1996b). Crystal structures of the binary and ternary complexes of 7α-hydroxysteroid dehydrogenase from *E.coli*. *Biochemistry* **35,** 7715–7730.

Ulick, S., Levine, L. S., Gunczler, P, Zanconato, G., Ramierz, L. C., Rauh, W., Rosler, A., Bradlow, H. L., and New, M. I. (1979). A syndrome of apparent mineralocorticoid excess associated with defects in the peripheral metabolism of cortisol. *J. Clin. Endocrinol. Metab.* **49,** 757–764.

Varughese, K. I., Skinner, M. M., Whiteley, J. M., Matthews, D. A., and Xuong, N. H. (1992). Crystal structure of rat liver dihydropteridine reductase. *Proc. Natl. Acad. Sci. U.S.A.* **89,** 6080–6084.

Watson, H. C., and Banaszak, L. J. (1964). Structure of glyceraldehyde-3-phosphate dehydrogenase. *Nature (London)* **204,** 918–920.

Structure-Based Inhibitor Design

SYDNEY P. CRAIG III AND ANN E. EAKIN

Laboratory of Molecular Parasitology & Drug Design, University of North Carolina School of Pharmacy, Chapel Hill, North Carolina 27599-7360 USA

Time and costs associated with the discovery of new drugs have been significantly reduced by enzyme structure-based approaches to the discovery of new chemotherapeutic agents. However, fundamental components of the overall approach continue to rely on technologies which, by their nature, involve relatively random processes (i.e., combinatorial chemistry and high-throughput screening). Thus, the efficiency of the drug discovery process potentially could be further improved through better use of structural information. In this regard, three-dimensional structures of enzymes are now being solved at high resolution and/or in conformations that provide data that should be more useful for inhibitor design or discovery. Scientists are beginning to appreciate the importance of water as a possible competitor of inhibitors for binding to target enzymes. New computational algorithms are improving the efficiency of identifying flexible inhibitors from among the large numbers of compounds in chemical databases. Also, tools of molecular genetics together with structures of target enzymes are likely to be used more frequently in dealing with the development of resistance to novel chemotherapeutic agents. Instead of detailing success stories in structure-based drug discovery, the following article considers how future efforts to discover or design new drugs may increasingly rely on information about molecular targets and less on data acquired via approaches involving random methodologies.

0083-6729/00 $30.00

I. Introduction

The target-based approach to drug design, yielding inhibitors that selectively bind to specific enzymes, can be compared to the process of designing keys to fit locks (Fischer, 1894). Until recently, inhibitor design largely involved the indirect evaluation of target enzymes via the accumulation of structure activity relationships (SAR) for a series of compounds derived from chemical modifications of lead compounds (those which yield a measurable biological or therapeutic activity). For SAR, the modified compounds are tested to assess the effects of the modifications on biological activity. This indirect approach to the discovery of inhibitors is analogous to making many slightly different keys in an effort to discover a key that will fit the lock without ever having an exact detailed image of the lock. A more direct approach to designing keys would be to analyze the lock in detail before attempting to manufacture a key that will be guaranteed to be specific for the lock. Similarly, it should be possible to utilize the structural details of a molecular target, such as an enzyme, to design and synthesize compounds that, at relatively low concentrations, are guaranteed to inhibit the activity of the enzyme and thus provide good leads for drug design.

Recently, the three-dimensional (3-D) structures of target enzymes have been used in the discovery process leading to a variety of novel chemotherapeutic agents. Examples of success stories, involving the structure-based design of inhibitors include the development of protease inhibitors for the treatment of AIDS; carbonic anhydrase inhibitors for the treatment of glaucoma; angiotensin-converting enzyme inhibitors for the treatment of hypertension and congestive heart failure; neutral endopeptidase inhibitors for regulating electrolyte levels; purine nucleoside phosphorylase inhibitors as therapeutic agents in host-graft rejection and T-cell proliferative diseases; sialidase inhibitors as antiviral agents; and thymidylate synthase inhibitors as antitumor agents (Bohacek *et al.,* 1996; Skulnick *et al.,* 1997; Wlodawer and Vondrasek, 1998). These discoveries have been made possible because of technological advances in computational chemistry, combinatorial chemistry, screening, and robotics, together with the availability of high-resolution 3-D structures of target enzymes.

The efficiency of the structure-based approach to drug design depends on the quality of the 3-D structures for the target enzymes and how the structural information is used. Also, X-ray crystallographic structures provide only static views of enzymes and available structures may not be in the most favorable conformation for efficient inhibitor design or discovery. Improvements in the resolution of existing structures and the solution of structures for enzymes in multiple conformations will facilitate the overall process of inhibitor design or discovery.

During recent years, a number of reviews have appeared on topics related to enzyme structure-based design of drugs (Verlinde and Hol, 1994; Gait and Karn, 1995; Kleinberg and Wanke, 1995; Bohacek *et al.*, 1996; Jackson, 1997; Marrone *et al.*, 1997; Salemme *et al.*, 1997; Skulnick *et al.*, 1997; Wade, 1997; Amzel, 1998; Wlodawer and Vondrasek, 1998), international conferences that include sections devoted to structure-based drug design are regularly convened, and it is not unusual to learn through the popular media of ongoing research involving structure-based drug design. This leaves the impression that we have arrived at an unprecedented period of human history with respect to the rate for the discovery of new drugs to treat diseases of both human and agricultural importance. The 3-D structures of target enzymes clearly have played a significant role in influencing the rate of development of novel chemotherapeutic agents. However, it is likely that technological advances in areas such as combinatorial chemistry, high-throughput screening, and robotics have played an equal, if not greater role. Unfortunately, contemporary methods of drug discovery still rely heavily on relatively random processes (i.e., methods of combinatorial chemistry and high-throughput screening), and structures of target enzymes are often used only as a reference point to be consulted repeatedly during the drug discovery process. Thus, the synthesis of thousands of compounds continues to be required for many contemporary drug discovery efforts (Li *et al.*, 1996; Salemme *et al.*, 1997; Kick *et al.*, 1997; Murray *et al.*, 1997; Whittaker, 1998).

The phrase "designing a key to fit a lock" has been used to describe enzyme structure-based drug design but examples are relatively rare for the *de novo* design of inhibitors using information coming largely from the target (i.e. Bugg *et al.*, 1993). In contrast, there are numerous examples of looking retrospectively at 3-D structures of drugs cocrystallized with their target enzymes to discover the molecular basis for specific tight binding interactions of potent inhibitors. Therefore, it is legitimate to ask why more inhibitors have not been designed directly, thereby avoiding random methods of synthesis and screening. Although the process of drug discovery is less labor intensive and more cost effective today than in the past, structure-based drug design continues to be a less efficient process than originally envisioned.

In this article, we suggest how and why structural information about therapeutic targets is likely to play a greater role in future efforts to discover or design new drugs. Particular attention will be paid to the resolution and conformations of target structures, the possible importance of crystallographic waters associated with target proteins, and the development of algorithms that more accurately simulate the molecular flexibilities of ligands and their enzyme targets. Also considered will be the possible role of molecular biology and enzyme structure in anticipating and dealing with the development of resistance to new drugs.

II. Target Selection, Cloning, and Determination of 3-D Structures

A scheme for the overall process of enzyme structure-based drug design appears in Fig. 1. A fundamental aspect of a target-based approach to drug design is the identification of an appropriate molecular target that, if inhibited, will block proliferation of the disease-causing agent.

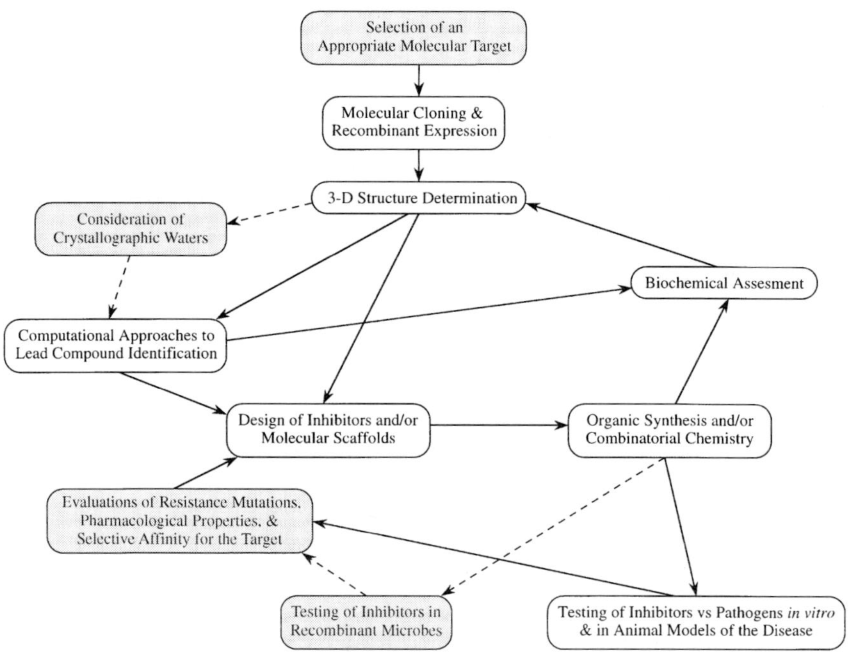

Fig. 1. Enzyme structure-based inhibitor design. Arrows show where information coming from one area of research is used to drive or guide work in another area. Areas illustrated with gray backgrounds indicate research fields that may play larger roles in future efforts to discover or design new drugs.

In some cases, merely slowing proliferation may enable immune systems to control or eliminate the disease condition. To minimize problems associated with toxicity, it is helpful if the molecular target is uniquely expressed by the pathogen or tumor cell. The existence of alternate metabolic pathways in normal host cells also may contribute to lessening the chances of toxicity. However, where the activity of a homologous host enzyme is essential for the survival of host tissues, the design or discovery of a selective inhibitor of the enzyme of the pathogen may provide a relatively nontoxic, but effective therapeutic agent.

Before the present decade, the cloning and recombinant expression of target proteins were somewhat problematical and could be major rate-limiting steps in the overall process of structure-based inhibitor design or discovery. However, rapidly expanding DNA databases (partly due to genome sequencing projects), the widespread use of polymerase chain reactions (PCR) for amplifying specific DNA sequences, and the commercial availability of a variety of expression vectors, including those with built-in strategies for purifying expressed proteins (i.e., the inclusion of removable metal binding tags), have facilitated this part of the process.

The crystallization of proteins also can be a rate-limiting step for the determination of 3-D structures of proteins. However, companies like Hampton Research of Laguna Nigel, California, now provide kits that facilitate the process of screening for conditions suitable for growing crystals in any protein biochemistry laboratory. After diffraction quality crystals have been generated, solving the structure of a protein can often be accomplished by methods of molecular replacement using the 3-D structure of a related protein (Tickle and Driessen, 1996; Turkenburg and Dodson, 1996). As the database of 3-D structures expands, the availability of potential molecular replacement probes increases for given protein targets. Alternatively, more complex methods may need to be employed, such as those requiring the generation of heavy atom derivatives via crystal soaking or by using crystals with protein that was generated in microbes that enable the substitution of the heavy atom, selenium, for sulfur atoms (Price and Nagai, 1995). The high electron density of the heavy metals in crystals yields anomalies in the diffraction of X rays that can be used to determine the positions of the small number of heavy atoms in the crystallized protein. This information can then be used to solve the structure of the entire protein. Additionally, the use of selenium-containing proteins enables the rapid solution of a protein structure from data collected from a single crystal at a synchrotron source using the method of multiwavelength anomalous diffraction (MAD; Hendrickson and Ogata, 1997).

The majority of the available 3-D structures for enzymes are in open conformations, either without ligands or with substrates, products, and/or analogs of the natural substrates or products in their active sites. Numerous structures of enzymes in open conformations with ligands (inhibitors) bearing little resemblance to their natural substrates have been used to guide structure-based drug design projects. As suggested in Section V, ligands that mimic the binding interactions occurring in the transition state of an enzyme-catalyzed reaction have the potential to be extraordinarily potent inhibitors. Thus, closed structures of enzymes, such as that reported by Focia *et al.* (1998b) (Fig. 2),

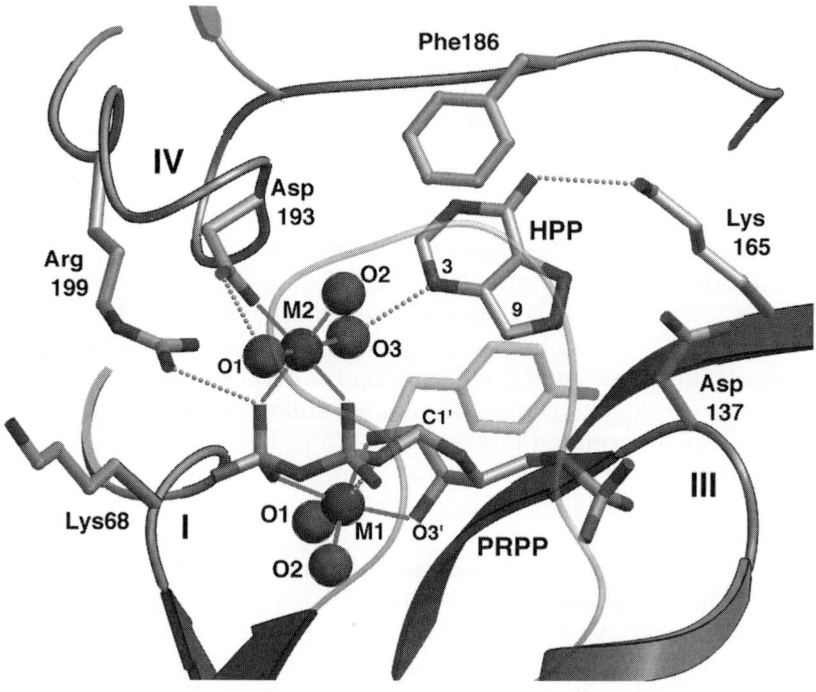

FIG. 2. The active site of a hypoxanthine phosphoribosyltransferase with bound substrate (phosphoribosylpyrophosphate: PRPP), substrate analog (7-hydroxy [4,3-*d*]pyrazolopyrimidine: HPP), and hydrated metal ions viewed through the closed flexible loop (transparent). Non-main-chain protein–ligand hydrogen bonds are displayed, as well as metal (M1 and M2) coordination interactions with substrates, waters, and the protein. [Reprinted with permission from Focia, P. J., Craig, S. P., III, and Eakin, A. E. (1998). Approaching the transition state in the crystal structure of a phosphoribosyltransferase. *Biochemistry,* **37,** 17120–17127. Copyright © 1998 American Chemical Society.]

are likely to be more useful than open structures as guides for structure-based inhibitor design.

Low-resolution crystal structures may suffer from inaccuracy in the positioning of atomic coordinates. Thus, high-resolution 3-D structures provide better guides for designing ligands that can interact favorably with the active sites of target enzymes. Different crystal forms of the same enzyme will occasionally diffract X rays with improved resolution (Price and Nagai, 1995; Nieves-Alicea *et al.*, 1998). Also, the use of X rays generated at synchrotron sources can enable the solution of structures at significantly improved resolution compared with those solved using data collected by more traditional methods (Focia *et al.*, 1998a; Kuhn *et al.*, 1998). Several structures have been solved in recent years with resolutions at or below 1 Å. Structures at this resolution not only enable the accurate positioning of coordinates for carbon, nitrogen, oxygen, etc. atoms, but distortions in the electron density maps may sometimes show the probable positions for hydrogen atoms (Kuhn *et al.*, 1998). Structures at this level of resolution will facilitate the design of inhibitors and increase the accuracy of predicting force field interactions with potential ligands.

The initial 3-D structure of a target enzyme is more difficult to solve than are subsequent structures which can usually be solved by methods of molecular replacement using the original structure as the search probe. Subsequent 3-D structures of target enzymes with novel ligands in their active sites can often be solved in a matter of weeks. These iterative structures provide reference points that can and should be consulted at every step of ligand modification during the process of inhibitor design. A very important reason for this iterative process is that both the target enzyme and the ligand are usually flexible molecules, enabling interactions that might not be predicted from static crystal structures.

III. Crystallographic Water Molecules

Water is typically responsible for about half of the mass of a protein crystal. The majority of these water molecules are disordered. However, the molecular motions of some waters are restricted by their interactions with either the protein or ligands associated with the protein. Thus, reported crystal structures usually include the 3-D coordinates for many ordered “crystallographic” water molecules.

Tightly bound water molecules have been suggested to affect the binding of the inhibitor, methotrexate, to dihydrofolate reductase (Nagy, 1988; Marrone *et al.*, 1997). Additionally, water molecules may be present that participate in active site chemistry whose displacement could be exploited in a strategy to interfere with the enzyme-catalyzed reaction, as in the case for inhibition of scytalone dehydratase of a plant fungus (Chen *et al.,* 1998). Similarly, the design of inhibitors of the HIV protease has, in some cases, involved displacement of a water molecule that was tetrahedrally coordinated between the enzyme and inhibitors (Wlodawer and Vondrasek, 1998).

Ringe (1995) has suggested that some water molecules may resist displacement by other ligands and thus their presence should be considered in enzyme structure-based inhibitor design strategies intended to maximize ligand interactions with the target enzymes. The question is this: Of the many ordered water molecules in crystal structures, which are those that will resist displacement? One approach to answering this question is to individually evaluate each water molecule in or near the active site. Many of the water molecules may be predicted to move in and out of the active site along with substrates, products, or metal ions (see Focia *et al.,* 1998b). Other water molecules, particularly those that form multiple hydrogen bonds with protein atoms, might be more permanently associated with the enzyme and thus may be more difficult to displace in aqueous solutions with the concentration of water at 55 *M*.

Allen *et al.* (1996) describe an experimental approach to mapping the binding surfaces of protein crystals of porcine pancreatic elastase. Their strategy involved solving the structure of crystals that have been soaked in specific organic solvents. This approach has the potential of distinguishing those water molecules that can be displaced by an organic solvent, like acetonitrile, from those water molecules that interact too tightly with the protein surface to be displaced (Fig. 3). The results from solvent mapping, together with information from peptide-inhibited complexes of the elastase, were used in designing a peptidomimetic inhibitor (a lead compound) that binds to the elastase in the same fashion as the class of dipeptides from which its design was derived (Peisach *et al.,* 1995). In this case, the molecular scaffold for the inhibitor was an aminimide that could form the basis for a combinatorial chemistry approach to inhibitor refinement. These results indicate that high-resolution structures of crystallized proteins in a variety of solvents may permit the identification of water molecules whose presence should be considered in the design of enzyme inhibitors.

IV. Computational Approaches to Lead Compound Discovery

Computational algorithms for use in screening databases of compounds have been in use for a number of years in support of efforts to

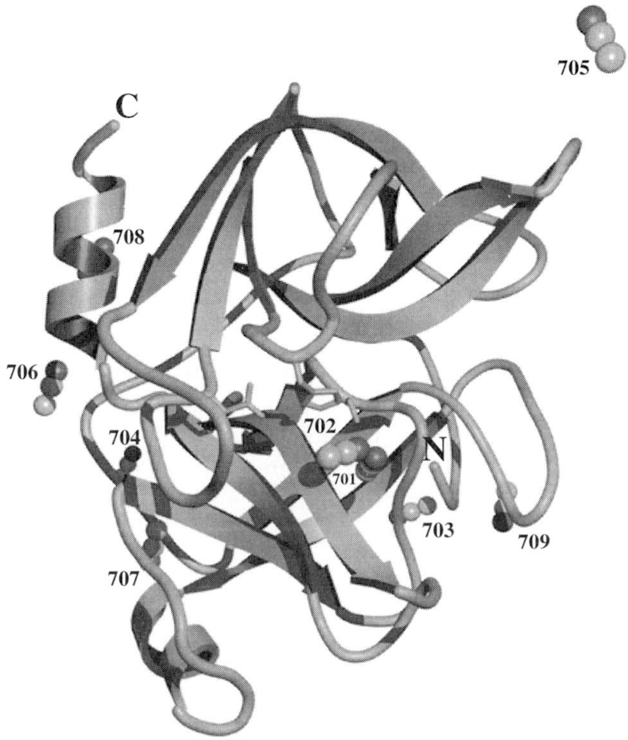

FIG. 3. Elastase structure represented by a ribbon diagram where the overall secondary structural features are shown as a solid helix, strand, or coil. Elastase contains 240 amino acid residues, and it is composed of two β-barrel domains. The catalytic triad (Asp 108, His 60, and Ser 203) and nine acetonitrile molecules are shown explicitly. The nitrogen atom of the acetonitrile molecule is in dark gray, and the carbons are in a lighter gray. [Reprinted with permission from Allen, K. N., *et al.* (1996). An experimental approach to mapping the binding surfaces of crystalline proteins. *Journal of Physical Chemistry,* **100,** 2605–2611. Copyright © 1996 American Chemical Society.]

identify those that will bind to the active site of a target enzyme. Computational algorithms originally intended to fit rigid shapes (Kuntz *et al.,* 1982), have been modified to include potential electrostatic interactions (Shoichet and Kuntz, 1993), and although the record of efficiency of this approach to inhibitor discovery has been somewhat unimpressive, new leads for drug design have been identified (i.e. Shoichet *et al.,* 1993; Strynadka *et al.,* 1996a,b). Computational methods have suffered from the fact that both the target and most potential ligands are flexible rather than rigid molecules. Unfortunately, most computational algorithms that account for flexibility require too much CPU time for their practical application in screening large chemical databases (i.e. Goodsell *et al.*, 1996; Wasserman and Hodge, 1996), although some may be extremely useful for evaluating interactions of specific flexible ligands with their molecular targets (McMartin and Bohacek, 1997). By default, many of the inhibitors identified by computationally screening large chemical databases are relatively rigid molecules, while flexible compounds usually are overlooked because they dock as rigid structures in unsuitable conformations.

Recently, Lorber and Shoichet (1998) reported a method for rapidly screening multiple conformations of flexible compounds in large chemical databases. For this method, the rigid body program, DOCK (Kuntz *et al.*, 1982), was modified to allow rapid fitting of multiple conformations of ligands. Conformations of a given molecule were precalculated in the same frame of reference so that each conformer shared a common rigid fragment with all other conformations. The ligand conformers were then docked together, as an ensemble of as many as 500 conformations, into the receptor binding sites (see Fig. 4). The advantage over previously reported algorithms using Monte Carlo dynamics simulations and simulated annealing is the much more efficient use of CPU time required for the practical screening of large chemical databases (Lorber and Shoichet, 1998).

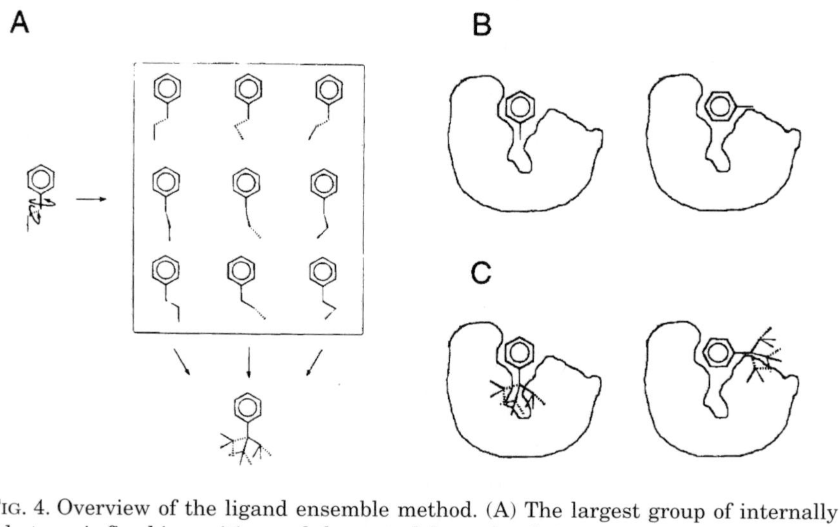

FIG. 4. Overview of the ligand ensemble method. (A) The largest group of internally rigid atoms is fixed in position and the rest of the molecule is systematically sampled at 60° or 120° increments. (B) The rigid fragment common to all conformations of the molecule is oriented in the binding site. (C) All flexible fragments of the molecule are scored in the orientation of the rigid fragment. (From Lorber and Shoichet, 1998. Reprinted with the permission of Cambridge University Press.)

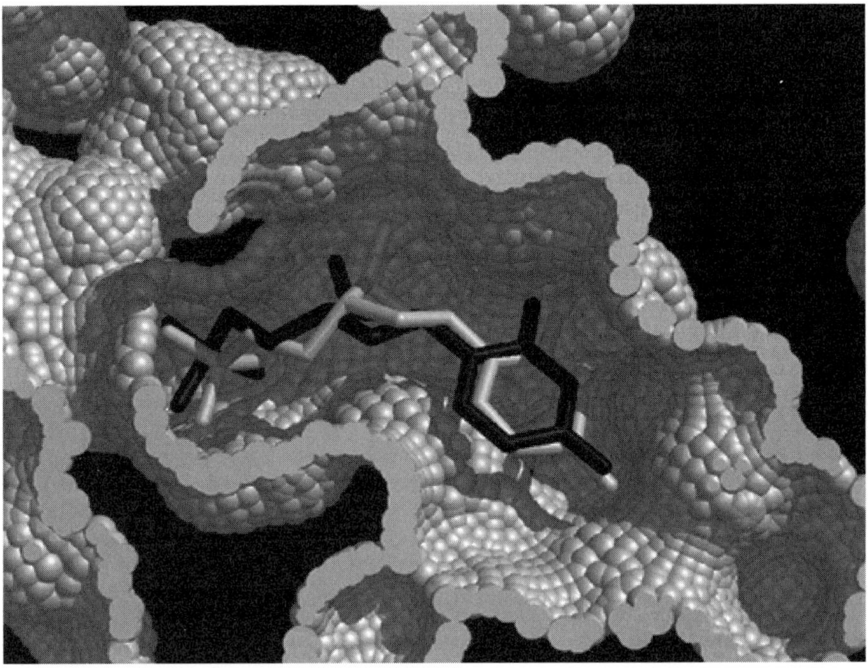

FIG. 5. The experimental (white) and the best fit from ensemble docking (black) orientations of dUMP in the binding site of thymidylate synthase represented by its molecular surface (Connolly, 1983). (From Lorber and Shoichet, 1998. Reprinted with the permission of Cambridge University Press.)

Three-dimensional structures have been determined for a number of enzymes with potent inhibitors as ligands. These structures were used by Lorber and Shoichet (1998) as a reference to test the accuracy of their algorithm for flexible ligand docking in predicting the binding interactions for three different inhibitor/enzyme systems (Figs. 5 and 6; see color plate for Fig. 6). The results of these tests show that the most favorable predicted binding conformations and orientations of the ligands are not identical with those of the inhibitors in the enzymes for which crystal structures have been determined, but they are close enough to warrant optimism for the future of flexible ligand docking in identifying lead compounds from among available chemical databases. This optimism may be particularly justified if the docking surface used is derived from a closed conformation of a target enzyme (see Section V). In this regard, together with Jun Feng, Doug Freymann, and Brian Shoichet, we have used the flexible ligand docking algorithm with a closed conformation of a hypoxanthine phosphoribosyltransferase (Focia *et al.*, 1998b) to screen an available chemicals database of more than 150,000 compounds. Although the results are preliminary, the method has been extraordinarily efficient in revealing the identity of inhibitors, including several that yield binding constants (K_i's) for the target enzyme that are significantly lower than those of the best inhibitors formerly reported for this class of enzymes (unpublished observations).

The lack of specificity of inhibitors for a molecular target in a pathogen can be a significant factor in toxicity to host tissues, and the ability to identify lead compounds that are specific for a target enzyme can be a key to therapeutic success. Gschwend *et al.* (1997) reported using DOCK together with structures of dihydrofolate reductases from humans and *Pneumocystis carinii* (a fungal opportunist that causes pneumonia in immunocompromised individuals) to screen an available chemicals database for selective inhibitors of the enzyme of the pathogen. Subsequent postdocking refinement, directed at discerning subtle structural and chemical features that might reflect species specificity, resulted in the successful identification of compounds (10 of 13) that bind preferentially to the fungal enzyme.

The development of computational methods to deal with the problems of ligand flexibility and to identify selective inhibitors of target enzymes will facilitate future efforts to discover novel inhibitors of enzymes for which 3-D structures have been determined. Furthermore, although target flexibility may contribute to potential inhibitors being overlooked even by flexible ligand docking algorithms, Knegtel *et al.* (1997) have tried to address this problem by docking small molecules to ensembles of DOCK surfaces generated from various 3-D structures of a target enzyme. However, if only a single structure is selected, a closed conformation that mimics the transition state of the target enzyme is likely to provide a more suitable surface for docking inhibitors (see Section V).

V. Estimating the Theoretical Limits for Reversible Inhibitor Binding

Most therapeutic agents coming from structure-based drug design are competitive inhibitors of specific enzyme targets. An early step in a structure-based drug design/discovery process is the identification of a compound (a lead) that is a competitive inhibitor with a low inhibition constant (K_i) with the target enzyme. A low K_i provides no guarantee that the compound will be nontoxic or yield favorable pharmacological properties, but the effectiveness of a "lead" compound as an inhibitor provides a performance standard against which modified versions of the lead can be compared. Subsequent modifications of lead compounds are usually intended to lessen toxicity, improve pharmacological properties, and enhance binding interactions with the target enzyme. In starting with a lead compound with a very low K_i with a target enzyme, chances are improved for eventually obtaining an effective drug.

The success of lead compounds should be considered in the context of the theoretical limits of favorable binding interactions with a target enzyme. Recently, Wolfenden (1999) has attempted to estimate the theoretical limits for binding affinities by comparing the rates of various enzyme-catalyzed reactions with those for the same noncatalyzed reactions (Table I). The idea is that the strong binding interactions associated with stabilization of the transition state are directly responsible for an enzyme's ability to lower the activation energy for a chemical reaction, and the strength of these interactions can be extrapolated from comparisons of the rates of the catalyzed and noncatalyzed reactions.

The last column of Table I shows the calculated values for the dissociation constants of substrates from each of the enzymes in the transition state. These values indicate that extremely high affinities of inhibitors for enzymes should be attainable, in principle, if perfect analogs of the altered substrate in the transition state can be devised (Wolfenden, 1999). Thus, if the delicate conformational requirements of

TABLE I

DISSOCIATION CONSTANTS (K^{tx}) OF SUBSTRATES FROM ENZYMES IN THE TRANSITION STATE[a]

Uncatalyzed reaction	$k_{25°C}$ sec^{-1}	k_{cat}/k_{non}	$K_{tx}M$
MP decarboxylation	3×10^{-6}	1.4×10^{17}	5×10^{-24}
hosphomonoester hydrolysis	5×10^{-14}	1×10^{15}	$<3 \times 10^{-20}$
Iandelate racemization	3×10^{-13}	1.7×10^{15}	2×10^{-19}
-Methyl glucoside hydrolysis	5×10^{-15}	$>10^{17}$	$>10^{-22}$
umarate hydration	3×10^{-14}	3.5×10^{15}	1×10^{-21}
hosphodiester hydrolysis	2×10^{-13}	1×10^{17}	$>10^{-22}$
eptide hydrolysis (average)	5×10^{-11}	1.3×10^{13}	3×10^{-18}
horismate mutation	3×10^{-5}	1.9×10^{6}	4×10^{-10}
arbonic anhydrase	2×10^{-3}	7.7×10^{6}	9.2×10^{-8}

Source: Reprinted from *Bioorganic and Medicinal Chemistry*, Volume 7, Wolfenden, R. Conrmational aspects of inhibitor design: Enzyme–substrate interactions in the transition state, ɔ. 647–652. Copyright © 1999, with permission from Elsevier Science.

[a]Rates for noncatalyzed reactions at 25°C were extrapolated from Arrhenius plots of rates dermined in neutral solution in sealed tubes at high temperature.

the powerful interactions involved in stabilizing the transition state can be satisfied, major improvements should be possible in the design of stable enzyme inhibitors. Even partial satisfaction of these requirements, such as the addition of a single positive charge that mimics the predicted transition state, can result in very significant changes in the binding affinities of an inhibitor for a target enzyme (Schramm *et al.*, 1994).

VI. Connectivity, Flexibility, and Conformational Entropy

Page and Jencks (1971) estimated that the binding of two substrates in their ground states could in principle lead to rate enhancements as large as 10^8 for enzyme catalyzed reactions. Similarly, Frick and Wolfenden (1989) predicted that the entropic advantage of combining the binding determinants of two molecules into one has the potential of enhancing their binding affinity for an enzyme by 8 orders of magnitude. This enormous decrease in the predicted K_i for an inhibitor of this type results from the entropic advantage associated with presenting the two binding determinants in a conformation that allows for simultaneous favorable interactions with the active site of the target enzyme. The advantage of connectivity has been demonstrated for inhibitors of adenosine deaminase (Fig. 7) where the covalent linkage of ligands with dissociation constants of 3.3×10^{-5} *M* and 0.25 *M* yields an inhibitor with a K_i of 2.5×10^{-12} *M*. In this case, the failure to achieve a full 8 orders of magnitude reduction in K_i by connecting the two ligands ($3.3 \times 10^{-5} \times 0.25 \times 10^{-8} = 8.2 \times 10^{-14}$ which is roughly 30-fold lower than the empirically determined value of 2.5×10^{-12}) was likely due to the fact that the bond between these two ligands is flexible. Eliminating this flexibility can be difficult, particularly in the absence of accurate 3-D structures of the target enzyme cocrystallized with the inhibitor as a guide because there is a risk that the two ligands will be locked in an orientation that prevents the simultaneous favorable interactions of both binding determinants with the binding domains of the enzyme. Nevertheless, a synthetic strategy might be possible for eliminating the remaining flexibility and further reducing the binding affinity for the target enzyme.

Bohacek *et al.* (1996) reviewed studies of a conformationally restrained cyclosporin analog (TCsA) that has a threefold lower binding affinity for cyclophilin and calcineurin than does Cyclosporin A (CsA). This lower binding affinity was believed to be due entirely to entropic effects. Bohacek *et al.* (1996) further speculated that the reason for the

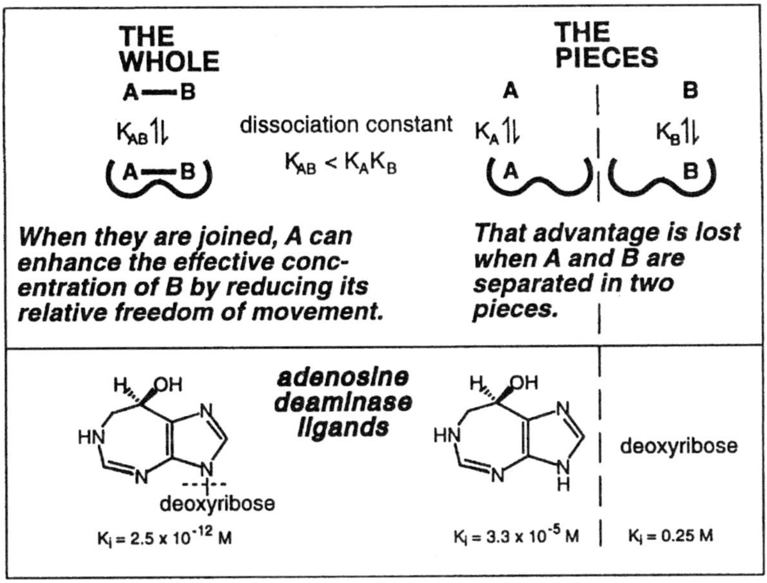

FIG. 7. The advantage of connectivity in ligand binding, as proposed by Page and Jencks (1971) and as measured for 2′deoxycoformycin, a transition state analog inhibitor of adenosine deaminase (Kati *et al.,* 1992). (Reprinted from *Bioorganic and Medicinal Chemistry,* Volume 7, Wolfenden, R. Conformational aspects of inhibitor design: Enzyme–substrate interactions in the transition state, pp. 647–652. Copyright © 1999, with permission from Elsevier Science.)

modest (three-fold) "increase in potency could be attributed to a hypothesis that the bound conformation of the more flexible Cyclosporin A is probably similar to its conformation in aqueous solution and that rigidifying a conformation already stabilized in solution may not gain much in binding free energy."

The role of conformational entropy was well-demonstrated in a study by Khan *et al.* (1998) using methods of X-ray crystallography to show that a flexible inhibitor of penicillopepsin binds to the target enzyme in the same conformation and with similar interactions as for a conformationally restrained analog of the inhibitor. Since the conformationally restrained analog possessed a 420-fold lower binding affinity for penicillopepsin, the authors were able to claim that the results provide the first rigorous measure of the extent to which reducing conformational flexibility in a ligand contributes to the binding energy in a protein–ligand complex.

The entropic advantage of combining the binding determinants of two molecules into one inhibitor will not be fully realized unless molecular flexibility is eliminated. However, if this is achieved, available data support the prediction by Frick and Wolfenden (1989) that binding energies will be enhanced by roughly 8 orders of magnitude if the binding determinants of two molecules are successfully combined in a single rigid molecule. An inescapable conclusion from this prediction and those of Section V is that, with rare exception, the majority of drugs in use today, with inhibition constants no better than the low nanomolar range, fall far short of the theoretical limits for the potential binding affinities of ligands for their molecular targets.

VII. Using Enzyme Structures in Dealing with Resistance Mutations

The use of drugs at concentrations that are less than 100% effective in killing the pathogen or affected cells provides a window of opportunity for mutations that can result in drug resistance. Analysis of the mechanisms of drug resistance is a large and active field of research. Molecular procedures [i.e., first cycle reverse transcription–polymerase chain reactions (RT-PCR)] are well established for cloning DNA encoding target enzymes from cells that have developed resistance to chemotherapeutic agents, and it is possible to rapidly determine whether mutations responsible for resistance to drugs are within the target enzyme. Also, if the 3-D structure of a target enzyme has been determined, it is possible to analyze the molecular basis for the development of resistance for those mutations that are in the target enzyme by modeling the affected amino acid into the 3-D structure of the target enzyme (i.e., Foote *et al.,* 1990; Peterson *et al.,* 1990). After the mechanisms of resistance have been determined, it should be possible to design or discover inhibitors that will bind favorably to mutant forms of the target enzyme that are resistant to binding the original drug.

Drug preparations composed of more than one inhibitor of a target enzyme have the potential to decrease the probability of the development of resistance to one or the other of the inhibitors. The development of resistance to the inhibitor of the HIV-reverse transcriptase (3TC) was discovered to slow the development of resistance to AZT (Larder *et al.,* 1995; Nijhuis *et al.,* 1997). Also, AZT-resistant virus, in which resistance to 3TC is selected for, can regain phenotypic sensitivity to AZT (Eron, 1996). Amino acids at position 184 of the HIV-reverse transcriptase have been identified as being responsible for the 3TC/AZT resistance mechanism. The trick from a drug design perspective would be in predicting the synergistic properties of the two inhibitors in advance of the serendipitous discovery that 3TC could reduce the probability of the development of resistance to AZT. But the evidence is clear: It should be possible to design a mixture or cocktail of compounds that would be effective against resistant and wild-type forms of the target enzyme. In fact, this is the basis of contemporary approaches to the therapy of HIV infections (De Clerk, 1998; also see Balzarini *et al.*, 1998).

Structure-based inhibitor design need not wait for the appearance of resistance mutations in natural populations. Resistance mutations can be generated by culturing pathogens in the presence of sublethal concentrations of inhibitors. With adequately large populations of cells, those that possess mutations conferring resistance to binding an inhibitor will eventually appear and overtake the cells that are not resistant to the inhibitor. This approach may not even require the *in vitro* culture of the pathogen. Klein and Geary (1997) and Canyuk *et al.* (1998) have used bacteria complemented by the activity of recombinant enzymes to test enzyme–ligand interactions and/or to screen for inhibitors of recombinant enzymes. Canyuk *et al.* (1998) suggested that this type of recombinant selection system could be set up for many different metabolic target enzymes. Taken one step further, recombinant screening systems could be adapted to select for resistance mutations in the target enzymes via maintaining the recombinant microbes under the pressure of inhibitors that severely impede the growth of the bacterial cells. Obviously, some resistance mutations may occur that are not within the target enzyme, but subsequent sequencing of the DNA encoding the target enzyme from the cloned microbes that have developed resistance to the inhibitor would reveal the identity of those amino acid substitutions that are in the target enzyme. Thus, it should be possible to generate and analyze resistance mutations in target enzymes before drugs that inhibit the enzymes are tested in natural populations. If the mechanisms of resistance conferred by these mutations can be determined, then it should be possible to design modified inhibitors to compensate for the resistance mutations. A cocktail of inhibitors designed to bind to the resistant, as well as the wild-type form, of a target enzyme could constitute a drug preparation that potentially could be used in large populations with a reduced risk of the future development of drug resistance.

VIII. Conclusions

Although the 3-D structures of target enzymes are presently being used to guide the structure-based design of drugs, technology in this field is still being developed. However, future directions and applications for structure-based approaches to drug discovery can be envisioned. Reliance on detailed structural information is likely to greatly increase the overall efficiency of the drug discovery process and provide a means to deal with the occurrence of resistance mutations in target enzymes. Thus, the future is likely to see the availability of drug preparations that are a mixture of compounds that are effective, as well as nontoxic chemotherapeutic agents, to which the development of resistance is unlikely.

REFERENCES

Allen, K. N. Bellamacina, C. R., Ding, X., Jeffery, C. J., Mattos, C., Petsko, G. A., and Ringe, D. (1996). An experimental approach to mapping the binding surfaces of crystalline proteins. *J. Phys. Chem. BB100,* 2605–2611.

Amzel, L. M. (1998). Structure-based drug design. *Curr. Opin. Biotechnol.* **9,** 366–369.

Balzarini, J., Pelemans, H., Riess, G., Roesner, M., Winkler, I., De Clercq, E., and Kleim, J. P. (1998). Retention of marked sensitivity to (S)-4-isopropoxycarbonyl-6-methoxy-3-(methylthiomethyl)-3,4-dihydroquin oxaline-2 (1H)-thione (HBY 097) by an azidothymidine (AZT)-resistant human immunodeficiency virus type 1 (HIV-1) strain subcultured in the combined presence of quinoxaline HBY 097 and 2',3'-dideoxy-3'-thiacytidine (lamivudine). *Biochem. Pharmacol.* **55,** 617–625.

Bohacek, R. S., McMartin, C., and Guida, W. C. (1996). The art and practice of structure-based drug design: A molecular modeling perspective. *Med. Res. Rev.* **16,** 3–50.

Bugg, C. E., Carson, W. M., and Montgomery, J. A. (1993). Drugs by design: Structure-based design, an innovative approach to developing drugs, has recently spawned many promising therapeutic agents, including several now in human trials for treating AIDS, cancer and other diseases. *Scientific Amer.* **269,** 92–98.

Canyuk, B., Craig, S. P., and Eakin, A. E. (1998). Bacterial complementation as a means to test enzyme-ligand interactions. *Appl. Microbiol. Biotechnol.* **50,** 181–186.

Chen, J. M., Xu, S. L., Wawrzak, Z., Basarab, G. S., and Jordan, D. B. (1998). Structure-based design of potent inhibitors of scytalone dehydratase: Displacement of a water molecule from the active site. *Biochemistry* **37,** 17735–17744.

Connolly, M. L. (1983). Analytical molecular surface calculations. *J. Appl. Crystallogr.* **16,** 548–558.

De Clerk, E. (1998). New perspectives for the treatment of HIV infections. *Verh.–K. Acad. Geneeskd. Belg.* **60,** 13–41.

Eron, J. J., Jr. (1996). The treatment of antiretroviral-naive subjects with the 3TC/zidovudine combination: A review of North American (NUCA3001) and European (NUCB) 3001) trials. *AIDS, Suppl.* **5,** S11–S19.

Fischer, E. (1894). Einfluss der configuration auf die wirkung der enzyme. *Chem. Ber.* **27,** 2985–2993.

Focia, P. J., Craig, S. P., III, Nieves-Alicea, R., Fletterick, R. J., and Eakin, A. E. (1998a). A 1.4 Å crystal structure for the hypoxanthine phosphoribosyltransferase of *Trypanosoma cruzi. Biochemistry* **37,** 15066–15075.

Focia, P. J., Craig, S. P., III, and Eakin, A. E. (1998b). Approaching the transition state in the crystal structure of a phosphoribosyltransferase. *Biochemistry* **37,** 17120–17127.

Foote, S. J., Galatis, D., and Cowman, A. F. (1990). Amino acids in the dihydrofolate reductase-thymidylate synthase gene of *Plasmodium falciparum* involved in cycloguanil resistance differ from those involved in pyrimethamine resistance. *Proc. Natl. Acad. Sci. U.S.A.* **87,** 3014–3017.

Frick, L., and Wolfenden, R. (1989). General approaches to the design of inhibitors. *In* "Design of Enzyme Inhibitors as Drugs" (M. Dandler and H. J. Smith, eds.), pp. 19–48. Oxford University Press, Oxford.

Gait, M. J., and Karn, J. (1995). Progress in anti-HIV structure-based drug design. *Trends. Biotechnol.* **13,** 430–438.

Goodsell, D. S., Morris, G. M., and Olson, A. J. (1996). Automated docking of flexible ligands: Applications of AutoDock. *J. Mol. Recognition* **9,** 1–5.

Gschwend, D. A., Sirawaraporn, W., Santi, D. V., and Kuntz, I. D. (1997). Specificity in structure-based drug design: Identification of a novel, selective inhibitor of *Pneumocystis carinii* dihydrofolate reductase. *Proteins: Struct., Funct., Genet.* **29,** 59–67.

Hendrickson, W. A., and Ogata, C. (1997). Phase determination by multiwavelength anomalous diffraction measurements. *In* "Methods in Enzymology" (C. W. Carter and R. M. Sweet, eds.), Vol. 276, pp. 494–523. Academic Press, San Diego, CA.

Jackson, R. C. (1997). Contributions of protein structure-based drug design to cancer chemotherapy. *Semin. Oncol.* **24,** 164–172.

Kati, W. M., Acheson, S. A., and Wolfenden, R. (1992). A transition state in pieces: Major contributions of entropic effects to ligand binding by adenosine deaminase. *Biochemistry* **31,** 7356–7366.

Khan, A. R., Parrish, J. C., Fraser, M. E., Smith, W. W., Bartlett, P. A., and James, M. N. G. (1998). Lowering the entropic barrier for binding conformationally flexible inhibitors to enzymes. *Biochemistry* **37,** 16839–16845.

Kick, E. K., Roe, D. C., Skillman, A. G., Liu, G., Ewing, T. J., Sun, Y. Kuntz, I. D., and Ellman, J. A. (1997). Structure-based design and combinatorial chemistry yield low nanomolar inhibitors of cathepsin D. *Chem. Biol.* **4,** 297–307.

Klein, R. D., and Geary, T. G. (1997). Recombinant microorganisms as tools for high throughput screening for nonantibiotic compounds. *J. Biomol. Screen.* **2,** 41–49.

Kleinberg, M. L., and Wanke, L. A. (1995). New approaches and technologies in drug design and discovery. *Am. J. Health Syst. Pharm.* **52,** 1323–1336.

Knegtel, R. M. A., Kuntz, I. D., and Oshiro, C. M. (1997). Molecular docking to ensembles of protein structures. *J. Mol. Biol.* **266,** 424–440.

Kuhn, P. Knapp, M., Soltis, S. M., Ganshaw, G., Thoene, M., and Bott, R. (1998). The 0.78 Å structure of a serine protease: *Bacillus lentus* subtilisin. *Biochemistry* **37,** 13446–13452.

Kuntz, I. D., Blaney, J. M., Oatley, S. J., Langridge, R., and Ferrin, T. E. (1982). A geometric approach to macromolecule-ligand interactions. *J. Mol. Biol.* **161,** 269–288.

Larder, B. A., Kemp, S. D., and Harrigan, P. R. (1995). Potential mechanism for sustained antiretroviral efficacy of AZT-3TC combination therapy. *Science* **269,** 696–699.

Li, R., Chen, X., Gong, B., Selzer, P. M., Li, Z., Davidson, E., Kurzban, G., Miller, R. E., Nuzum, E. O., McKerrow, J. H., Fletterick, R. J., Gillmor, S. A., Craik, C. S., Kuntz, I. D., Cohen, F. E., and Kenyon, G. L. (1996). Structure-based design of parasitic protease inhibitors. *Bioorg. Med. Chem.* **4,** 1421–1427.

Lorber, D. M., and Shoichet, B. K. (1998). Flexible ligand docking using conformational ensembles. *Protein Sci.* **7,** 938–950.

Marrone, T. J., Briggs, J. M., and McCammon, J. A. (1997). Structure-based drug design: computational advances. *Annual Rev. Pharmacol. Toxicol.* **37,** 71–90.

McMartin, C., and Bohacek, R. S. (1997). QXP: Powerful, rapid computer algorithms for structure-based drug design. *J. Comput.-Aided Mol. Des.* **11,** 333–344.

Murray, C. W., Clark, D. E., Auton, T. R., Firth, M. A., Li, J., Sykes, R. A., Waszkowycz,

B., Westhead, D. R., and Young, S. C. (1997). PRO:SELECT: Combining structure-based drug design and combinatorial chemistry for rapid lead discovery. *J. Comput.-Aided Mol. Des.* **11,** 193–207.

Nagy, P. (1988) Orientation and structure-guilding role of the water molecules bound at the contact surface of the dihydrofolate reductase-methotrexate complex. *J. Comput.-Aided Mol. Des.* **2,** 65–76.

Nieves-Alicea, R., Focia, P. J., Craig, S. P., III, and Eakin, A. E. (1998). Limited proteolysis of a trypanosomal phosphoribosyltransferase yields crystals that diffract X-rays to near atomic resolution. *Biochim. Biophys. Acta* **1388,** 500–505.

Nijhuis, M., Schuurman, R., de Jong, D., van Leeuwen, R., Lange, J., Danner, S., Keulen, W., de Groot, T., and Boucher, C. A. (1997). Lamivudine-resistant human immunodeficiency virus type 1 variants (184V) require multiple amino acid changes to become co-resistant to zidovudine *in vivo*. *J. Infect. Dis.* **176,** 398–405.

Page, M. I., and Jencks, W. P. (1971). Entropic contributions to rate accelerations in enzymic and intramolecular reactions and the chelate effect. *Proc. Natl. Acad. Sci. U.S.A.* **68,** 1678–1683.

Peisach, E., Casebier, D., Gallion, S. L., Furth, P., Petsko, G. A., Hogan, J. C., and Ringe, D. (1995). Interaction of a peptidomimetic aminimide inhibitor with elastase. *Science* **269,** 66–69.

Peterson, D. S., Milhous, W. K., and Wellems, T. E. (1990). Molecular basis of differential resistance to cycloguanil and pyrimithamine in *Plasmodium falciparum* malaria. *Proc. Natl. Acad. Sci. U.S.A.* **87,** 3018–3022.

Price, C. R., and Nagai, K. (1995). Protein engineering as a tool for crystallography. *Curr. Opin Biotechnol.* **6,** 425–430.

Ringe, D. (1995). What makes a binding site a binding site? *Curr. Opin. in Struct. Biol.* **5,** 825–829.

Salemme, F. R., Spurlino, J., and Bone, R. (1997). Serendipity meets precision: The integration of structure-based drug design and combinatorial chemistry for efficient drug discovery. *Structure* **5,** 319–324.

Schramm, V. L., Horenstein, B. A., and Kline, P. C. (1994). Transition state analysis and inhibitor design for enzymatic reactions. *J. Biol. Chem.* **269,** 18259–18262.

Shoichet, B. K., and Kuntz, I. D. (1993). Matching chemistry and shape in molecular docking. *Protein Eng.* **6,** 723–732.

Shoichet, B. K., Stroud, R. M., Santi, D. V., Kuntz, I. D., and Perry, K. M. (1993). Structure-based discovery of inhibitors of thymidylate synthase. *Science* **259,** 1445–1450.

Skulnick, H. I., Johnson, P. D., Aristoff, P. A., Morris, J. K., Lovasz, K. D., Howe, W. J., Watenpaugh, K. D., Janakiraman, M. N., Anderson, D. J., Reischer, R. J., Schwartz, T. M., Banitt, L. S., Tomich, P. K., Lynn, J. C., Horng, M. M., Chong, K. T., Hinshaw, R. R., Dolak, L. A., Seest, E. P., Schwende, F. J., Rush, B. D., Howard, G. M., Toth, L. M., Wilkinson, K. R., Kakuk, T. J., Johnson, C. W., Cole, S. L., Zaya, R. M., Zipp, G. L., Possert, P. L., Dalga, R. J., Zhong, W.-Z., Williams, M. G., and Romines, K. R. (1997). Structure-based design of nonpeptidic HIV protease inhibitors: the sulfonamide-substituted cyclooctylopyramones. *J. Med. Chem.* **40,** 1149–1164.

Strynadka, N. C. J., Eisenstein, M., Katchalski-Katzir, E., Shoichet, B. K., Kuntz, I. D., Abazyan, R., Totrov, M., Janin, J., Cherfils, J., Zimmerman, R., Olson, A., Duncan, B., Rao, M., Jackson, R., Sternberg, M., and James, M. N. G. (1996a). Molecular docking programs successfully predict the binding of a beta-lactamase inhibitory protein to tem-1 beta-lactamase. *Nat. Struct. Biol.* **3,** 233–239.

Strynadka, N. C. J., Jensen, S. E., Alzari, P. M., and James, M. N. G. (1996b). A potent new mode of β-lactamase inhibition revealed by the 1.7 Å X-ray crystallographic structure of the TEM-1-β-lactamase complex. *Nat. Struct. Biol.* **3,** 290–297.

Tickle, I. J., and Driessen, H. P. (1996). Molecular replacement using known structural information. *Methods Mol. Biol.* **56,** 173–203.

Turkenburg, J. P., and Dodson, E. J. (1996). Modern developments in molecular replacement. *Curr. Opin. Struct. Biol.* **6,** 604–610.

Verlinde, C. L. M. J., and Hol, W. G. J. (1994). Structure-based drug design: Progress, results and challenges. *Structure* **2,** 577–587.

Wade, R. C. (1997). 'Flu' and structure-based drug design. *Structure* **5,** 1139–1145.

Wasserman, Z. R., and Hodge, C. N. (1996). Fitting an inhibitor into the active site of thermolysin: A molecular dynamics study. *Proteins* **24,** 227–237.

Whittaker, M. (1998). Discovery of protease inhibitors using targeted libraries. *Curr. Opin. Chem. Biol.* **2,** 386–396.

Wlodawer, A., and Vondrasek, J. (1998). Inhibitors of HIV-1 protease: A major success of structure-assisted drug design. *Annual Rev. Biophys. Biomol. Struct.* **27,** 249–284.

Wolfenden, R. (1999). Conformational aspects of inhibitor design: Enzyme-substrate interactions in the transition state. *Bioorg. Med. Chem.* **7,** 647–652.

VITAMINS AND HORMONES, VOL. 58

Protein Folding Using Contact Maps

MICHELE VENDRUSCOLO AND EYTAN DOMANY

Department of Physics of Complex Systems, Weizmann Institute of Science, Rehovot 76100, Israel

---

We discuss the problem of representations of protein structure and give the definition of contact maps. We present a method to obtain a three-dimensional polypeptide conformation from a contact map. We also explain how to deal with the case of nonphysical contact maps. We describe a stochastic method to perform dynamics in contact map space. We explain how the motion is restricted to physical regions of the space. First, we introduce the exact free energy of a contact map and discuss two simple approximations to it. Second, we present a method to derive energy parameters based on perceptron learning. We prove in an extensive number of situations that the pairwise contact approximation both when alone and when supplemented with a hydrophobic term is unsuitable for stabilizing proteins' native states.

0083-6729/00 $30.00

I. Introduction

Computational approaches to protein folding are divided into two main categories. In *energy minimization* methods the native state is identified with the ground state of a suitable energy function (Brooks, 1998). In *fold recognition* methods, the native state is selected as the most compatible structure among those present in a library (Bowie *et al.,* 1992; Jones *et al.,* 1992; Fisher *et al.,* 1996). Both approaches depend on three important choices: (1) the representation of protein structure, (2) the set of alternative structures among which the native fold is sought for, and (3) a bias toward the "best" conformation. In energy minimization approaches such a bias is an energy function, and the best conformation is the one of lowest energy. On the other hand, in fold recognition methods, a compatibility function for a sequence on a structure is used. The compatibility is often expressed in terms of database-derived properties and restraints.

In this review we analyze the attempt to use contact maps to perform protein fold prediction efficiently. Contact maps are a particularly manageable representation of protein structure which has been already applied in the past to the study of protein conformation (Chan and Dill, 1990), structure comparison (Holm and Sander, 1939), interaction patterns in proteins (Lifson and Sander, 1979; Godzik *et al.,* 1993), and correlated mutations (Olmea and Valencia, 1997; Ortiz *et al.,* 1998). The possibility of performing energy minimization in the space of contact maps has been proposed by Mirny and Domany (1996). We present here the consistent development of their idea, discussing successes, failures, and perspectives.

II. Structure Representation

Following Anfinsen's thermodynamic hypothesis (Anfinsen, 1973), the native state of a protein is commonly assumed to be the minimum of a free-energy function. This is a powerful assumption and molecular dynamics is the most direct method to implement it (Brooks, 1998). The structure is represented by listing the coordinates of all the atoms and Newton's equations of motion are solved in a suitable force field tuned for molecular systems. Unfortunately, present computers cannot follow the trajectory of a protein all the way down to its native state. The best result to date is the simulation of the 1-μs trajectory of villin headpiece subdomain in water, which allowed the detection of the hydrophobic collapse and of the formation of secondary structures (Duan and Kollman,

1998). Furthermore, the method depends crucially on the determination of suitable energy parameters. An incorrect energy function, which does not assign the lowest energy to the native conformation, leads a careful energy minimization procedure to some misfolded conformation, as for example shown by Karplus and collaborators in the case of hemerythrin and the immunoglobulin VL domain (Novotny *et al.*, 1984).

We believe that perhaps one can solve the problem without going to such microscopic detail and we set about to investigate this assumption. The problem of structure representation is to find the best trade-off between computability and accuracy of the predictions. Our inclination is that good predictions can be obtained by constructing simplified models.

Lattice models offer the possibility to gradually turn on the complexity of the representation of the structure. Usually a protein is represented as a chain of monomers occupying lattice sites and representing C_α atoms. The complexity can be measured by the number of states available to each monomer (Park and Levitt, 1995. The lowest possible complexity is that of tetrahedral and simple cubic lattices, where 3 and 5 states per monomer, respectively, are available. Lattices of high coordination number, up to 55 states, have been studied (Ortiz *et al.*, 1998). The main motivation to study lattice models is that at low complexity it is possible to effectively solve the problem of searching the ground state, either exactly, by enumerating all the conformations, or approximately, by Monte Carlo methods. By using Monte Carlo simulations, solutions can be routinely obtained for polymers up to length 125 (Šali *et al.*, 1994; Dinner *et al.*, 1996). Using pairwise contact energy functions, evidence for important features of the folding process has been produced, most notably, the hydrophobic collapse. It is possible to consider more detailed lattice models which still retain some of the advantage in computability and allow a more realistic representation of protein structure. For example, it is possible to represent side chains by an additional virtual atom. Advances in this direction have been recently reported by Skolnick and collaborators (Ortiz *et al.*, 1998). Interestingly, the accuracy increases very slowly with the complexity, and the typical resolution of structures in the Protein Data Bank (PDB) (Bernstein *et al.*, 1977) can be obtained with models with 10 to 20 states (Park and Levitt, 1995). The C_α model can be simulated also off-lattice. For short chains and simple interactions it is possible to identify the ground state with reasonable reliability (Clementi *et al.*, 1998; Irbäck and Potthast, 1995).

A minimalistic representation of a protein's structure is given by its *contact map* (Lifson and Sander, 1979; Chan and Dill, 1990; Godzik *et al.,* 1993; Holm and Sander, 1993; Mirny and Domany, 1996; Vendruscolo *et al.,* 1997). The contact map of a protein with N residues is an $N \times N$ matrix **S,** whose elements are defined as

$$S_{ij} = \begin{cases} 1 & \text{if residues } i \text{ and } j \text{ are in contact} \\ 0 & \text{otherwise} \end{cases}. \tag{1}$$

One can define contact between two residues in different ways; one is to consider two amino acids in contact when their two C_α atoms are closer than some threshold R_c ("C_α" definition) (Vendruscolo *et al.,* 1997). Another definition is based on the minimal distance between two atoms that belong to the two residues (Hinds and Levitt, 1994; Mirny and Domany, 1996) ("all-atom" definition). In a contact map α-helices appear as thick bands of contacts along the diagonal, β-sheets as bands running parallel or perpendicular to the diagonal. Given all the inter-residue contacts or even a subset of them, it is possible to reconstruct quite well a protein's structure, by means of either distance geometry (Crippen and Havel, 1988), molecular dynamics (Brünger *et al.,* 1986), or Monte Carlo (Vendruscolo *et al.,* 1997) methods.

In contrast to Cartesian coordinates, the map representation of protein structure is independent of the coordinate frame. This property made contact maps attractive for protein structure comparisons and for *searching a limited database* for similar structures (Chan and Dill, 1990; Godzik *et al.,* 1993; Holm and Sander, 1993).

One of the main reasons for selecting the contact map representation of structure is our expectation and hope that we may be able to search the space of contact maps in an efficient manner and find low-energy maps (that hopefully correspond to conformations close to the native one). In particular, one hopes that relatively simple changes on a map may generate very substantial coherent moves of the corresponding polypeptide chain conformation, moves that would have taken much longer to achieve by working with the chain itself.

To actually implement such moves in the space of contact maps, one has to overcome two important problems. The main and foremost one is the need to ensure that the map **S** that has been generated is *physical.* Our definition of "physical" will be given in detail later; broadly speaking, we mean that there exists a chain conformation whose contact map is indeed our proposed **S.** Arbitrary changes performed on a map yield, with very high probability, nonphysical maps. The reason is that the total number of possible $N \times N$ contact maps is $O(2^{aN^2})$, whereas the number of physical maps is much smaller, of order $O(2^{bN})$. The need for a procedure that limits the search to the small subspace of physical maps was identified by Mirny and Domany (1996), who proposed a restricted set of moves. The hope was that when moves selected from this set are performed on a physical map, the new map is also physical. These rules were, however, heuristic and no clear proof of the validity of this hope could be given.

Subsequently, the problem of generating physical maps was dealt with by means of a *reconstruction procedure* (Vendruscolo *et al.,* 1997). For any proposed contact map **S** one generates a chain, with **S** serving as the guide of the construction process. The procedure stops when the contact map **S**′ of the resulting chain is close to the target map; this way one generates a map which is physical *and* close to the target.

The second important issue is the manner in which low-energy maps are generated from an existing one (Vendruscolo and Domany, 1998a). The procedure has to be such that the resulting map is "protein-like," that is, has secondary structure elements and the corresponding chain has the right density, bond-angle distribution, chirality, etc. The final map should be physical and the decorrelation time with the starting map should be short.

In what follows we sketch how these two problems were addressed; for a more detailed description of these procedures the reader is referred to the original publications (Vendruscolo *et al.,* 1997; Vendruscolo and Domany, 1998a).

III. The Reconstruction Procedure

The aim is to find an efficient procedure, which can be performed "on line" and in parallel with the dynamics in the space of contact maps, which will "project" any map onto a nearby one that is guaranteed to be physically realizable. The protein is represented as a "string of beads," in which each bead stands for an amino acid—the coordinates of the center of a bead are identified with those of the corresponding C_α atom. For a given target contact map **S,** the algorithm searches for a conformation that this string of beads can take, such that the contact map **S**′ of our string is familiar (or close) to **S.** If there exists a chain conformation whose contact map is identical with **S,** this contact map is, by definition, physical. In general, our method aims at converting a possibly ill-defined, nonphysical set of contacts to a legitimate one. The three-dimensional structure is in our case a means, rather than the end.

Note that related, previously developed methods (Havel *et al,.* 1979; Brünger *et al.,* 1986; Bohr *et al,.* 1993; Nilges, 1995; Lund *et al.,* 1996; Aszodi and Taylor, 1996; Mumenthaler and Braun, 1996; Skolnick *et al.*, 1997), had a different aim; to construct three-dimensional structures from measured distance information, using various forms of distance geometry (Crippen and Havel, 1988), supplemented by restricted molecular dynamics (Scheek *et al.*, 1989) or simulated annealing (Brünger *et al.*, 1997).

One should emphasize here the distinction between a contact map and a distance map. In a contact map a minimal amount of information is available—given a pair of amino acids, we know only if they are in contact or not. That is, only lower and upper bounds on their separation are given. A distance matrix, on the other hand, presents real-valued distances between pairs of amino acids. The method presented here is not restricted to contact maps and has been generalized to distance maps (M. Vendruscolo and E. Domany, unpublished). The deviations between different structures that were reconstructed from the same contact map are typically much higher than those between two structures derived from a distance matrix.

The proposed algorithm is divided into two parts. The first part, *growth,* consists of adding one monomer at a time, that is, a step-by-step growth of the chain. The second part, *adaptation,* is a refinement of the structure, obtained as a result of the growth stage, by local moves. In both stages, to bias the dynamics, we introduce cost functions defined on the basis of the contact map. Such cost functions contain only geometric constraints, and do not resemble the true energetics of the polypeptide chain.

A. Growth

The first element of the growth is *single monomer addition,* which is carried out in the spirit of the Rosenbluth method (Rosenbluth and Rosenbluth, 1955). To add monomer i to the growing chain we generate at random N_t trial positions (typically $N_t = 10$):

$$\mathbf{r}_i^{(j)} = \mathbf{r}_{i-1} + \mathbf{r}^{(j)}, \tag{2}$$

where $j = 1, \ldots, N_t$. The length and the directions of $r^{(j)}$ are set from a statistical analysis of PDB. One out of the N_t trials is chosen according to the probability

$$p^{(j)} = \frac{e^{-E_g^{(j)}/T_g}}{\sum_{j=1}^{N_t} e^{-E_g^{(j)}/T_g}}, \tag{3}$$

where E_g is a cost function that rewards contacts that should be present, according to the given contact map, and discourages contacts that should not be there.

The second element of the procedure is *chain growth.* The step-by-step growth just discussed optimizes the position of successive amino acids along the sequence. The main difficulty in the method is that the single step of the growing chain has no information on the contacts that should be realized many steps (or monomers) ahead. To solve this problem, we carry out several attempts (typically 10) to reconstruct the structure, choosing the best one. In practice, this is done as follows.

For each attempt, when position $\mathbf{r}^{[j(i)]}$ is chosen for monomer i according to Eq. (3), its probability is accumulated in the weight

$$W_i = \prod_{k=1}^{i} p_k^{[j(k)]}. \tag{4}$$

When we have reached the end of the chain we store the weight W_N. The trial chain with the highest W_N is chosen.

The *cost function* for growth is

$$E_{\mathrm{g}}^{(j)} = \sum_{k=1}^{i-1} d \cdot a_{\mathrm{g}}(S_{ik}) \cdot \vartheta(d_{\mathrm{t}} - r_{ik}^{(j)}), \tag{5}$$

where $r_{ik}^{(j)} = |\mathbf{r}_i^{(j)} - \mathbf{r}_k|$. The enhancing factor $d = i - k$ is introduced to guide the growth toward contacts that are long ranged along the chain; ϑ is the Heaviside step function; and the constant a_{g} can take two values; $a_{\mathrm{g}}(S_{ik} = 0) \geq 0$ and $a_{\mathrm{g}}(S_{ik} = 1) \leq 0$. That is, when a contact is identified in the chain, that is, $r_{ik} < d_{\mathrm{t}}$, it is either rewarded (when the target map has a contact between i and k) or penalized.

B. Adaptation

When we have grown the entire chain of N points, we refine the structure according to the following scheme. We choose a point i at random and try, using a crankshaft move (Šali *et al.,* 1994), to displace it to $\mathbf{r}'_i$, keeping fixed the distances from both points $i - 1$ and $i + 1$. We use a local cost function $E_{\mathrm{a}}^{(i)}$:

$$E_{\mathrm{a}}^{(i)} = \sum_{k=1}^{N} f_{\mathrm{a}}(r'_{ik}) = a_{\mathrm{a}}(S_{ik}) \cdot \vartheta(d_{\mathrm{t}} - r'_{ik}), \tag{6}$$

where $r'_{ik} = [\mathbf{r}'_i - \mathbf{r}_k]$. Note that the enhancing factor d has been omitted, so that f_{a} does not favor contacts between monomers that are distant along the chain. The displacement is accepted with probability π, according to the standard Metropolis prescription:

$$\pi = \min(1, \exp(-\Delta E_a / T_a), \tag{7}$$

where ΔE_a is the change int he cost function E_a induced by the move, and T_a is a temperature-like parameter, used to control the acceptance ratio of the adaptation scheme.

A key ingredient of our method is annealing (Kirkpatrick *et al.*, 1983). As in all annealing procedures, the temperature-like parameter T_a is decreased slowly during the simulation to help the system find the ground state in a rugged energy landscape. In our method, however, instead of using simulation time as a control parameter on the temperature, we chose the number n of missing contacts. Two regimes were roughly distinguished. In the first regime many contacts are missed and the map is very different from the target one. In the second regime few contacts are missed, and the map is close to the target. The parameters a_a and T_a are interpolated smoothly between values suitable for these two limiting cases. In the first regime, we strongly favor the recovery of contacts that should be realized, whereas in the second regime we strongly disfavor contacts that are realized but should not be present. We set, as shown in Fig. 1,

$$a_a^{(n)}(S) = a^f(S) + [a^i(S) - a^f(S)]\sigma(n), \tag{8}$$

and

$$T_a^{(n)} = T_a^f + (T_a^i - T_a^f)\sigma(n). \tag{9}$$

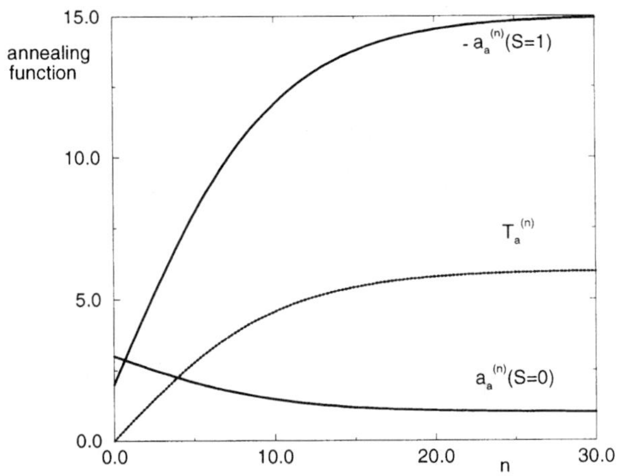

FIG. 1. Annealing functions for the parameters used in adaptation. See Eqs. (8) and (9).

The function $\sigma(n)$ interpolates between the initial value a^{i} and the final value a^{f},

$$\sigma(n) = \frac{2}{1 + e^{-\alpha_g n}} - 1. \tag{10}$$

By choosing a^{i}, a^{f}, $T^{\mathrm{i}}_{\mathrm{a}}$, $T^{\mathrm{f}}_{\mathrm{a}}$, and α_{g} we define the two regimes, far from and close to the target map.

We have tried several alternatives to each of the components of the method outlined. For a detailed description of these, we refer the reader to Vendruscolo *et al.* (1997). We present here a brief description of some selected results that were obtained using the algorithm presented.

A. Results

1. *Experimental Contact Maps*

In this section we present results about the reconstruction of experimental contact maps as taken from PDB. Since our purpose, as explained in the introduction, is to use the reconstruction in connection with dynamics, we chose the contact length $R_c = 9$ Å to obtain the most faithful representation of the energy of the protein (Mirny and Domany, 1996). Such a threshold is determined by the requirement that the average number of $C_\alpha - C_\alpha$ contacts for each amino acid is roughly equal to the respective numbers obtained with the all-atom definition of contacts.

The most commonly used dissimilarity measure between structures is the root mean square (RMS) distance D (McLachlan, 1979):

$$D = \sqrt{\frac{1}{N}\sum_{i=1}^{N}(\mathbf{r}_i - \mathbf{r}_i')^2}, \tag{11}$$

where one structure is translated and rotated to get a minimal D.

The dissimilarity measure between contact maps is defined as the Hamming distance:

$$D^{\mathrm{map}} = \sum_{j>i} |S_{ij} - S'_{ij}|, \tag{12}$$

which counts the number of mismatches between maps $\mathbf{S}$ and $\mathbf{S}'$.

We present in Fig. 2a the results over 100 reconstruction runs for chains for the protein 6pti (bovine pancreatic trypsin inhibitor).

In Fig. 2b we show the contact map for the protein 6pti, $N = 56$, as taken from PDB, that was used as a target to construct a chain. The

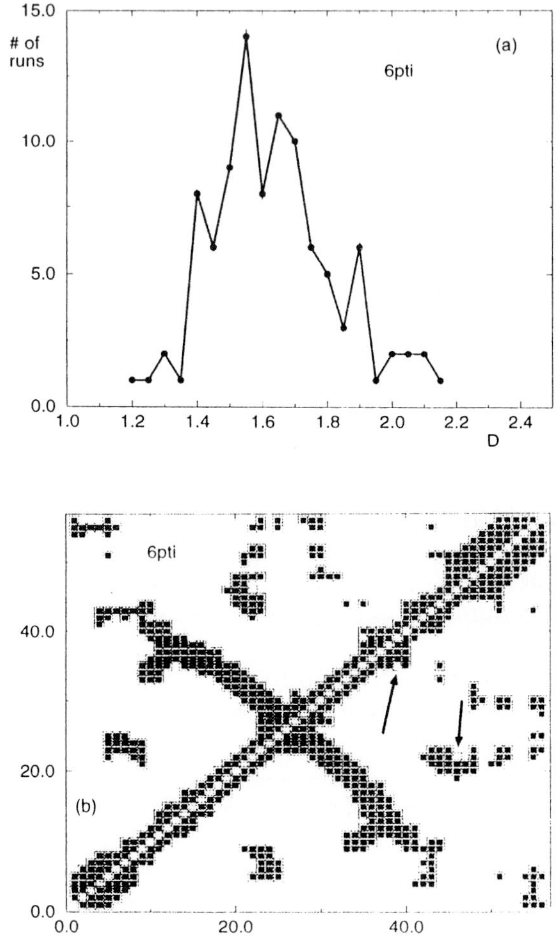

FIG. 2. (a) Histogram of the distance D for the 100 runs used to test the reconstruction procedure. Data are presented for protein 6pti. (b) Contact maps for protein 6pti for a threshold $d_t = 9$ Å. Full squares are the PDB data, open squares the output of the reconstruction procedure. None of the target contacts is missed and two spurious ones are added (the arrows point at their locations). [Adapted with permission from Vendruscolo, M., Kussell, E., and Domany, E. (1997). Recovery of protein structure from contact maps. *Folding & Design,* **2**(5), 295–306. Copyright © 1997 Current Biology.]

contact map of a typical reconstructed chain is also shown. In this particular case none of the 342 original contacts was missed and only two false-positive contacts were added. These are close to clusters of correct contacts, indicating slight local differences with the crystallographic structure. The distance recorded in this case was $D = 1.56$.

We carried out extensive similar tests for other proteins of various lengths (Vendruscolo *et al*,. 1997). Our method produces, using a native contact map as target, a structure whose contact map is in nearly perfect agreement with the target. Furthermore, the distance of this reconstructed chain from the native structure is quite close to the resolution that can be obtained from the information contained in contact maps.

2. *Nonphysical Contact Maps*

Our main purpose is to develop a strategy to construct a three-dimensional structure, starting from a given set of contacts, even if these contacts are not physical, that is, not compatible with any conformation allowed by a chain's topology. In such a case we require our procedure to yield a chain whose conformation is as "close" as possible to the contact map we started with. The exact measure of such closeness depends on the source of nonphysicality, as will be demonstrated in the examples that follow.

Our first examples of nonphysical contact maps were obtained by randomizing a native contact map; this was done by flipping M randomly chosen entries. Contacts between consecutive amino acids (neighbors along the chain) were conserved.

A typical contact map with noise is shown in Fig. 3. The protein is 1trm chain A, whose contact map has 1595 contacts, when the threshold is set to 9 Å. We show the native map and the target map obtained by flipping at random $M = 400$ entries of the native map, together with the map produced by our method. For the particular case shown, the distance to the crystallographic structure $D = 2.4$ Å. The most important conclusion that can be drawn from Fig. 3 is that isolated nonphysical contacts are identified as such and ignored and the underlying physical contact map is recovered.

The dependence of this recovery on the noise level is shown in Fig. 3, where we present the average distance of the final structure from the uncorrupted 1trm A contact map, for various values of M. Averages were taken over 10 different realizations of the noise, and over 10 reconstruction runs for each realization. The distance for totally unrelated structures for 1trm A is around 15 Å. It is remarkable that up to $M <$ 1000 a fair resemblance to the experimental structure is still found.

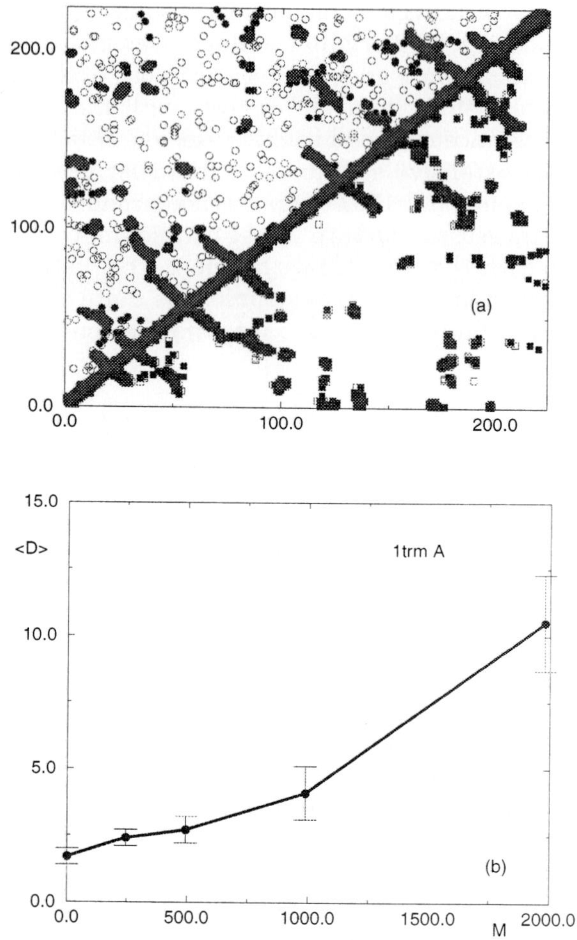

FIG. 3. (a) Above diagonal: reference map (open circles) obtained by randomizing the underlying physical map (full squares) of protein 1trm chain A. Below diagonal: reconstructed contact map (open square) obtained using the noise-corrupted map as target. (b) Average distances D versus noise M for protein 1trm A. [Adapted with permission from Vendruscolo, M., Kussell, E., and Domany, E. (1997). Recovery of protein structure from contact maps. *Folding & Design,* **2**(5), 295–306. Copyright © 1997 Current Biology.]

Even with the addition of a noise that is around 60% of the signal, the reconstruction procedure works. We have found similar results for the smaller protein 6pti, which has 342 contacts, and can be fairly well reconstructed with a noise of up to 200 flipped contacts.

To summarize this section, for physically realizable target contact maps our method is very fast and reliable for finding a chain conformation whose contact map is nearly identical to the target. Moreover, the method is able to find a good candidate structure even when the target map has been corrupted with nonphysical contacts. The information contained in a known native contact map suffices to reconstruct a conformation, which is relatively close to that of the original structure, as was already observed by Havel *et al.* (1979). There is, however, an intrinsic limit to the resolution of a contact map. We used a threshold of 9 Å between C_α atoms to define contact; for this threshold the distance between two typical structures, both compatible with the contact map, is about 1 Å. The threshold of 9 Å is relevant for our purpose, of working with contact energies in a scheme to derive structure from sequence.

IV. Dynamics in Contact Map Space

Our aim is to generate a large number of contact maps that can serve as candidates for the native structure. Such maps are necessary for protein folding by means of energy minimization, as well as in order to generate decoys needed to test properties of various energy functions. Hence, the requirements from any procedure that generates such maps are as follows:

- The generated maps should be physical.
- The maps should be "protein-like"; for example, they should have secondary structure elements.
- The maps should have low values of the energy (defined in terms of the sequence and the contact map).
- Efficiency is necessary in order to generate large numbers of independent maps in reasonable computing times.

The requirement of physicality is addressed by the method described in Section III; whenever a new candidate map is generated, we use it as the target map of the reconstruction procedure, and obtain, in this way, a contact map that corresponds to a physical "chain of beads."

To move efficiently in contact map space in a way that satisfies the requirements just listed, we introduced a four-step procedure, delined next. For further details we refer to Vendruscolo and Domany (1998a).

1. *Nonlocal dynamics.* Starting from an existing map, we perform largescale "cluster" moves. Clusters are in approximate corre-

spondence with secondary structure elements. At this stage, no attempt is made to preserve physicality. The contact map obtained by this procedure is typically uncorrelated to the starting one.

2. *Local dynamics.* The resulting map is refined by using local moves of different kinds. Secondary structure formation can be viewed as a "growth" process. Starting from a random coil, an α-helix is formed by twisting one turn at a time (Chakrabartty and Baldwin, 1995). Analogously, a helix can translate locally by untwisting a turn at one extreme and reforming it at the opposite end, in a movement reminiscent of the reptation dynamics of polymers (de Gennes, 1979). A β-sheet is created and removed by zipping and unzipping two β-strands (Muñoz *et al.,* 1997). We also use the conservative dynamics introduced by Mirny and Domany (1996) to further refine the resulting map.
3. *Reconstruction.* We use the previously introduced reconstruction algorithm (Vendruscolo *et al.,* 1997) to restore physicality by projecting the map obtained from the second step onto the physical subspace.
4. *Refinement.* We perform further optimization by energy minimization in *real space* using standard crankshaft moves (Š *et al.,* 1994; Vendruscolo and Domany, 1998a).

The projection procedure from a contact map to its three-dimensional counterpart is the bottleneck of the method. The dynamic rules that we introduce are aimed at generating uncorrelated starting points for this reconstruction. In this way, after each four-step move, we obtain a good candidate map for the native state. The contact energy [Eq. (15)] with some standard parametrization [i.e., choice of the $w(a,b)$] is used in steps 1, 2, and 4 following the standard Metropolis prescription for the acceptance of a move.

A. Nonlocal Dynamics

Rules of nonlocal dynamics have been introduced in the context of the simulation of equilibrium properties of spin systems (Kandel and Domany, 1991) and of surfaces (Evertz *et al.,* 1991). Under suitable conditions, systems with a large number of degrees of freedom arrange themselves in conformations where the degrees of freedom are "coherently" grouped together. Using an incoherent dynamic procedure the time it takes to go from one coherent conformation to another can be prohibitively long. Physical intuition guides the choice of nonlocal rules to obtain an efficient dynamics. Since in our case we are developing a minimization algorithm, we are not concerned with detailed balance, and we can optimize the dynamics by choosing moves that minimize the energy.

We now present the physical considerations that guided our choice of nonlocal moves. The unknown interactions between amino acids dictate the rules that determine the stability of protein folds. Such rules govern the chain topology in a rather stringent way. The overall number of existing protein families is estimated to be around 1000 (Chothia, 1992; Orengo *et al.,* 1994). Protein families are characterized by particular arrangements of secondary structures. Secondary structure elements are easily identified also in a contact map as clusters of points of characteristic size, shape, and location. In the contact map representation, secondary structures can be handled very efficiently by binary operations. For example a parallel β sheet is created by turning from 0 to 1 a set of contacts forming a cluster with the shape of a thin band parallel to the main diagonal. To turn a 2-α bundle from an up–down topology to the alternative up–up one, a rotation of a cluster of points is required. The operations described provide only a scaffold, which is nonphysical, and must be rectified by the other two steps of the dynamics. Every time the procedure yields a completely new topology. A MD simulation could obtain the same result only by completely (or at least partially) unfolding and refolding the protein.

In contact maps of experimentally determined protein structures, clusters of contacts can be divided in four classes. Thick bands of adjacent contacts along the main diagonal represent α-helices (see region 1 in Fig. 4a). Thin bands represent parallel β-sheets if they are parallel to the main diagonal (region 2 in Fig. 4a), and antiparallel β-sheets if they are antiparallel (region 3 in Fig. 4a). Small clusters or isolated points represent structurally relevant contacts between amino acids that are well separated along the sequence. These features characterize protein-like contact maps and should be preserved by the dynamics in contact map space.

As preliminary information we determine the expected number N_c^* of contacts and the number N_s^* of clusters that are expected to appear in the contact map. These numbers will be stochastically conserved during the dynamics. We have already presented evidence to the effect (Vendruscolo *et al.,* 1997) that $N_c^* = aN^\nu$, where N is the length of the protein, $\nu \simeq 1$ and a depends on the threshold that is used to define a contact. As for N_s^*, there are algorithms to predict the secondary structure content, like the PHD (Rost and Sander, 1993) or the GOR (Garnier *et al.,* 1996) algorithms. Alternatively, having a good starting guess for the native contact mpa, one can directly determine N_c^* and N_s^*. The

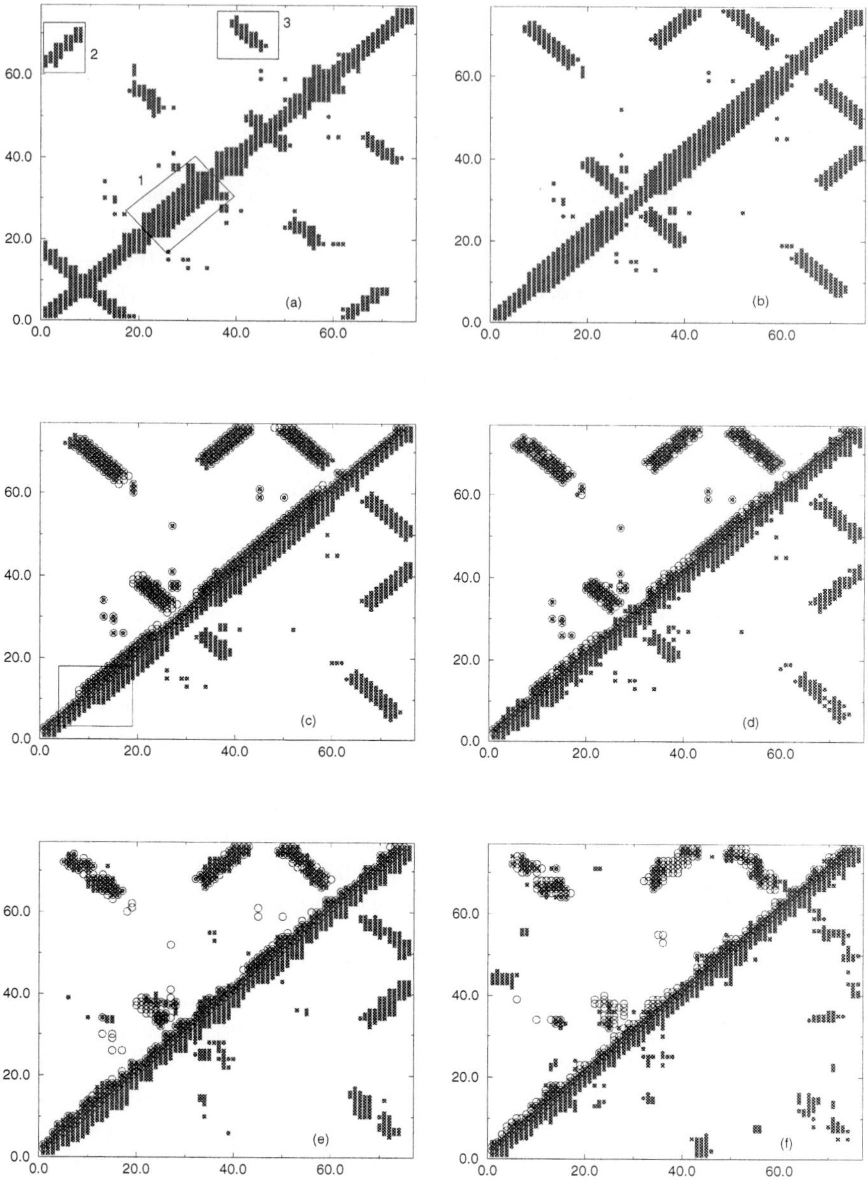

FIG. 4. (a) Contact map for the native state of protein ubiquitin (1ubq). There are 292 non-nearest-neighbor contacts. Region 1 is an α-helix, region 2 a parallel β-sheet, and region 3 an antiparallel β-sheet. (b) Contact map after a step of the non-local dynamics. (c) After a step of the local growth dynamics. The untwisting of a helix is shown in the box. (d) After a step of the local conservative dynamics. (e) After reconstruction. (f) After final minimization in real space. In (c–f) squares represent the current map and circles the previous one. [Reprinted with permission from Vendruscolo, M., and Domany, E. (1998). Efficient dynamics in the space of contact maps. *Folding & Design,* **3**(5), 329–336. Copyright © 1998 Current Biology.]

cluster algorithm is divided into three steps: labeling, deletion and creation.

1. *Labeling.* Starting from an existing map, the first step is to identify the clusters that are present, which is done using the Hoshel-Kopelman algorithm (Stauffer and Aharony, 1992). With our definition of contact, contacts S_{ij} with $|i-j| \leq 2$ are always present due to chain connectivity (see Fig. 4a). Our dynamic rules do not violate these topological constraints. Cluster labeling is made in the upper triangle, excluding the first three diagonals. By symmetry, it is sufficient to perform all the dynamics in this region. At this stage we calculate the number N_c of contacts and the number N_s of secondary structures. Secondary structures are defined as cluster of more than $H = 10$ points. After this labeling procedure, each point (i, j) in the map has label $L(i, j) = (C, K)$ with a class C and a number K inside the class. Five classes are considered. In class α we put the clusters that are formed by bands along the main diagonal. In class $\beta_{\parallel}$ we put the clusters that constitute bands parallel to the main diagonal but apart from it. In class $\beta_{\perp}$ we group clusters that are in the form of bands perpendicular to the main diagonal. In the fourth class we gather all the remaining irregular clusters. In the last class we put all the points that do not belong to any cluster (e.g., isolated points).

2. *Destruction.* N^- existing clusters are deleted from the map. N^- is chosen from a uniform random distribution between 1 and N_s. Destruction is simply realized by choosing at random a label (C, K) and by turning contacts in the corresponding cluster from 1 to 0.

3. *Creation.* N^+ clusters are created in the map. N^+ is drawn from a Gaussian distribution of mean $N_s^* + (N_s - N^-)$ and variance 1. If $N_s - N^- > N_s^*$ then $N^+ = 0$. Each time we make M attempts to create a cluster (typically $M = 100$), and we choose the one with the more favorable energy, according to Eq. (15). At each creation we first decide with probability $P(\alpha)$, $P(\beta_{\parallel})$, and $P(\beta_{\perp})$ whether to grow an α, a $\beta_{\parallel}$, or a $\beta_{\perp}$. Typically $P(\alpha) = P(\beta_{\parallel}) = P(\beta_{\perp}) = 1/3$. The length of the created band is a uniform random number in [5,30] for α, in [5,12] for $\beta_{\parallel}$ and $\beta_{\perp}$. Creation starts by selecting randomly a seed point on the map. For α-clusters this point is chosen on the principal diagonal, for $\beta_{\parallel}$ at a point displaced on the map. For α-clusters this point is chosen on the principal diagonal, for $\beta_{\parallel}$ at a point displaced from the principal diagonal by more than the length of the cluster. No restrictions are imposed on the seed of $\beta_{\perp}$. From this point we lay down a cluster in the form of a band as shown in Fig. 4b. We do not allow secondary structures to overlap or to be closer than four spacings on the map, since

this is not commonly observed in actual contact maps. If while growing the cluster we encounter a point which already has a label that violates this condition we restart the condition. The result of a nonlocal move for 1ubq, starting from the native map shown in Fig. 4a, is shown in Fig. 4b.

B. Local Dynamics

The principal aim of these moves is to allow local rearrangements of the secondary structures that have been placed by the nonlocal dynamics. We first give the dynamic rules to deal with α-helices. Consider an helix of n amino acids, which we have previously identified as starting from amino acid i and ending on $i + n$. Typically, n ranges from 5 to 30. To increase the size from the head, we add the two contacts $(i + n + 1, i + n - 2)$, $(i + n + 1, i + n - 3)$. The tail is increased by adding the two contacts above the diagonal $(i - 1, i + 2)$ and $(i - 1, i + 3)$. To decrease the size of the helix, one removes the contacts $(i + n, i + n - 3)$ and $(i + n, i + n - 4)$ on the head and $(i, i + 3)$ and $(i, i + 4)$ on the tail. To translate the helix, one performs a reptation-like move in which one turn is removed from one end and added to the other, by using the same rules.

Similar rules govern the growth and the translation of sheets. Consider first an antiparallel β-sheet formed by two strands. The first strand extends from amino acids i to $i + n$ and the second from j to $J + m$. By unzipping amino acids $i + n$ and j, we reduce the size at the end closer to the main diagonal. This move is realized by setting to 0 (irrespective of their state) the five contacts $(i + n, j + 2)$, $(i + N - 1, j + 2)$, $(i + n - 1, j + 1)$, $(i + n - 2, j + 1)$ and $(i + n - 2, j)$. Opening the sheet from the other side is realized by setting to 0 the contacts $(i + 2, j + m)$, $(i + 2, j + m - 1)$, $(i + 1, j + m - 1)$, $(i + 1, j + m - 2)$, $(i, j + m - 2)$. Zipping together the ends is realized by setting the corresponding contacts to one. Translating the sheets, as in the case of helices, is realized by opening one end while closing the other. Rules that are entirely similar are applied to parallel β-sheets. In the general case, the sheet might present irregularities that would appear as supplementary contacts at the extremities. Since we do not attempt here to realize a physical map we implement these simple rules and rely on the reconstruction procedure to take care of the local structural details. The resulting contact map after this step is shown in Fig. 4c. We use the conservative dynamics introduced by Mirny and Domany (1996) to further refine the resulting map; the result is shown in Fig. 4d. Typically minor local rearrangements take place.

C. Reconstruction

As already observed, a generic contact map is not guaranteed to correspond to any real chain conformation in space. This is likely the case of the contact map obtained after the first two steps of the dynamics. By using the reconstruction method presented earlier, we project this contact map onto its closest physical counterpart, that is, we create a contact map which is "close" to the starting one, as measured by the Hamming distance, and is guaranteed to be physical, that is, there is a real chain conformation which has that contact map. To achieve this result, we construct a backbone conformation in Cartesian space and try to force it to have the contacts specified in the input contact map. If an input contact map is nonphysical, existing contacts are discarded and possibly new ones are introduced. However, because any difference in the number and locations of contacts is penalized, the contact map of the resulting conformation is necessarily close to the starting one (Vendruscolo *et al.*, 1997). Monomers are not allowed to invade each other's space. This is ensured by introducing a lower threshold R_L, below which they experience a hard-core repulsion. The lowest C_α–C_α distance found in PDB proteins is around 3.5 Å. We chose $R_L = 5.0$ Å. With such a value, it is still possible to reconstruct all the PDB proteins and the tendency to create too compact structures, typical of the contact energy approximation, is minimized. Result of the reconstruction is shown in Fig. 4e.

D. Refinement

We perform further optimization by an energy minimization in real space using a standard Metropolis crankshaft technique (Šali *et al.*, 1994; Vendruscolo and Domany, 1998a). Results of the minimization are shown in Fig. 4f. In this calculation we used a set $\mathbf{w}_{153}$ of contact energy parameters, which was derived using the method presented by Mirny and Domany (1996), applied to the database of 153 proteins listed in Vendruscolo *et al.* (1998) with the present definition of contact. The initial energy of the native fold of 1ubq is 25.72 and the energy of the final map is much lower, −84.20.

V. Approximation for Free Energy

A. The (Free) Energy Associated with a Contact Map

As explained earlier, many microscopic configurations of a protein with sequence $\boldsymbol{A} = (a_1, a_2, a_3, \ldots, a_N)$ are characterized by the same contact map **S**. We now show that one can define an *exact free energy*, $\mathcal{H}(\mathbf{A},\mathbf{S})$, for the assignment of **S** to the sequence **A**. Denote by C a microstate of the system, specified by the coordinates of all atoms of the protein (and of the solvent and any other relevant molecules). The true, microscopic energy of this configuration is $E(C)$. In thermal equilibrium each microstate appears with a probability proportional to the corresponding Boltzmann weight $e^{-\frac{1}{kT}E(C)}$.

The *free energy* $\mathcal{H}(\mathbf{A},\mathbf{S})$ (to which we refer simply as energy) associated with sequence **A** and map **S** is defined as follows:

$$\text{Prob}(\boldsymbol{S}) \propto e^{-\mathcal{H}(\mathbf{A},\boldsymbol{S})} = \sum_{C} e^{-\frac{1}{kT}E(C)} \Delta(C,\boldsymbol{S}), \tag{13}$$

where

$$\Delta(C,\boldsymbol{S}) = \begin{cases} 1 & \text{if } \boldsymbol{S} \text{ consistent with } C \\ 0 & \text{otherwise} \end{cases},$$

and $\Delta(C,\boldsymbol{S})$ is a "projection operator" that ensures that only those configurations C whose contact map is **S** contribute to the sum [Eq. (13)]. In other words, only those microstates whose contact map is **S** contribute to the sum and hence to $\mathcal{H}(\mathbf{A},\mathbf{S})$.

This definition of the (free) energy of a map is exact: it is nothing but the negative log of the *probability* of observing the map **S** for sequence **A**. Therefore $\mathcal{H}(\mathbf{A},\mathbf{S})$ has an important property; inasmuch as the native fold's contact map, $\mathbf{S}_0$, has the highest probability of appearing the corresponding (free) energy, $\mathcal{H}(\mathbf{S}_0,\mathbf{A})$ is the *lowest* among all possible $\mathcal{H}(\mathbf{A},\mathbf{S})$, that is,

$$\mathcal{H}(\mathbf{A},\mathbf{S}_0) < \mathcal{H}(\mathbf{A},\mathbf{S}) \qquad \forall \mathbf{S} \neq \mathrm{S}_0. \tag{14}$$

The main problem with this exact energy is that the sum [Eq. (13)] is impossible to carry out. Therefore, one takes various phenomenologically motivated guesses for the form of $\mathcal{H}(\mathbf{A},\mathbf{S})$, that presumable would have been obtained had the sum been carried out. This approach is related in spirit to the phenomenological Landau-Ginzburg type of free energy used in several areas of condensed matter physics. We also start from the simplest approximate form for this complicated function—that of the *pairwise contact energy:*

$$\mathcal{H}^{\text{pair}}(\mathbf{A},\boldsymbol{S}) = \sum_{i<j}^{N} S_{ij} w(a_i, a_j). \tag{15}$$

That is, if there is a contact between residues i and j, the parameter $w(a_i,a_j)$, which represents the energy gained by bringing amino acids a_i and a_j in contact is added to the energy.

Another term that has been introduced (Mirny and Domany, 1996) in the same spirit is a *hydrophobic* (or solvation) term, of the form

$$\mathcal{H}^{\text{hydro}}(\mathbf{A},\mathbf{S}) = \sum_{i}^{N}\left[\beta(a_i)\left(\sum_{k\neq i}^{N}\mathbf{S}_{ik} - n(a_i)\right)^2\right]. \tag{16}$$

Here $n(a)$ is the desired number of contacts that amino acid a should have—as obtained from the PDB, for example. A hydrophobic residue will have a larger value of n than a hydrophilic one. The term penalizes the configuration for deviations of the actual number of contacts of a_i, as read from the map **S,** from the desired $n(a_i)$. Whereas the terms in $\mathcal{H}^{\text{pair}}$ are specific to the residue with which a_i is in contact, here a_i does not "care" who its partners are. There may be some "double counting" in using both energy terms, but this should be compensated for by the way one determines the parameters $w(a,b)$ and $\beta(a)$, the latter of which measures the strength of these hydrophobic terms. In our approach we determine the parameters by *learning,* with no recourse to a microscopic interpretation, so that we should not worry about double counting.

B. Optimization

In the energy function of the form of Eq. (15) one has to decide which set of parameters **w** to use. The first idea was to derive a particular set of energy parameters from amino acid pairing frequencies observed in available crystallographic structures (Tanaka and Scheraga, 1976; Miyazawa and Jernigan, 1985).

An alternative method was proposed later by Maiorov and Crippen (1992). Energy parameters are derived by requiring that the observed native structures be the lowest in energy among an ensemble of alternative conformations. Given the native contact map $\mathbf{S}_0$, for each alternative map **S** they wrote an inequality:

$$\mathcal{H}^{\text{pair}}(\mathbf{A},\mathbf{S}_0,\mathbf{w}) < \mathcal{H}^{\text{pair}}(\mathbf{A},\mathbf{S},\mathbf{w}), \tag{17}$$

and obtained the set **w** from the solution of such inequalities. It was also realized that parameter derivation can be formulated as an optimization problem. For a fixed set of sequences with their native maps and for a fixed set of alternative contact amps, one defines a cost function in parameter space and looks for the set of parameters of "minimum cost." Goldstein *et al.* (1992) maximized the ratio R between the width of the distribution of the energy and the average energy difference between the native state and the unfolded ones. More recently Mirny and Shakhnovich (1996) expressed the Z score as a function of the energy parameters which were then derived by optimization.

C. Learning the Energy Parameters by a Perceptron

We set out to determine the energy parameters on the basis of the basic requirement of inequality (14). The basic requirement expresses the condition that the energy should attain its lowest value at the *true native map*. It can be stated by posing the following question:

> Is it possible choose energy parameters such that for all the proteins in the database the native states have the lowest energy among all possible decoys?

We choose one protein (or more) with a known native map $\mathbf{S}_0$ and generate for it a large set of decoys $\mathbf{S}_\mu$; then we look for 210 contact parameters for which the inequalities (17) hold for all $\mathbf{S}_\mu$.

The basic attribute of $\mathcal{H}^{\mathrm{pair}}$ that we use is its linearity in the contact energies $w(a,b)$; for any map $\mathbf{S}_\mu$ the energy (15) is a linear function of the 210 contact energies that can appear and it can be written as

$$\mathcal{H}^{\mathrm{pair}}(\mathbf{A},\mathbf{S}_u,\mathbf{w}) = \sum_{c=1}^{210} N_c(\mathbf{S}_\mu)w_c. \tag{18}$$

Here the index $c = 1,2, \ldots, 210$ labels the different contacts that can appear and $N_c(\mathbf{S}_\mu)$ is the total number of contacts of type c that actually appear in map $\mathbf{S}_\mu$. The difference between the energy of this map and the native one is therefore

$$\Delta\mathcal{H}_\mu^{\mathrm{pair}} = \sum_{c=1}^{210} w_c x_c^\mu = \mathbf{w}\cdot\mathbf{x}_\mu, \tag{19}$$

where we used the notation

$$x_c^\mu = N_c(\mathbf{S}_\mu) - N_c(\mathbf{S}_0) \tag{20}$$

and $\mathbf{S}_0$ is the native map.

Also the difference in the hydrophobic energy between the decoy $\mathbf{S}_\mu$ and the native map can be cast in a linear form. To see this, note that

$$\mathcal{H}^{\mathrm{hydro}}(\mathbf{a},\mathbf{S}_\mu,\mathbf{v}) = \sum_{c=1}^{20} M_c(\mathbf{S}_\mu)v_c. \tag{21}$$

In this case the index c runs over the 20 species of amino acids and

$$M_c(\mathbf{S}_\mu) = \sum_{i=1}^{N}\left[\sum_{k\neq 1}^{N} \mathbf{S}_{ik,\mu} - n(a_c)\right]^2 \delta(a_i, c), \tag{22}$$

where $\delta(a_i,c) = 1$ if $a_i = c$ and 0 otherwise.

The difference between the hydrophobic energy of this map and the native one is therefore

$$\Delta\mathcal{H}_\mu^{\text{hydro}} = \sum_{c=1}^{20} v_c y_c^\mu = \mathbf{v}\cdot\mathbf{y}_\mu, \tag{23}$$

where we used the notation

$$y_c^\mu = M_c(\mathbf{S}_\mu \mathrm{M}) - M_c(\mathbf{S}_0). \tag{24}$$

In the presence of the hydrophobic term, inequality (14) can then be written as

$$\Delta\mathcal{H}_\mu^{\text{pair}} + \Delta\mathcal{H}_\mu^{\text{hydro}} = \mathbf{w}\cdot\mathbf{x}_\mu + \mathbf{v}\cdot\mathbf{y}_\mu = \mathbf{u}\cdot\mathbf{z}_\mu > 0, \tag{25}$$

where we introduced the following 230-component vectors:

$$\mathbf{z} = (x_1, \ldots, x_{210}, y_1, \ldots, y_{20}),$$

$$\mathbf{u} = (w_1, \ldots, w_{210}, v_1, \ldots, v_{20}). \tag{26}$$

The total energy is linear in the parameters **w** and **v.**

We denote the normalization factor of the vector $\mathbf{z}_\mu$ by

$$Z_\mu = \sum_{c=1}^{230} (z_c^\mu)^2, \tag{27}$$

and from here on we set $Z_c^\mu \leftarrow z_c^\mu / Z_\mu^{1/2}$, so that both **z** and **u** are normalized to 1.

Once requirement (17) has been expressed using an energy in the form of Eq. (25), the question of whether it does or does not have a solution reduces to deciding whether a set of examples is learnable by a perceptron (Rosenblatt, 1962). Every candidate contact map provides a pattern for the training session.

The vector **u** is "learned" in the course of a training session. The P patterns are presented cyclically; after presentation of pattern μ the weights **u** are updated according to the following learning rule:

$$\mathbf{u}' = \begin{cases} (\mathbf{u} + \eta\mathbf{z}_\mu)/|\mathbf{u} + \eta\mathbf{z}_\mu| & \text{if } \mathbf{u}\cdot\mathbf{z}_\mu < 0 \\ \mathbf{u} & \text{otherwise} \end{cases}. \tag{28}$$

This procedure is called *learning* because when the present **u** misses the correct "answer" $h_\mu > 0$ for example μ, all weights are modified in a manner that reduces the error. No matter what initial guess is made for u, a convergence theorem guarantees that if a solution **u** exists, it will be found in a finite number of training steps. Since the parameters **v** have to be positive, we introduced the following trick. We added 20 fictitious examples **z** which are vectors of zeros except the component y_i, $i = 1, \ldots, 20$ which is set to 1.

Different choices are possible for the parameter η. Here we use the learning rule introduced by Nabutovsky and Domany (1991) since it allows, at least in principle, to assess whether the problem is learnable. The parameter η is given at each learning step by

$$\eta = \frac{-h_\mu + 1/d}{1 - h_\mu/d}, \tag{29}$$

where the parameter d (called despair) evolves during learning according to

$$d' = \frac{d + \eta}{\sqrt{1 + 2\eta h_\mu + \eta^2}}. \tag{30}$$

Initially one sets $d = 1$.

The training session can terminate with only two possible outcomes. Either a solution is found [that is, no pattern that violates condition (25) is found in a cycle], or unlearnability is detected. The problem is *unlearnable* if the despair parameter d exceeds a critical threshold:

$$d > d_c = \sqrt{M}[2Z_{\max}]^{M/2}, \tag{31}$$

where M is the number of components of **w,** and $Z_{\max}$ is the maximal value of the normalization factors [see Eq. (27)]. This value of d_c is easily derived from examples of the type of Eq. (20), using the same method as given by Nabutovsky and Domany (1991).

It is easy to see that $\mathbf{u}^* = (\mathbf{w}^*, \mathbf{v}^*)$ is a solution of the system (25) if and only if $\mathbf{u}^*_\lambda = (\mathbf{w}^*, \lambda\mathbf{v}^*)$ is a solution of

$$\mathbf{w} \cdot \mathbf{x}_\mu + \frac{1}{\lambda}\mathbf{v} \cdot \mathbf{y}_\mu > 0. \tag{32}$$

The outcome of the learning process does not depend on the initial choice of λ. However, the learning time usually depends on λ, which has to be chosen conveniently. A useful condition to set λ is to obtain on average $|\mathbf{x}| \sim |\mathbf{y}/\lambda|$.

VI. Results

Whether or not basic requirement (14) can be satisfied depends on several factors:

1. The definition of contact
2. The assignment of the contact length R_c
3. The number M_p of proteins in the database
4. The method used to produce decoys

A. Threading

We first discuss the dependence of learnability on M_p and R_c using the all-atom definition of contact and producing decoys by gapless threading. Once the contact maps of the M_p proteins in the database are obtained, decoys are generated for a given sequence of length N from the structures of proteins of lengths N' ($> N$) by selecting submaps of size $N \times N$ along the main diagonal of the contact map of the longer protein. For each definition of contact that we considered we found two "phases." There is a region in (R_c,M_p) in which the problem is *learnable*; that is, there is a set $\mathbf{w}$ of pairwise contact energy parameters that stabilizes simultaneously all the native maps in the set. On the other hand, outside this region (e.g., for fixed R_c and large enough M_p) the problem is *unlearnable,* and no set $\mathbf{w}$ exists.

Without doing any calculation, we can predict a few general features of the (R_c,M_p) phase diagram. Having set a definition of contact (e.g., the all-atom one) one can plot the distribution of distances between amino acid pairs. Choosing R_c smaller than the smallest observed distance would result in contact maps with no contacts, independently of the conformation. No set of energy parameters can then discriminate the native map from the decoys. Similarly, choosing R_c larger than the largest observed distance would result in contact maps with all the entries set to 1. In this case again no discrimination is possible. Thus we expect to find a window of learnability in R_c. It is also reasonable to expect that such a window will shrink with increasing M_p. The problem is thus reduced to investigate whether such window remains open for an arbitrary large value of M_p, or it closes for M_p large enough.

The boundaries of the region of learnability must be interpreted in a probabilistic sense. At given M_p and R_c, learnability depends on the particular choice of the proteins in the database. In principle, one should define $P(R_c,M_p)$, the probability for a randomly extracted set to be learnable at (R_c,M_p). The boundary is then defined by $P(R_c,M_p)$ = const. In the present study, we chose not to give a precise numerical estimation of $P(R_c,M_p)$, which would require many extractions of sets of M_p proteins from the PDB and would be numerically intensive. We are interested in establishing the *existence* of the boundary rather than in its precise determination. Hence we will show that it is possible to find unlearnable datasets when their size is large enough.

The approximate boundaries of the region of learnability are shown in Fig. 5. The window of learnability shrinks to zero for M_p above 270. The precise value of the limit of learnability depends on the particular choice of proteins, and the fluctuations in M_p are of the order of a few tens of proteins. Such fluctuations are larger than expected; they are due to a few proteins that are markedly more difficult to learn than others, and their inclusion in the dataset lowers the chances of learnability. For example, 1vdfA and 1gotG are two such "hard to learn" proteins. 1vdfA is a chain of 46 amino acids that forms a single α-helix; 1gotG is a chain formed by 2 α-helices hinging at 90 degrees.

Each point shown in Fig. 5 is derived from a few (1–3) randomly generated sets of M_p proteins in each. A full circle indicates that all the sets considered were learnable; open circles signal that at least one set was

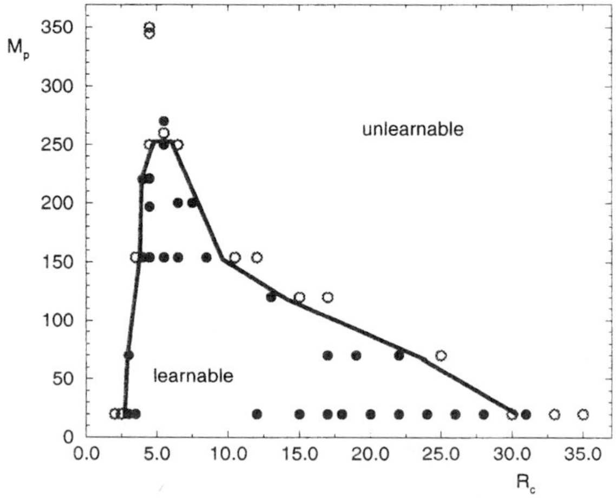

FIG. 5. Region of learnability for the all-atom definition of contact. Several sets of M_p proteins were generated for each value of R_c. Full circles indicate that all the sets considered for a particular value of (R_c, M_p) were learnable; otherwise we use open circles.

unlearnable. The largest fluctuations are found for small M_p at the right boundary. For example, for different sets with $M_{p_0} = 20$ we found that the maximal R_c of learnability varies between 28 Å and 31 Å.

In this study, we selected five sets of protein structures from the PDB. The sets ranged from 154 to 945 proteins. The PDB is an archive of experimentally derived structures of proteins. To date, 8295 coordinate entries have been deposited. it is known that the information contained in the entire PDB is redundant and some is incorrect. Routine methods are available to select subsets of nonhomologous proteins whose experimental structures are determined reliably (Heringa *et al.,* 1992; Hobohm and Sander, 1994; Hooft *et al.,* 1996).

Correlations between solutions at different M_p and R_c ranged from 0.22 (for a solution at $R_c = 7.5$ Å and $M_p = 200$ with a solution at $R_c =$ 4.5 Å and $M_p = 154$) to 0.94, which is the typical correlation between two solutions of maximal stability at $R_c = 4.5$ Å and $M_p > 150$.

These findings indicate that with the all-atom definition of contact and the physically motivated choice of $R_c = 4.5$ Å (Mirny and Domany, 1996) it is possible to stabilize simultaneously, with high probability, about 200 randomly selected nonhomologous proteins against decoys generated by gapless threading.

Which decoys are more challenging? It is instructive to consider the overlap Q between the contact map of the native state and of the decoys, which is defined as

$$Q = \frac{1}{\max(N_c^0, N_c^\mu)} \sum_{h>k+1}^{N} \mathbf{S}_{hk}^0 \mathbf{S}_{hk}^\mu , \tag{33}$$

where N is the length of protein and N_c^0 and N_c^μ are the number of contacts in $\mathbf{S}_0$ and $\mathbf{S}_\mu$, respectively. We considered, for the C_α definition and $R_c = 8.5$ Å, two sets of proteins. The first, of 123 proteins, is learnable and the second, of 141 proteins, unlearnable. First, for each decoy we calculated, using as initial weights the solution of an independent set of 197 proteins, the energy difference $\Delta\mathcal{H}$ with the native state and the overlap Q. In Figs. 6a and 6b we present scatter plots of $\Delta\mathcal{H}$ on Q for our decoys. The first question we asked is whether decoys of high overlap are the more challenging ones. To answer this question, we repeated the learning procedure by considering only decoys with $Q < Q_t$, where Q_t is a threshold value for the overlap. The set of 141 proteins was still unlearnable for $Q_t = 0.6$. It seems that the "difficult" decoys are spread over the entire range of Q. Including all the decoys in the learning procedure we were able to learn the set of 123 proteins, but not

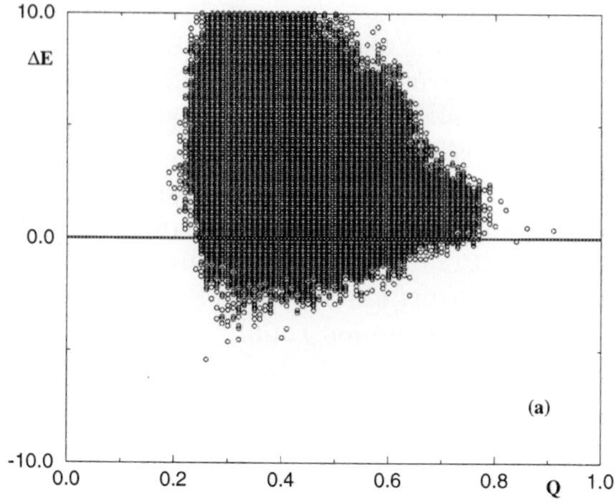

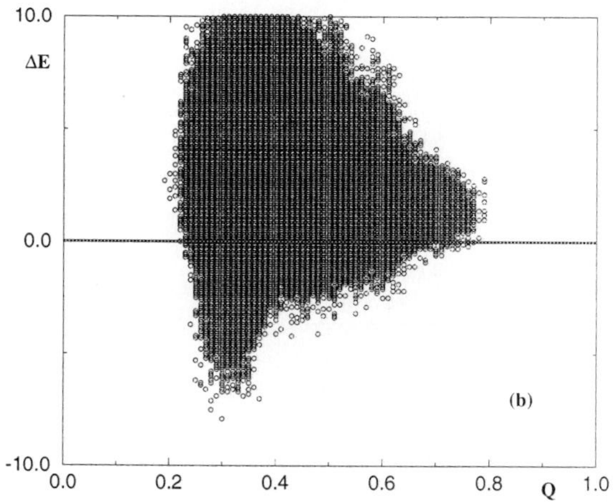

FIG. 6. Scatter plot of the energy difference $\mathcal{H}$ between decoys and native state and the overlap Q with the native state. (a) Learnable set of 123 proteins for which the energy $\mathcal{H}^{pair}$ was calculated using the solution of an independent set of 197 proteins; (b) unlearnable set of 141 proteins with the same initial set of energy parameters; (c) the set of 123 proteins with the energy parameters obtained from learning; and (d) the set of 141 proteins with the energy parameters arrived at when unlearnability has been established.

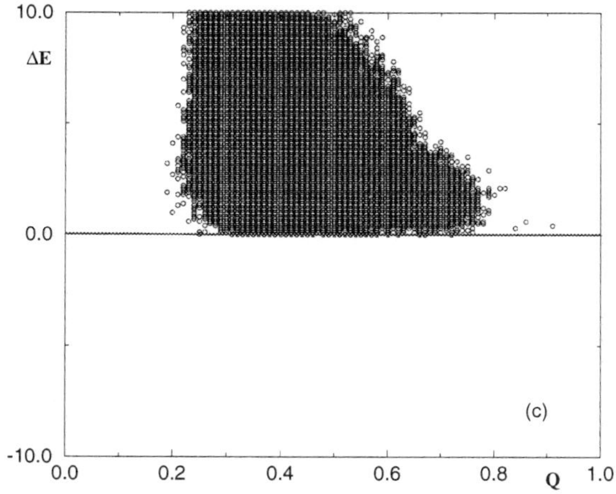

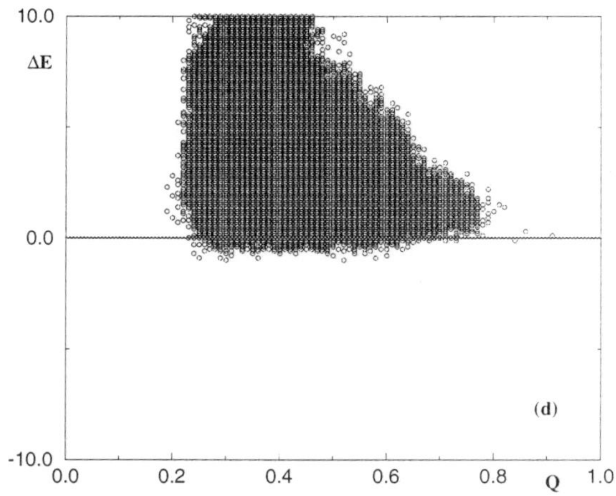

FIG. 6. Continued

the set of 141 proteins. As shown in Fig. 6c, after learning the set of 123 proteins, decoys of low energy are present in the approximate range 0.2 $> Q >$ 0.8. The important finding is that the unlearnable case is not qualitatively different (see Fig. 6d). also in this case decoys of low energy are present in the range $0.2 < Q < 0.8$, although now some of them have $\Delta\mathcal{H} < 0$. The difference is that with the set of 123 proteins there are 805,938 decoys, whereas for the set of 141 proteins, 1,071,753. With 210 energy parameters it is possible to satisfy the smaller set of inequalities but not the larger. Decoys of arbitrary Q enter in the learning process and one can argue that the lack of correlation between $\Delta\mathcal{H}$ and Q is a major cause that renders the problem unlearnable for large enough M_p and, further, that an improved energy function should first of all provide such a correlation.

To investigate the dependence of learnability on the definition of contact we repeated the same analysis for two other definitions of contacts. The first based on C_α and the second on all carbon atoms—a contact was assumed if any two C atoms of the two amino acids were closer than R_c. In both cases we found regions of learnability qualitatively similar to that shown in Fig. 5.

To summarize, we showed that, no matter which definition of contact and which R_c are chosen, there is always a maximal number (of the order of few hundreds) of proteins that can be stabilized together. Is this number large or small? A definite answer is outside the limits of this study. It is small when compared with the total number of proteins existing in nature (hundreds of thousands), but it is fairly large when compared with the number of representative proteins in the PDB (also of the order of few hundreds; Hobohm and Sander, 1994). We choose to generate decoys by gapless threading since it is a very efficient way to obtain decoys. To further generalize our conclusion, in the following sections we consider a more refined way to obtain decoys, based on energy minimization in the space of contact maps.

B. Crambin

The conclusion drawn from threading is that there is a maximal number of proteins that can be stabilized together using the pairwise contact approximation to the energy. One can ask a more limited question, namely, if it is possible to fine-tune energy parameters to stabilize just one protein, or possibly a set of proteins belonging to the same structural family. Threading offers poor contenders to the native state. Better candidates for the ground state are produced by contact map dynamics, the method presented in Section IV, which explores in an efficient way the space of contact maps. For a particular protein—crambin—we show in Fig. 7 that the decoys produced by threading are far less a challenge than those produced by contact map dynamics. The set of decoys obtained by threading can be learned—it is possible to produce a set of pairwise contact energy parameters that stabilizes crambin in a threading experiment. Here we ask the same question for decoys obtained by contact map dynamics (see point 4 at the beginning of this section).

Crambin (Teeter *et al.*, 1993) is a protein of length $N = 46$. We con-

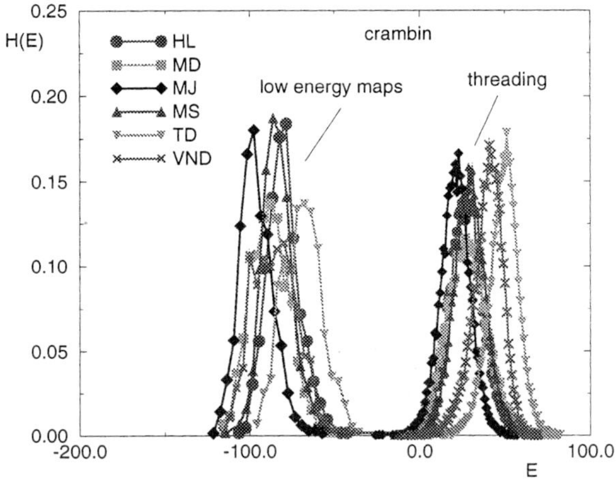

FIG. 7. Histograms that demonstrate the difference in energy between ensembles of contact maps obtained by threading and by energy minimization, shown for different contact energy parameters: VND: as obtained in this work, by finding a solution for a threading example; HL: Hinds and Levitt, 1994; MD: Mirny and Domany, 1996; MJ: Miyazawa and Jernigan, 1996; MS: Mirny and Shakhnovich, 1996; TD: Thomas and Dill, 1996. The energy parameter sets were shifted and rescaled to obtain $\langle \mathbf{w} \rangle = 0$ and $\langle \mathbf{w}^2 \rangle - \langle \mathbf{w} \rangle^2 = 1$ (averages are over the 210 energy parameters). Energies were shifted to set the native state to $E = 0$.

structed its native map by taking the coordinates of the C_α atoms from the PDB and using a threshold of 8.5 Å to define contacts. In the crambin chain, 5 out of the 20 amino acids do not appear and 3 appear only once. Thus, among the corresponding 210 possible contact energies, only 117 parameters can effectively enter the energy [Eq. (15)] for any set of candidate maps. These parameters form a 117-component vector **w**. The native map contains 187 non-nearest-neighbor contacts, which involve only 72 of the 117 possible contact energy parameters. We summarize here our main result about the question we have addressed in the present work. We will present below decisive evidence supporting our main conclusion:

> The problem of fine-tuning the pairwise contact energy parameters to stabilize the native state of crambin is *unsolvable*.

1. *Learning the Pairwise Contact Term*

In an unlearnable case there exist sets of examples for which no solution can be found; for large enough P the training set will include, with nonvanishing probability, such an unlearnable subset.

The condition that has to be met (Nabutovsky and Domany, 1991) to establish unlearnability is that the despair *d*, recorded during the learning process, should exceed the critical value d_c. According to Eq. (31), for $M = 117$, we typically get a critical despair $d_c \simeq 10^{163}$.

In perceptron learning of $P > 2M$ randomly generated examples [which is an unlearnable problem (Cover, 1965; Gardner, 1988) for large M], d grows *exponentially* with learning time (Nabutovsky and Domany, 1991). Here, due to the specific nonrandom nature of the learning task, we need much more than $P = 2M = 234$ examples. Learning is realized in an iterative manner. Starting with an initial set of energy parameters we generate a set of examples that is then learned; the new set of energy parameters is used to generate new examples and so on. We will refer to each such iteration as an *epoch*. It was necessary to generate $P = 298{,}710$ examples in 11 epochs. We underscore the fact that each example was obtained by an energy minimization procedure, using the method discussed in Section IV. Learning this set of examples is an unsolvable task, proved by the divergence of the despair d. We reached $d > d_c$ after approximately 37,500 learning cycles. To speed up the procedure, without affecting the final result, we selected the 10,000 examples at epoch 11 that had the lowest energy according to the energy parameters obtained from learning at epoch 10.

2. *Energies of the False Ground States*

Even though the problem is unlearnable, our procedure produces contact energies that have several appealing features. The first is that whereas for the existing contact potentials it is very easy to find maps whose energy is below that of the native map, with the **w** obtained after several training epochs this becomes a difficult (albeit possible) task. With the initial energy parameters the vast majority of the contact maps that are generated have a lower energy than the native contact map. As can be seen from Fig. 8, where the energy scale is shifted so that the native contact map has always zero energy, for increasing epoch index, the energy distribution shifts to the right and becomes narrower. Learning is thus accompanied by an improvement of the Z-score, which is a commonly used estimator of the quality of a set of energy parameters (Mirny and Shakhnovich, 1996). Hence our learning procedure flattens the energy landscape, reducing both the number of violating examples and their energy difference with the true native state. Another relevant question concerns the “location” of these false minima, that is, how different are the corresponding structures from the native one? To study this, we reconstructed the three-dimensional conformations corresponding to the violating examples and measured their average dis-

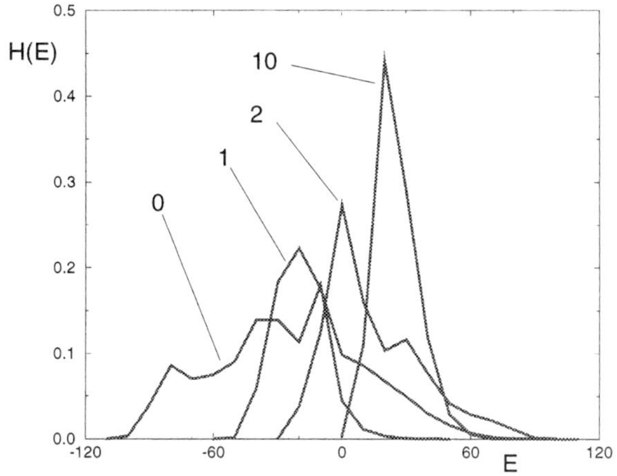

FIG. 8. Normalized histogram $H(\mathcal{H})$ of the energies of the contact maps at epochs t = 0, 1, 2, and 10. The energy scale is shifted so that the native contact map energy is 0. [Reprinted with permission from Vendruscolo, M., and Domany, E. (1998). Pairwise contact potentials are unsuitable for protein folding. *J. Chem. Phys.* **109,** 11101–11108. Copyright © 1998 American Institute of Physics.]

tance D [see Eq. (11)] from the native conformation. We found that D does not decrease with the epoch index; moreover false minima are found at an approximate average D of 8 Å at *any* epoch. Only their number decreased significantly. Two compact uncorrelated conformations are typically found at a distance of 15 Å, which is also the result we obtained using other parameter sets taken from the literature.

Here, we define the overlap Q between contact maps in a slightly different way as than in Eq. (33):

$$Q = \frac{N_p}{N_c}, \tag{34}$$

where N_p is the number of contacts present in the native map *and* in the predicted one and N_c is the total number of contacts of the native map [contacts $(i, i + 1)$ and $(i, i + 2)$ are not counted]. We generated 1000 low-energy contact maps, using the set of energy parameters obtained at epoch 10. The histogram of the fraction of contact maps with a given overlap Q with the native state is shown in Fig. 5. The distribution is peaked at around 0.40, which means that typically we are able to correctly recover 40% of the native contacts. We analyzed the dis-

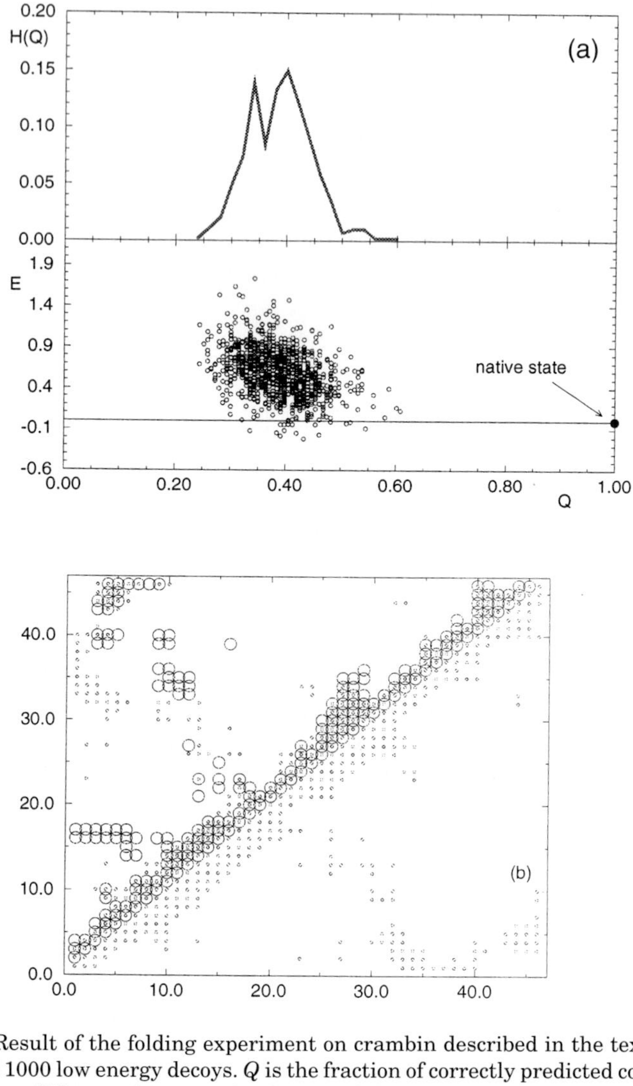

FIG. 9. (a) Result of the folding experiment on crambin described in the text in which we generated 1000 low energy decoys. Q is the fraction of correctly predicted contacts and $f9$ is the energy difference between the decoy and the native state. [Reprinted with permission from Vendruscolo, M., and Domany, E. (1998). Pairwise contact potentials are unsuitable for protein folding. *J. Chem. Phys.* **109,** 11101–11108. Copyright © 1998 American Institute of Physics.] (b) Typical low energy map obtained during the simulation. Small dots indicate contacts in the native map, open circles contacts in the simylated map.

tance D of the conformations corresponding to the maps of Fig. 9. We found that in the worst cases conformations are found at an approximate average D of 8 Å. It is not possible to improve this result, since the contact energy parameters cannot be optimized further.

Consistent with our conclusion that the pairwise contact energy approximation is unsuitable for protein folding, Hao and Scheraga (1996) showed that a more accurate parametrization of the energy is capable of improving the performance in a folding simulation. They found that "when the crambin model is simulated to fold to the low-energy state with the optimized energy parameters, the folded structure is generally close to the target native structure," although "the energy of the target structure was never able to separate completely from the continuous density of the nonnative states through the optimization." They do find nonnative conformations with energy lower than the native state, but the contact maps of these have, on the average, high overlap with the native map.

One should not be misled by similar work on *model* proteins where, using a pairwise contact energy function, it is possible to discriminate the native state, either by optimization (Hao and Scheraga, 1996, 1998; Mirny and Shakhnovic, 1996) or by imposing conditions analogous to Eq. (17) (Van Mourik *et al.,* 1998). In this case, a database of foldable sequences is designed using a pairwise contact potential and subsequently a set of contact energy parameters is recovered. Success in this case is possible because the contact energy of Eq. (15) is the *exact* form of the free energy of the model.

3. *Learning the Hydrophobic Term*

We proved that the pairwise contact approximation is unsuitable to stabilize the native contact map of crambin against a set of decoys obtained by contact map dynamics. The next step we take is to improve it by adding the hydrophobic term of Eq. (16) and ask again if the native state of crambin can be assigned with the lowest energy.

We found that the previously used set of 298,710 decoys was learnable with the hydrophobic term. We increased the number of decoys to 390,117 in 15 epochs and in this way we obtained an unlearnable set. After some trial we choose $\lambda = 15$ in Eq. (32) since we found that the despair d grows faster for this choice.

Introducing the hydrophobic term does not make the problem learnable. However, it improves the situation depicted in Fig. 9. We repeated the folding experiment previously done (see Section VI,B,2). We found that the typical value of Q is larger, as shown in Fig. 10a. The av-

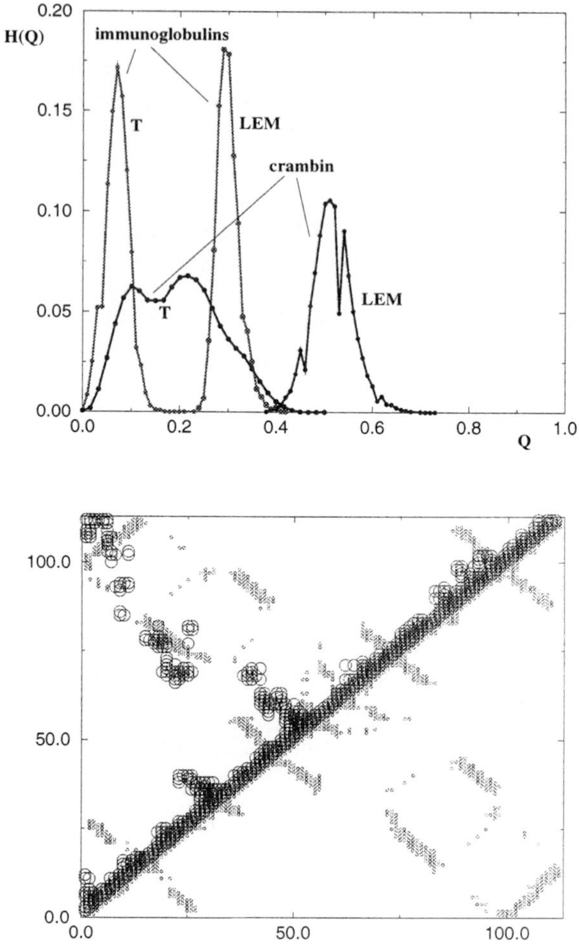

FIG. 10. (a) Histogram of the distribution of the overlap Q between maps of low energy (LEM) and the mative map found in the simulation with the hydrophobic term for immunoglobulins and for crambin. For comparison, the distributions obtained by threading are also shown (T). (b) Typical low-energy contact map (open circles) generated during the simulation of 1dba. For comparison the native map is also shown (full dots).

erage value of Q moves from 0.4 (pairwise term only) to 0.5 (hydrophobic term). These two values should be compared with 0.2, which is the value obtained for decoys produced by threading crambin into a set of 456 proteins (the distribution is marked by a T in the figure).

C. Immunoglobulins

Crambin could be a peculiar case because it is a short protein, it has only 15 species of amino acids, etc. To what extent do the results about unlearnability extend to other proteins? To answer this question, we undertook a study of a set of proteins of the family of immunoglobulins.

An immunoglobulin molecule is formed by two light chains and two heavy ones (Branden and Tooze, 1991) held together by disulfide bridges and by the packing of their respective domains. The light chain folds into two domains, the variable V_L and the constant C_L and the heavy chain folds into four domains, one variable V_H and three constant C_{H1}, C_{H2}, and C_{H3}. Sequence similarity in constant domains is typically above 35% and around 10–20% in variable domains.

We focused our study on the variable domains of the light chain V_L, which is formed by two β-sheets, one of four strands and the other of six strands. The reason to choose immunoglobulins is mainly due to the extremely rich amount of information available. The sequence database of Kabat *et al.* (1991) now contains about 5700 different variable domains and the structure of about 140 are available from the PDB.

We asked the question if it is possible to stabilize 6 V_L domains simultaneously. The domains chosen were 8fab A, 1baf L, 1cbv L, 1dba L, 2f19 L, and 1fdl L, whose respective lengths were 104, 107, 112, 107, 107, and 107 amino acids.

1. *Learning the Pairwise Contact Term*

As opposed to the case of crambin, where only 117 pairwise contact species are present, all the 210 possible ones are present for immunoglobulins. To prove unlearnability, we adopted the same iterative procedure used for crambin. We generated 97,309 examples, in 6 epochs. To speed the procedure, we considered only a subset of decoys, selected according to the following procedure. Consider the component c of the vector $\mathbf{x}^{\mu}$ of a particular decoy μ. If the numbers $N_c(S_{\mu})$ and $N_c(S_0)$ are equal then $x_c^{\mu} = 0$ and this particular decoy can never be used to update w_c. If we initialize $w_c = 0$ and use only decoys with $x_c = 0$ then $w_c = 0$ during the entire learning procedure. We excluded all the decoys having contacts of 13 species, in this way we remained with 35,677 decoys with 197 contact species. Obviously, for such a subset, adding the 13 energy parameters cannot change the result, since they never enter in the calculation of the energy. We further selected the 5000 decoys of lowest energy (according to the energy parameters learned at epoch 5) and performed the learning on this subset. In this way the critical value of the despair was surpassed in an amenable computer time.

2. *Learning the Hydrophobic Term*

The set of 97,309 decoys was learnable by adding the hydrophobic term. We enlarged it to 110,576. Using the same trick explained earlier, we excluded from the learning 13 pairwise contact energy terms and 2 hydrophobic ones. This selection procedure identified a subset of 42,798 decoys for which we proved unlearnability.

The failure to assign the lowest energy to the native contact map is more severe in the case of immunoglobulins. The low-energy contact maps in the case of crambin did resemble the native map to some extent, as shown in Fig. 9. In the case of immunoglobulins there is very little similarity between low-energy maps and the native ones (see Fig. 10b).

We think that this result deserves an explanation, which we propose in the following argument. As already done for crambin, we obtained a set of decoys by threading for immunoglobulins. Such a distribution is narrower and shifted to the left with respect to the corresponding distribution for crambin. We assume that threading offers random protein-like conformations. Therefore, the shift is originated by the exponential increase of the number of physical contact maps (Vendruscolo *et al.*, 1999) with the length of the proteins. The improvement, as measured by the right-shift of the distribution of Q, obtained by learning energy parameters, is comparable for crambin and for immunoglobulins. However, since immunoglobulins are longer, one would have to work harder to obtain energy parameters suitable for folding—many more decoys are to be weeded out. Since we proved that it is impossible to tune better the energy parameters for both approximations of the energy that we discussed, the only choice is to develop better forms for the energy. This will be the object of future study.

VII. Conclusions

We presented the results of our attempt to perform protein folding in the space of contact maps.

- We demonstrated the feasibility of the original idea of performing energy minimization in contact map space.
- We proved that two simple phenomenological approximations to the free energy cannot possibly be tuned to assign the lowest energy to the observed native conformation of even one single protein.
- We leave open the possibility of using better approximations for the free energy and more detailed representations of the chain underlying a contact map.

- We also leave open the possibility of predicting native states by other methodologies that do not use energy minimization.

ACKNOWLEDGMENTS

We are grateful to Ron Elber for discussing with us a similar approach, based on Eq. (14) (unpublished), and to Gaddy Getz, Ido Kanter, Edo Kussell, Rafael Najmanovich, and Kibeom Park for their contributions to different parts of the work presented. This research was supported by grants from the Minerva Foundation, the Germany-Israel Science Foundation, and by a grant from the Israeli Ministry of Science.

REFERENCES

Anfinsen, C. (1973). Principles that govern the folding of protein chains. *Science* **181,** 223–230.

Aszodi, A., and Taylor, W. R. (1996). Homology modelling by distance geometry. *Fold. Des.* **1,** 325–334.

Bernstein, F. C., Koetzle, T. F., Williams, G. J. B., Meyer, E. F., Jr., Brice, M. D., Rodgers, J. R., Kennard, O., Shimanouchi, T., and Tasumi, M. (1977). The protein data bank: A computer-based archival file for macromolecular structures. *J. Mol. Biol.* **112,** 535–542.

Bohr, J., Bohr, H., Brunak, S., Cotterill, R. M., Fredholm, H., Lautrup, B., and Petersen, E. F. (1993). Protein structures from distance inequalities. *J. Mol. Biol.* **231,** 861–869.

Bowie, J. U., Luthy, R., and Eisenberg, D. (1992). A method to identify protein sequences that fold into a known 3-dimensional structure. *Science* **253,** 164–170.

Branden, C., and Tooze, J. (1991). "Introduction to Protein Structure." Garland Publishing, New York and London.

Brooks, C. L. (1998). Simulation of protein folding and unfolding. *Curr. Opin. Struct. Biol.* **8,** 222–226.

Brünger, A. T., Clore, G. M., Gronenborn, A. M., and Karplus, M. (1986). Three-dimensional structure of proteins determined by molecular dynamics with interproton distance restraints: Application to crambin. *Proc. Natl. Acad. Sci. U.S.A.* **83,** 3801–3805.

Brünger, A. T., Adams, P. D,. and Rice, L. M. (1997). New applications of simulated annealing in X-rays crystallography and solution NMR. *Curr. Opin. Struct. Biol.* **5,** 325–336.

Chakrabartty, A., and Baldwin, R. L. (1995). α-helix stability. *Adv. Protein Chem.* **46,** 141–176.

Chan, H. S., and Dill, K. A. (1990). Origins of structure in globular proteins. *Proc. Natl. Acad. Sci. U.S.A.* **87,** 6388–6392.

Chothia, C. (1992). One thousand families for the molecular biologist. *Nature (London)* **357,** 543–544.

Clementi, C., Maritan, A., and Banavar, J. R. (1998). Folding, design, and determination of interaction potentials using off-lattice dynamics of model heteropolymers. *Phys. Rev. Lett.* **81,** 3287–3290.

Cover, T. (1965). Geometrical and statistical properties of systems of linear inequalities with applications in pattern recognition. *IEEE Trans. Electron. Comput.* **14,** 326.

Crippen, G., and Havel, T. F. (1988). "Distance Geometry and Molecular Conformation." Wiley, New York.

de Gennes, P. G. (1979). "Scaling Concepts in Polymer Physics." Cornell University Press, Ithaca, NY.

Dinner, A. R., Šali, A., and Karplus, M. (1996). The folding mechanism of larger model proteins: The role of native structure. *Proc. Natl. Acad. Sci. U.S.A.* **93,** 8356.

Duan, Y., and Kollman, P. A. (1998). Pathways to a protein folding intermediate observed in a 1-microsecond simulation in water solution. *Science* **282,** 740–744.

Evertz, H. G., Hasenbush, M., Marcu, M., Pinn, K., and Solomon, S. (1991). High precision measurement of the SOS surface thickness in the rough phase. *J. Phys. I (France)* **1,** 1669–1674.

Fisher, D., Rice, D., Bowie, J. U., and Eisenberg, D. (1996). Assigning amino acid sequences to 3-dimensional protein folds. *FASEB J.* **10,** 126–136.

Gardner, E. (1988). The space of interactions in neural networks models. *J. Phys. A* **21,** 257.

Garnier, J., Gibrat, J. F., and Robson, B. (1996). GOR method for predicting protein secondary structure from amino acid sequence. *In* "Methods in Enzymology" (R. F. Doolittle, ed.), Vol. 266, pp. 540–553. Academic Press, San Diego, CA.

Godzik, A., Skolnik, J., and Kolinski, A. (1993). Regularities in interaction patterns of globular proteins. *Protein Eng.* **6,** 801–810.

Goldstein, R., Luthey-Schulten, Z. A., and Wolynes, P. G. (1992). Optimal protein folding codes from spin glass theory. *Proc. Natl. Acad. Sci. U.S.A.* **89,** 4918–4922.

Hao, M.-H., and Scheraga, H. A. (1996). How optimization of potential function affects protein folding. *Proc. Natl. Acad. Sci. U.S.A.* **93,** 4984–4989.

Hao, M.-H., and Scheraga, H. A. (1998). Molecular mechanisms for cooperative folding of proteins. *J. Mol. Biol.* **277,** 973–983.

Havel, T. F., Crippen, G. M., and Kuntz, I. D. (1979). Effect of distance constraints on macromolecular conformation. II. Simulation of experimental results and theoretical predictions. *Biopolymers* **18,** 73–81.

Heringa, J., Sommerfeldt, H., Higgins, D., and Argos, P. (1992). OBSTRUCT: A program to obtain largest cliques from a protein sequence set according to structural resolution and sequence similarity. *CABIOS* **8,** 599–600.

Hinds, D. A., and Levitt, M. (1994). Exploring conformational space with a simple lattice model for protein structure. *J. Mol. Biol.* **243,** 668–682.

Hobohm, U., and Sander, C. (1994). Enlarged representative set of protein structure. *Protein Sci.* **3,** 522–524.

Holm, L., and Sander, C. (1993). Protein structure comparison by alignment of distance matrices. *J. Mol. Biol.* **233,** 123–138.

Hooft, R. W., Vriend, G., Sander, C., and Abola, E. E. (1996). Errors in protein structures. *Nature (London)* **381,** 272.

Irbäck, A., and Potthast, F. (1995). Studies of off-lattice model for protein folding: Sequence dependence and improved sampling at finite temperature. *J. Chem. Phys.* **103,** 10298–10305.

Jones, D. T., Taylor, W. R., and Thornton, J. M. (1992). A new approach to protein fold recognition. *Nature (London)* **358,** 86–89.

Kabat, E., Wu, T., Perry, H., Gottesman, K., and Foeller, C. (1991). "Sequences of Proteins of Immunological Interest," 5th ed. NIH Publ. No.91-3242. U.S. Department of Health and Human Services, Public Health Service, National institute of Health, Bethesda, MD.

Kandel, D., and Domany, E. (1991). General cluster Monte Carlo dynamics. *Phys. Rev. B* **43,** 8539–8555.

Kirkpatrick, S., Gelatt, C. D., Jr., and Vecchi, M. P. (1983). Optimization by simulated annealing. *Science* **220,** 671–680.

Lifson, S., and Sander, C. (1979). Antiparallel and parallel beta-strands differ in amino acid residue preferences. *Nature (London)* **282,** 109–111.

Lund, O., Hansen, J., Brunak, S., and Bohr, J. (1996). Relationship between protein structure and geometrical constraints. *Protein Sci.* **5,** 2217–2225.

Maiorov, V. N., and Crippen, G. M. (1992). Contact potential that recognizes the correct folding of globular proteins. *J. Mol. Biol.* **227,** 876–888.

McLachlan, A. D. (1979). Gene duplications in the structural evolution of chymotrypsin. *J. Mol. Biol.* **128,** 49–79.

Mirny, L., and Domany, E. (1996). Protein fold recognition and dynamics in the space of contact maps. *Proteins* **26,** 391–410.

Mirny, L., and Shakhnovich, E. I. (1996). How to derive a protein folding potential? A new approach to an old problem. *J. Mol. Biol.* **264,** 1164–1179.

Miyazawa, S., and Jernigan, R. L. (1985). Estimation of effective interresidue contact energies from protein crystal structures: Quasi-chemical approximation. *Macromolecules* **18,** 534–552.

Miyazawa, S., and Jernigan, R. L. (1996). Residue-residue potentials with a favorable contact pair term and an unfavorable high packing density term, for simulation and threading. *J. Mol. Biol.* **256,** 623–644.

Mumenthaler, C., and Braun, W. (1996). Automated assignment of simulated and experimental NOESY spectra of proteins by feedback filtering and self-correcting distance geometry. *J. Mol. Biol.* **254,** 465–480.

Muñoz, V., Thompson, P. A., Hofrichter, J. I., and Eaton, W. A. (1997). Folding dynamics and mechanism of β-hairpin formation. *Nature (London)* **390,** 196–199.

Nabutovsky, D., and Domany, E. (1991). Learning the unlearnable. *Neural Comput.* **3,** 604.

Nilges, M. (1995). Calculation of protein structure with ambiguous distance restraints. Automated assignment of ambiguous NOE crosspeaks and disulfide connectivists. *J. Mol. Biol.* **245,** 645–660.

Novotny, J., Bruccoleri, R., and Karplus, M. (1984). An analysis of incorrectly folded protein models. Implications for structure prediction. *J. Mol. Biol.* **177,** 787–818.

Olmea, O., and Valencia, A. (1997). Improving contact predictions by the combination of correlated mutations and other sources of sequence information. *Fold. Des.* **2,** S25–S32.

Orengo, C. A., Jones, D. T., and Thornton, J. M. (1994). Protein superfamilies and domain superfolds. *Nature (London)* **372,** 631–634.

Ortiz, A. R., Kolinski, A., and Skolnick, J. (1998). Nativelike topology assembly of small proteins using predicted restraints in Monte Carlo simulations. *Proc. Natl. Acad. Sci. U.S.A.* **95,** 1020.

Park, B. H., and Levitt, M. (1995). Energy functions that discriminate x-ray and near-native folds from well-constructed decoys. *J. Mol. Biol.* **249,** 493 (1995).

Rosenblatt, F. (1962). "Principles of Neurodynamics." Spartan Books, New York.

Rosenbluth, M. N., and Rosenbluth, A. W. (1955). Monte Carlo calculation of the average extension of molecular chains. *J. Chem. Phys.* **23,** 356–359.

Rost, B., and Sander, C. (1993). Improved prediction of protein secondary structure by use of sequence profile and neural networks. *Proc. Natl. Acad. Sci. U.S.A.* **90,** 7558–7562.

Šali, A., Shakhnovich, E. I., and Karplus, M. (1994). Kinetics of protein folding. A lattice model study of the requirements for folding to the native state. *J. Mol. Biol.* **235,** 1614–1636.

Scheek, R. M., van Gunsteren, W. F., and Kaptein, R. (1989). Molecular dynamics simulation techniques for determination of molecular structures from nuclear magnetic resonance data. *In* "Methods in Enzymology" (N. J. Oppenheimer and T. L. James, eds.), Vol. 177, pp. 204–218. Academic Press, Orlando, FL.

Skolnick, J., Kolinski, A., and Ortiz, A. R. (1997). MONSSTER: A method for folding globular proteins with a small number of distance restraints. *J. Mol. Biol.* **265,** 217–241.

Stauffer, D., and Aharony, A. (1992). "Introduction to Percolation Theory," 2nd ed. Taylor & Francis, London.

Tanaka, S., and Scheraga, H. A. (1976). Medium- and long-range interaction parameters between amino acids for predicting three-dimensional structures of proteins. *Macromolecules* **9,** 945–950.

Teeter, M. M., Roe, S. M., and Heo, N. H. (1993). Atomic resolution (0.83 Angstroms) crystal structure of the hydrophobic crambin at 130 K. *J. Mol. Biol.* **230,** 292–311.

Thomas, P. D., and Dill, K. A. (1996). An iterative method for extracting energy-like quantities from protein structures. *Proc. Natl. Acad. Sci. U.S.A.* **93,** 11628–11633.

Van Mourik, J., Clementi, C., Maritan, A., Seno, F., and Banavar, J. R. (1998). Determination of interaction potentials of amino acids from native protein structures: Test on simple lattice models. E-print cond-mat/9801137.

Vendruscolo, M., and Domany, E. (1998a). Efficient dynamics in the space of contact maps. *Fold. Des.* **3,** 329–336.

Vendrascolo, M., and Domany, E. (1998b). Pairwise contact potentials are unsuitable for protein folding. *J. Chem. Phys.* **109,** 11101–11108.

Vendruscolo, M., Kussell, E., and Domany, E. (1997). Recovery of protein structure from contact maps. *Fold. Des.* **2,** 295–306.

Vendruscolo, M., Najmanovich, R., and Domany, E. (1999). A novel method to derive contact energy parameters from large databases obtained by threading. *Proteins* in press.

Vendruscolo, M., Subramanian, B., Kanter, I., Domany, E., and Lebowitz, J. L. (1999). Statistical properties of contact maps. *Phys. Rev. E* **59,** 977–984.

VITAMINS AND HORMONES, VOL. 58

HIV Protease: Enzyme Function and Drug Resistance*

SERGEI GULNIK, JOHN W. ERICKSON, AND DONG XIE

Structural Biochemistry Program, SAIC Frederick, National Cancer Institute, Frederick Cancer Research and Development Center, Frederick, Maryland 21702–1201

HIV protease is responsible for processing of the gag and gag-pol polyproteins during virion maturation. The activity of this enzyme is essential for virus infectivity, rendering the protein a major therapeutic target for AIDS treatment. This articles reviews the biochemical and biophysical properties of the enzyme. The clinical and *in vitro* observations of resistance to protease inhibitors are discussed from the perspective of drug resistance mechanisms of HIV protease mutants.

---

*The content of this publication does not necessarily reflect the views or policies of the Department of Health and Human Services, nor does mention of trade names, commercial products, or organization imply endorsement by the U.S. Government.

0083-6729/00 $30.00

I. INTRODUCTION

Human immunodeficiency virus (HIV) encodes three enzymes essential for its life cycle: integrase, RNA-dependent DNA polymerase (reverse transcriptase), and protease. HIV protease (PR) is responsible for posttranslational processing of the polyprotein gene products of *gag* and *gag-pol* to yield structural proteins of the viral core and enzymes including the protease itself (Fig. 1). HIV PR is absolutely required for viral infectivity since mutation of the catalytic residues as well as chemical inhibition of the enzyme leads to the production of immature noninfectious viral particles (Kohl *et al.,* 1988; Seelmeier *et al.,* 1988; McQuade *et al.,* 1990; Kaplan *et al.,* 1993). Therefore, HIV PR is considered an attractive target for the design of drugs against AIDS. For this reason HIV PR has been the subject of numerous investigations aimed at characterizing its biochemical properties, the mechanism of its action, and its three-dimensional structure. Within a relatively short period of time, the enzyme was cloned, expressed in *Escherichia coli* (Kramer *et al.,* 1986; Debouck *et al.,* 1987), purified (see Debouck, 1992, for references), and intensively characterized biochemically. The molecular structures of recombinant and chemically synthesized protein were determined by X-ray crystallography (Navia *et al.,* 1989; Wlodawer *et al.,* 1989; Lapatto *et al.,* 1989). Based on this knowledge, a large number of potent HIV PR inhibitors have been designed and prepared (for recent reviews, see Wlodawer and Erickson, 1993; Vacca, 1994; Vacca and Condra, 1997; Flexner, 1998; Wlodawer and Vondrasek, 1998; Erickson and Eissenstat, 1999). Some of these compounds are currently the most powerful agents for blocking viral replication and HIV infection. However, the rapid selection of drug-resistant mutations limits the long-term efficacy of antiretroviral agents. In this review we address some functional properties of HIV PR that

FIG. 1. Schematic structure of HIV-1 gag and gag-pol polyprotein precursors and PR cleavage sites. (A) p55 Gag and p160 gag-pol polyproteins are translated from single spliced mRNA. Minus 1 ribosomal frame shift at the very C terminus of NC occurs at a frequency of ~5% and results in the gag-pol synthesis. Abbreviations: MA, matrix protein; CA, capsid protein; NC, nucleocapsid protein; PR, protease; RT, reverse transcriptase; RH, RNAse H domain of RT; IN, integrase; P1, P2, and P6, small proteins with unknown function. (B) P4–P4′ sequence of HIV-1 PR cleavage sites in gag-pol polyprotein (consensus B) and kinetic constants of peptide hydrolysis reported by Maschera *et al.,* (1996). Amino acids are numbered P1, P2, etc., from the scissile bond toward the N terminus of peptide and P1′, P2′, etc., from the scissile bond toward the C terminus based on the nomenclature of Schecter and Berger (1967). The position of the scissile bond is marked by * . ND, no data.

A

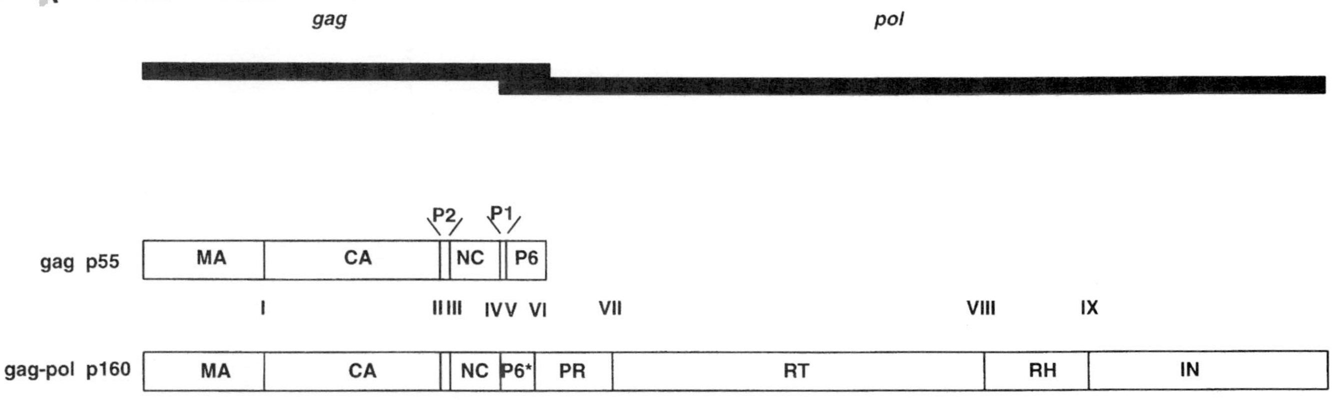

B

	Cleavage site	Sequence (P4 P3 P2 P1 * P1' P2' P3' P4')	k_{cat} (s^{-1})	K_m (mM)	k_{cat}/K_m (s^{-1} mM^{-1})
I	MA-CA	S Q N Y * P I V Q	23	3.75	6.2
II	CA-P2	A R V L * A E A M	2.6	0.37	6.9
III	P2-NC	A T I M * M Q R G	41	3.0	14.0
IV	NC-P1	R Q A N * F L G K	ND	ND	ND
V	P1-P6	P G N F * L Q S R	<0.25	>5	0.05
VI	P6*-PR	S F N F * P Q I T	43	4.0	11.0
VII	PR-RT	T L N F * P I S P	20	3.0	7.0
VIII	RT-RH	A E T F * Y V D G	1.8	0.19	10.0
IX	RH-IN	R K I L * F L D G	5.1	1.15	4.4

might help researchers to understand the possible molecular basis for drug resistance.

II. HIV PR Is a Member of the Aspartic Proteinase Family

HIV PR, as well as all other retroviral proteinases, belongs to the family of aspartic proteinases. This was postulated in 1985 by Toh and coworkers, based on the presence of the signature sequence, DTG, containing an active site aspartic acid residue as well as some other regions of sequence homology with eukaryotic members of the family, such as pepsin. Further evidence from biochemical inhibition and modeling studies supported this idea (Pearl and Taylor, 1987; Katoh *et al.,* 1987). The retroviral protease sequences, however, are much smaller—less than half the size of their mammalian relatives. They contain only one active site aspartic acid residue while two of them are required to make a catalytic diad, suggesting that retroviral enzymes function as homodimers.

Interestingly, Tang and coworkers predicted in 1978 that mammalian aspartic proteinases, which function as monomers with two copies of the active site consensus sequence, originated by gene duplication. From this perspective, retroviral proteinases that consist of two identical monomers can be considered as ancestors of eukaryotic aspartic proteinases. The inhibition of HIV PR by pepstatin A and 1,2-epoxy-3-(4-nitrophenoxy) propane (Seelmeier *et al.,* 1988; Meek *et al.,* 1989)—defining inhibitors of aspartic proteinases—and inactivation of the enzyme by mutagenesis of the catalytic aspartates (Kohl *et al.,* 1988) provided further evidence in support of the assignment of HIV PR to the aspartic proteinase family. Finally the three-dimensional structures of HIV PR clearly reveal a high degree of homology with other aspartic proteinases in their active site architecture.

III. HIV PR Structure

The design of suitable recombinant expression systems (see Stebbins and Debouck, 1994, for a recent review) allowed sufficient production of HIV PR for structural investigation. The crystal structure of HIV PR was solved in 1989 (Navia *et al.,* 1989; Wlodawer *et al.,* 1989; Lapatto *et al.,* 1989). The molecule represents a symmetrical homodimer with one-half of the dimer related to the other by a crystallographic two-fold axis (Fig. 2A). The dimer interface is formed primarily by a four strand-

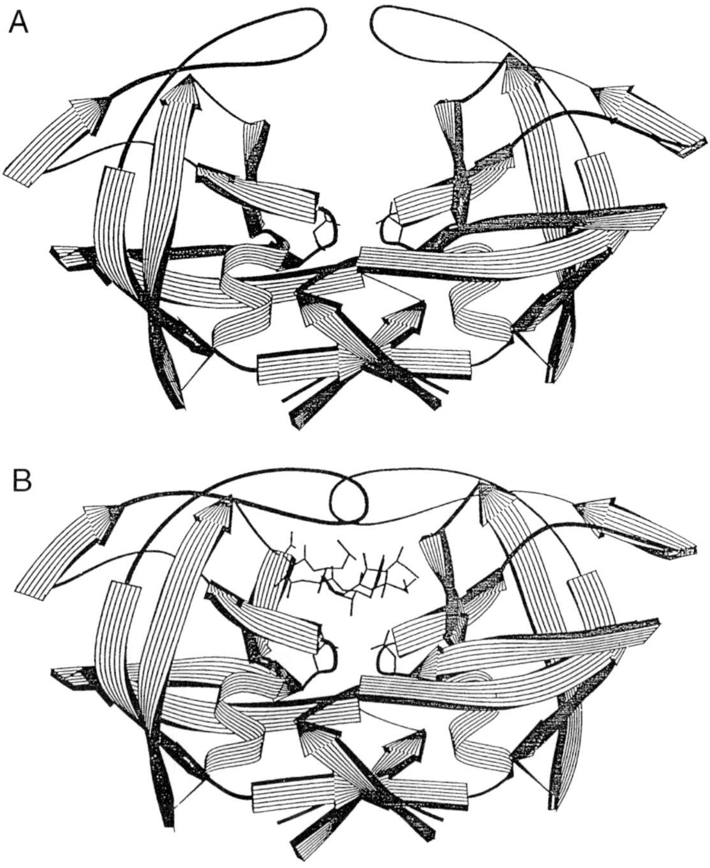

FIG. 2. Chain tracings of HIV PR from crystal structures of (A) unbound and (B) inhibited forms of the enzyme. Inhibitor is drawn as stick figure.

ed antiparallel β-sheet involving both the amino and carboxyl termini of each monomer. The architecture of the HIV PR monomer is similar to that of the single domain of the bilobal cellular aspartic proteinases. The secondary structure of the HIV PR monomer contains predominantly β-strands and only one short helix. The conserved D25-T26-G27 active site triad in each monomer is positioned in a loop that forms part of the catalytic site. The carboxylate groups of D25, D25' from both monomers are nearly coplanar and involved in hydrogen bonding interactions with the "catalytic" water molecule. A fairly long and primarily hydrophobic substrate binding cleft is located opposite the catalytic aspartates and can accommodate peptides of 7–8 amino acids in length. The entrance to the binding cleft is covered by two long β-hairpin loops termed *flaps,* which participate in the binding of inhibitors and presumably of substrates. Unlike mammalian aspartic proteases, which have only one flap, a pair of twofold symmetry-related flaps is present in HIV PR and other retroviral aspartic proteases. The flap is one of the most mobile elements in the aspartic protease structure (Davies, 1990). However, the magnitude of flap movement is significantly greater in HIV PR and other retroviral proteases: Both flaps move up to 15 Å upon inhibitor binding, establishing the so-called "closed" conformation that helps to lock inhibitor in the active site (Wlodawer and Erickson, 1993) (Figure 2B).

IV. Mechanism of Substrate Cleavage and Protonation State of the Catalytic Aspartates

A general acid–base mechanism of catalysis by aspartic proteases has been proposed based on a number of biochemical and crystallographic studies (Davies, 1990). One of the active site aspartic acids is protonated and acts as a general acid, while the other is deprotonated and plays the role of a general base. It is generally agreed that there is no stable covalent acyl or amide intermediate present during catalysis. Nevertheless, the exact mechanism of action of aspartic proteases still remains controversial in its fine details (Meek *et al.,* 1994, and references therein).

Recently Silva *et al.* (1996) proposed a detailed elaboration of the general acid–base mechanism based on the crystal structure of HIV PR in complex with a pseudo-C2 symmetric inhibitor which contains a central difluoroketone motif. The suggested chain of events during substrate catalysis includes asymmetric substrate binding and subsequent hydration of the scissile bond carbonyl by the active site water molecule (step 1); conformational reorientation of nitrogen resulting in the gauche conformation of hydrated peptide bond (step 2); simultaneous proton exchange in which one of the catalytic aspartic acids donates a proton to the nitrogen lone pair, and the other charged aspartate accepts one proton from the diol hydroxyl (step 3); and peptide bond breakage and restoration of the initial protonation state of the active site (step 4).

The ionization state of the active site will influence the polarization of the catalytic water molecule, which is the attacking nucleophile in step 1. Thus, defining the protonation state of the two catalytic aspartic acid residues is critical in elucidating the catalytic mechanism and also for designing potent inhibitors. Theoretically, there are four possible charge states for D25 and D25′: (−1, −1), (−1, 0), (0, −1), and (0, 0), where −1 and 0 represent the deprotonated and protonated state for the carboxyl groups, respectively.

Determining the exact charge state for the protease active site has been the subject of numerous computational and experimental investigations. Based on the crystallographic structures of HIV-1 PR complexed with two substrate-based inhibitors, Harte and Beveridge (1993) performed a study of the protonation state using molecular dynamics simulations. Both catalytic aspartates were predicted to be protonated (0, 0) for the U-85548e inhibitor, and ionized (−1, −1) for the MVT-101 inhibitor, despite the fact that the crystals for both complexes were grown at pH 5.4. The differences in the charge state of catalytic aspartates in these two complexes can be rationalized based on the charge of the inhibitor cores: The hydroxyethylene peptide isostere of U-85548e contains no charged groups, whereas a secondary amine in MVT-101 will be positively charged (Fig. 3). Theoretical studies on HIV-1 PR in complex with several other inhibitors, using semiempirical quantum chemical analysis or a free-energy perturbation method, suggested that the two aspartates are in a singly charged state (Ferguson *et al.,* 1991; Baldwin *et al.,* 1995a; Silva *et al.,* 1996; Tawa *et al.,* 1998).

The experimental characterizations were carried out by measuring the NMR chemical shift of ^{13}C at different pHs. For the uninhibited enzyme, Smith *et al.* (1996) found that D25 and D25' were chemically indistinguishable. At pH 5.9, the catalytic carboxyls were both deprotonated. This is in contrast to the assumption that carboxyls in the free enzyme have a total charge of -1 in order to act as general acid and general base (Hyland *et al.,* 1991). In the protease–pepstatin complex, the two aspartate residues display clearly different titration profiles. One of the carboxyl groups was shown to be protonated and have a pK_a above 6.5, while the other exhibited a pK_a below 2.5 and remained deprotonated in the pH range of 2.5 to 6.5 (Smith *et al.,* 1996). This charge state of the active site is energetically favorable for the accommodation of pepstatin, which mimics the tetrahedral intermediate (Fig. 3). In the protease–pepstatin complex, the protonated carboxyl group forms a hydrogen bond with the hydroxyl group of the statine moiety of the inhibitor.

Similar results for the protonation state of the protease active site were demonstrated by a nuclear magnetic resonance (NMR) study of HIV-1 PR complexed with KNI-272, another substrate-based peptidomimetic transition state analog (Wang *et al.,* 1996). In agreement with the NMR results, a recent isothermal titration calorimetry study of pep-

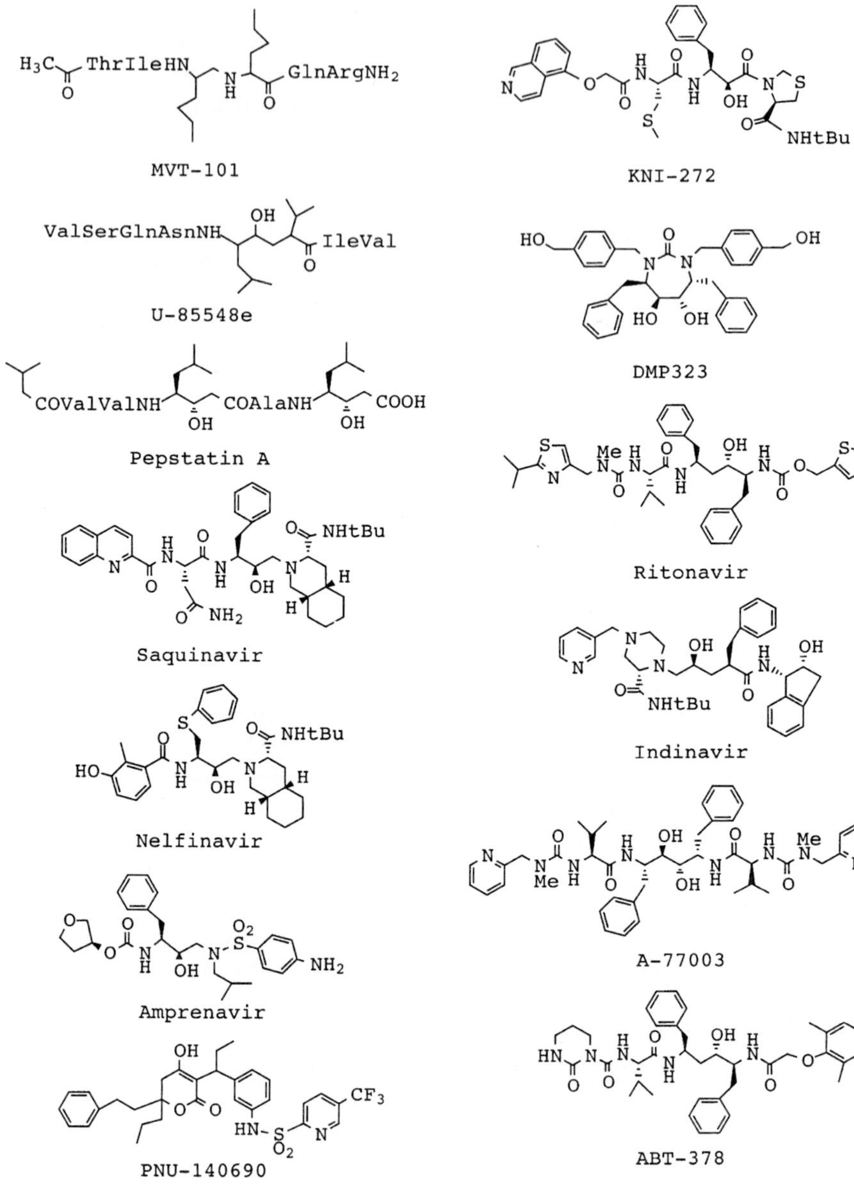

FIG. 3. Chemical structures of the HIV-1 PR inhibitors discussed in this review.

statin binding to HIV-1 PR demonstrated that 0.4 proton was transferred from the buffer to one of the catalytic aspartates upon pepstatin binding at pH 6.5 (Xie *et al.*, 1998). This indicates the increase of pK_a of one catalytic carboxyl group to a value above 6.5 when pepstatin binds, while the other aspartate remains deprotonated.

Another NMR study of HIV-1 PR in complex with a nonpeptide cyclic urea-based inhibitor provided a very different picture for the protonation state of the active site (Yamazaki *et al.*, 1994). Both catalytic aspartyl groups were shown to be protonated. The DMP323 inhibitor used in this work, in contrast to pepstatin, has two hydroxyl groups (Fig. 3). It has been suggested that these two hydroxyl groups form a hydrogen-bond network with the protonated catalytic aspartates, resulting in the most stable protease–inhibitor complex.

Taken together, both computational and experimental studies demonstrate that the protonation state of the two catalytic aspartates in HIV-1 PR can be profoundly influenced by the presence or absence of inhibitor or substrate, as well as their electrostatic properties.

V. Substrate Specificity

HIV PR is responsible for cleavage of at least nine peptide bonds in the p55 gag and p160 gag-pol polyproteins. The kinetic properties and specificity of HIV PR have been analyzed for the cleavage of peptide substrates encompassing the natural cleavage sites (Fig. 1) (see reviews in Dunn *et al.*, 1994; Tomasselli and Heinrikson, 1994). It seems clear that HIV PR interacts with amino acids from P4 through P4′ of the substrate.

Based on sequences around the scissile bond, substrates can be classified into three groups (Henderson *et al.*, 1988). The first group is characterized by the presence of asparagine in P2, an aromatic amino acid in P1, and proline in P1′ (peptides I, VI, VII, Fig. 1B). The proline residue in that P1′ position is rarely observed for other endoproteases, making the specificity of retroviral proteases unique. Substrates in the second group have arginine in P4 and phenylalanine and leucine in P1′ and P2′ (peptide IV, IX, Fig. 1B). Peptides having glutamic acid or glutamine in P2′ belong to the third group (peptides II, III, V, VI, Fig. 1B).

Later Griffiths *et al.* (1992) proposed another classification of substrates based on P1-P1′ amino acids: hydrophobic–hydrophobic and aromatic–proline. Analysis of HIV PR cleavage sites in heterologous proteins revealed, however, a number of exceptions to both classifications (Poorman *et al.*, 1991). For example, small apolar and charged amino acids were reported in both P1 (G and E) and P1′ (G, E, and R) positions. Interestingly, glutamic acid is the most frequently observed amino acid in the P2′ position of heterologous cleavage sites, and its presence is correlated with less discrimination toward P1 and P1′ substituents by HIV PR. By analyzing kinetic data obtained from a series of synthetic peptides, and performing molecular modeling of enzyme–substrate complexes, Griffiths *et al.* (1992) concluded that extensive interactions at long distances of the scissile peptide bond are responsible for the ability of HIV PR to selectively and efficiently cleave a variety of unrelated sequences.

HIV PR has been reported to cleave *in vitro* a variety of cellular proteins, including calmodulin (Tomasselli *et al.,* 1991a), pro-interleukin 1β (Tomasselli *et al.,* 1991b), NF-κB (Riviere *et al.,* 1991), several cytoskeletal proteins (Shoeman *et al.,* 1993), and the C3 component of the complement system (Kisselev *et al.,* 1997; for a recent review, see Snasel and Pichova, 1996). The cleavage of cellular antiapoptosis protein bcl-2 was proposed to be responsible for the high cytotoxicity of HIV PR (Korant *et al.,* 1998). However, the importance of these observations for HIV pathogenesis remains unclear.

HIV PR can undergo autolysis; that is, the enzyme can serve as a substrate for itself. Autolysis leads to enzyme inactivation, and can introduce significant errors in experimental enzyme kinetics data. There are three major autodigestion sites in HIV PR: L5-W6, L33-E34, and L63-I64 (Strickler *et al.,* 1989; Hui *et al.,* 1993; Rose *et al.,* 1993). Because none of the cleavage sites is readily accessible in the folded dimer molecule, it was proposed that the disordered monomers serve as better substrates for autolysis (Mildner *et al.,* 1994). Analysis of HIV PR specificity revealed that several amino acids are generally excluded from certain positions in the substrate. For example, introduction of β-branched amino acids at the P1 position or lysine at P2′ dramatically decreases the ability of HIV PR to efficiently process peptide substrates (Konvalinka *et al.,* 1990; Phylip *et al.,* 1990; Griffiths *et al.,* 1992; Tozser *et al.,* 1992; Dunn *et al.,* 1994). This observation led to the construction of a triple mutant, Q7K/L33I/L63I, which has a significantly lower autolysis rate, but otherwise the same biochemical properties, as the wild-type HIV PR (Mildner *et al.,* 1994). This mutant has been widely used in biochemical and biophysical studies.

VI. Assay Methods

It is important to have a sensitive, relatively simple and reliable assay for HIV PR activity that can be used for biochemical and enzymological studies and for screening and characterization of inhibitors. Assays for HIV PR activity can be divided into two major groups: discontinuous and continuous (see Hellen, 1994, for a recent review). The first group requires withdrawal of samples for analysis at various times of substrate cleavage, whereas the second allows continuous monitoring of enzymatic reaction. SDS gel electrophoresis in conjunction with immunoblotting or autoradiography and HPLC separation of peptide products are examples of the most commonly used discontinuous assays for analyzing polyprotein and peptide substrate cleavage, respectively. Although discontinuous assays often provide simultaneous information about cleavage specificity, they are usually time consuming and suffer from low detection sensitivity. Continuous enzyme assays are usually fast, sensitive, simple, and can be easily automated for high-throughput screening of inhibitors. As with other aspartic proteinases, HIV PR does not have esterase and amidase activities, which are most commonly exploited for proteinase assays. Several different approaches have been developed for continuous monitoring of enzyme activity. In one such technique the substrate contains a *p*-nitrophenylalanine (Nph) residue in the P1′ position. Below pH 6.5, protonation of the N-terminal Nph, generated during cleavage, shifts its absorbance maximum from 278 to 272 nm. This makes it possible to follow product formation continuously by monitoring the absorbance decrease at 300–310 nm (Dunn *et al.*, 1994).

Another technique is based on the use of fluorogenic peptides containing fluorophor and quencher groups in the same molecule. The fluorescence is quenched in substrate due to either collision between the fluorophor and quencher or resonance energy transfer. Cleavage results in an increase in fluorescence due to the liberation of fluorescent product, and allows proteolytic activity to be continuously monitored (Krafft and Wang, 1994).

In studying enzyme substrate specificity, usually many peptides of different sequences need to be investigated. It is difficult to make all studied peptides chromogenic or fluorogenic. Moreover, the addition of chromophore onto a substrate might make the assay less biologically relevant. In search of an alternative way to study peptide substrate specificity of HIV-1 PR, we recently developed a competition assay using spectrophotometric measurements (Xie *et al.*, 1999b). In this assay, the kinetic constants for the cleavage of a fluorogenic substrate are first determined independently. Then the initial cleavage rate of the fluorogenic substrate is measured in the presence of a second substrate. The second substrate can be any peptide that competes for the enzyme active site and is not required to have any optical property change upon cleavage. Using unique experimental design and data analysis for a two-substrate system, we were able to determine both K_m and k_{cat} values for cleavage of the second substrate. This method provides fast, precise, and continuous measurement of the cleavage rate of unmodified oligopeptide. In general, it can be applied to other enzyme systems.

VII. pH Optima

The pH dependence of the specificity constant k_{cat}/K_m is characterized by a bell-shaped curve with a maximum at pH 4.5–5.5. Ionization of the two carboxyl groups of catalytic aspartyl residues D25 and D25′ was suggested to account for this pH dependency (Billich *et al.,* 1990; Richards *et al.,* 1990; Ido *et al.,* 1991; Konvalinka *et al.,* 1992; Meek *et al.,* 1994; Polgár *et al.,* 1994; Szeltner and Polgar, 1996). Notwithstanding, the estimated pK_a values for the active site aspartates vary significantly: 3.0–5.1 for the more acidic aspartate, and 5.0–6.9 for the more basic one. These values appear in contrast to results of the NMR studies discussed earlier, which showed that in the HIV PR–pepstatin A complex, the pK_a of one catalytic carboxyl group is >6.5, while the other is <2.5 (Smith *et al.,* 1996). In addition to the two catalytic aspartates, there are several other titratable residues in this pH range, D29, D29′, D30, and D30′. These residues are located in the vicinity of the active site and may contribute to the pH dependence of the kinetic constants. There is little agreement in the literature on the dependence of individual kinetic constants on pH as well. Richards *et al.* (1990) reported that the rate constant k_{cat} was independent of pH, while K_m showed strong pH dependence. In contrast Billich *et al.* (1990) found K_m to be pH independent. It appears that the pH profile of the kinetic constants may depend on the substrate (Polgár *et al.,* 1994).

Ionic strength is another important factor that modulates HIV PR activity. It is commonly accepted that HIV PR is more active at higher ionic strength (Billich *et al.,* 1990; Richards *et al.,* 1990; Meek *et al.,* 1994; Polgár *et al.,* 1994). This activation may result from enhancement of favorable hydrophobic interactions between the enzyme and the mainly hydrophobic peptide substrates or inhibitors (“salting in” of peptides to the active site of protease) (Meek *et al.,* 1994). Our results show that the dependence of activity on ionic strength for HIV PR may be somewhat more complicated. For example, the k_{cat}/K_m ratio for hydrolysis of KARVYNphEANleNH_2 does not depend on ionic strength at pH near 4.5 (Fig. 4). At lower pH, the activity of HIV PR decreases at higher ionic strength, whereas at pH above 4.5 the activity increases.

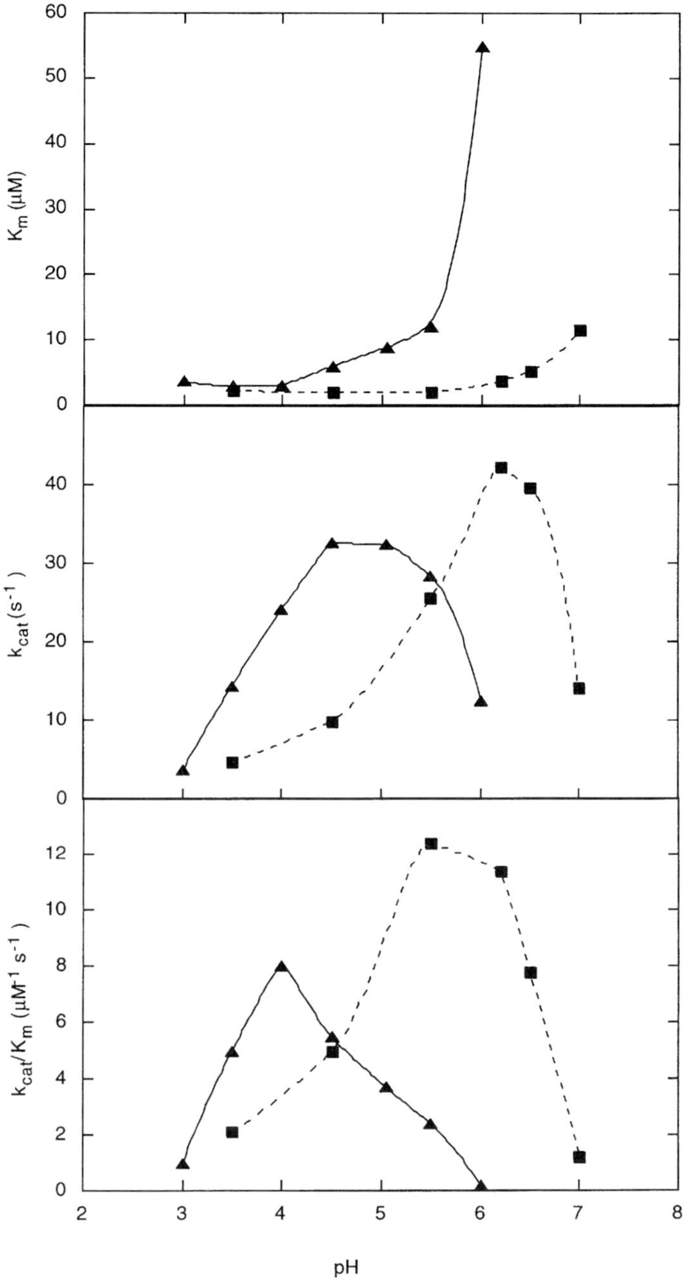

FIG. 4. pH Dependence of K_m (upper), k_{cat} (middle), and k_{cat}/K_m (lower) for the hydrolysis of KARVYNphEANleNH$_2$ by HIV-1 PR. Kinetic constants were measured in buffers, containing 0.2 *M* sodium chloride (triangles) or 1.25 *M* ammonium sulfate (squares).

VIII. Dimer Stability

An interesting feature of HIV PR is its dimeric functional form in contrast to the monomeric eukaryotic aspartic proteases. The dimer stability of HIV PR has been the subject of extensive investigations because the monomer–dimer equilibrium might provide a mechanism for regulating protease activity (see Darke, 1994, for review and references therein). The dimer dissociation constant (K_D) has been determined using different approaches based on measurements of enzyme activity (Cheng *et al.*, 1990; Zhang *et al.*, 1991; Jordan *et al.*, 1992; Kuzmic *et al.*, 1993), fluorescent inhibitor binding (Darke *et al.*, 1994), protein folding/unfolding induced by urea or temperature (Grant *et al.*, 1992; Todd *et al.*, 1998), and sedimentation equilibrium (Grant *et al.*, 1992; Xie *et al.*, 1999a).

The reported K_D values differ substantially. For example, at pH 5.0, an indirect enzyme kinetic study yielded a K_D of 3.6 n*M* (Zhang *et al.*, 1991), whereas a value of 23 p*M* was obtained by characterizing the urea-induced dimer dissociation of HIV-1 PR (Todd *et al.*, 1998). Although the experimental conditions were slightly different, for example, 37°C and 1 *M* NaCl for the kinetics experiment and 25°C and no salt for the urea denaturation measurement, the discrepancy cannot be easily reconciled since the dimeric enzyme is known to be equally or more stable at higher ionic strength (Kuzmic *et al.*, 1993; Pargellis *et al.*, 1994; Darke *et al.*, 1994).

One factor that could theoretically lead to erroneous measurements of K_D is the overestimation of dimer stability due to substrate binding. This most likely happens in assays where K_D values are estimated from the nonlinear dependence of enzyme activity on total enzyme concentration. When the substrate concentration is higher than or near K_m, substrate binding should shift the equilibrium toward dimer formation. However, even at substrate concentrations significantly lower than K_m, substrate stabilization effects may be significant. This is evident from the definition of K_m:

$$K_m = k_{-1}/k_1 + k_{cat}/k_1,$$

where k_{cat} is the catalytic rate constant, and k_1 and k_{-1} are the on- and off-rate constants for substrate binding. The first term on the right-hand side of the equation represents the equilibrium constant for substrate binding, or K_S. Hence, K_m will always be greater than or equal to K_S. The extent of the difference between these two parameters is determined by k_{cat}/k_1. The reported values of k_{cat} for cleavage of most substrates by HIV-1 PR are in the range of 1–40 s^{-1} (Dunn *et al.*, 1994).

The on-rate for a peptidomimetic inhibitor binding to HIV-1 PR measured by stopped-flow spectrofluorometry was $1.6 \times 10^5\ M^{-1}s^{-1}$ (Furfine *et al.,* 1992). Hence, assuming a similar k_1 for substrate binding, the k_{cat}/k_1 ratio can be estimated to be 6–250 μ*M*. This value is of the same magnitude as K_m, making estimates of K_S essentially indeterminate; they could be substantially lower than K_m. Thus, substrate at concentrations well below K_m but still in the range of K_S would stabilize the dimer.

Most of the studies on HIV PR dimer stability were carried out at acidic pH or high-salt concentration where the protease exhibits optimum activity, and the dimer is most stable, with K_D values in the low nanomolar or subnanomolar range. Although the exact environment of viral polyprotein precursor processing by HIV PR is unresolved, it appears to occur during or subsequent to viral budding (see reviews in Vogt, 1996; Coffin *et al.,* 1997), possibly under physiologic conditions of neutral pH and low-salt concentration. Indeed, in bacterial expression systems, HIV PR can be autoactivated from recombinant precursor proteins at neutral pH, and accurately cleaves the gag and gag-pol polyprotein processing sites (Kotler *et al.,* 1992, 1997). Cytoplasmic HIV PR activity has also been observed in virus-infected cells *in vitro,* but this usually leads to a reduction in infectious particle formation (Kaplan and Swanstrom, 1991).

The K_D of HIV-1 PR at neutral pH was reported to be 58 n*M* (Darke *et al.,* 1994). In that work, the time course of the binding of a fluorescent inhibitor to PR was monitored. The dimer concentrations and subsequently the K_D values were calculated using an independently measured equilibrium dissociation constant (K_i) and on and off rate constants for inhibitor binding. However, the measurements of K_i and the on rate were carried out at protease concentrations near the reported K_D value. Under such conditions, inhibitor-induced dimerization may result in an underestimated inhibitor affinity and overestimated dimer concentration and stability. Recently, we have directly determined the K_D value of HIV PR at neutral pH using sedimentation equilibrium (Xie *et al.,* 1999a). The triple mutant, Q7K/L33I/L63I, which exhibits a significantly low autolysis rate as discussed earlier, was used as a surrogate of the wild-type enzyme. At pH 7.0, the K_D was determined to be 5.8 μ*M,* an almost 1000-fold increase from the value at acidic pHs.

The relatively low dimer stability at neutral pH provides a possible mechanism for regulating dimer formation and virus maturation during virus infection. For example, assuming a radius of 5 μm for an infected T cell, the cellular HIV PR concentration due to the budding of 1–10 viral particles can be estimated to be 1–10 n*M*. In a 50-nm-radius virus particle, on the other hand, the protease concentration should be in the range of 1 m*M*. Thus, according to this hypothesis, the low protease concentration in infected cells prevents immature precursor processing before viral budding; and the local increase in protease concentration in the virion ensures dimer formation and rapid maturation.

IX. Thermodynamic Characteristics of Inhibitor Binding

For the purpose of structure-based drug design, the knowledge of thermodynamic parameters other than K_i, such as the binding enthalpy ΔH and entropy ΔS, is important for understanding the mechanism of enzyme–drug interactions. Despite the intensive effort to design potent HIV PR inhibitors, relatively few thermodynamic studies have been reported for this system. Isothermal titration calorimetry (ITC) is usually the technique of choice for high-sensitivity thermodynamic characterization of molecular recognition processes (Freire *et al.*, 1990; Ladbury and Chowdhry, 1996).

The thermodynamic parameters for pepstatin A binding to HIV-1 PR have been determined using ITC (Xie *et al.*, 1998). At pH 6.5 the equilibrium dissociation constant was 1.5 μ*M*. A distinguishing feature of this binding process is its positive binding enthalpy. At 25°C, the ΔH was 5.3 kcal/mol. Similar results were obtained for the binding of acetyl pepstatin to HIV PR (Luque *et al.*, 1998). The binding of a tripeptide analog inhibitor to simian immunodeficiency virus PR and a V32I/I47V/V82I triple mutant of HIV-1 PR also exhibited positive ΔH values of 6.4 and 2.5 kcal/mol, respectively (Hoog *et al.*, 1996). The large, positive ΔH is highly unfavorable for the binding reaction, and indicates that the complexation of reactants is an entropy-driven process. These thermodynamic properties are the characteristics of a binding reaction involving burial of hydrophobic groups and concomitant release of solvent molecules (Murphy and Freire, 1992). A further dissection of the binding energetics for acetyl pepstatin was performed using a semi-empirical structure-based method (Luque *et al.*, 1998). The entropic contribution to the binding Gibbs free energy from the burial of hydrophobic groups and solvent release was shown to be −37 kcal/mol.

X. Inhibitors of HIV PR

Identification of HIV PR as a member of the aspartic proteinase family allowed investigators to take advantage of the extensive research on inhibitors of renin—a human aspartic proteinase involved in blood pressure regulation. Unique features of HIV PR such as the symmetry of homodimer and the unusual substrate specificity (ability to accommodate proline in the S1′ binding site, which is not found for mammalian proteinases), together with the availability of numerous crystallographic data, greatly accelerated the process of design and improvement of the selectivity and potency of HIV PR inhibitors. Structural, synthetic, and clinical aspects of HIV PR inhibitors have been the subject of several recent reviews (Wlodawer and Erickson, 1993; Vacca, 1994; Winslow and Otto, 1995; West and Fairlie, 1995; Moyle and Gazzard, 1996; Barry *et al.,* 1997; Deeks *et al.,* 1997; Korant and Rizzo, 1997; Flexner, 1998; Wlodawer and Vondrasek, 1998; Erickson and Eissenstat, 1999).

Four inhibitors of HIV PR have been approved as drugs for the treatment of patients with AIDS: saquinavir (Roberts *et al.,* 1990), indinavir (Dorsey *et al.,* 1994; Vacca *et al.,* 1994), ritonavir (Kempf *et al.,* 1995), and nelfinavir (Kaldor *et al.,* 1997) (Fig. 3). These drugs, as well as most of the compounds now in clinical trials, represent a class of substrate-based peptidomimetic inhibitors with the scissile bond replaced by a nonhydrolyzable transition state analog. The fine tuning of most of these compounds was achieved by using a combination of chemical synthesis, crystal structure determination, and computer-aided drug design. All of these inhibitors contain a central hydroxy group that mimics the diol transition state intermediate and makes hydrogen bond interactions with the catalytic aspartic acids of HIV PR. Ritonavir has a P1-P1′ symmetric core and originated from the C2 symmetry-based design concepts of Erickson and coworkers (1990).

A new approach based on the substitution of a structurally buried “flap” water molecule in HIV PR was proposed by DuPont Merck researchers. The novel cyclic urea inhibitors [DMP323, DMP450 and analogs (Lam *et al.,* 1994; Hodge *et al.,* 1996)] contain a rigid seven-membered ring system with the carbonyl oxygen making hydrogen bonding interactions of the replaced structural “flap” water molecule (Ala *et al.,* 1998). A similar approach was used by Gilead in its design of nonpeptidic cyclic sulfones (Kim *et al.,* 1996). Both of these designs incorporated C2 symmetry.

Two series of nonpeptidic, non-transition-state mimic compounds are now in development by Pharmacia-Upjohn and Parke-Davis (Prasad *et al.,* 1995, 1996; Thaisrivongs *et al.,* 1996). Both series are based on 4-hydroxypyrone derivatives discovered from high-throughput enzyme-based screening and optimized by structure-based drug design.

Irreversible inhibitors, which inactivate aspartic proteinases by alkylation of the active site aspartic acids (Yu *et al.,* 1996), become attractive in light of the growing problem of resistance to existing HIV PR inhibitors. Theoretically, even low potency compounds might be able to inactivate mutant enzymes since lower occupancy of the active site can only delay but not necessarily prevent complete inhibition. Generally, the major concern in developing these compounds is the possible toxicity associated with lack of specificity of the reactive epoxide moiety. However, several recent reports on new, highly potent epoxide-based irreversible inhibitors of HIV PR suggest a low level of toxicity in cell culture and a high degree of specificity (Park *et al.,* 1995; Lee *et al.,* 1996).

The dimer interface of HIV PR represents another potential target for inhibition that has been studied by a number of investigators. Dimer interface-based peptides inhibit HIV PR in a noncompetitive manner and inhibitors based on cross-linked interfacial peptides have been proposed (Zhang *et al.,* 1991; Schramm *et al.,* 1996; Zutshi *et al.,* 1997, and reference therein). Recently, nonpeptidic dimerization inhibitors, polycyclic triterpenes, have been found (Quere *et al.,* 1996). However, unlike competitive inhibitors which target the active site of HIV PR and have potencies in the subnanomolar range, dimerization inhibitors exhibit only micromolar K_i's.

XI. Introduction to Resistance to HIV PR Inhibitors

One of the most striking features of HIV is the large number of genetic variants found in virus isolates. For retroviruses, genetic errors are introduced both during reverse transcription as well as during proviral DNA transcription into progeny RNA genomes by the cellular RNA polymerase. Preexistent, low populated resistant species can become selected in the presence of drug. This phenomenon generates a serious problem in AIDS therapy: drug resistance. The fast development of viral resistance to nucleoside and non-nucleoside inhibitors of HIV reverse transcriptase makes these compounds much less effective.

Theoretically, HIV PR has several advantages as a target for therapy. The enzyme consists of two identical monomers so that each single genetic variation generates in fact a double mutant for the protease dimer. The relatively small size of the enzyme might lead to some restrictions in mutational possibility. However, despite the high expectations, mutations within the protease gene conferring resistance have emerged for each of the four FDA-approved HIV PR inhibitors, both in cell cultures and in patients (for recent reviews, see Winslow and Otto, 1995; Ridky and Leis, 1995; Erickson, 1995; Roberts *et al.,* 1998; Molla *et al.,* 1998; Boden and Markowitz, 1998; Erickson *et al.,* 1999).

The main factors responsible for the fast generation of resistant variants are the error-prone reverse transcription and the dynamic viral turnover at steady state. It has been estimated that each possible single mutation in the HIV PR gene can occur 10^4–10^5 times daily in infected individuals (Coffin, 1995). If, in the presence of a drug, any mutant is more fit than the wild-type HIV PR and therefore can provide a selective advantage to the virus, it will be selected quickly. Genotypic changes—mutations in the HIV PR gene—usually coincide with clinical manifestations of resistance: viral RNA rebound and decrease in CD4 cell count. In this review we discuss primarily the resistance data for HIV PR inhibitors approved as drugs for AIDS treatment.

To date, more than 87 changes have been observed in at least 47 positions in the 99-amino-acid monomer. A compilation of resistance mutations for HIV PR and reverse transcriptase is maintained at the Los Alamos National Laboratory, HIV Sequence Database (Hammond *et al.,* 1997). Some of the most representative mutations found *in vivo* are shown in Fig. 5. Mutations in HIV PR can be classified as active site or nonactive site depending on whether they are located within or outside the inhibitor binding subsites (Erickson, 1995; Erickson and Burt, 1996). There are several possible strategies by which the virus could evolve resistance to protease inhibitors (Erickson, 1995): active site mutations which interfere directly with binding; nonactive site mutations, which interfere with inhibitor binding through long-range perturbations; mutations, which reduce the dimer stability of HIV PR, thus decreasing its apparent drug affinity; mutations in the cleavage sites of HIV PR polyprotein substrates; and up-regulatory mutations in or outside the protease gene, for example, those resulting in a more efficient frame shift event.

XII. Active Site Mutations

The substrate–inhibitor binding cleft in HIV PR contains a number of defined pockets—"specificity subsites"—that accommodate the side chains of inhibitor and presumably substrate. The interactions of inhibitor and enzyme side-chain atoms in these subsites define the specificity and potency of a inhibitor. Thus, enzyme residues in specificity subsites are obvious candidates for mutations that could directly affect inhibitor binding. Clinical data indicate the presence of a certain order in the selection process. Usually active site mutations represent primary mutations, temporally and quantitatively associated with the initial loss of suppression of viral replication (Eastman *et al.,* 1998). Therefore, the degree of sensitivity of different inhibitors to HIV PR active

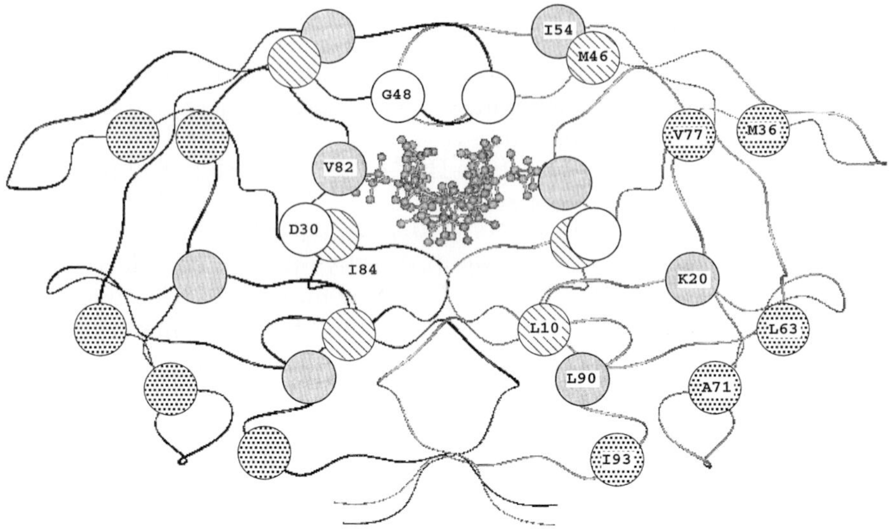

FIG. 5. Position of some important protease mutations selected *in vivo* by the four clinically approved PR inhibitors. Mutations represented by dotted, striped, gray, or open circles were observed for all four drugs, three, two, and one drug, respectively. The monomer subunits are shown in gray and black. The mutation sites are labeled by their residue type and sequence number, with the active and nonactive site mutations labeled on the left- and right-hand sides of the dimer, respectively. A difluoroketone-containing inhibitor, A79285, is shown in ball-and-stick colored by atom type. The PDB file is 1dif (Silva *et al.*, 1996).

site mutants may have an important predictive value in an ongoing effort to create resistance-proof inhibitors and to find a proper combination of inhibitors to block the development of drug resistance.

In vitro susceptibility of different mutants to inhibition may be characterized by two approaches: measuring the inhibition constant (K_i) of purified mutant HIV PR; or determining IC_{50} or IC_{90} values for the inhibition of replication of recombinant viruses containing mutant protease in cell culture (phenotypic assay). Because HIV PR inhibitors are highly optimized for the wild-type enzyme, it is not surprising that active site mutations, by themselves, may cause a significant drop in inhibitor potency. In general, a given inhibitor exhibits elevated K_i and IC_{50} values toward mutants selected using that compound (Kaplan *et al.,* 1994b; Gulnik *et al.,* 1995; Partaledis *et al.,* 1995; Patick *et al.,* 1995; Pazhanisamy *et al.,* 1996; Maschera *et al.,* 1996; Nillroth *et al.,* 1997; Croteau *et al.,* 1997; Jadhav *et al.,* 1997). For example, two active site mutations, I84V and V82A, have been found in multisubstituted clinical isolates from patients subjected to ritonavir, saquinavir, or indinavir monotherapy (Condra *et al.*, 1995, 1996; Molla *et al.*, 1996; Eastman *et al.*, 1998; Winters *et al.*, 1998; Zennou *et al.*, 1998). A 7-, 6-, and 10-fold increase in K_i over wild type has been reported for ritonavir, saquinavir, and indinavir, respectively, for the I84V mutant protease; and 10-, 4-, and 22-fold increases were found for the V82A mutant (Gulnik *et al.*, 1995).

IC_{90} ratios over wild type were reported to be 3.7, 0.7, and 1.0 for the I84V mutant and 4.7, 0.5, and 2.3 for the V82A mutant for these three drugs, respectively (Klabe *et al.*, 1998). Another clinically relevant mutant—V82F, which has been found for ritonavir and indinavir—causes a 24- and 85-fold increase in K_i, respectively, and a 5.9- and 2.3-fold increase in IC_{90}. This mutation has never been selected using saquinavir, which can be explained by the 2-fold *decrease* in K_i and 4-fold *decrease* in IC_{90}. The combination of two active site mutations, V82F/I84V and V82T/I84V, occasionally found in clinical isolates during indinavir monotherapy (Condra *et al.*, 1995, 1996), caused a large increase in K_i for this inhibitor: 41- and 48-fold, respectively (L. Suvorov and S. Gulnik, unpublished results). Although K_i values are usually positively correlated with IC_{50} values, there are cases where a correlation was not observed. In several cases, IC_{50} determinations in cell culture failed to detect a significant level of resistance for singly substituted active site mutants found in patient isolates (Condra *et al.*, 1995; Molla *et al.*, 1996; Klabe *et al.*, 1998) even though elevated K_i's were observed in the PR assay. Because these mutations afford a selective advantage to the virus in the presence of drug, the more sensitive K_i analysis may have greater predictive value.

XIII. Nonactive Site Mutations

Genotypic analyses of viral isolates at different stages of antiretroviral therapy revealed an ordered accumulation of mutations associated with the progression of resistance (Condra *et al.*, 1995; Molla *et al.*, 1996; Zhang *et al.*, 1997). It is not uncommon that HIV PR in viral isolates from heavily drug-experienced individuals contain 10–14 mutations (H. Mitsuya, personal communication; Condra *et al.*, 1996). Among them only 1–3 are active site mutations, while the majority are located outside of the active site. While the effect of active site mutations can often be rationalized based on crystallographic data or on modeling experiments (see later discussion), understanding the effect of nonactive site mutations, which do not directly interfere with inhibitor binding, is less straightforward. Experimental data suggest that at least some of the nonactive site mutations decrease the binding affinity of inhibitor. An example is L90M—the most clinically relevant mutation for saquinavir—which exhibits a 3.7-fold increase in K_i (L. Suvorov and S. Gulnik, unpublished results). It has been reported that mutations at codons 36 and 54, M36I and I54V, increase K_i and IC_{50} values in both enzymatic and cell culture-based antiviral assays (Nijhuis *et al.*, 1999). Nonactive site mutation N88S has been described as a major drug resistance-conferring mutation for the hydroxyethyl-urea inhibitor SC-55389A (Smidt *et al.*, 1997). The M46I mutation on the flaps appears to be critical for the manifestation of resistance against indinavir (Condra *et al.*, 1995). Molecular dynamics simulations on flap movement suggest that this mutation may affect the dynamics of flap closure (Collins *et al.*, 1995), thus altering the enzyme kinetics and affecting K_i.

In contrast to active site drug resistant mutants which always exhibit a reduction in the sensitivity toward inhibitors, it has been shown that many nonactive site mutations do not significantly alter inhibitor binding (Gulnik *et al.*, 1995; Pazhanisamy *et al.*, 1996; Schock *et al.*, 1996). On the other hand, all active site mutants have decreased catalytic efficiency (k_{cat}/K_m) compared to wild-type HIV PR (Gulnik *et al.*, 1995; Pazhanisamy *et al.*, 1996; Wilson *et al.*, 1997; Ridky *et al.*, 1998). In some cases, reduction in enzyme efficiency was reflected by impaired virus replication in cell culture.

Ho and colleagues (1994) first reported that the M46I mutation, while not changing sensitivity to A-77003, appeared to improve replication kinetics of the severely deficient mutant virus that contained the active site R8Q mutation. The M46I mutation was identified as a "compensatory mutation," because it at least partially restored the viral replication rate and/or the enzyme catalytic efficiency introduced by active site mutations. The existence of other compensatory mutations was later confirmed by many *in vitro* studies.

Markowitz *et al.*, (1995) have shown that the L63P mutation can partially compensate for the deficient replication of the V82F/I84V double mutant and suggested that A71V may play a similar role. The compensatory effect of A71V has been later clearly demonstrated by both enzymatic assay and viral replication assay for the M36I/I54V/V82T triple mutant found in clinical isolates from an individual treated with ritonavir (Nijhuis *et al.*, 1999). Upon addition of the A71V mutation, the k_{cat}/K_m ratio of the quadruple mutant increased by 10-fold. The indinavir-selected M46L/L63P mutation improved the catalytic efficiency of HIV PR and was able to partially compensate for the decrease in activity of the V82T/I84V mutant (Schock *et al.,* 1996). It has been also demonstrated that L63P can improve the replication capacity of the L90M and D30N mutants (Martinez-Picado *et al.,* 1999), and that M46I and I47V mutations partially compensate for the poor efficiency of the amprenavir selected I50V mutant (Pazhanisamy *et al.,* 1996). The addition of L10I increases the replication rate of the virus containing a triple mutant, G48V/A71T/V82A, by more than 25-fold; addition of the A71T mutation significantly improved the activity of G48V/L90M at the enzyme level (Rose *et al.,* 1996).

Interestingly, it has been noted that the L90M mutation accumulated *in vivo* with a higher frequency in the proteinase genes that carry the mutations L10I, M36I, L63P, and A71T (Jacobsen *et al.,* 1996). These four positions represent major sites for the natural polymorphism of HIV PR (Barrie *et al.,* 1996; Lech *et al.,* 1996; Kozal *et al.,* 1996; Cornelissen *et al.,* 1997). It is possible, therefore, that the initial presence of these background mutations in HIV PR genes of drug-naive individuals may increase the potential for resistance development.

XIV. Cleavage Site Mutants

As discussed earlier, HIV PR recognizes and cleaves the gag-pol precursor polyprotein at nine specific sites (Fig. 1A). The cleavage of polyprotein precursor apparently requires precise timing (Kaplan *et al.,* 1994a) and occurs in an ordered fashion (Pettit *et al.,* 1994; Kräusslich *et al.,* 1995). Because HIV PR active site mutations generally reduce the catalytic efficiency of the enzyme, it might be advantageous for the virus to select for mutations at or near the cleavage sites to restore the efficiency of hydrolysis. In a sense, such mutations can be defined as compensatory mutations but outside the protease region. Mutants of this type were first observed by Doyon *et al.* (1996) during *in vitro* selection experiments using two potent substrate analog protease inhibitors, BILA 2185 and BILA 1906. Multiply substituted HIV PR clones generated in the presence of these inhibitors displayed very high levels of resistance, up to a 1500-fold increase in IC_{50} values. Besides the mutations in HIV PR, additional substitutions were found in two gag cleavage sites, in the P3 and P2 residues of the p7–p1 junction (QA to RV), and in the P1′ residue of the p1–p6 junction (L to F). Mutation at the p1–p6 cleavage site appeared early during selection, while the p7–p1 mutation was found only in very late passages.

Reconstructed viral molecular clones with mutations in the protease region had severely impaired growth kinetics, which could be at least partially restored by addition of the gag mutations. The early appearance of the p1–p6 mutation and late appearance of the p7–p1 mutation have also been reported recently from *in vitro* selection studies using ABT-378 (Carrillo *et al.,* 1998).

In an *in vivo* study, Salzman and colleagues (Zhang *et al.,* 1997) found mutations in the P2 position of the p7–p1 gag cleavage site (A to V) in all six patients who developed resistance to indinavir. In contrast to the *in vitro* results, the changes in the gag p7–p1 cleavage site occurred very early, approximately at the same time as the emergence of the first drug resistance mutation in the protease gene. The p7–p1 and p1–p6 gag cleavage site mutations were found more occasionally in patients treated with saquinavir or ritonavir (Mammano *et al.,* 1998). Although these cleavage site mutations were present together with several mutations in HIV PR, no common correlation between particular gag and protease mutations was reported. In the same study, mutations at other gag cleavage sites have been detected, including MA-CA, CA-p2, and p2-NC. It should be noted, however, that changes at these sites involved residues which are variable in inhibitor-naive patients.

It is possible that cleavage at the p1–p6 and p7–p1 sites for which mutations are more frequently observed is a rate-limiting step in polyprotein processing in virions. Indeed, the p1–p6 site is cleaved by HIV PR with the slowest rate as shown by using the *in vitro* translated p55 gag polyprotein (Pettit *et al.,* 1994) or synthetic peptides (Schock *et al.,* 1996; Maschera *et al.,* 1996; Ermolieff *et al.,* 1997). Interestingly, the p1–p6 and p7–p1 cleavage site mutations result in about a 10-fold increase in the catalytic efficiency with the wild-type HIV PR (Doyon *et al.,* 1996). The fact that they have not been selected without drug pressure may indicate that precisely balanced proteinase efficiency toward the cleavage of different sites in gag and gag-pol precursors is required to maintain accurate timing and coordination in the process of virion maturation. The presence of the cleavage site mutations highlights the ability of the virus to overcome restraints on PR variability.

XV. Reduced Dimer Stability

Previously we have mentioned that the dimer stability of HIV PR may play a role in regulating enzyme activity and precursor activation. Recently, we have determined the dimer stability of four drug-resistant PR mutants at pH 7.0 (Xie *et al.,* 1999a). Interestingly, all mutants were shown to be weaker dimers than the Q7K/L33I/L63I mutant, a surrogate of the wild-type enzyme (Fig. 6). These results indicate that in an

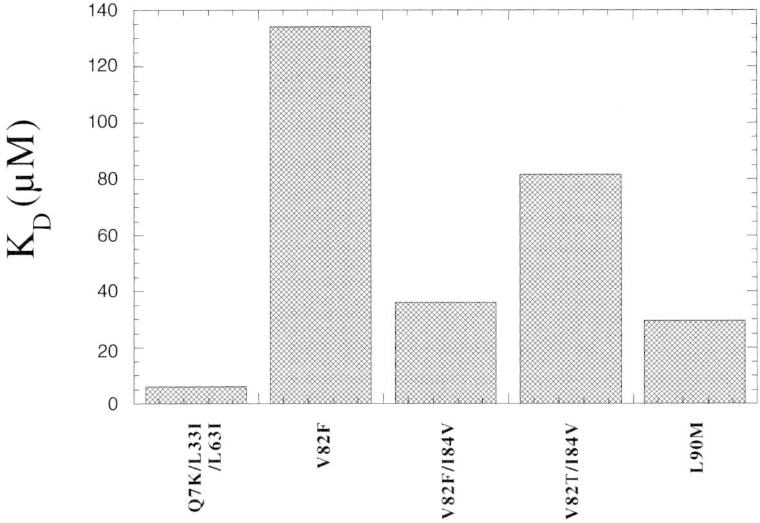

FIG. 6. Dimer dissociation constant of HIV-1 PR mutants at pH 7.0 and 4°C, determined by analytical centrifugation method.

infected cell where PR concentration is below the dimer dissociation constant, the studied mutants have less of a tendency to form the functional dimer than does the wild-type enzyme. It is important to note that the drugs used for mutant selections were designed to bind to the dimeric form of protease. In situations where the enzyme is primarily in the monomeric form, the monomer–dimer equilibrium of PR will be shifted toward dimer by inhibitor binding. The apparent drug-binding affinity of PR will be reduced due to the linkage of drug binding and the induced dimer formation. Thus, the observation of weaker dimer stability of drug-resistant PR mutants suggests an additional mechanism for reduced drug affinity.

XVI. STRUCTURAL BASIS OF DRUG RESISTANCE

The loss of sensitivity of HIV PR mutants toward inhibitors has been rationalized from the specific structural changes introduced by mutations (Erickson, 1995; Erickson and Burt, 1996; Wlodawer and Vondrasek, 1998). In some cases, the effects were modeled using the crystal structures of corresponding wild-type enzyme-inhibitor complexes.

However, direct determination of the crystal structures of mutant–inhibitor complexes gives more reliable results.

Several crystal structures of drug-resistance mutants complexed with inhibitors have been reported (Baldwin *et al.,* 1995b; Chen *et al.,* 1995; Ala *et al.,* 1997). The structure of the V82A mutant of HIV PR complexed with the C2-symmetric inhibitor, A-77003, revealed localized structural changes that resulted in asymmetric repacking of protein and inhibitor atoms in the S1 and S1′ subsites (Baldwin *et al.,* 1995b). Consequently, deletion of the Cγ methyl group in the mutant enzyme resulted in the loss of van der Waals interactions only in the S1', but not the S1 subsite. This was a rather unexpected result that was not predicted by modeling.

Some mutations exhibit additive or even synergistic effects when combined in the active site of HIV PR. For example, for KNI-272, a 5.3- and 32.0-fold increase of K_i over wild type has been reported for the V32I and I84V mutants, respectively, while a combination of these mutations, V32I/I84V, resulted in a 156.0-fold increase in K_i (Gulnik *et al.,* 1995). Nevertheless, K_i values are not always additive. In the case of the cyclic urea inhibitor, DMP323, an I84V mutation causes a 25-fold increase in K_i, which was attributed to a reduction in the number of van der Waals contacts with the inhibitor (Ala *et al.,* 1997). In contrast, the V82F mutant actually binds DMP323 more tightly, presumably as a result of additional van der Waals interactions of the mutant phenylalanine side chain with the P1 group of the inhibitor. In the structure of the double mutant V82F/I84V, however, the side chain of F82 rotates out of the S1/S1′ subsite, and this results in a 1000-fold increase in K_i.

In several instances modeling studies of active site mutants made predictions that were later confirmed by X-ray structure determination. For example, the dramatic reduction of A-77003 potency toward the R8Q mutant (Gulnik *et al.,* 1995) was explained by a loss of stacking interactions of the guanidinium groups of R8/8′ with the pyridine rings of the inhibitor (Ho *et al.,* 1994). This has been confirmed by crystallographic analysis (T. N. Bhat and J. Erickson, unpublished data).

In contrast to active site mutations, the effect of nonactive site mutations in many cases cannot be easily rationalized, even from the crystal structure of the mutant enzyme. Although it has been reported that M46I and L63P can induce small structural changes in residues distal to the active site (Chen *et al.,* 1995), the significance of these changes is unclear. Modeling studies of a hextuple mutant, selected using KNI-272, revealed that mutations far from the active site can influence inhibitor binding through a concerted "domino effect" (Erickson and Burt, 1996).

By comparing SIV and HIV PR structures, Rose and colleagues (1998) identified five domains in the protease dimer that move as rigid bodies against one another: two active site domains, two flap domains, and a terminal domain. It was noted that many of the nonactive site mutations were located at interdomain interfaces. The authors hypothesized that such mutations may alter the equilibrium between closed and open conformations of protease in favor of the open conformation, which should increase the off-rate of inhibitor and reduce binding affinity.

On the other hand, the closed conformation of HIV PR was predicted from molecular dynamics simulations of the M46I mutant (Collins *et al.,* 1995). Nonactive site mutations that are involved in resistance pathways generally have little influence on inhibitor K_i's in contrast to their enhancing effects on substrate catalysis. A reduction of sensitivity to saquinavir for G48V, L90M, and G48V/L90M mutants was previously shown to be a result of increased dissociation rate constants (Maschera *et al.,* 1996). However, these mutations are believed to directly or indirectly influence subsite–inhibitor interactions. It appears that nonactive site mutations may change the enzyme flexibility in a manner that is more readily manifested in substrate catalysis than inhibitor binding.

XVII. Biochemical Basis of Viral Fitness

Viral fitness can be defined as a relative property that reflects the replicative advantage of a particular genotype over others in a population. The rate of emergence of a mutant will depend on the absolute replication rate as well as its relative fitness. Considering the high viral turnover rate during HIV infection, even a small increase in viral fitness caused by drug selection pressure may result in rapid outgrowth of the wild-type virus by the mutant. Several methods have been used to estimate the relative fitness of viruses in cell culture, including the comparison of kinetics of viral antigen production in parallel cultures (which is the most commonly used technique), and direct measurement of relative replication capacity based on clonal analysis of mixed competitive infections of cells by two viruses (Martinez-Picado *et al.,* 1999; Nijhuis *et al.,* 1999). The latter method seems to be more sensitive in its ability to discriminate small fitness differences but requires precise control over input viral ratios. For example, it has been shown that the D30N mutant is less fit than wild type or L90M in the absence of drug (Martinez-Picado *et al.,* 1999). However, it is most likely that since D30N has been selected in the presence of nelfinavir *in vitro* (Patick *et al.*, 1996) it would be most fit in the presence of a certain concentration of this particular drug. Thus, it is important to emphasize that the most valuable information can be obtained from competition assays *in the presence of inhibitor*, but such experiments have not been reported.

On the biochemical level, fitness of wild-type or mutant HIV PR in the presence of inhibitor can be modeled using the available data on inhibition constants of inhibitors and catalytic efficiencies of enzymes. The biochemical resistance of a mutant can be defined by the reduction of binding affinity toward a given inhibitor and is described by the $K_{i,mutant}/K_{i,wild\text{-}type}$ ratio. Besides having a reduced sensitivity to inhibitor, the mutant enzyme should also maintain a certain level of activity in order for the virus to be functional. The proteolytic efficiency of an enzyme is described by its k_{cat}/K_m ratio. If we assume that the activity of wild-type HIV PR is optimized for the virus, any reduction of k_{cat}/K_m will lead to a weaker, less functional virus. Given two different protease mutants that have the same K_i ratio toward a particular inhibitor, the one with higher k_{cat}/K_m, that is, the more efficient enzyme, is more likely to be selected under the pressure of this inhibitor. To be able to compare the selective advantage of different mutants in the presence of a particular inhibitor, we introduced an empirical parameter, called vitality, described by the following equation (Gulnik *et al.*, 1995):

$$\text{vitality} = (K_i \times k_{cat}/K_m)_{mutant}/(K_i \times k_{cat}/K_m)_{wild\text{-}type}.$$

The higher the vitality, the more likely it is that the mutant will be able to maintain the virus function in the presence of inhibitor and the more probable it is that this mutation will be selected as a resistant variant. Thus, the vitality value can be considered as a measure of biochemical or enzymatic fitness. The magnitude of vitality value in general correlates with the appearance of mutants during *in vitro* drug-resistance selection studies (Gulnik *et al.*, 1995; Pazhanisamy *et al.*, 1996; Wilson *et al.*, 1997; Nillroth *et al.*, 1997). There are only a limited number of specific primary mutations for every drug. Among them, drug-selected active site mutants usually have the highest vitality values. For example, mutations at position 82 (V82A, T, F) are characteristic for ritonavir and indinavir (Condra *et al.*, 1995, 1996; Molla *et al.*, 1996; Eastman *et al.*, 1998). This observation is in agreement with the high vitality values for these mutants (Fig. 7).

In contrast, the V82I mutant, which was not selected by these inhibitors, exhibits a significantly lower vitality value. The most common mutation for saquinavir, L90M (Jacobsen *et al.*, 1996), has the highest vitality value among the active site mutations tested. For saquinavir,

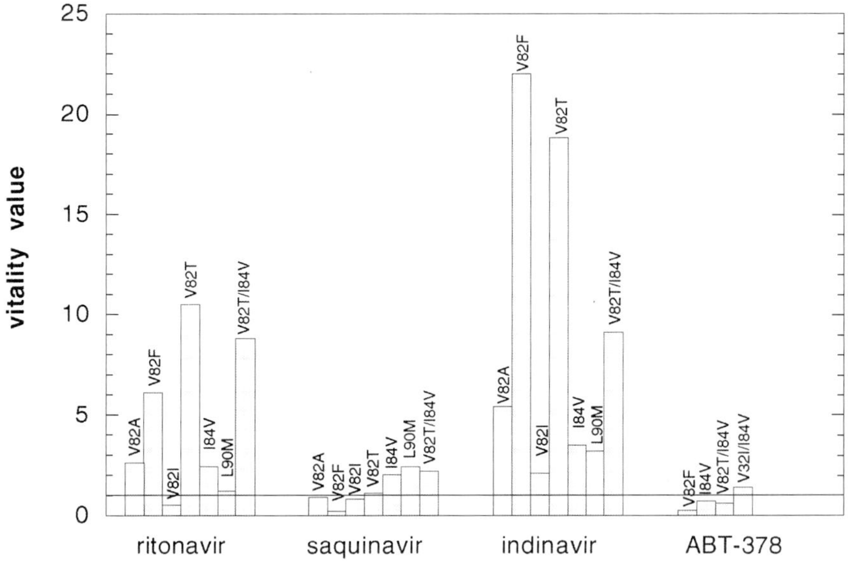

FIG. 7. Vitality values for mutant HIV PRs. Mutants with vitality values below the threshold of 1.0 are predicted to be less fit than wild-type in the presence of drug.

the I84V mutant, which has also been found *in vivo,* has the second highest vitality value among the single mutants tested. Overall, the vitality values for saquinavir are much lower than those for ritonavir and indinavir but this is mainly a reflection of K_i ratio, since the k_{cat}/K_m ratio is independent of inhibitor. Unfortunately, saquinavir has low oral bioavailability apparently due to poor absorption and rapid metabolism by the cytochrome *P*450 system (Noble and Faulds, 1996). A new Abbott inhibitor, ABT-378, which is in clinical development, is structurally related to ritonavir and retains high antiviral activity against ritonavir-resistant strains (Carrillo *et al.,* 1998). The vitality value of this inhibitor is below 1 for many of the mutants tested (L. Suvorov and S. Gulnik, unpublished data).

Nijhuis *et al.* (1999) followed the evolution of HIV PR in one individual treated with ritonavir. Initially, during the antiviral therapy, two drug-resistant viral variants were selected: V82T and M36I/I54V/V82T. Recombinant mutant viruses displayed 8- and 48-fold increases in IC_{50} values over the wild type, respectively. The addition of the A71V mutation to the triple mutant did not further increase the IC_{50} value. The biochemical resistance paralleled phenotypic resistance. Recombinant mutant enzymes showed K_i increases of 17-fold for V82T and around 400-fold for both the triple and quadruple mutants. Thus, the addition of A71V did not improve resistance in either enzymatic or cell culture assays. However, the vitality value for the quadruple mutant enzyme was about 10-fold higher than that for the triple mutant, due to the increase in catalytic efficiency upon addition of the A71V mutation. These results show the importance of measuring biochemical fitness and not just resistance to explain mutant evolution.

It is important to emphasize that vitality can be used for comparison of selective advantage of different mutants in the presence of a particular inhibitor. Vitalities values for different inhibitors calculated using the same mutant, as was done by Klabe *et al.* (1998), will reflect just the K_i ratios because the k_{cat}/K_m ratio is constant for a given mutant and should be independent of inhibitors. Another important point is that the vitality value need not necessarily correlate with antiviral IC_{50} value. If any enzymatic parameter correlates with the IC_{50} ratio, that should be the K_i ratio.

Another kinetic model for comparison of inhibitor resistance potential of mutant HIV-1 PRs was proposed by Tang and Hartsuck (1995). They introduced the term *mutant processing activity*, a_{MI}, which is defined by the equation:

$$a_{MI} = \sigma(k_{cat}/K_m)/(1 + I/K_i),$$

where σ is determined by the initial substrate and mutant enzyme concentrations, and I is the inhibitor concentration. The parameter a_{MI} is essentially the initial rate of substrate cleavage by a mutant enzyme. Mutation modulated activity (MMA) was then introduced as the ratio of a_{MI} for a particular mutant to the processing activity of uninhibited wild-type enzyme.

This kinetic model has been used to rationalize the generation of saquinavir-resistant mutants G48V, L90M, and G48V/L90M (Ermolieff *et al.*, 1997). The ability to simulate proteolytic activity at different inhibitor concentrations is an important and useful feature of this kinetic model. However, the model is only valid under processing conditions where the effective substrate concentration is much less than K_m. The analysis of a given mutant's fitness requires the assumption of a threshold MMA value for viability. For example, significantly different thresholds of catalytic efficiency of protease at which infectious virions can still be generated have been reported at 2–25% (Rose *et al.*, 1994) and 0.045% (Croteau *et al.*, 1997).

The catalytic efficiency of mutant and wild-type HIV PR (k_{cat}/K_m) is a component of both biochemical models of viral fitness discussed earlier. Therefore, both vitality values and MMA depend on the substrate used for the k_{cat}/K_m determination. It might be possible to pick one or several peptides representative of natural gag and gag-pol cleavage sites which will give the best correlation between the model and experimental observations. A peptide corresponding to the p6*–PR cleavage site in the gag-pol polyprotein, for example, gives such good correlation (Ermolieff *et al.,* 1997; Nijhuis *et al.,* 1999).

XVIII. Cross-Resistance to Protease Inhibitors

Enzymatic studies of HIV PR active site mutants revealed a high degree of cross-resistance to different inhibitors (Gulnik *et al.,* 1995; Schock *et al.,* 1996; Partaledis *et al.,* 1995; Rose *et al.,* 1996; Klabe *et al.,* 1998). This observation can be explained by the fact that all HIV PR inhibitors are optimized to inhibit the wild-type enzyme. Therefore, any change in enzyme active site subsites introduced by mutations may be expected to affect inhibitor binding. Mutations at positions 82 and 84, for example, introduce cross-resistance to ritonavir and indinavir. The I84V mutation significantly increases inhibition constants for all approved PR inhibitors and many of those in clinical trials, including amprenavir (Partaledis *et al.,* 1995) and PNU-140690 (L. Suvorov and S. Gulnik, unpublished data). The G48V mutant, selected by saquinavir, is cross-resistant to ritonavir and indinavir as well (Maschera *et al.,* 1996; Wilson *et al.,* 1997).

Phenotypic cross-resistance patterns for mutants selected *in vivo* are more complicated and in some instances failed to reveal any obvious sequence determinants for cross-resistance (Condra *et al.,* 1996; Winters *et al.,* 1998). It seems that the evolution of resistance may be unique in each individual and may depend on the preexisting frequencies of mutant variants as well as the presence of compensatory background polymorphisms prior to treatment. Generally, the level of cross-resistance correlates with the number of accumulated mutations, which usually increases with time after the original manifestation of resistance. Most of the viral isolates from patients who failed long-term protease inhibitor treatment are highly cross-resistant to all four clinically available inhibitors (Condra *et al.,* 1996; Winters *et al.,* 1998; Shafer *et al.,* 1998). The most prevalent mutations are found at positions 10, 36, 46, 54, 71, 77, 82, and 90, often in combination with cleavage site mutations in gag (Hertogs *et al.,* 1998). Analyzing the data from clinical isolates, Roberts *et al.* (1998) ranked the four clinically approved inhibitors in order of frequency and breadth of cross-resistance as indinavir > ritonavir > saquinavir = nelfinavir.

In search of better treatment strategies for HIV PR inhibitor-naive patients, several considerations are important. It is obvious that maximal suppression of viral replication will reduce the chance for the selection of resistant virus. Indeed, it has been reported that the duration of viral suppression during ritonavir therapy may be predicted by the plasma HIV RNA levels at nadir (Kempf *et al.,* 1998; Eastman *et al.,* 1998). In general, protease inhibitors are dosed to achieve the highest plasma concentration to IC_{50} ratio possible for the maximal reduction of viral replication rate.

Improvement of the bioavailability of existing compounds has been achieved by inhibiting enzymes involved in their metabolism. For example, combination therapy with a low concentration of ritonavir enhances the plasma levels of saquinavir and other protease inhibitors by inhibiting their metabolism through the cytochrome *P*450 system (see review in Molla *et al.,* 1998). Combination therapy with one or two protease inhibitors and one or more reverse transcriptase inhibitors achieves the most effective and long-lasting viral suppression. From the perspective of cross-resistance, an inhibitor that selects for a primary set of mutations that do not result in significant cross-resistance with other available inhibitors should be the first choice. At the first sign of phenotypic resistance, it may be wise to switch to or add another inhibitor.

XVIX. Conclusions

The availability of potent HIV PR inhibitors gives new promise to AIDS therapy. The combination of these drugs with reverse transcriptase inhibitors can dramatically suppress viral replication in HIV-infected patients. Despite the current success, the issue of complete eradication of HIV virus is still the subject of debate. The emergence of a drug-resistant virus to every approved drug is especially alarming and demands continuous efforts to elucidate the precise role of HIV PR in the virus biology and underscores the importance of studies on drug-resistance mechanisms.

Several unaddressed HIV PR-related issues remain. Little is known, for example, about the regulatory mechanisms of HIV PR activity. What triggers the activation of the enzyme? What are the critical steps for viral maturation? Is it necessary to shut off the enzyme activity after maturation and how does it happen? What is the effective environment (pH, ionic strength, etc.) in which protease acts *in vivo* and what is the best way of mimicking these conditions *in vitro*? Several critical questions also remain unanswered regarding the action of protease inhibitors.

Where does the inhibition occur, in cell or in virus? What is the target for inhibition? Is it the processed protease or the enzyme still in context of the dimeric gag-pol precursor? Why are some inhibitors, such as saquinavir and nelfinavir, less affected by many active site mutations? Further biochemical and biophysical investigations, in combination with clinical studies, could shed light on these issues and assist in designing a new generation of resistance-proof inhibitors.

Note added in proof: Since this paper was prepared (December 1998), the fifth HIV PR inhibitor, Amprenavir (also known as VX-478) was approved for clinical use.

ACKNOWLEDGMENTS

We thank Drs. T. N. Bhat and L. Suvorov for sharing unpublished data, Michael Eissenstat and T. N. Bhat for many helpful discussions and figure preparation, Ms. Maritta P. Grau for manuscript editing, and Ms. Christine Ray for manuscript preparation. This project has been funded in whole or in part with federal funds from the National Cancer Institute, National Institutes of Health, under contract NO1-CO-56000.

REFERENCES

Ala, P. J., Huston, E. E., Klabe, R. M., McCabe, D. D., Duke, J. L., Rizzo, C. J., Korant, B. D., DeLoskey, R. J., Lam, P. Y., Hodge, C. N., and Chang, C. H. (1997). Molecular basis of HIV-1 protease drug resistance: Structural analysis of mutant proteases complexed with cyclic urea inhibitors. *Biochemistry* **36,** 1573–1580.

Ala, R., DeLoskey, R., Huston, E., Jadhav, P., Lam, P., Eyermann, C., Hodge, C., Schadt, M., Lewandowski, F., Weber, P., McCabe, D., Duke, J., and Chang, C. (1998). Molecular recognition of cyclic urea HIV-1 protease inhibitors. *J. Biol. Chem.* **273,** 12325–12331.

Baldwin, E. T., Bhat, T. N., Gulnik, S., Liu, B., Topol, I. A., Kiso, Y., Mimoto, T., Mitsuya, H., and Erickson, J. W. (1995a). Structure of HIV-1 protease with KNI-272, a tight-binding transition-state analog containing allophenylnorstatine. *Structure* **3,** 581–590.

Baldwin, E. T., Bhat, T. N., Liu, B., Pattabiraman, N., and Erickson, J. W. (1995b). Structural basis of drug resistance for the V82A mutant of HIV-1 proteinase. *Nat. Struct. Biol.* **2,** 244–249.

Barrie, K. A., Perez, E. E., Lamers, S. L., Farmerie, W. G., Dunn, B. M., Sleasman, J. W., and Goodenow, M. M. (1996). Natural variation in HIV-1 protease, Gag p7 and p6, and protease cleavage sites within gag/pol polyproteins: Amino acid substitutions in the absence of protease inhibitors in mothers and children infected by human immunodeficiency virus type 1. *Virology* **219,** 407–416.

Barry, M., Gibbons, S., Back, D., and Mulcahy, F. (1997). Protease inhibitors in patients with HIV disease. Clinically important pharmacokinetic considerations. *Clin. Pharmacokinet.* **32,** 194–209.

Billich, A., Hammerschmid, F., and Winkler, G. (1990). Purification, assay and kinetic features of HIV-1 proteinase. *Biol. Chem. Hoppe-Seyler* **371,** 265–272.

Boden, D., and Markowitz, M. (1998). Resistance to human immunodeficiency virus type 1 protease inhibitors. *Antimicrob. Agents Chemother.* **42,** 2775–2783.

Carrillo, A., Stewart, K. D., Sham, H. L., Norbeck, D. W., Kohlbrenner, W. E., Leonard, J. M., Kempf, D. J., and Molla, A. (1998). *In vitro* selection and characterization of human immunodeficiency virus type 1 variants with increased resistance to ABT-378, a novel protease inhibitor. *J. Virol.* **72,** 7532–7541.

Chen, Z., Li, Y., Schock, H. B., Hall, D., Chen, E., and Kuo, L. C. (1995). Three-dimensional structure of a mutant HIV-1 protease displaying cross-resistance to all protease inhibitors in clinical trials. *J. Biol. Chem.* **270,** 21433–21436.

Cheng, Y.-S. E., Yin, F. H., Foundling, S., Blomstrom, D., and Kettner, C. A. (1990). Stability and activity of human immunodeficiency virus protease: Comparison of the natural dimer with a homologous, single-chain tethered dimer. *Proc. Natl. Acad. Sci. U.S.A.* **87,** 9660–9664.

Coffin, J. M. (1995). HIV population dynamics *in vivo*: Implications for genetic variation, pathogenesis and therapy. *Science* **267,** 483–489.

Coffin, J. M., Hughes, S. H., and Varmus, H. E., eds. (1997). "Retroviruses." Cold Spring Harbor Lab. Press, Cold Spring Harbor, NY.

Collins, J. R., Burt, S. K., and Erickson, J. W. (1995). Flap opening in HIV-1 protease simulated by 'activated' molecular dynamics. *Nat. Struct. Biol.* **2,** 334–338.

Condra, J. H., Schleif, W. A., Blahy, O. M,. Gabryelski, L. J., Graham, D. J., Quintero, J. C., Rhodes, A., Robbins, H. L., Roth, E., Shivaprakash, M., Titus, D., Yang, T., Teppler, H., Squires, K. E., Deutsch, P. J., and Emini, E. A. (1995). *In vivo* emergence of HIV-1 variants resistant to multiple protease inhibitors. *Nature (London)* **374,** 569–571.

Condra, J. H., Holder, D. J., Schleif, W. A., Blahy, O. M., Danovich, R. M., Gabryelski, L. J., Graham, D. J., Laird, D., Quintero, J. C., Rhodes, A., Robbins, H. L., Roth, E., Shivaprakash, M., Yang, T., Chodakewitz, J. A., Deutsch, P. J., Leavitt, R. Y., Massari, F. E., Mellors, J. W., Squires, K. E., Steigbigel, R. T., Teppler, H., and Emini, E. A. (1996). Genetic correlates of *in vivo* viral resistance to indinavir, a human immunodeficiency virus type 1 protease inhibitor. *J. Virol.* **70,** 8270–8276.

Cornelissen, M., van den Burg, R., Zorgdrager, F., Lukashov, V., and Goudsmit, J. (1997). *pol* gene diversity of five human immunodeficiency virus type 1 subtypes: Evidence for naturally occurring mutations that contribute to drug resistance, limited recombination patterns, and common ancestry for subtypes B and D. *J. Virol.* **71,** 6348–6358.

Croteau, G., Doyon, L., Thibeault, D., McKercher, G., Pilote, L., and Lamarre, D. (1997). Impaired fitness of human immunodeficiency virus type 1 variants with high-level resistance to protease inhibitors. *J. Virol.* **71,** 1089–1096.

Darke, P. L. (1994). Stability of dimeric retroviral proteases. *In* "Methods in Enzymology" (L. C. Kuo and J. A. Shafer, eds.), Vol. 241, pp. 104–127. Academic Press, San Diego, CA.

Darke, P. L., Jordan, S. P., Hall, D. L., Zugay, J. A., Shafer, J. A., and Kuo, L. C. (1994). Dissociation and association of the HIV-1 protease dimer subunits: Equilibria and rates. *Biochemistry* **33,** 98–105.

Davies, D. R. (1990). The structure and function of the aspartic proteinases. *Annu. Rev. Biophys. Biomol. Struct.* **19,** 189–215.

Debouck, C. (1992). The HIV-1 protease as a therapeutic target for AIDS. *AIDS Res. Hum. Retroviruses* **8,** 153–164.

Debouck, C., Gorniak, J. G., Strickler, J. E., Meek, T. D., Metcalf, B. W., and Rosenberg, M. (1987). Human immunodeficiency virus protease expressed in *Escherichia coli* exhibits autoprocessing and specific maturation of the gag precursor. *Proc. Natl. Acad. Sci. U.S.A.* **84,** 8903–8906.

Deeks, S. G., Smith, M., Holodniy, M., and Kahn, J. O. (1997). HIV-1 protease inhibitors. A review for clinicians. *JAMA, J. Am. Med. Assoc.* **277,** 145–153.

Dorsey, B. D., Levin, R. B., McDaniel, S. L., Vacca, J. P., Guare, J. P., Darke, P. L., Zugay, J. A., Emini, E., Schleif, W. A., Quintero, J. C., Lin, J. H., Chen, I.-W., Holloway, M. K., Fitzgerald, P., M. D., Axel, M. G., Ostovic, D., Anderson, P. S., and Huff, J. R. (1994). L-735,524: The design of a potent and orally bioavailable HIV protease inhibitor. *J. Med. Chem.* **37,** 3443–3451.

Doyon, L., Croteau, G., Thibeault, D., Poulin, F., Pilote, L., and Lamarre, D. (1996). Second locus involved in human immunodeficiency virus type 1 resistance to protease inhibitors. *J. Virol.* **70,** 3763–3769.

Dunn, B. M., Gustchina, A., Wlodawer, A., and Kay, J. (1994). Subsite preferences of retroviral proteinases. *In* "Methods of Enzymology" (L. C. Kuo and J. A. Shafer, eds.), Vol. 241, pp. 254–278. Academic Press, San Diego, CA.

Eastman, P., Mittler, J., Kelso, R., Gee, C., Boyer, E., Kolberg, J., Urdea, M., Leonard, J., Norbeck, D., Mo, H., and Markowitz, M. (1998). Genotypic changes in human immunodeficiency virus type 1 associated with loss of suppression of plasma viral RNA levels in subjects treated with ritonavir (norvir) monotherapy. *J. Virol.* **72,** 5154–5164.

Erickson, J., Neidhart, D. J., VanDrie, J., Kempf, D. J., Wang, X. C., Norbeck, D. W., Plattner, J. J., Rittenhouse, J. W., Turon, M., Wideburg, N., Kohlbrenner, W. E., Simmer, R., Helfrich, R., Paul, D. A., and Knigge, M. (1990). Design, activity, and 2.8 Å crystal structure of a C_2 symmetric inhibitor complexed to HIV-1 protease. *Science* **249,** 527–533.

Erickson, J. W. (1995). The not-so-great escape. *Nat. Struct. Biol.* **2,** 523–529.

Erickson, J. W., and Burt, S. K. (1996). Structural mechanisms of HIV drug resistance. *Annu. Rev. Pharmacol. Toxicol.* **36,** 545–571.

Erickson, J. W., and Eissenstat, M. E. (1999). HIV protease as a target for the design of antiviral agents for AIDS. *In* "Proteases of Infectious Agents" (B. Dunn, ed.), pp. 1–60. Academic Press, San Diego, CA.

Erickson, J. W., Gulnik, S. V., and Markowitz, M. (1999). Protease inhibitors: Resistance, cross-resistance, fitness and the choice of initial and salvage therapies. *AIDS* **13** (Suppl A), S189–S204.

Ermolieff, J., Lin, X., and Tang, J. (1997). Kinetic properties of saquinavir-resistant mutants of human immunodeficiency virus type 1 protease and their implications in drug resistance *in vivo*. *Biochemistry* **36,** 12364–12370.

Ferguson, D. M., Radmer, R. J., and Kollman, P. A. (1991). Determination of the relative binding free energies of peptide inhibitors to the HIV-1 protease. *J, Med. Chem.* **34,** 2654–2659.

Flexner, C. (1998). HIV-protease inhibitors. *N. Engl. J. Med.* **338,** 1281–1292.

Freire, E., Mayorga, O. L., and Straume, M. (1990). Isothermal titration calorimetry. Direct thermodynamic characterization of biological molecular interactions. *Anal. Chem.* **62,** 950A-959A.

Furfine, E. S., D'Souza, E., Ingold, K. J., Leban, J. J., Spector, T., and Porter, D. J. (1992). Two-step binding mechanism for HIV protease inhibitors. *Biochemistry* **31,** 7886–7891.

Grant, S., Deckman, I., Culp, J., Minnich, M., Brooks, I., Hensley, P., Debouck, C., and Meek, T. (1992). Use of protein unfolding studies to determine the conformational and dimeric stabilities of HIV-1 and SIV protease. *Biochemistry* **31,** 9491–9501.

Griffiths, J. T., Phylip, L. H., Konvalinka, J., Strop, P., Gustchina, A., Wlodawer, A., Davenport, R. J., Briggs, R., Dunn, B. M., and Kay, J. (1992). Different requirements for productive interaction between the active site of HIV-1 proteinase and substrates containing -hydrophobic*hydrophobic- or -aromatic*pro- cleavage sites. *Biochemistry* **31,** 5193–5200.

Gulnik, S. V., Suvorov, L. I., Liu, B., Yu, B., Anderson, B., Mitsuya, H., and Erickson, J. W. (1995). Kinetic characterization and cross-resistance patterns of HIV-1 protease mutants selected under drug pressure. *Biochemistry* **34,** 9282–9287.

Hammond, J., Larder, B. A., Schinazi, R. F., and Mellors, J. W. (1997). Mutation in retroviral genes associated with drug resistance. *In* "Human Retroviruses and AIDS" (Korber, B., Foley, B., Leither, T., McCutchan, F., Hahn, B., Mellors, J. W., Myers, G., Kuiken, C., eds.), pp. III-207–249. Los Alamos, NM.

Harte, W. E., Jr., and Beveridge, D. L. (1993). Prediction of the protonation state of the active site aspartyl residues in HIV-1 protease-inhibitor complexes via molecular dynamics simulation. *J. Am. Chem. Soc.* **115,** 3883–3886.

Hellen, C. U. T. (1994). Assay methods for retroviral proteases. *In* "Methods of Enzymology" (L. C. Kuo and J. A. Shafer, eds.), Vol. 241, pp. 46–58. Academic Press, San Diego, CA.

Henderson, L. E., Benveniste, R. E., Sowder, R., Copeland, T. D., Schultz, A. M., and Oroszlan, S. (1988). Molecular characterization of *gag* proteins from simian immunodeficiency virus (SIV_{Mne}). *J. Virol.* **62,** 2587–2595.

Hertogs, K., Kemp, S., Bloor, S., Miller, V., Staszewski, S., Mellors, J., Van den Eynde, C., Peeters, F., Larder, B., and Pauwels, R. (1998). Patterns of cross-resistance among protease inhibitors in over 1500 clinical HIV-1 isolates. Comparison of genotypic and phenotypic resistance profiles. *Int. Workshop HIV Drug Resist. Treat. Strategies, 2nd,* 24–27 June, 1998, Lake Maggiore, Italy. Abstract 70, p. 49.

Ho, D. D., Toyoshima, T., Mo, H., Kempf, D. J., Norbeck, D., Chen, C.-M., Wideburg, N. E., Burt, S. K., Erickson, J. W., and Singh, M. K. (1994). Characterization of human immunodeficiency virus type 1 variants with increased resistance to a C_2-symmetric protease inhibitor. *J. Virol.* **68,** 2016–2020.

Hodge, C. N., Aldrich, P. E., Bacheler, L. T., Chang, C. H., Eyermann, C. J., Garber, S., Grubb, M., Jackson, D. A., Jadhav, P. K., Korant, B., Lam, P. Y., Maurin, M. B., Meek, J. L., Otto, M. J., Rayner, M. M., Reid, C., Sharpe, T. R., Shum, L., Winslow, D. L., and Erickson-Viitanen, S. (1996). Improved cyclic urea inhibitors of the HIV-1 protease: Synthesis, potency, resistance profile, human pharmacokinetics and X-ray crystal structure of DMP 450. *Chem. Biol.* **3,** 301–314.

Hoog, S. S., Towler, E. M., Zhao, B., Doyle, M. L., Debouck, C., and Abdel-Meguid, S. S. (1996). Human immunodeficiency virus protease ligand specificity conferred by residues outside of the active site cavity. *Biochemistry* **35,** 10279–10286.

Hui, J. O., Tomasselli, A. G., Reardon, I. M., Lull, J. M., Brunner, D. P., Tomich, C. S., and Heinrikson, R. L. (1993). Large scale purification and refolding of HIV-1 protease from *Escherichia coli* inclusion bodies. *J. Protein Chem.* **12,** 323–327.

Hyland, L. J., Tomaszek, T. A., Jr., and Meek, T. D. (1991). Human immunodeficiency virus-1 protease. 2. Use of pH rate studies and solvent kinetic isotope effects to elucidate details of chemical mechanism. *Biochemistry* **30,** 8454–8463.

Ido, E., Han, H. P., Kezdy, F. J., and Tang, J. (1991). Kinetic studies of human immunodeficiency virus type 1 protease and its active-site hydrogen bond mutant A28S. *J. Biol. Chem.* **266,** 24359–24366.

Jacobsen, H., Hanggi, M., Ott, M., Duncan, I. B., Owen, S., Andreoni, M., Vella, S., and Mous, J. (1996). *In vivo* resistance to a human immunodeficiency virus type 1 proteinase inhibitor: Mutations, kinetics, and frequencies. *J. Infect. Dis.* **173,** 1379–1387.

Jadhav, P. K., Ala, P., Woerner, F. J., Chang, C. H., Garber, S. S., Anton, E. D., and Bacheler, L. T. (1997). Cyclic urea amides: HIV-1 protease inhibitors with low nanomolar potency against both wild type and protease inhibitor resistant mutants of HIV. *J. Med. Chem.* **40,** 181–191.

Jordan, S., Zugay, J., Darke, P., and Kuo, L. (1992). Activity and dimerization of human immunodeficiency virus protease as a function of solvent composition and enzyme concentration. *J. Biol. Chem.* **267,** 20028–20032.

Kaldor, S. W., Kalish, V. J., Davies, J. F., 2nd, Shetty, B. V., Fritz, J. E., Appelt, K., Burgess, J. A., Campanale, K. M., Chirgadze, N. Y., Clawson, D. K., Dressman, B. A., Hatch, S. D., Khalil, D. A., Kosa, M. B., Lubbehusen, P. P., Muesing, M. A., Patick, A. K., Reich, S. H., Su, K. S., and Tatlock, J. H. (1997). Viracept (nelfinavir mesylate, AG1343): A potent, orally bioavailable inhibitor of HIV-1 protease. *J. Med. Chem.* **40,** 3979–3985.

Kaplan, A. H., and Swanstrom, R. (1991). Human immunodeficiency virus type 1 Gag proteins are processed in two cellular compartments. *Proc. Natl. Acad. Sci. U.S.A.* **88,** 4528–4532.

Kaplan, A. H., Zack, J. A., Knigge, M., Paul, D. A., Kempf, D. J., Norbeck, D. W., and Swanstrom, R. (1993). Partial inhibition of the human immunodeficiency virus type 1 protease results in aberrant virus assembly and the formation of noninfectious particles. *J. Virol.* **67,** 4050–4044.

Kaplan, A. H., Manchester, M., and Swanstrom, R. (1994a). The activity of the protease of human immunodeficiency virus type 1 is initiated at the membrane of infected cells before the release of viral proteins and is required for release to occur with maximum efficiency. *J. Virol.* **68,** 6782–6786.

Kaplan, A. H., Michael, S. F., Wehbie, R. S., Knigge, M. F., Paul, D. A., Everitt, L., Kempf, D. J., Norbeck, D. W., Erickson, J. W., and Swanstrom, R. (1994b). Selection of multiple human immunodeficiency virus type 1 variants that encode viral proteases with decreased sensitivity to an inhibitor of the viral protease. *Proc. Natl. Acad. Sci. U.S.A.* **91,** 5597–5601.

Katoh, I., Yasunaga, T., Ikawa, Y., and Yoshinaka, Y. (1987). Inhibition of retroviral protease activity by an aspartyl proteinase inhibitor. *Nature (London)* **329,** 654–656.

Kempf, D. J., Marsh, K. C., Denissen, J. F., McDonald, E., Vasavanonda, S., Flentge, C. A., Green, B. E., Fino, L., Park, C. H., Kong, X. P., *et al.* (1995). ABT-538 is a potent inhibitor of human immunodeficiency virus protease and has high oral bioavailability in humans. *Proc. Natl. Acad. Sci. U.S.A.* **92,** 2484–2488.

Kempf, D. J., Rode, R. A., Xu, Y., Sun, E., Heath-Chiozzi, M. E., Valdes, J., Japour, A. J., Danner, S., Boucher, C., Molla, A., and Leonard, J. M. (1998). The duration of viral suppression during protease inhibitor therapy for HIV-1 infection is predicted by plasma HIV-1 RNA at the nadir. *AIDS* **12,** F9–F14.

Kim, C. U., McGee, L. R., Krawczyk, S. H., Harwood, E., Harada, Y., Swaminathan, S., Bischofberger, N., Chen, M. S., Cherrington, J. M., Xiong, S. F., Griffin, L., Cundy, K. C., Lee, A., Yu, B., Gulnik, S., and Erickson, J. W. (1996). New series of potent, orally bioavailable, non-peptidic cyclic sulfones as HIV-1 protease inhibitors. *J. Med. Chem.* **39,** 3431–3434.

Kisselev, A. F., Mentele, R., and von der Helm, K. (1997). Cleavage of the complement system C3 component by HIV-1 proteinase. *Biol. Chem. Hoppe-Seyler* **378,** 439–442.

Klabe, R. M., Bacheler, L. T., Ala, P. J., Erickson-Viitanen, S., and Meek, J. L. (1998). Resistance to HIV protease inhibitors: A comparison of enzyme inhibition and antiviral potency. *Biochemistry* **37,** 8735–8742.

Kohl, N. E., Emini, E. A., Schleif, W. A., Davis, L. J., Heimbach, J. C., Dixon, R. A. F., Scolnick, E. M., and Sigal, I. S. (1988). Active human immunodeficiency virus protease is required for viral infectivity. *Proc. Natl. Acad. Sci. U.S.A.* **85,** 4686–4690.

Konvalinnka, J., Strop, P., Velek, J., Cerna, V., Kostka, V., Phylip, L. H., Richards, A. D., Dunn, B. M., and Kay, J. (1990). Sub-site preferences of the aspartic proteinase from the human immunodeficiency virus, HIV-1. *FEBS Lett.* **268,** 35–38.

Konvalinka, J., Horejsi, M., Andreansky, M., Novek, P., Pichova, I., Blaha, I., Fabry, M., Sedlacek, J., Foundling, S., and Strop, P. (1992). An engineered retroviral proteinase from myeloblastosis associated virus acquires pH dependence and substrate specificity of the HIV-1 proteinase. *EMBO J.* **11,** 1141–1144.

Korant, B. D., and Rizzo, C. J. (1997). The HIV protease and therapies for AIDS. *Adv. Exp. Med. Biol.* **421,** 279–284.

Korant, B. D., Strack, P., Frey, M. W., and Rizzo, C. J. (1998). A cellular anti-apoptosis protein is cleaved by the HIV-1 protease. *Adv. Exp. Med. Biol.* **436,** 27–29.

Kotler, M., Arad, G., and Hughes, S. (1992). Human immunodeficiency virus type 1 Gag-protease fusion proteins are enzymatically active. *J. Virol.* **66,** 6781–6783.

Kotler, M., Simm, M., Zhao, Y. S., Sova, P., Chao, W., Ohnona, S. F., Roller, R., Krachmarov, C., Potash, M. J., and Volsky, D. J. (1997). Human immunodeficiency virus type 1 (HIV-1) protein Vif inhibits the activity of HIV-1 protease in bacteria and *in vitro*. *J. Virol.* **71,** 5774–5781.

Kozal, M. J., Shah, N., Shen, N., Yang, R., Fucini, R., Merigan, T. C., Richman, D. D., Morris, D., Hubbell, E., Chee, M., and Gingeras, T. R. (1996). Extensive polymorphisms observed in HIV-1 clade B protease gene using high-density oligonucleotide arrays. *Nat. Med.* **2,** 753–759.

Krafft, G. A., and Wang, G. T. (1994). Synthetic approaches to continuous assays of retroviral proteases. *In* "Methods in Enzymology" (L. C. Kuo and J. A. Shafer, eds.), Vol. 241, pp. 70–86. Academic Press, San Diego, CA.

Kramer, R. A., Schaber, M. D., Skalka, A. M., Ganguly, K., Wong, S. F., and Reddy, E. P. (1986). HTLV-III gag protein is processed in yeast cells by the virus pol-protease. *Science* **231,** 1580–1584.

Kräusslich, H.-G., Fäcke, M., Heuser, A.-M., Konvalinka, J., and Sentgraf, H. (1995). The spacer peptide between human immonodeficiency virus capsid and nucleocapsid proteins is essential for ordered assembly and viral infectivity. *J. Virol.* **69,** 3407–3419.

Kuzmic, P., Garcia-Echeverria, C., and Rich, D. H. (1993). Stabilization of HIV proteinase dimer by bound substrate. *Biochem. Biophys. Res. Commun.* **194,** 301–305.

Ladbury, J., and Chowdhry, B. (1996). Sensing the heat: The application of isothermal titration calorimetry to the thermodynamic studies of biomolecular interactions. *Chem. Biol.* **3,** 791–801.

Lam, P. Y. S., Jadhav, P. K., Eyermann, C. J., Hodge, C. N., Ru, Y., Bacheler, L. T., Meek, J. L., Otto, M. J., Rayner, M. M., Wong, Y. N., Chang, C.-H., Weber, P. C., Jackson, D. A., Sharpe, T. R., and Erickson-Viitanen, S. (1994). Rational design of potent, bioavailable, nonpeptide cyclic ureas as HIV protease inhibitors. *Science* **263,** 380–384.

Lapatto, R., Blundell, T., Hemmings, A., Overington, J., Wilderspin, A., Wood, S., Merson, J. R., Whittle, P. J., Danley, D. E., Geoghegan, K. F., Hawrylik, S. J., Lee, S. E., Scheld, K., and Hobart, P. M. (1989). X-ray analysis of HIV-1 proteinase at 2.7Å resolution confirms structural homology among retroviral enzymes. *Nature (London)* **342,** 299–302.

Lech, W. J., Wang, G., Yang, Y. L., Chee, Y., Dorman, K., McCrae, D., Lazzeroni, L. C., Erickson, J. W., Sinsheimer, J. S., and Kaplan, A. H. (1996). *In vivo* sequence diversity of the protease of human immunodeficiency virus type 1: Presence of protease inhibitor-resistant variants in untreated subjects. *J. Virol.* **70,** 2038–2043.

Lee, C. S., Choy, N., Park, C., Choi, H., Son, Y. C., Kim, S., Ok, J. H., Yoon, H., and Kim, S. C. (1996). Design, synthesis, and characterization of dipeptide isostere containing cis-epoxide for the irreversible inactivation of HIV protease. *Bioorg. Med. Chem. Lett.* **6,** 589–594.

Luque, I., Todd, M., Gomez, J., Semo, N., and Freire, E. (1998). Molecular basis of resistance to HIV-1 protease inhibition: A plausible hypothesis. *Biochemistry* **37,** 5791–5797.

Mammano, F., Petit, C., and Clavel, F. (1998). Resistance-associated loss of viral fitness in human immunodeficiency virus type 1: Phenotypic analysis of protease and gag coevolution in protease inhibitor-treated patients. *J. Virol.* **72,** 7632–7637.

Markowitz, M., Mo, H., Kempf, D. J., Norbeck, D. W., Bhat, T. N., Erickson, J. W., and Ho, D. D. (1995). Selection and analysis of human immunodeficiency virus type 1 variants with increased resistance to ABT-538, a novel protease inhibitor. *J. Virol.* **69,** 701–706.

Martinez-Picado, J., Savara, A., Sutton, L., and D'Aquila, R. (1999). Replicative fitness of protease inhibitor resistant mutants of human immunodeficiency virus type 1. *J. Virol.* **73,** 3744–3752.

Maschera, B., Darby, G., Palu, G., Wright, L. L., Tisdale, M., Myers, R., Blair, E. D., and Furfine, E. S. (1996). Human immunodeficiency virus. Mutations in the viral protease that confer resistance to saquinavir increase the dissociation rate constant of the protease-saquinavir complex. *J. Biol. Chem.* **271,** 33231–33235.

McQuade, T. J., Tomasselli, A. G., Liu, L., Karacostas, V., Moss, B., Sawyer, T. K., Heinrikson, R. L., and Tarpley, W. G. (1990). A synthetic HIV-1 protease inhibitor with antiviral activity arrests HIV-like particle maturation. *Science* **247,** 454–456.

Meek, T. D., Dayton, B. D., Metcalf, B. W., Dreyer, G. B., Strickler, J. E., Gorniak, J. G., Rosenberg, M., Moore, M. L., Magaard, V. W., and Debouck, C. (1989). Human immunodeficiency virus 1 protease expressed in *Escherichia coli* behaves as a dimeric aspartic protease. *Proc. Natl. Acad. Sci. U.S.A.* **86,** 1841–1845.

Meek, T. D., Rodriguez, E. J., and Angeles, T. S. (1994). Use of steady state kinetic methods to elucidate the kinetic and chemical mechanisms of retroviral proteases. *In* "Methods in Enzymology" (L. C. Kuo and J. A. Shafer, eds.), Vol. 241, pp. 127–156. Academic Press, San Diego, CA.

Mildner, A. M., Rothrock, D. J., Leone, J. W., Bannow, C. A., Lull, J. M., Reardon, I. M., Sarcich, J. L., Howe, W. J., Tomich, C.-S. C., Smith, C. W., Heinrikson, R. L., and Tomasselli, A. G. (1994). The HIV-1 protease as enzyme and substrate: Mutagenesis of autolysis sites and generation of a stable mutant with retained kinetic properties. *Biochemistry* **33,** 9405–9413.

Molla, A., Korneyeva, M., Gao, Q., Vasavanonda, S., Schipper, P. J., Mo, H. M., Markowitz, M., Chernyavskiy, T., Niu, P., Lyons, N., Hsu, A., Granneman, G. R., Ho, D. D., Boucher, C. A., Leonard, J. M., Norbeck, D. W., and Kempf, D. J. (1996). Ordered accumulation of mutations in HIV protease confers resistance to ritonavir. *Nat Med.* **2,** 760–766.

Molla, A., Granneman, G. R., Sun, E., and Kempf, D. J. (1998). Recent developments in HIV protease inhibitor therapy. *Antiviral Res.* **39,** 1–23.

Moyle, G., and Gazzard, B. (1996). Current knowledge and future prospects for the use of HIV protease inhibitors. *Drugs* **51,** 701–712.

Murphy, K. P., and Freire, E. (1992). Thermodynamics of structural stability and cooperative folding behavior in proteins. *Adv. Protein Chem.* **43,** 313–361.

Navia, M. A., Fitzgerald, P. M. D., McKeever, B. M., Leu, C.-T., Heimbach, J. C., Herber, W. K., Sigal, I. S., Darke, P. L., and Springer, J. P. (1989). Three-dimensional structure of aspartyl protease from human immunodeficiency virus HIV-1. *Nature (London)* **337,** 615–620.

Nijhuis, M., Schuurman, R., Jong, D., Erickson, J., Gustchina, E., Albert, J., Schipper, P., Gulnik, S., and Boucher, C. A. B. (1999). Increased fitness of drug resistant HIV-1 variants as a result of continued replication during suboptimal therapy. *Submitted for publication.*

Nillroth, U., Vrang, L., Markgren, P., Hulten, J., Hallberg, A., and Danielson, H. (1997). Human immunodeficiency virus type 1 proteinase resistance of symmetric cyclic urea inhibitor analogs. *Antimicrob. Agents Chemother.* **41,** 2383–2388.

Noble, S., and Faulds, D. (1996). Saquinavir. A review of its pharmacology and clinical potential in the management of HIV infection. *Drugs* **52,** 93–112.

Pargellis, C. A., Morelock, M. M., Graham, E. T., Kinkade, P., Pav, S., Lubbe, K., Lamarre, D., and Anderson, P. C. (1994). Determination of kinetic rate constants for the binding of inhibitors to HIV-1 protease and for the association and dissociation of active homodimer. *Biochemistry* **33,** 12527–12534.

Park, C., Koh, S. J., Son, Y. C., Choi, H., Lee, C. S., Cho, N., Moon, K. Y., Jung, W. H., Kim, S. C., and Yoon, H. (1995). Rational design of irreversible, pseudo-C_2-symmetric HIV-1 protease inhibitors. *Bioorg. Med. Chem. Lett.* **5,** 1843–1848.

Partaledis, J. A., Yamaguchi, K., Tisdale, M., Blair, E. E., Falcione, C., Maschera, B., Myers, R. E., Pazhanisamy, S., Futer, O., Cullinan, A. B., Stuver, C. M., Byrn, R. A., and Livingston, D. J. (1995). *In vitro* selection and characterization of human immunodeficiency virus type 1 (HIV-1) isolates with reduced sensitivity to hydroxyethylamino sulfonamide inhibitors of HIV-1 aspartyl protease. *J. Virol.* **69,** 5228–5235.

Patick, A. K., Rose, R., Greytok, J., Bechtold, C. M., Hermsmeier, M. A., Chen, P. T., Barrish, J. C,. Zahler, R., Colonno, R. J., and Lin, P.-F. (1995). Characterization of a human immunodeficiency virus type 1 variant with reduced sensitivity to an aminodiol protease inhibitor. *J. Virol.* **69,** 2148–2152.

Patick, A. K., Mo, H., Markowitz, M., Appelt, K., Wu, B., Musick, L., Kalish, V., Kaldor, S., Reich, S., Ho, D., and Webber, S. (1996). Antiviral and resistance studies of AG1343, an orally bioavailable inhibitor of human immunodeficiency virus protease. *Antimicrob. Agents Chemother.* **40,** 292–297.

Pazhanisamy, S., Stuver, C. M., Cullinan, A. B., Margolin, N., Rao, B. G., and Livingston, D. J. (1996). Kinetic characterization of human immunodeficiency virus type-1 protease-resistant variants. *J. Biol. Chem.* **271,** 17979–17985.

Pearl, L. H., and Taylor, W. R. (1987). A structural model for the retroviral proteases. *Nature (London)* **329,** 351–354.

Pettit, S., Moody, M., Wehbie, R., Kaplan, A., Nantermet, P., Klein, C., and Swantstrom, R. (1994). The p2 domain of human immunodeficiency virus type 1 Gag regulates sequential proteolytic processing and is required to produce fully infectio virions. *J. Virol.* **68,** 8017–8027.

Phylip, L. H., Richards, A. D., Kay, J., Konvalinka, J., Strop, P., Bláha, I., Velek, J., Kostka, V., Ritchie, A. J., Broadhurst, A. V., Farmerie, W. G., Scarborough, P. E., and Dunn, B. M. (1990). Hydrolysis of synthetic chromogenic substrates by HIV-1 and HIV-2 proteinases. *Biochem. Biophys. Res. Commun.* **171,** 439–444.

Polgár, L., Szeltner, Z., and Boros, I. (1994). Substrate-dependent mechanisms in the catalysis of human immunodeficiency virus protease. *Biochemistry* **33,** 9351–9357.

Poorman, R. A., Tomasselli, A. G., Heinrikson, R. L., and Kézdy, F. J. (1991). A cumulative specificity model for proteases from human immunodeficiency virus types 1 and 2, inferred from statistical analysis of an extended substrate data base. *J. Biol. Chem.* **266,** 14554–14561.

Prasad, J. V., Lunney, E. A., Ferguson, D., Tummino, P. J., Rubin, R. J., Reyner, E. L., Stewart, B. H., Guttendorf, R. J., Domagala, J. M., Suvorov, L. I., Gulnik, S. V., Topol,

I. A., Bhat, T. N., and Erickson, J. W. (1995). HIV protease inhibitors possessing a novel, high affinity and achiral P1/P2 ligand with unique pattern of *in vitro* resistance. Importance of conformationally-restricted template in the design of enzyme inhibitors. *J. Am. Chem. Soc.* **117,** 11070–11074.

Prasad, J. V., Lunney, E. A., Para, K. S., Tummino, P. J., Ferguson, D., Hupe, D., Domagala, J. M., and Erickson, J. W. (1996). Nonpeptidic potent HIV-1 protease inhibitors. *Drug Des. Discovery* **13,** 15–28.

Quere, L., Wenger, T., and Schramm, H. J. (1996). Triterpenes as potential dimerization inhibitors of HIV-1 protease. *Biochem. Biophys. Res. Commun.* **227,** 484–488.

Richards, A. D., Phylip, L. H., Farmerie, W. G., Scarborough, P. E., Alverez, A., Dunn, B. M., Hirel, P.-H., Konvalinka, J., Strop, P., Pavlickova, L., Kostka, V., and Kay, J. (1990). Sensitive, soluble chromogenic substrates for HIV-1 proteinase. *J. Biol. Chem.* **265,** 7733–7736.

Ridky, T., and Leis, J. (1995). Development of drug resistance to HIV-1 protease inhibitors. *J Biol Chem.* **270,** 29621–29623.

Ridky, T. W., Kikonyogo, A., Leis, J., Gulnik, S., Copeland, T., Erickson, J., Wlodawer, A., Kurinov, I., Harrison, R. W., and Weber, I. T. (1998). Drug-resistant HIV-1 proteases identify enzyme residues important for substrate selection and catalytic rate. *Biochemistry* **37,** 13835–13845.

Riviere, Y., Blank, V., Kourilsky, P., and Israel, A. (1991). Processing of the precursor of NF-kappa B by the HIV-1 protease during acute infection. *Nature (London)* **350,** 625–626.

Roberts, N. A., Craig, J. C., and Sheldon, J. (1998). Resistance and cross-resistance with saquinavir and other HIV protease inhibitors: Theory and practice. *AIDS* **12,** 453–460.

Roberts, N. A., Martin, J. A., Kinchington, D., Broadhurst, A. V., Craig, J. C., Duncan, I. B., Galpin, S. A., Handa, B. K., Kay, J., Kröhn, A., Lambert, R. W., Merrett, J. H., Mills, J. S., Parkes, K. E. B., Redshaw, S., Ritchie, A. J., Taylor, D. L., Thomas, G. J., and Machin, P. J. (1990). Rational design of peptide-based HIV proteinase inhibitors. *Science* **248,** 358–361.

Rose, J. R., Salto, R., and Craik, C. S. (1993). Regulation of autoproteolysis of the HIV-1 and HIV-2 proteases with engineered amino acid substitutions. *J. Biol. Chem.* **268,** 11939–11945.

Rose, J. R., Babe, L. M., and Craik, C. S. (1994). Defining the level of human immunodeficiency virus type 1 (HIV-1) protease activity required for HIV-1 particle maturation and infectivity. *J. Virol.* **69,** 2751–2758.

Rose, R., Craik, C., and Stroud, R. (1998). Domain flexibility in retroviral proteases: Structural implications for drug resistant mutations. *Biochemistry* **37,** 2606–2621.

Rose, R. E., Gong, Y. F., Greytok, J. A., Bechtold, C. M., Terry, B. J., Robinson, B. S., Alam, M., Colonno, R. J., and Lin, P. F. (1996). Human immunodeficiency virus type 1 viral background plays a major role in development of resistance to protease inhibitors. *Proc. Natl. Acad. Sci. U.S.A.* **93,** 1648–1653.

Schechter, I., and Berger, A. (1967). On the size of the active site in proteases. I. Papain. *Biochem. Biophys. Res. Commun.* **27,** 157–162.

Schock, H. B., Garsky, V. M., and Kuo, L. C. (1996). Mutational anatomy of an HIV-1 protease variant conferring cross-resistance to protease inhibitors in clinical trials. Compensatory modulations of binding and activity. *J. Biol. Chem.* **271,** 31957–31963.

Schramm, H. J., Boetzel, J., Buttner, J., Fritsche, E., Gohring, W., Jaeger, E., Konig, S., Thumfart, O., Wenger, T., Nagel, N. E., and Schramm, W. (1996). The inhibition of human immunodeficiency virus proteases by 'interface peptides.' *Antiviral Res.* **30,** 155–170.

Seelmeier, S., Schmidt, H., Turk, V., and von der Helm, K. (1988). Human immunodeficiency virus has an aspartic-type protease that can be inhibited by pepstatin A. *Proc. Natl. Acad. Sci. U.S.A.* **85,** 6612–6616.

Shafer, R. W., Winters, M. A., Palmer, S., and Merigan, T. C. (1998). Multiple concurrent reverse transcriptase and protease mutations and multidrug resistance of HIV-1 isolates from heavily treated patients. *Ann. Intern. Med.* **128,** 906–911.

Shoeman, R. L., Sachse, B., Honer, E., Mothes, E., Kaufmann, M., and Traub, P. (1993). Cleavage of human and mouse cytoskeletal and sarcomeric proteins by human immunodeficiency virus type 1 protease: Actin, desmin, myosin, and tropomyosin. *Am. J. Pathol.* **142,** 221–230.

Silva, A. M., Cachau, R. E., Sham, H. L., and Erickson, J. W. (1996). Inhibition and catalytic mechanism of HIV-1 aspartic protease. *J. Mol. Biol.* **255,** 321–346.

Smidt, M. L., Potts, K. E., Tucker, S. P., Blystone, L., Stiebel, T. R., Jr., Stallings, W. C., McDonald, J. J., Pillay, D., Richamn, D. D., and Bryant, M. L. (1997). A mutation in human immunodeficiency virus type 1 protease at position 88, located outside the active site, confers resistance to the hydroxyethylurea inhibitor SC-55389A. *Antimicrob. Agents Chemother.* **31,** 515–522.

Smith, R., Brereton, I. M., Chai, R. Y., and Kent, S. B. (1996). Ionization states of the catalytic residues in HIV-1 protease. *Nat. Struct. Biol.* **3,** 946–950.

Snasel, J., and Pichova, I. (1996). The cleavage of host cell proteins by HIV-1 protease. *Folia Biol.* **42,** 227–230.

Stebbins, J., and Debouck, C. (1994). Expression systems for retroviral proteases. *In* "Methods in Enzymology" (L. C. Kuo and J. A. Shafer, eds.), Vol. 241, pp. 3–16. Academic Press, San Diego, CA.

Strickler, J. E., Gorniak, J., Dayton, B., Meek, T., Moore, M., Magaard, V., Malinowski, J., and Debouck, C. (1989). Characterization and autoprocessing of precursor and mature forms of human immunodeficiency virus type 1 (HIV 1) protease purified from *Escherichia coli. Proteins* **6,** 139–154.

Szeltner, Z., and Polgár, L. (1996). Rate-determining steps in HIV-1 protease catalysis. The hydrolysis of the most specific substrate. *J. Biol. Chem.* **271,** 32180–32184.

Tang, J., and Hartsuck, J. A. (1995). A kinetic model for comparing proteolytic processing activity and inhibitor resistance potential of mutant HIV-1 proteases. *FEBS Lett.* **367,** 112–116.

Tang, J., James, M. N. G., Hsu, I. N., Jenkins, J. A., and Blundell, T. L. (1978). Structural evidence for gene duplication in the evolution of the acid proteases. *Nature (London)* **271,** 618–621.

Tawa, G. J., Topol, I., Burt, S., and Erickson, J. W. (1998). Calculation of relative binding free energies of peptidic inhibitors to HIV-1 protease and its I84V mutant. *J. Am. Chem. Soc.* **120,** 8856–8863.

Thaisrivongs, S., Skulnick, H. I., Turner, S. R., Strohbach, J. W., Tommasi, R. A., Johnson, P. D., Aristoff, P. A., Judge, T. M., Gammill, R. B., Morris, J. K., Romines, K. R., Chrusciel, R. A., Hinshaw, R. R., Chong, K. T., Tarpley, W. G., Poppe, S. M., Slade, D. E., Lynn, J. C., Horng, M. M., Tomich, P. K., Seest, E. P., Dolak, L. A., Howe, W. J., Howard, G. M., *et al.* (1996). Structure-based design of HIV protease inhibitors: Sulfonamide-containing 5,6-dihydro-4-hydroxy-2-pyrones as non-peptidic inhibitors. *J. Med. Chem.* **39,** 4349–4353.

Todd, M. J., Semo, N., and Freire, E. (1998). The structural stability of the HIV-1 protease. *J. Mol. Biol.* **283,** 475–488.

Toh, H., Ono, M., Saigo, K., and Miyata, T. (1985). Retroviral protease-like sequence in the yeast transposon Ty1. *Nature (London)* **315,** 691–692.

Tomasselli, A. G., and Heinrikson, R. L. (1994). Specificity of retroviral proteases: An analysis of viral and nonviral protein substrates. *In* "Methods in Enzymology" (L. C. Kuo and J. A. Shafer, eds.), Vol. 241, pp. 279–301. Academic Press, San Diego, CA.

Tomasselli, A. G., Howe, W. J., Hui, J. O., Sawyer, T. K., Reardon, I. M., DeCamp, D. L., Craik, C. S., and Heinrikson, R. L. (1991a). Calcium-free calmodulin is a substrate of proteases from human immunodeficiency viruses 1 and 2. *Proteins* **10,** 1–9.

Tomasselli, A. G., Hui, J. O., Adams, L., Chosay, J., Lowery, D., Greenberg, B., Yem, A., Deibel, M. R., Zürcher-Neely, H., and Heinrikson, R. L. (1991b). Actin, troponin C, Alzheimer amyloid precursor protein and pro-interleukin 1β as substrates of the protease from human immunodeficiency virus. *J. Biol. Chem.* **266,** 14548–14553.

Tozser, J., Weber, I. T., Gustchina, A., Blaha, I., Copeland, T. D., Louis, J. M., and Oroszlan, S. (1992). Kinetic and modeling studies of S3-S3' subsites of HIV proteinases. *Biochemistry* **31,** 4793–4800.

Vacca, J. P. (1994). Design of tight-binding human immunodeficiency virus type 1 protease inhibitors. *In* "Methods in Enzymology" (L. C. Kuo and J. A. Shafer, eds.), Vol. 241, pp. 311–334. Academic Press, San Diego, CA.

Vacca, J. P., and Condra, J. H. (1997). Clinically effective HIV-1 protease inhibitors. *Drug Des. Discovery* **2,** 261–272.

Vacca, J. P., Dorsey, B. D., Schleif, W. A., Levin, R. B., McDaniel, S. L., Darke, P. L., Zugai, J., Quintero, J. C., Blahy, O. M., Roth, E., Sardana, V. V., Schlabach, A. J., Graham, P. I., Condra, J. H., Gotlib, L., Holloway, M. K., Lin, J., Chen, I.-W., Vastag, K., Ostovic, D., Anderson, P. S., Emini, E. A., and Huff, J. R. (1994). L-735, 524: An orally bioavailable human immunodeficiency virus type 1 protease inhibitor. *Proc. Natl. Acad. Sci. U.S.A.* **91,** 4096–4100.

Vogt, V. M. (1996). Proteolytic processing and particle maturation. *Curr. Top. Microbiol. Immunol.* **214,** 95–131.

Wang, Y. X., Freedberg, D. I., Yamazaki, T., Wingfield, P. T., Stahl, S. J., Kaufman, J. D., Kiso, Y., and Torchia, D. A. (1996). Solution NMR evidence that the HIV-1 protease catalytic aspartyl groups have different ionization states in the complex formed with the asymmetric drug KNI-272. *Biochemistry* **35,** 9945–9950.

West, M. L., and Fairlie, D. P. (1995). Targeting HIV-1 protease: A test of drug-design methodologies. *Trends Pharmacol. Sci.* **16,** 67–75.

Wilson, S. I., Phylip, L. H., Mills, J. S., Gulnik, S. V., Erickson, J. W., Dunn, B. M., and Kay, J. (1997). Escape mutants of HIV-1 proteinase: Enzymic efficiency and susceptibility to inhibition. *Biochim. Biophys. Acta* **1339,** 113–125.

Winslow, D. L., and Otto, M. J. (1995). HIV protease inhibitors. *AIDS* **9** (Suppl. A), S183–S192.

Winters, M. A., Schapiro, J. M., Lawrence, J., and Merigan, T. C. (1998). Human immunodeficiency virus type 1 protease genotypes and *in vitro* protease inhibitor susceptibilities of isolates from individuals who were switched to other protease inhibitors after long-term saquinavir treatment. *J. Virol.* **72,** 5303–5306.

Wlodawer, A., and Erickson, J. W. (1993). Structure-based inhibitors of HIV-1 protease. *Annu. Rev. Biochem.* **62,** 543–585.

Wlodawer, A., and Vondrasek, J. (1998). Inhibitors of HIV-1 protease: A major success of structure-assisted drug design. *Annu. Rev. Biophys. Biomol. Struct.* **27,** 249–284.

Wlodawer, A., Miller, M., Jaskólski, M., Sathyanarayana, B. K., Baldwin, E., Weber, I. T., Selk, L. M., Clawson, L., Schneider, J., and Kent, S. B. H. (1989). Conserved folding in retroviral proteases: Crystal structure of a synthetic HIV-1 protease. *Science* **245,** 616–621.

Xie, D., Gulnik, S., Collins, L., Gustchina, E., Bhat, T. N., and Erickson, J. W. (1998). Thermodynamics and proton uptake for pepstatin binding to retroviral and eukaryotic aspartic proteases. *Adv. Exp. Med. Biol.* **436,** 381–386.

Xie, D., Gulnik, S., Gustchina, E., Yu, B., Shao, W., Qoronfleh, W., Nathan, A., and Erickson, J. W. (1999a). Drug resistant mutations can affect dimer stability of HIV-1 protease at neutral pH. *Protein Sci.* **8,** 1702–1707.

Xie, D., Suvorov, L., Gustchina, E., Erickson, J., and Gulnik, S. (1999b). Real-time measurements of dark substrate catalysis. *Protein Sci.* (in press).

Yamazaki, T., Nicholson, L., Torchia, D., Wingfield, P., Stahl, S., Kaufman, J., Eyermann, C., Hodge, N., Lam, P., Ru, Y., Jadhav, P., Chang, C.-H., and Weber, P. (1994). NMR and X-ray evidence that the HIV protease catalytic aspartyl groups are protonated in the complex formed by the protease and a non-peptide cyclic urea-based inhibitor. *J. Am. Chem. Soc.* **116,** 10791–10792.

Yu, Z., Caldera, P., McPhee, F., De Voss, J. J., Jones, P. R., Burlingame, A. L., Kuntz, I. D., Craik, C. S., and Ortiz de Montellano, P. R. (1996). Irreversiible inhibition of the HIV-1 protease: Targeting alkylating agents to the catalytic aspartate groups. *J. Am. Chem. Soc.* **118,** 5846–5856.

Zennou, V., Mammano, F., Paulous, S., Mathez, D., and Clavel, F. (1998). Loss of viral fitness associated with multiple Gag and Gag-Pol processing defects in human immunodeficiency virus type 1 variants selected for resistance to protease inhibitors *in vivo*. *J. Virol.* **72,** 3300–3306.

Zhang, Y. M., Imamichi, H., Imamichi, T., Lane, H. C., Falloon, J., Vasudevachari, M. B., and Salzman, N. P. (1997). Drug resistance during indinavir therapy is caused by mutations in the protease gene and in its Gag substrate cleavage sites. *J. Virol.* **71,** 6662–6670.

Zhang, Z.-Y., Poorman, R. A., Maggiora, L. L., Heinrikson, R. L., and Kézdy, F. J. (1991). Dissociative inhibition of dimeric enzymes. *J. Biol. Chem.* **266,** 15591–15594.

Zutshi, R., Franciskovich, J., Shultz, M., Schweitzer, B., Bishop, P., Wilson, M., and Chmielewski, J. (1997). Targeting the dimerization interface of HIV-1 protease-inhibition with corss-linked interfacial peptides. *J. Am. Chem. Soc.* **119,** 4841–4845.

VITAMINS AND HORMONES, VOL. 58

Ultrastructure of Cells Undergoing Apoptosis

KENJI KIMURA,* HIRONOBU SASANO,* TOORU SHIMOSEGAWA,† SHIZUE MOCHIZUKI,* HIROSHI NAGURA,* AND TAKAYOSHI TOYOTA†

*Department of Pathology and †Department of Medicine, Tohoku University School of Medicine, 2–1 Seiryou-machi, Aoba-ku, Sendai-shi, Miyagi-ken, Japan

I. Overview of Apoptosis

Analysis of cell turnover, the balance between cell proliferation and cell death, can provide important information about our understanding of the biological characteristics of various tissues and their functions. Apoptosis or programmed cell death is a selective process of physiologic cell deletion (Kerr *et al.,* 1972; Wyllie *et al.,* 1980; Russel, 1981; Wyllie, 1981; Duvall and Wyllie, 1986; Smith *et al.,* 1989). The process of apoptosis plays a critical role in development and/or morphogenesis, in determining the size and shape of developing tissues, in maintenance of dynamic steady state in turnover of certain cell lineages in normal tissues, and in defining the number of tumor cell populations in the course of neoplastic development. Apoptosis is also important in various pathologic conditions, such as the response to irradiation (Yamada and Ohyama, 1998) stimulation by glucocorticoid (Wyllie,1980), withdrawal of glucocorticoid (Kyprianou and Isaacs, 1988), cytotoxic damages to gastric epithelial cells following Helicobacter pylori infection (Yabuki *et al.,* 1997), and changes in some kinds of growth hormones (Solviter *et al.,* 1993; Wyllie, 1993).

Apoptosis is generally considered to be a form of cell death that occurs as a result of intentional suicide based on a genetic mechanism (Majno and Joris, 1995; Sasano, 1995; Kimura *et al.,* 1997). Apoptosis is also a form of cell death characterized by morphologic as well as biochemical criteria and can be considered as a counterpart of mitosis (Majno and Joris, 1995). Apoptosis is characterized morphologically by

0083-6729/00 $30.00

shrinkage of the cells resulting in cytoplasm condensation and the fragmentation of the cell nucleus, and membrane blebbing. Apoptosis is also defined biochemically by increased endogenous endonuclease activity, chromatin cleavage, and DNA fragmentation, that is, the DNA is broken down into segments that are multiples of approximately 185 base pairs, due to specific cleavage between nucleosomes (Kerr *et al.,* 1987). In contrast, accidental cell death or oncosis or necrosis, which represents all cell death by murders, is generally defined as cell death caused by various cytotoxic agents or a pathologic environment, and is morphologically characterized by cytoplasmic swelling without the nuclear changes just described (Majno and Joris, 1995). It is also important to note that the whole process of apoptosis or programmed cell death is under genetic control and can be initiated by an internal clock or by extracellular agents. In addition, the duration of the slow process of apoptosis can be very fast, even in minutes (Majno and Joris, 1995). Apoptosis has been demonstrated to be induced by glucocorticoids, ceramide, wild-type p53, various monoclonal antibodies against cell-surface antigens including Apo-1 or Fas, and others to be inhibited by *bc1–2* protein encoded by the protooncogene *bc1–2* (Majno and Joris, 1995).

Biochemical detection of DNA cleavage into oligonucleosomal fragments of multiples of 180–200 base pairs, or the ladder pattern of nucleosomal DNA fragments on agarose gel electrophoresis, has been generally recognized as the hallmarks of apoptosis (Wyllie, 1980; Wyllie *et al.,* 1984; Majno and Joris, 1995). However, as expected, this method can by no means identify the individual cells that are undergoing apoptosis. In addition, it has been recently demonstrated that these ladder patterns are not necessarily considered as specific for the process of apoptosis (Rink *et al.,* 1995). On the other hand, the concept of "apoptosis" itself, which Kerr and Wyllie originally proposed was based on morphologic and ultrastructural examination (Kerr *et al.,* 1972; Wyllie *et al.,* 1980; Wyllie, 1981). It is therefore very important for those involved in the study of apoptosis to obtain sufficient knowledge of ultrastructural features of the cells undergoing apoptosis and incorporate morphologic observation or examination, especially that using electron microscopy into the study of apoptosis in his or her laboratory. These features are discussed in the following sections.

II. Ultrastructural Features of Cells Undergoing Apoptosis

Kerr implied in 1971, shrinkage necrosis or apoptosis as a distinct mode of cellular death. The major morphologic feature of the cells un-

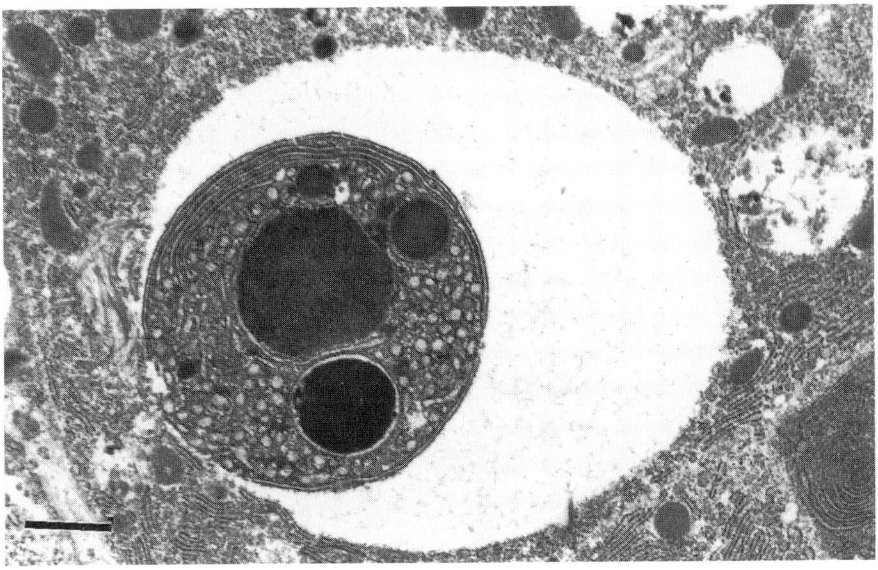

FIG. 1. Ultrastructures of acinar cells in the pancreas of adrenalectomized rats treated with cerulein. Male Wistar strain rats weighing approximately 250 g underwent bilateral adrenalectomy and received 5 μg/kg per hour cerulein, intravenously, for 6 h. The cell shrinks and nuclei condense. The cytoplasm becomes dense but various intracellular organellae including mitochondria and endoplasmic reticula are still discernible. Bar = 1 μm.

dergoing apoptosis is the shrinkage of the cells. The cell becomes small in size and the cytoplasm becomes dense but various cell organellae including mitochondria are relatively normal (Fig. 1). Progression of the cytoplasmic condensation then results in marked crowding of organellae, which characteristically retain their integrity. However, the most characteristic ultrastructural feature of the cells undergoing apoptosis is chromatin condensation. The chromatin becomes very pyknotic, aggregating in the periphery of the nucleus to form dense smooth masses applied against the nuclear membrane (Fig. 2). These well-delimited dense masses of various shapes and sizes have been morphologically described as half-moon-, horseshoe-, lancet-, and navicular-shaped nuclei. The nucleus itself sometimes breaks up, resulting in one or two fragments of these dense smooth intranuclear masses (Fig. 2). These nuclear changes may be discernible in light microscopic examination but the need for ultrastructural confirmation of light microscopic identification of dying cells is required.

The other ultrastructural features detected in the cells undergoing

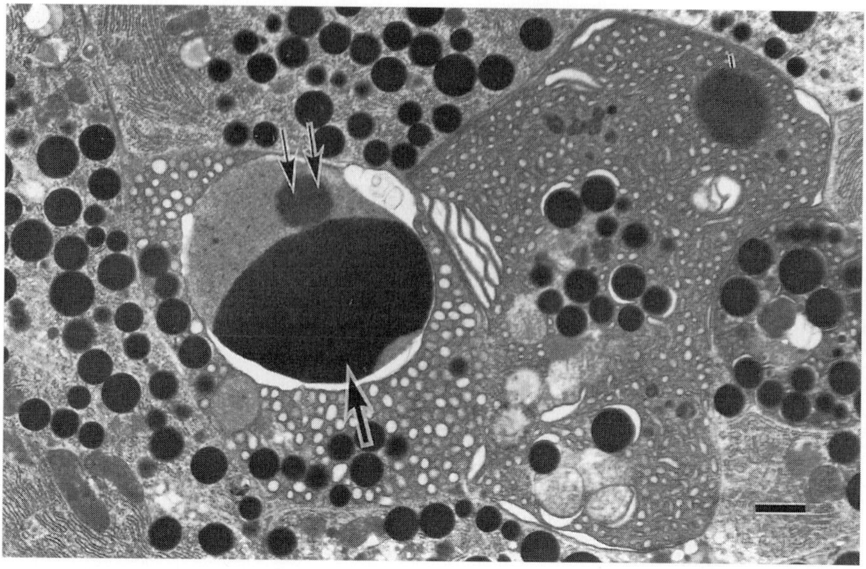

FIG. 2. Ultrastructures of acinar cells in the pancreas of adrenalectomized rats treated with cerulein. The arrow indicates apoptotic acinar cells. In the nucleus, condensed chromatin aggregates close to the nuclear membrane and demonstrates an appearance of crescent-like clumping (arrow). The nucleus also breaks up and forms relatively dense smooth intranuclear masses (double arrows). Zymogen granules are preserved in these cells. The polarity of zymogen granules in adjacent nonapoptotic acinal cells is relatively preserved. Bar = 1 μm.

apoptosis is formation of cytoplasmic blebs and so-called "apoptotic bodies." These processes emitted by the cells often contain pyknotic nuclear fragments (the budding phenomenon). These apoptotic bodies also contain cytoplasms and densely packed cell organellae. Sometimes, these apoptotic bodies may contain only cytoplasmic elements. These apoptotic bodies may be phagocytized by adjacent macrophages or parenchymal cells. Therefore, the cells undergoing apoptosis can be deleted from the tissue through these processes and the adjacent cells may migrate or proliferate to replace the space occupied by the deleted apoptotic cells.

In contrast to these ultrastructural features of the cells undergoing apoptosis or programmed cell death, the cells undergoing the process of oncosis or accidental cell death may demonstrate cellular swelling, organellae swelling, and blebbing not accompanied by the nuclear changes described earlier (Majno and Joris, 1995).

III. CORRELATION BETWEEN ULTRASTRUCTURAL AND BIOCHEMICAL FEATURES OF THE CELLS UNDERGOING APOPTOSIS

When examining the presence or absence of the cells undergoing apoptosis in the tissue specimen or cultured cells, it is very important to study these tissues and cells using election microscopy in order to examine whether ultrastructural features described earlier are detected or not. However, it is also true that these ultrastructural features of the cells undergoing apoptosis or programmed cell death are observed only for a limited period in the course of apoptosis (Arends *et al.,* 1990; Ansari *et al.,* 1993; Gorczyca *et al.,* 1993; Hockenbery, 1995). In particular, these ultrastructural features may not be discernible in the early stages of cells undergoing apoptosis. In addition, examination by electron microscopy is not necessarily practical when screening for the presence or absence of apoptosis in tissue specimens or cell lines due to the labor-intensive and time-consuming nature of the technique. Apoptosis has been recognized as an essential component of tissue dynamics for many years. However, the relative difficulty in assessing apoptotic cells has prevented its study in both normal and pathologic processes (Ansari *et al.,* 1993). Consequently, the development of other *in situ* methods to study apoptosis has been desired.

In 1992, Gavrieli *et al.* demonstrated that DNA fragmentation can be detected *in situ* by labeling 3′-OH ends with biotinylated deoxyuridine triphosphate (dUTP) through the action of terminal deoxynucleotidyl transferase (TdT). Since then, this method, subsequently termed 3′-OH nick end labeling, or TdT-mediated dUTP-biotin nick end labeling (TUNEL), has been widely used to detect cells with DNA fragmentation, especially in surgical pathology materials of human disorders. The relatively good correlation of results obtained by the TUNEL method and DNA ladder formation by agarose gel electrophoresis has been demonstrated (Ansari *et al.,* 1995; Abe *et al.,* 1995). The TUNEL method has been widely applied to the analysis of apoptosis in a wide variety of materials including tissue specimens and cell lines (Hiraishi *et al.,* 1993; Sasano *et al.,* 1995; Kimura *et al.,* 1998). With the introduction of various commercially available staining kits, the TUNEL method can now be easily and rapidly performed in properly fixed and processed tissue materials. The TUNEL method is now being incorporated into many pathology laboratories throughout the world. However, increasing evidence suggests that the TUNEL method is by no means specific for the detection of apoptosis (Ansari *et al.,* 1995) and this method may also detect cells that are committed to, but not yet in the process of, apoptosis (Wyllie *et al.,* 1984; Ansari *et al.,* 1995). Therefore it becomes very important to examine the ultrastructural or morphologic features of TUNEL-positive cells.

We have recently examined ultrastructural features of TUNEL-positive cells in continuous mirror image semithin and ultrathin sections of Epon-Araldite-embedded tissues of pancreas from bilaterally adrenalectomized rats with cerulein-induced pancreatitis (Kimura *et al.*, 1997). Application of this method through continuous mirror sections can directly demonstrate ultrastructural features of the TUNEL-positive cells, although the sensitivity of the TUNEL method may be lower in Epon-Araldite-embedded tissue sections than in regular 10% formalin or 8% paraformaldehyde fixed and paraffin-embedded tissue sections possibly due to the interruption of the binding of dUTP to nick by Epon-Araldite plastic and osmium.

Results of this study revealed that all ultrastructurally discernible apoptotic cells were positive by the TUNEL method (Fig. 3), but de-

A

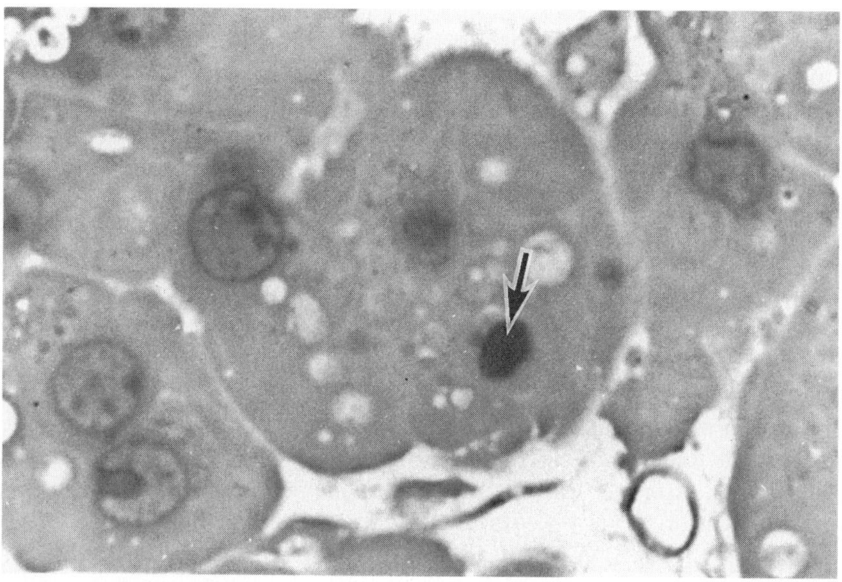

FIG. 3. (A) A TUNEL-positive nucleus of an acinar cell in an Epon-Araldite semithin section. The arrow indicates an intensely labeled acinar cell nucleus. The other nuclei are stained by hematoxylin (× 300). (B) Ultrastructural features of the same acinus in a serial adjacent ultrathin section. The arrow indicates condensed chromatin with creascentic clumping. Bar = 2 μm. (C) At higher magnification, mitochondria and zymogen granules are observed to be almost intact. Rough endoplasmic reticulum revealed a vesicular pattern. Bar = 2 μm.

B

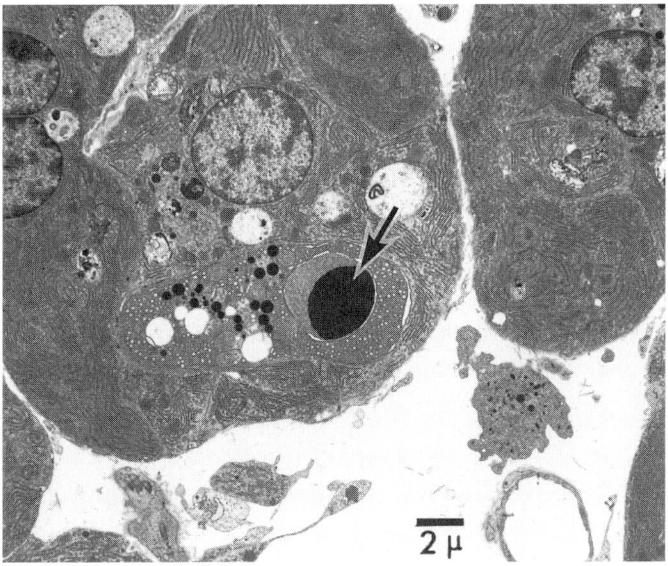

C

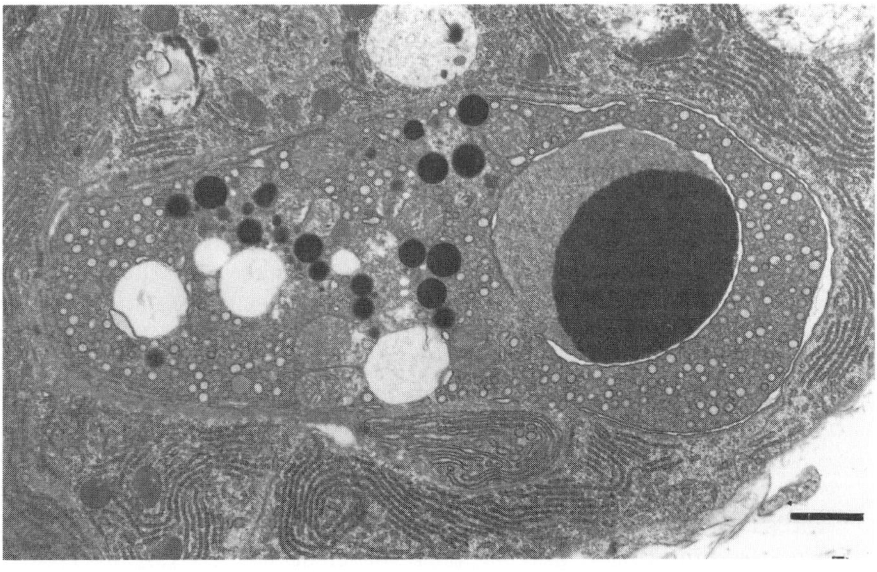

FIG. 3. Continued

generated nuclei that may represent the end stage of cell death, either programmed or accidental, are also strongly labeled by the TUNEL method. Ansari *et al.* (1993) suggested that the TUNEL method may identify not only the apoptotic nuclei, but also the nuclei in necrotic tissues; the staining of necrotic cells could be explained either by the activation of an apoptotic pathway in damaged cells by the degradation of DNA by lysozomal enzymes. However, these two patterns can be distinguished by the distribution of TUNEL-positive nuclei; apoptotic nuclei are usually scattered, without localized aggregation. Several investigators have proposed the possibility that cells which have not yet begun the process of apoptosis and which lack its morphologic features can also be detected by the TUNEL method (Gorczyca *et al.,* 1993; Migheli *et al.,* 1995), which may therefore be able to detect apoptotic cells much earlier than electron microscopy. In our study employing continuous mirror image semithin and ultrathin sections, however, nearly all TUNEL-positive cells exhibited the characteristic ultrastructural features of apoptosis.

Migheli *et al.* (1995) also applied the TUNEL method at the electron microscopic level, using an immunogold staining technique in embryonic mouse dorsal root ganglia. They reported that nuclear localization of immunogold was observed not only in typical apoptotic cells at various stages of cell death, but also in some apparently normal cells, which were structurally indistinguishable from adjacent unstained cells. Morphologically viable TUNEL-positive cells, including the "pre-apoptotic" cells (Migheli *et al.,* 1995) may represent experimentally induced artefactual DNA damage, leading to TUNEL positivity. The possibility that the relatively rapid process of apoptosis in pancreatic acinar cells in a cerulein-induced pancreatitis model made it difficult to detect pre-apoptotic cells with TUNEL cannot be completely ruled out. However, it is also true that ultrastructural examination of TUNEL-positive cells in continuous mirror image semithin and ultrathin sections of Epon-Araldite-embedded tissue sections employed in our study (Kimura *et al.,* 1997) has the following technical advantages over the immunogold technique reported by Migheli *et al.* (1995): (1) Much more ultrastructural detail can be obtained than in a single sections, (2) TUNEL-positive cells can be identified in a much wider area by light microscopy, and (3) TUNEL-positive cells are much more clearly visualized due to the colorimetric reaction in contrast to relatively small dots in the immunogold electron microscopy method. Despite these minor differences, results of our study and Migheli's clearly demonstrated that the ultrastructural features of apoptosis are unequivocally associated with DNA fragmentation and support the close association of DNA fragmentation detected by TUNEL with ultrastructurally identified apoptosis.

IV. Conclusion

Ultrastructural examination using conventional electron microscopy is still a very important method of analyzing apoptosis or programmed cell death. It is therefore very important for those involved in the study of tissue dynamics or cell proliferation and death to be familiar with these ultrastructural features of apoptosis.

REFERENCES

Abe, R., Shimosegawa, T., Kimura, K., Abe, T., Kashimura, J., Koizumi, M., and Toyota, T. (1995). The role of endogenous glucocorticoids in rat experimental models of acute pancreatitis. *Gastroenterology* **109,** 933–943.

Ansari, B., Coates, P. J., Greenstein, B. D., and Hall, P. A. (1993). *In situ* end-labeling detects DNA strand breaks in apoptosis and other physiological and pathological states. *J. Pathol.* **170,** 1–8.

Arends, M. J., Morris, R. G., and Wyllie, A. H. (1990). Apoptosis. The role of endonuclease *Am. J. Pathol.* **136,** 593–600.

Duvall, E., and Wyllie, A. H. (1986). Death and the cell. *Immunol. Today* **7,** 115–119.

Gavrieli, Y., Sherman, Y., and Ben-Sasson, S. A. (1992). Identification of programmed cell death *in situ* via specific labeling of nuclear DNA fragmentation. *J. Cell Biol.* **119,** 493–501.

Gorczyca, W., Gong, J., and Darzynkiewicz, Z. (1993). Detection of DNA strand breaks in individual apoptotic cells by the *in situ* terminal deoxynucleotidyl transferase and nick translation assays. *Cancer Res.* **53,** 1945–1951.

Hiraishi, K., Suzuki, K., Hakomori, S., and Adachi, M. (1993). Le^Y antigen expression is correlated with apoptosis (programmed cell death). *Glycobiology* **3,** 381–390.

Hockenbery, D. (1995). Review: Defining apoptosis. *Am. J. Pathol.* **146,** 16–19.

Kerr, J. F. R. (1971). Shrinkage necrosis: A distinct mode of cellular death. *J. Pathol.* **105,** 13–20.

Kerr, J. F. R., Wyllie, A. H., and Currie, A. R. (1972). Apoptosis: Basic biological phenomenon with wide-ranging implications in tissue kinetics. *Br. J. Cancer* **26,** 239–257.

Kerr, J. F. R., Searle, J., Harmon, B. V., and Bishop, C. J. (1987). Apoptosis in perspectives on mammalian cell ing of death. *In* "Oxford Ageing" (C. S. Potten, ed.), pp. 93–98. Oxford University Press, Oxford.

Kimura, K., Sasano, H., Shimosegawa, T., Kato, K., Noguchi, T., Mochizuki, S., Sawai, T., Koizumi, M., Toyota, T., and Nagura, H. (1997). Ultrastructural and confocal laser scanning microscopic examination of tunel-positive cells. *J. Pathol.* **181,** 235–242.

Kimura, K., Shimosegawa, T., Sasano, H., Abe, R., Satoh, A., Masamune, A., Koizumi, M., Nagura, H., and Toyota, T. (1998). Endogenous glucocorticoids decrease the acinar cell sensitivity to apoptosis during cerulein pancreatitis in rats. *Gastroenterology* **114,** 372–381.

Kyprianou, N., and Isaacs, J. T. (1988). Activation of programmed cell death in the rat ventral prostate after castration. *Endocrinology (Baltimore)* **122,** 552–562.

Majno, G., and Joris, I. (1995). Apoptosis, oncosis, and necrosis. An overview of cell death. *Am. J. Pathol.* **146,** 3–15.
Migheli, A., Attanasio, A., and Shiffer, D. (1995). Ultrastructural detection of DNA strand breaks in apoptotic numeral cells by *in situ* end-labeling techniques. *J. Pathol.* **176,** 27–35.
Rink, A., Fung, K. M., Trajanowski, J. Q., Lee, Y. M., Neugebauer, E., and McIntosh, T. K. (1995). Evidence of apoptotic cell death after experimental traumatic brain injury in the rat. *Am. J. Pathol.* **147,** 1575–1583.
Russel, J. H. (1981). Internal disintegration model of cytotoxic lymphocyte-mediated cytotoxicity. *Biol. Rev. Cambridge Philos. Soc.* **56,** 153–197.
Sasano, H. (1995). *In situ* end labeling and its application to the study of endocrine disease: How can we study programmed cell death in surgical pathology materials? *Endocr. Pathol.* **6,** 1–3.
Sasano, H., Imatani, A., Shizawa, S., Suzuki, T., and Nagura, H. (1995). Cell proliferation and apoptosis in normal and pathologic human adrenal. *Mod. Pathol.* **8,** 11–17.
Smith, C. A., Williams, G. T., Kingston, R., Jendinson, E. J., and Owen, J. J. (1989). Antibodies to CD3/T-cell receptor complex induce death by apoptosis in immature T cells in thymic cultures. *Nature (London)* **337,** 181–184.
Solviter, R. A., Sollas, A. L., and Dean, E. (1993). Adrenalectomy-induced granule cell degeneration in the rat hippocampal dentate gyrus: Characterization of an in vitro model of controlled neuronal death. *J. Comp. Neurol.* **330,** 324–336.
Wyllie, A. H. (1980). Glucocorticoid-induced thymocyte apoptosis is associated with endogenous endonuclease activation. *Nature (London)* **284,** 555–556.
Wyllie, A. H. (1981). Cell death: A new classification separating apoptosis from necrosis. *In* "Cell Death in Biology and Pathology" (I. D. Bowen and R. A. Lockshin, eds.) pp. 9–34. Chapman & Hall, London.
Wyllie, A.H. (1993). Apoptosis. (The 1992 Frank Rose Memorial Lecture) *Br. J. Cancer* **67,** 205–208.
Wylie, A. H., Kerr, J. F. R., and Currie, A. R. (1980). Cell death: The significance of apoptosis. *Int. Rev. Cytol.* **68,** 251–306.
Wylie, A. H., Morris, R. G., Smith, A. L., and Dunlop, D. (1984). Chromatin cleavage in apoptosis: Association with condensed chromatin morphology and dependence on macromolecular synthesis. *J. Pathol.* **142,** 67–77.
Yabuki, N., Sasano, H., Tobita, M., Imatani, A., Hoshi, T., Kato, K., Ohara, S., Asaki, S., Toyota, T., and Nagura, H. (1997). Analysis of cell damage and proliferation in Helicobacter pylori-infected human gastric mucosa from patients with gastric adenocarcinoma. *Am. J. Pathol.* **151,** 821–829.
Yamada, T., and Ohyama, H. (1998). Radiation-induced interphase death of rat thymocytes is internally programmed (apoptosis). *Int. J. Radial. Biol.* **53,** 65–75.

VITAMINS AND HORMONES, VOL. 58

Role of Cytoskeleton in Apoptosis

R. ATENCIA, A. ASUMENDI, AND M. GARCÍA-SANZ

Departamento de Biologia Celular, Facultad de Medicina, Universidad del País Vasco / EHU, 48940 Leioa, España

Apoptosis is a form of cell death that takes place under physiologic conditions, and plays a key role in the control of biological processes such as embryonic development, tissue remodelation and renewal, or regulation of cell populations. Since its discovery in the early 1970s, there have been many relevant advances in the knowledge of the biochemical and molecular events involved in apoptosis. However, although the apoptotic process was defined on the basis of morphologic observations, only recently have we started to elucidate the molecular mechanisms that drive the structural changes observed in cells undergoing apoptosis. The article reviews current knowledge about the implications of cytoskeleton components (microfilaments, intermediate filaments, microtubules, and other cytoskeleton-related proteins) in the dynamics of apoptosis. © 2000 Academic Press.

0083-6729/00 $30.00

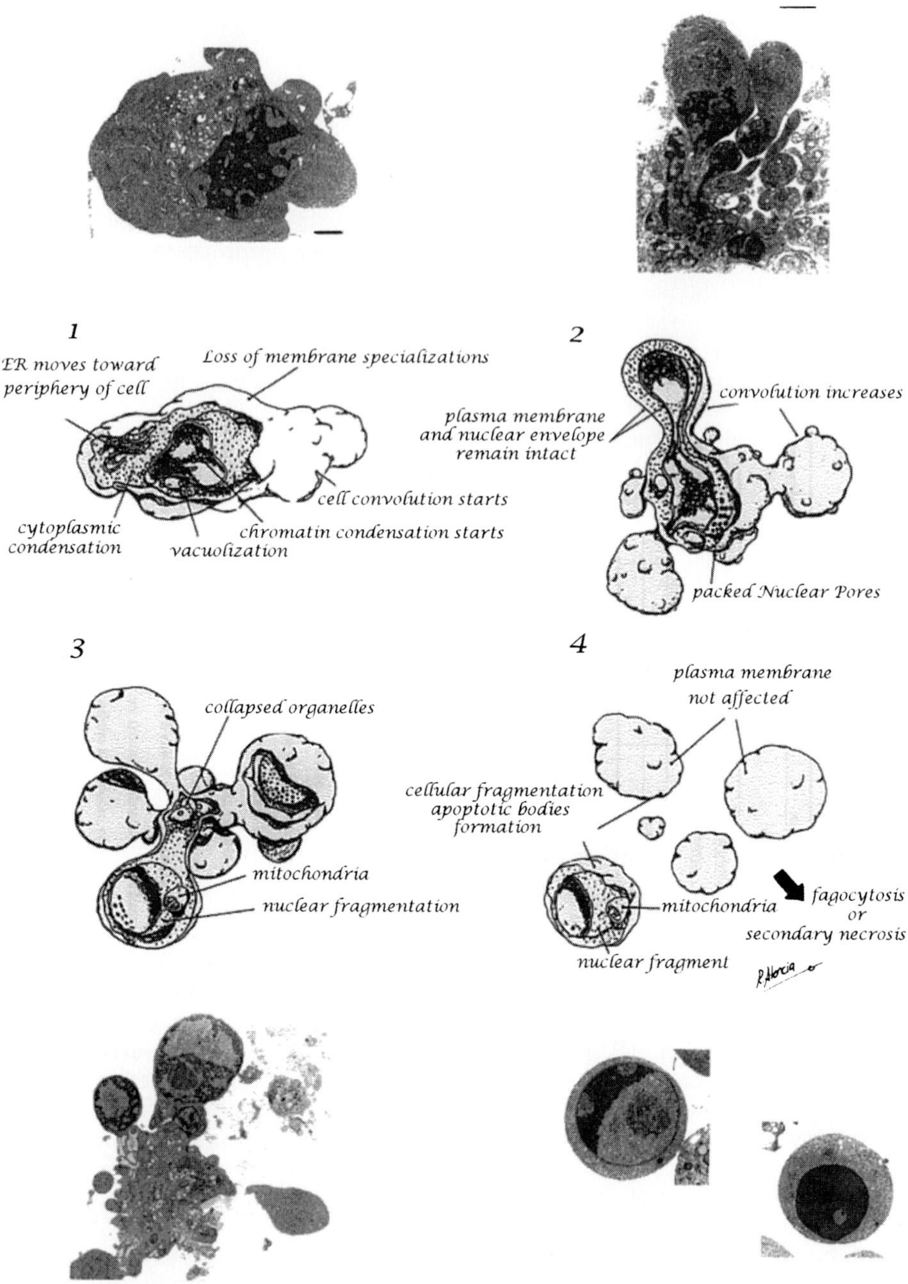

FIG. 1. Morphologic changes during the execution phase of apoptosis. This picture shows an idealization of the main structural changes that take place during the onset of apoptosis. The included micrographs show these changes observed on apoptotic cells when using a transmission electron microscope. Bar = 2 μm.

I. Introduction

In the last two decades, we have witnessed the discovery of a new kind of cellular function that has been defined as *apoptosis* or physiologic cell death. This term was first applied (Kerr *et al.,* 1972) for a mechanism of controlled cell deletion that plays a complementary role to mitosis in the regulation of cellular populations and, at present, is accepted as a basic biological activity of cells. The first descriptions of apoptosis were made under the basis of morphologic observations. As defined by Kerr and Wyllie (Kerr, 1971; Kerr *et al.,* 1972; Wyllie *et al.,* 1980), the most outstanding feature of the apoptotic process consists of a set of structural changes that comprise the following steps: loss of cell junctions and other specialized plasma membrane structures such as microvilli, cytoplasmic condensation and cellular convolution, chromatin hypercondensation and margination toward the nuclear periphery, and cellular fragmentation into membrane-bounded apoptotic bodies (Fig. 1).

Twenty-eight years later, the most relevant advances in the biochemical characterization of this process have been focused around the discovery of an endonuclease activity that cleaves the nuclear DNA during the onset of apoptosis (Wyllie, 1980; Arends *et al.,* 1990). We currently know that there is a family of proteases (recently joined under the term *caspases*) that is responsible for the activation of this endonuclease activity and other apoptotic-related phenomena such as lamin disassembly (for reviews, see Martins and Earnshaw, 1997; Porter *et al.,* 1997). This field of research has attracted many researchers, thus leaving untouched for years other points of interest such as the study of the key morphologic changes that define the apoptotic process. For these reasons, little was known about the biochemistry underlying apoptotic morphology until the last 5 years. However, in this period we have started to identify some of the molecular mechanisms that control the structural changes characteristic of apoptosis. We discuss in this review the current knowledge about the implications of the cytoskeleton in the dynamics of apoptosis.

II. Actin Skeleton and Apoptosis

Actin is the most abundant protein in many eukaryotic cells, and its ubiquity is completely justified. The actin skeleton (formed by the actin filaments and a myriad of actin-binding proteins) is involved in many cellular functions. The actin skeleton performs almost any cellular activity that requires movement or changes in the cellular shape, at least partially. We can include among these activities cell motility, chemotaxis, endocytosis, secretion, cell division, and, as we will see in the following lines, physiologic cell death or apoptosis. The organization of actin filaments inside the cell is highly regulated, providing the cell with an engine for rearranging its cytoskeleton in response to external or internal stimuli. Numerous proteins have been identified as parts of this cellular engine. These proteins control the assembly and length of actin filaments, link these filaments into bundles or networks, and establish connections between actin and cytoplasmic and membrane components, thus defining the complex three-dimensional network of the microfilament system. To date, we have much information about how actin and actin-binding proteins interact and how these interactions are regulated (for reviews, see Bretscher, 1991, Condeelis, 1993; Fechheimer and Zigmond, 1993; Carlier, 1998; Ayscough, 1998; Tanaka and Takai, 1998) in relation to the normal activity of nonmuscle cells.

A. Reorganization of Microfilament Framework during Apoptosis

As we have previously commented, the microfilament network helps to regulate overall cellular morphology and to coordinate cellular motility and orientation. For this reason, it was a logical presumption that actin skeleton could play an important role in the sequence of morphologic events that lead a cell to die by apoptosis. In this sense, the first experimental approaches in the search for a role of microfilaments in apoptosis relied on the activity of cytochalasins. Cytochalasins are fungal metabolites that can prevent the polymerization of actin and disrupt microfilament networks *in vivo* (Selden *et al.,* 1980; Schliwa, 1982; Cooper, 1987). In this sense, several studies have shown that cytochalasins B, C, and D prevent the formation of apoptotic bodies but not chromatin condensation nor endonuclease activity in several cell lines induced to undergo apoptosis (Cotter *et al.,* 1992; Endressen *et al.,* 1995; Atencia *et al.,* 1997). These results confirm that either actin polymerization or the presence of intact microfilaments is a necessary requirement for the formation of apoptotic bodies; but how is the microfilament network organized at the stages that precede the cellular fragmentation? Several studies have focused on this question in the last 2 years. Using fluorescent phallotoxins (phalloidin—mainly—and phallacidin) as probes for actin on different cell lines undergoing apoptosis, researchers have obtained the distribution pattern of actin at the stage of cellular blebbing and convolution, previous to formation of apoptotic bodies. Their results (Levee *et al.,* 1996; Pitzer *et al.,* 1996; Laster and McKenzie, 1996; García-Sanz *et al.,* 1996) supported the idea that rearrangement of the microfilament network is necessary for the formation of apoptotic bodies. All four reports described a striking reorganization of the actin skeleton in the blebbing cells, defining zones of high concentration of microfilaments at the base of apoptotic blebs. Soon after these findings, we analyzed at our laboratory the structure of the actin skeleton in apoptotic teratocarcinoma stem cells. By using confocal microscopy and computerized three-dimensional reconstruction, we observed the actin microfilaments located at the center of apoptotic cells. This filament mesh formed a three-dimensional structure that defined the border of blebbing zones and showed sporadic, ringlike structures (Atencia *et al.,* 1997). Moreover, we found that several actin-binding proteins such as myosin II and α-actinin located at the same position in apoptotic cells. It is known that the myosin family is involved in many forms of cell motility. Specifically, conventional nonmuscle myosin (myosin II) has been involved in such basic cellular processes as cytokinesis, secretion of vesicles, or maintenance of the cortical actin layer (Maciver, 1996). Therefore, our results suggested the existence of fully functional contractile machinery that could be responsible for the formation of apoptotic bodies. Regarding this matter, in the last year a research group from the University of Pennsylvania has published a very interesting paper (Mills *et al.,* 1998) proposing a biochemical model for the action of actin skeleton during the execution phase of apoptosis.

Moreover, the authors of this work have set up an elegant way to obtain a synchronous population of cells at the execution phase of apoptosis by using the caspase inhibitor z-Val-Ala-Asp-fluoromethylketone (z-VAD-FMK). When they used this caspase inhibitor at a concentration of 100 μM on PC12 cells entering apoptosis by serum deprivation, apoptosis was blocked, thus generating a synchronous population of cells actively extruding or retracting membrane blebs. On these cells, the authors performed a screening of kinase inhibitors, and obtained evidence that only myosin light chain kinase inhibitors decreased blebbing. Furthermore, they obtained data about the role of small G protein, Rho (a mediator in the phosphorylation of myosin light chain), in the process of blebbing. Finally, the authors proposed a working model of how membrane blebbing might occur.

These findings are very valuable at both the technical and theoretical levels. The possibility of obtaining large amounts of cells synchronized at an apoptotic state overrides one of the most important problems when studying apoptosis, that is, the scattered nature of this process. For sure, the method of synchronizing cells at the final phase of apoptosis described by Mills *et al.* (1998) will provide a powerful tool to analyze other events characteristic of physiologic cell death. On the theoretical side, we have, 28 years after the first morphologic description of apoptosis, a biochemical model that may explain, at least partially, the dynamics of the structural changes associated with the active phase of apoptosis. Nevertheless, other interesting results have been published in the last 2 years that could help to complete the model. We discuss these results next.

B. The Cortical Cytoskeleton and Apoptosis

Underlying and associated with the plasma membrane of animal cells is a cortical skeleton that both determines the topology of the membrane and is involved in cellular functions related to the plasma membrane. We can find a highly specialized form of this cortical skeleton, as the spectrin-actin network of the erythrocyte, that endows the membrane with stability and flexibility (for a review, see Bretscher, 1991). Spectrin-like proteins can be found in virtually all eukaryotic cells, and have been called mainly spectrin and fodrin. Fodrin is thought to be responsible for coupling a variety of membrane-spanning cell surface proteins to cytoplasmic elements, mainly, actin. Indeed, it has been suggested that fodrin could also be associated with the maintenance of lipid asymmetry.

In relation to apoptosis, several studies showed evidence of a link between a selective proteolysis of fodrin and the appearance of membrane blebbing during the apoptotic process. Particularly, the α-subunit of fodrin has been shown to be cleaved to a 150-kDa fragment during the onset of apoptosis in several cell lines (Martin *et al.,* 1995; Vanags *et al.,* 1996). However, thymocytes and T-cell leukemia lines undergo selective proteolysis of the α-fodrin subunit that yields a 120-kDa fragment. (Martin *et al.,* 1995; Kouchi *et al.,* 1997). In the search for a protease for this selective cleavage of fodrin, calpain was the first candidate. The calcium-activated neutral protease, or calpain system, represents one of the few well-characterized nonlysosomal proteolytic systems of mammalian cells (for a good review, see Croall and Demartino, 1991).

Calpains play a regulatory role in cellular protein metabolism, modifying (and not degrading) numerous target proteins, including cytoskeletal proteins. However, the involvement of calpain in apoptosis seems not to be a general mechanism to cleave fodrin, since synthetic calpain inhibitors fail to prevent α-fodrin proteolysis, membrane blebbing, and formation of apoptotic bodies in several experimental systems (Vanags *et al.,* 1996; Kouchi *et al.,* 1997; Atencia *et al.,* 1997). An alternate candidate is an ICE-like protease, formerly named CPP32 and now known as caspase-3 (Porter *et al.,* 1997). This apoptosis-related protease was first considered as an activator of the calpain system but there seems to be experimental data supporting a direct role for this caspase-3 in the selective proteolysis of fodrin (Cryns *et al.,* 1996). In any case, although the direct consequences of fodrin proteolysis should be elucidated, it is conceivable, given its role as a major component of the cortical cytoskeleton, that fodrin proteolysis during apoptosis may contribute to the membrane blebbing and rearrangement of actin skeleton.

C. Actin Skeleton and the Loss of Cell Surface Structures during Apoptosis

In addition to the cortical cytoskeleton we have found that cell surface structures such as microvilli and membrane ruffles contain an actin-based skeleton. Specifically, microvilli have a microfilament core bundle that inserts into a second microfilament arrangement, the terminal web. The core of microvilli is bundled through the action of villin (tissue-specific) and fimbrin (found in essentially all nonmuscle cells) (Bretscher, 1991). Other proteins function as cross-linkers between actin filaments and specific groups of integral membrane proteins. Belonging to this class, the ezrin-radixin-moesin (ERM) family consists of three closely related proteins (Arpin *et al.,* 1994; Tsukita and Yonemura, 1997). These proteins are activated in the cytoplasm, translocate to the plasma membrane and function as general actin filament-plasma membranes crosslinking to form and stabilize microvilli.

As we have stated earlier, one of the early morphologic changes observed during the active phase of apoptosis is a loss of the specialized membrane structures such as microvilli, resulting in a smoothing of the cell surface. Nevertheless, the molecular mechanisms underlying these changes were unknown until the publication in 1997 of experimental data that start to clarify the biochemistry of microvillar breakdown during apoptosis (Kondo *et al.,* 1997). In this work, researchers induced apoptosis by adding human Fas ligand to mouse fibroblasts expressing human Fas, and observed that there was a translocation of ERM proteins from microvilli to cytoplasm, coupled with a dephosphorylation of these proteins. Moreover, they found that the cytoplasmic translocation of the ERM proteins resulted in a general dissociation of the actin-based cytoskeletal components from the plasma membrane. These data are consistent with previous results from the same authors that showed an alteration of several cellular processes such as cell-cell adhesion, cell–matrix adhesion, and microvillar formation when ERM expression was suppressed with antisense oligonucleotides (Takeuchi *et al.,* 1994). Therefore, we have now a first approach to the molecular mechanisms involved in the loss of cell surface structures during the execution phase of apoptosis.

D. Actin Skeleton and the Loss of Intercellular Junctions during Apoptosis

Next, we refer to another morphologic event of apoptosis that has started to be understood from a biochemical point of view: We are talking about the detachment of cells from their neighbors during the early stages of apoptosis. Here, a line of evidence points to cleavage of an actin-related protein such as catenin. Adherens junctions link cells by organizing actin filaments to plasma membrane. The transmembrane receptors are the cadherins, a family of Ca^{2+}-dependent cell–cell adhesion molecules (Takeichi, 1995). *In vivo,* the intercellular bonds depend on the association of the cadherin cytoplasmic domain with the central region of β-catenin (Nagafuchi and Takeichi, 1988). β-Catenin serves as an adapter that links cadherins to α-catenin and thereby to the actin skeleton, since α-catenin interacts with both actin and α-actinin (Kundsen *et al.,* 1995; Rimm *et al.,* 1995). Now it has been proved that, during apoptosis, β-catenin is proteolytically processed, and this proteolysis prevents its ability to bind α-catenin (Brancolini *et al.,* 1997). Removal of the β-catenin domain responsible for binding α-catenin should contribute to the loss of cell–cell interactions during apoptosis. Moreover, β-catenin is cleaved *in vitro* by caspase-3 in a form similar in size to that observed by Brancolini in apoptotic cells. In this context, cleavage of β-catenin seems to be related with the loss of its function as a regulator of adherens junctions during apoptosis.

E. Assembling the Pieces of the Apoptotic Engine

The most characteristic changes of cells when entering the active phase of apoptosis focus on the plasma membrane and cytoskeleton, as cells lose adherence, cell-surface structures, they bleb and fragment into apoptotic bodies. As we have seen, these morphologic changes depend on rearrangement of the actin skeleton; thus, proteolytic cleavage of several components of this actin framework, such as fodrin, β-catenin, or the ERM family, may contribute to the recruitment and re-

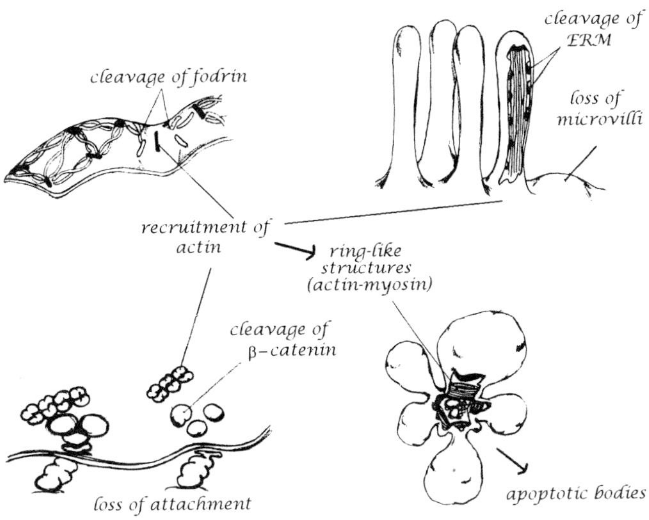

FIG. 2. Summary of the changes affecting the actin skeleton during apoptosis and their consequences.

organization of actin microfilaments to form, together with activated myosin, contractile machinery that leads the cell to convolute and finally fragment into apoptotic bodies (Fig. 2).

However, until now, we have only seen several frames of the whole sequence of events that drives the cell in its way to death. Further research will be needed to complete the biochemical map of reactions that control the structural aspects of apoptosis—and it will be an exciting challenge for the future.

III. Intermediate Filaments

A. General Features and Members

Intermediate filaments (IFs) form the structural framework of both the cytoskeleton and the nuclear lamina in most eukaryotic cells. IFs branch through the cytoplasm of many cells and associate with the nuclear and plasma membrane, as well as various membranous organelles (Georgatos *et al.,* 1994). Despite the fact that the existence of IFs has been known for nearly a century, when neurofilaments were revealed via silver staining of neurons, only during the 1980s did it become apparent that IFs are a family of proteins that share common sequences and structural features that have been identified only in pluricellular organisms. IFs are so called because of their diameter (10 nm), which emphasizes the intermediate width between actin filaments (6 nm) and microtubules (23 nm) when seen under transmission electron microscopy. IFs display a greater diversity in their number, sequence, and abundance than other cytoskeletal proteins (Fuchs and Weber, 1994), being differentially expressed in various cells and tissues of vertebrates or invertebrates.

Each IF protein subunit has the same multidomain secondary structure, consisting of three structural domains: a central, largely α-helical segment (rod) and two non-helical end regions (head and tail domain at the NH_2 and COOH terminals, respectively). The rod domain contains multiple repeats of a heptad sequence, is highly invariant in its amino acid sequence, and is responsible for the fact that all members of this family polymerize into common 10-nm filaments. The head and tail domains, in contrast, are extremely variable both in size and amino acid sequence and variable domains are primarily responsible for determining the unique characteristics of the individual subunit proteins (for reviews, see Fuchs and Cleveland, 1998; Klymkowsky, 1997).

IFs are highly dynamic structures that are regulated mainly by several regulatory posttranslational modifications such as phosphorylation. Phosphorylation plays a significant role in a variety of IF function-related aspects such as disassembly and reorganization, solubility, localization within specific cellular domains, and association with other cytoplasmic or membrane-associated proteins (for a review, see Ku *et al.,* 1997a). The interactions between IFs and other structures are mediated by intermediate filament-associated proteins (IFAPs), a growing family of structural proteins that anchors IFs to each other and to other cytoskeletal components, thus being essential for maintaining the macromolecular organization and integrity of the IF network (Chou *et al.,* 1997).

It is now assumed, as suggested in the first place by Weber and colleagues (1989), that nuclear lamins are the ancestors of cytoplasmic IFs. Based on their genomic structure, polymerization properties, and patterns of expression, vertebrate IF proteins are classified into six groups: acid keratins (type I), basic keratins (type II), vimentin-like IFs (type III), neurofilament IFs (type IV), nuclear lamins (type V), and lens IFs (type VI). All members share an important role structuring the cytoplasm or nucleus to maintain the mechanical integrity of cells. Moreover, in each case they have specific functions, many of which are yet unknown.

B. Cytoplasmic Intermediate Filaments during Apoptosis: Reorganization and Cleavage

Keratins, present in all epithelial cells, form some of the most stable protein–protein interactions known in nature. The use of transgenic mice, as well as the discovery of keratin mutations as the genetic basis of some pathologies such as epidermolysis bullosa simplex (EBS), provided definitive evidence of the fact that the normal function of the keratin network is to impart mechanical integrity to epidermal cells (for reviews, see Fuchs, 1994; Fuchs and Cleveland, 1998). EBS is a rare genetic skin disease typified by mechanical stress-induced intraepidermal blistering due to cytolysis within the basal layer. It has a phenotype similar to that of transgenic animals expressing keratin (K14) mutant genes. The known epithelial pathologies involving mutations in keratin genes are growing and to date encompass 14 keratins (Fuchs and Cleveland, 1998). All of them reflect the essential function of keratins in the maintenance of the integrity in the epithelia.

The type III family includes vimentin (the most widely expressed, produced by mesenchymal cells); desmin, present mainly in muscle cells; and gliar fibrillar acidic protein (GFAP) that form the gliar filaments (Fuchs, 1994). Desmin IFs play crucial roles in maintaining the structural integrity of muscle cells as revealed by null mice, which exhibit muscle fibers susceptible to damage during contraction. Mice lacking GFAP exhibit defects in the white matter of the brain and in the blood–brain barrier and show deficiency in memory for conditioned responses (Liedtke *et al.,* 1996). Vimentin is essential for assembly of a GAFP network in astrocytes and also in the myelin-producing cells in the peripheral nervous system (Galou *et al.,* 1997). By far the most abundant IFs in nerve cells are neurofilaments (type IV IFs), composed in mammalians of three subunits termed NF-L, NF-M, and NF-H. Many studies proved the important role of neurofilaments in structuring axons, specifically supporting the growth and maintenance of axonal caliber in normal neurons (Lee and Cleveland, 1996).

Few reports show the apoptosis-associated biochemical events that involve these cytoplasmic IFs. In epithelial cells, keratins undergo an early reorganization during apoptosis that precedes morphologic nuclear disruption (Caulin *et al.,* 1997), and at later stages they depolymerize completely (Tinnemans *et al.,* 1995; Caulin *et al.,* 1997). In simple-type epithelia after initiation of apoptosis, keratin filaments begin to reorganize first locally and then globally throughout the cell into discrete granular structures (Caulin *et al.,* 1997). The reorganization and disruption of keratin filaments is associated with a phosphorylation and cleavage at specific sites (Ku *et al.,* 1997b; Caulin *et al.,*1997). Both are apparently independent processes because phosphorylation is not necessary for cleavage to take place. Reorganization involved type II K18 phosphorylation at serine 53 and cleavage at sequence VEVD/A by caspases. Other phosphorylation sites on keratins associated with apoptosis have also been found (Liao *et al.,* 1997) to probably be involved in their dramatic reorganization. It is interesting to note that K18 is cleaved in a conserved non α-helical linker region, which is identical to that found in B lamins (Rao *et al.,* 1996) and is also conserved in other IFs family members such as other type I keratins, desmin, and NF-L (Caulin *et al.,* 1997).

In addition, some IFAPs that function to connect cytoskeletal components to each other could be involved in this apoptotic reorganization. For instance, filaggrin, an IFAP that aggregates epidermal keratin filaments, has been shown to contribute to the form of apoptosis associated with terminal differentiation in epidermis (Dale *et al.,* 1997). Other IFs such as vimentin filaments also form early aggregates that undergo proteolytical cleavage in a manner independent of that of apoptosis-inducing stimuli (van Engeland *et al.,* 1997). The fact that vimentin degradation is inhibited by the ICE-inhibitor z-VAD-FMK (Giese *et al.,* 1997) suggests that IF degradation shares common signaling pathways probably involving one or more ICE-like proteases or caspases. Thus, the similar behavior of the cytoplasmic IFs (cIFs) studied and the possibility of caspase-mediated common proteolysis at a highly conserved sequence may reflect a general model of cIFs processing during apoptosis.

TGases, which are involved in apoptotic events, are believed to contribute to cross linking of the membrane and intracellular proteins in the late stages of apoptosis (Fesus *et al.,* 1987; Schmidt *et al.,* 1988). In this sense, it has been shown that TGases or TGase-related antigens are closely associated with vimentin in fibroblasts (Trejo-Skalli *et al.,* 1995) and keratin IFs in keratinocytes (Clement *et al.,* 1997), but the role of this association is yet to be determined. *In vitro* vimentin and keratins are substrates of TGases (Trejo-Skali *et al.,* 1995). In terminal differentiation of keratinocytes (which is associated with apoptotic process), the cornified envelope formation is TGase dependent (Yaneda *et al.,* 1992). It would be interesting to find out if TGases are involved in the formation of IF cytoplasmic reorganization observed in apoptotic cells in addition to their phosphorylation and cleavage.

The significance of keratin and vimentin fragmentation, which seems to represent a general behavior of cIFs, remains unclear. It is likely that cells need to reorganize their cytoskeleton before apoptotic body formation occurs, mainly when this is abundant. For instance, highly polymerized keratin filament packing represents an added problem during apoptosis. It is likely that cytoskeleton reorganization by phosphorylation and caspase-mediated proteolysis at specific sites should facilitate apoptotic morphologic changes. However, to date there is no evidence to suggest an additional function to this general behavior of cIFs. Some human mutations such as the Weber-Cockayne form of EBS have been detected in the immediate proximity of the sequence found to be cleaved by caspases during apoptotic process (Ku *et al.,* 1997b). In these cases, the mutations could interfere with keratin IF proteolysis during apoptosis and subsequently their reorganization during cell death (Ku *et al.,* 1997b). Thus, a defect of keratin degradation could play an important role in the pathophysiology of some keratin disorders.

C. Nuclear Lamins: Their Involvement in the Apoptotic Nuclear Events

Lamins form the nuclear lamina (NL), a thin fibrous network underlying the inner nuclear membrane that constitutes a major structural framework for the nuclear envelope (Georgatos *et al.,* 1994). Two types of lamin polypeptides are characterized, type A and type B. In mammals, type A is represented by lamins A and C, which are transcript forms of the same gene; type B encompasses B1 and B2 lamins, encoded by distinct genes (Nigg, 1992). Lamins are expressed in a specific manner at different states of differentiation and development. AB-lamin isotype is found in all tissues examined to date. A-type lamins are expressed exclusively in differentiation tissues (Moir *et al.,* 1995). The growing knowledge about lamin dynamics and the lamin interactions with the nuclear envelope and chromatin allow a better understanding of NL. It has been assumed from many reports that lamins participate in some aspects of nuclear organization such as structural organization of chromatin (Luderus *et al.,* 1992; Dessev, 1992; McKeon, 1991), and disassembly and assembly of the nuclear envelope (Moir *et al.,* 1995). Lamins bind DNA (Paddy *et al.,* 1990) and inner nuclear membrane proteins (Georgatos *et al.,* 1994; Senior and Gerace, 1988), thus a role in nuclear envelope and chromatin organization could be expected. Moreover, it seems that A- and B-lamin behavior is different and interacts with chromatin and nuclear membrane proteins (Gerace and Burke, 1988) in different ways. A direct role for lamins in chromatin condensation at the nuclear periphery is indicated by the observation that lamin B binds to specific DNA sequence motifs, called *matrix attachment regions,* that mediate the interaction of chromatin with the nuclear matrix (Luderus *et al.,* 1992). Since lamins are involved in chromatin organization, they consequently influence DNA replication (Moir *et al.,* 1995) and possibly transcription (Csermely and Kahn, 1992). Finally, the classification of lamins as IFs together with their strategic localization suggested the mechanical function of lamins. In this sense, the alteration of NL composition becomes nucleus fragile and changes its morphology (Furukawa and Hotta, 1993), which reinforces the fact that lamins probably provide structural support for the nuclear envelope.

In some aspects, processes occurring during apoptosis, such as chromatin condensation, are similar to those occurring during mitosis even if they are opposite processes. This is also true for lamin behavior. In fact, nuclear changes observed in apoptotic cells are related, as we will later see, to the changes observed in lamins. The nuclear lamin undergoes disassembly during mitosis in the same manner as it happens in the active stage of apoptosis and, in both cases, is preceded by lamin phosphorylation (Peter *et al.,* 1990). However, this process takes place in different ways. During mitosis, the laminar structure is broken down as a result of lamin solubilization occurring by hyperphosphorylation at specific sites flanking the α-helical rod domain by $p34^{cdc2}$ kinase (Peter *et al.,* 1990; Ward and Kirschner, 1990; Oberhammer *et al.,* 1994), a process that is reversible. In apoptotic cells, by contrast, a small amount of lamin is solubilized before the onset of apoptosis and after that lamins are cleaved and consequently disassembled irreversibly (Oberhammer *et al.,* 1994).

This cleavage, but not phosphorylation, has been found to be indispensable for the lamina breakdown (Oberhammer *et al.,* 1994; Neamati *et al.,* 1995). Lamin cleavage seems to be a universal event in apoptosis, in the light of all studies proving so in different types of apoptotic cells, and in response to different apoptosis-inducing stimuli (Neamati *et al.,* 1995; Ucker *et al.,* 1992b; Kaufmann, 1989; Lazebnik *et al.,* 1995a; Oberhammer *et al.,* 1994, and references below).

The use of cell-free systems, initially approached by Lazebnik *et al.* (1993), has enormously facilitated the identification of the apoptotic proteases involved in the digestion of different substrates, lamins included. In most cases, a protease related to ICE (caspase) is involved in lamin cleavage (Lazebnik *et al.,* 1995a, 1995b; Orth *et al.,* 1996; Greidinger *et al.,* 1996; Takahashi *et al.,* 1996). A serine protease that is Ca^{2+}-dependent and is associated with the nuclear scaffold has also been implicated in lamin-B degradation (Clawson *et al.,* 1992; Zhivotovsky *et al.,* 1997), which could, in turn, be regulated by Bcl-2 (Zhivotovsky *et al.,* 1997). ICE-related protease Mch2 (now called caspase-6) has been described as lamin-A-specific protease (Takahashi *et al.,* 1996; Orth *et al.,* 1996). Cleavage by Mch2 occurs adjacent to a conserved asp residue in the α-helical rod domain of lamin A (Takahashi *et al.,* 1996). This region has been found to be important for lamin polymerization (McKeon, 1991) and, moreover, has specific chromatin-binding sites (Glass *et al.,* 1993). Taking the functions of the lamin cleavage sites together, it is reasonable to think that protease-mediated disruption of lamins allows nuclear lamin disassembly and chromatin release at specific sites which, in turn, facilitate apoptotic execution. The use of lamin proteases inhibitors, which prevent lamin cleavage, along with uncleavable lamin mutant proteins has led to a better knowledge of lamin participation in the nuclear morphologic changes observed in apoptotic cells and, thus, an understanding of their contribution to cell death.

In particular, some studies carried out in the early 1990s shed light on this specific question. Several conclusions can be drawn from those studies. In the first place, it is generally accepted that lamin degradation is a significant early step in apoptotic consecution (Neamati *et al.,* 1995; Oberhammer *et al.,* 1994; Rao *et al.,* 1996). Moreover, lamin disassembly facilitates, in some way, apoptotic events since in all cases the fact of preventing this cleavage delays apoptosis (Rao *et al.,* 1996; Oberhammer *et al.,* 1994) or interrupts it (Lazebnik *et al.,* 1995a), leaving way for an atypical apoptotic morphology in which typical condensation of chromatin cannot be observed. Instead, chromatin collapses against the nuclear periphery (Lazebnik *et al.,* 1995a; Rao *et al.,* 1996) and is localized at the intact nuclear lamina. Moreover, the nuclear shrinkage seen in normal apoptotic cells is absent (Rao *et al.,* 1996). The contribution of lamins to the final phase of apoptosis, that is, the actin-dependent formation of apoptotic bodies, is not clear and contradictory results have been obtained. It seems necessary to carry out more research works in relation to this issue. Considering the structural function of lamins as intermediate filaments, it is feasible that their degradation could contribute to making the nuclear envelope more malleable, thus facilitating their arrangement in apoptotic bodies. Kaufmann (1989) initially showed an association between oligonucleosomal DNA fragmentation and lamin degradation. Since then, many studies have reported this relationship (Neamati *et al.,* 1995; Oberhammer *et al.,* 1994; Ucker *et al.,* 1992a; Lazebnik *et al.,* 1993).

Lamins have also been associated with chromatin condensation and endonuclease activation occurring during apoptosis (Lazebnik *et al.,* 1993, 1995a). Even if lamin and chromatin degradation are independent processes carried out in different ways (Rao *et al.,* 1996; Lazebnik *et al.*, 1995a), when lamin degradation is prevented, a considerable delay in apoptosis consecution is noticeable. This can be explained if we assume that if independent processes, lamin degradation facilitates in some way DNA degradation, for instance, by making it more accesible to endonucleases. This assumption is supported by the fact that lamin B binds directly to regularly spaced matrix attachment regions (MARs) (Luderus *et al.*, 1992). Thus it is likely that lamin degradation releases MARs, leaving chromatin unfolded and, thus, allowing for endonuclease access. In addition, in the systems studied, lamin fragmentation precedes (Neamati *et al.*, 1995) or overlaps (Oberhammer *et al.*, 1994) with the appearance of DNA fragments.

It seems that generally lamin A and B do not function in different ways during apoptosis. Rao and collaborators (1996), using mutated A and B lamins that cannot be proteolized, proved that in both cases the apoptotic events are delayed in the same manner. In contrast, in Fas/CD95-mediated apoptosis, an early preferent degradation of B lamins occurs that precedes DNA degradation, and a later degradation of A lamins (Mandal *et al.*, 1996). The significance of the difference in behavior of lamin degradation is unknown, but it is possible that two different proteases may be involved in this case. However, to date there is no evidence to support this.

Even if apoptotic changes, as far as lamins are concerned, play a crucial role during cell death, it seems likely that the changes taking place in other components of the nuclear matrix play a definitive role as well. Among these nuclear matrix components, NuMa (nuclear mitotic apparatus protein) is phosphorylated and specifically cleaved early during dexamethasone-induced apoptosis in thymocytes (Weaver *et al.*, 1996). Degradation occurs at the same time as lamin B degradation before detecting DNA fragmentation. NuMa has been suggested to form core filaments of the nucleoskeleton or intranuclear matrix during interphase (Cleveland, 1995). As a result of its degradation in apoptotic cells, the structure of the cytoskeleton destabilizes and finally collapses. Also, this degradation occurs in other cellular types where induced apoptosis by different stimuli occurs (Weaver *et al.*, 1996), and therefore it seems likely to be a common phenomenon during this process (Fig. 3).

In summary, what we know is that some IFs undergo a reorganization and caspase-mediated specific proteolysis during apoptotic cell death. In addition, there are reasons to think (i.e, the common conserved sequence known to be cleaved in some IFs) that could be a general phenomenon. Whether the caspases are the same for lamins and cytoplasmic IFs remains unknown. However, the meaning of caspase-

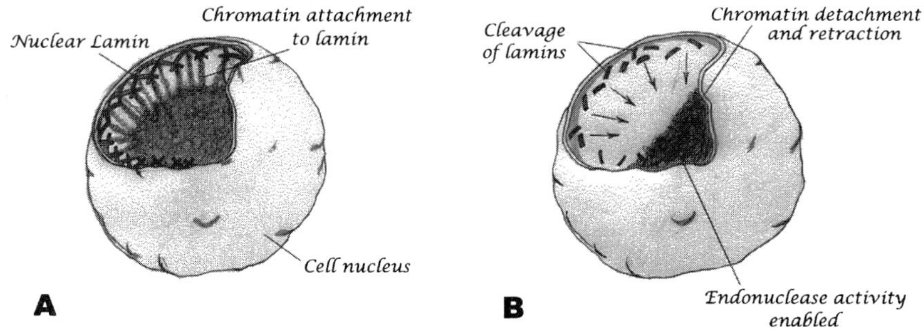

FIG. 3. (A) Normal disposition and (B) consequences of proteolytic cleavage of lamins during the execution phase of apoptosis.

mediated IF proteolysis or their contribution to the apoptotic events seem to be different on each case, but could only reflect the ignorance about this particular question. To date no IFs have been demonstrated to have any other meaning but to facilitate the packing of the cellular structures (cIFs included) into apoptotic bodies. The lamins, on the contrary, seem to contribute more so to the morphologic events that occur during the active phase of apoptosis (i.e, nuclear shrinkage, chromatin condensation and degradation), which may be due to their ubiquitous localization and their strategic link to both the chromatin and the inner nuclear membrane.

IV. Microtubules and Associated Proteins during Apoptosis

Microtubules represent one of the specific fiber systems of the eukaryotic cytoskeleton. This cytoplasmic network is essential, in association with actin filaments, for a wide variety of cellular functions, fundamentally cell motility, transport of molecules, translocation of organelles, acquisition and maintenance of cell shape, and polarity and establishment of the mitotic machinery during cell division.

Structurally, microtubules are formed by the assembly of tubulin, an acidic protein heterodimer composed of α- and β-subunits of approximately 55 kDa each. The dimer is polarized such that α-tubulin points to the plus end and β-tubulin points toward the minus end. This polarity enables β-tubulin to interact with γ-tubulin, which, forming rings or spirals templates, is an essential component of microtubule organizing centers (MOCs). Moreover, this structure can serve as anchorage of microtubules at the centrosomes. The microtubule network provides directional pathways for intercellular transport including chromosomes in cell division (for reviews, see Mandelkow and Mandelkow, 1995; Wade and Hyman, 1997)

The microtubular network interacts with different classes of proteins: structural associate proteins that bind, stabilize, and promote the assembly of dimers (MAPs, tau and isophorms of these proteins) and motor proteins (kinesins and dyneins) that bind to tubulin for intracellular transport and perhaps also affect tubulin polymerization/depolymerization dynamics (Joshi, 1997).

Microtubules are a key factor to cytoplasmic cell activity: They determine that polymeric structures show complex dynamics resulting from a continuous process of assembly and disassembly. This process is coordinated with the other filament networks and regulated at the spatial and temporal levels according to the functional demands of the cells.

Microtubules represent, by their high metabolic activity and their capital relations, one of the preferred cell targets, in pharmacologic terms, to act against degenerative and neoplasic diseases.

A. Microtubular Skeleton and Apoptosis

Despite the existence of extensive experimental evidence proving that drugs destabilizing the microtubular architecture trigger apoptosis, few studies show specific information on the spontaneous behavior of microtubules in the course of cell death.

The first change that can be detected during apoptosis is the exposure of phosphatidylserine residues at the external surface of the plasma membrane without loss of functional membrane activity (van Engeland *et al.*, 1996). Following that, a progressive change in the cytoskeletal organization occurs. In contrast to the intermediate filaments, which were disassembled and proteolytically cleaved as previously stated, microtubules were not degraded in this initial process, since evident tubulin structures were found to be present as aggregated, during the execution phase, inside the incipient apoptotic bodies (van Engeland *et al.*, 1997).

In several experimental studies, tublin has been described as partially depolymerized and reorganized in early apoptosis. This reorganization is similar to what we have previously reported for actin filaments and probably involves changes in phosphorylation and subsequent tubulin polymerization. This effect is not dependent on cell types or drug employed and has not been detected when apoptosis is induced in mitotic cells. The fact that the presence of the mitotic apparatus excludes the formation of apoptotic microtubules (S. M. Pittman *et al.,* 1994, Ireland and Pitman, 1995; S. Pittman *et al.,* 1997) may well be due to the hierarchical control for the use of monomers, necessary during the mitotic cell stage.

B. Microtubules Associate Proteins in Apoptosis

The disassembly of microtubules during apoptosis is accompanied by dephosphorylation of tau and other structural proteins. Tau is a microtubule-associated protein thought to stabilize fibrillar structure in axons and thus provide the basis for intraneuronal transport. This molecule has several effects: It binds to microtubules, promotes their nucleation and elongation, and protects them against disassembly. The interest in tau protein increased after the discovery that it is present as aggregate in nerve cells during neurodegenerative disorders (Delacourte and Buée, 1997). Several authors have proposed that closely before or in the execution phase of apoptosis, microtubules get disassembled, with the activation of phosphatases, particularly PP2A-like phosphatase that dephosphorylates tau in ser-pro residues as well as other substrates. This activation of PP2A and the subcellular distribution of tau are considered to be the first cytoplasmatic signs of apoptosis, or the debut of the execution phase of cell death (Nuydens *et al.,* 1997; Arendt *et al.,* 1988; Mills *et al.,* 1998). Dephosphorylation of tau can be correlated with the reduction of tau mRNA expression detected in neurons during early phases of the apoptotic process (Esclaire *et al.,* 1998). Tau protein has been isolated with the phospholipase Cγ, suggesting that this structural protein may be involved in other dynamic functions including the regulation of signal transduction pathways (Jenkins and Jonhson, 1998).

The microtubular network interacts with other structural proteins such as MAPs and motor proteins (kinesins and dyneins). Limited information exists in relation to the role of these proteins during apoptosis. Apparently, all structural proteins are desphosphorylated in the course of apoptosis and vice versa; they appear highly phosphorylated and/or overexpressed in cells that refuse to enter an apoptotic process or show increased resistance to drugs that induce apoptosis (Lesort *et al.,* 1998; Zhang *et al.,* 1998).

Given that kinesins (and, in general, motor proteins) are related to intracellular translation for organelles such the Golgi complex and endoplasmic reticulum, these motor proteins are expected to be involved in the organellar collapse observed in the execution phase of apoptosis. The membrane proximal region of TNFR55 has recently been described as essential for signaling to microtubule-dependent mitochondrial translocation and clustering. In this experimental system, in the absence of translocation, cell death responses were markedly delayed, suggesting a cooperative effect on cell death (De Vos *et al.,* 1998).

C. Apoptosis Induced by Drugs That Disturb the Microtubular Organization

The microtubular organization of the cell may be altered using cytotoxic drugs that enhance or limit the microtubule stability and disturb the polymerization rate. In all cases when cellular death was induced, cells showed apoptotic morphology. Similarly, other pathologic processes that change the cytoplasmatic homeostasis, such as the modification of Ca^{2+} or glutamate levels, may also induce apoptosis via cytoskeleton cleavage (Ankarcrona *et al.,* 1996).

Taxol is the most widely employed tubulin-like drug. Used at therapeutic concentrations, it is a paradigmatic example of the complex action of drugs that affect microtubular organization to induce apoptosis. Initially, it has been shown that taxol (and analogous products) enhances and stabilizes the polymerization of tubulin, resulting in arrest cell cycle in the G_2 or M phase followed by apoptotic cell death (Gangemi *et al.,* 1995; Ireland and Pittman, 1995).

Experimental evidence shows that in asynchronous treated cultures taxol induces different cell behaviors in relation to the cell cycle stage. Treated cells undergo apoptosis directly at all stages of the cell cycle except mitosis. Moreover, if cells are arrested in prophase, taxol induces rapid activation of a p53-independent pathway, whereas in cells arrested in G_1 the apoptotic pathway induced is p53 dependent. In both processes treated cells displayed morphologic hallmarks of apoptosis (Woods *et al.,* 1995). On the other hand, cells that were arrested in mitosis contained multiple asters and/or disordered mitotic spindles, but did not show apoptotic morphology. (Pittman *et al.,* 1994; Ireland and Pittman, 1995).

Using taxol concentration sufficiently to block mitosis, but insufficiently to induce apoptosis directly, it has been observed that several cytoplasmic pathways are activated. One of the first mechanisms detected associated with tubulin repolymerization is the activation of Raf-1 kinase, related with Bcl-2 phosphorylation and apoptosis. Raf-1 also plays an important role in stathmin phosphorylation that is closely connected with microtubule reorganization (Lovric *et al.,* 1998). Controversial information exists related to the mechanisms and consequences of induction of Raf-1 kinase activation. It has been suggested that this mechanism takes place directly after microtubule depolymerizaton, apparently via an intermediate protein (Blagosklonny *et al.,* 1997).

In contrast, in another experimental model, it has been reported that the induction of apoptosis by taxol is apparently independent of *de novo* protein synthesis (Gangemi *et al.,* 1995). It has also been described that activation of Raf-1 is an independent phenomenon showing no relationship with the apoptotic process induced by taxol (Ibrado *et al.,* 1997). In any case, the kinase activation is associated with this network since cells treated with other agents that interact with tubulin and microtubules, such as vincristine and vimblastine, also induce Raf-1 activation. However, drugs that directly induce DNA damage and apoptosis do not activate this metabolic pathway (Blagosklonny *et al.,* 1997).

Another early mechanism induced by taxol and drugs with an antagonistic effect such as vincristine and vimblastine is the hyperphosphorylation of Bcl-2. The characteristics of this protein have been extensively described, including its antiapoptotic activity (Red, 1997). Bcl-2 hyperphosphorylation may be induced by a drug after activation of a protein kinase A (PKA) or Raf-1 kinase as well as other cellular pathways during the apoptotic process. Additionally, reduced Bcl-2/Bax dimerization and increased free Bax proapoptotic protein have also been detected associated with (Sirivastava *et al.,* 1998) or independently of microtubules damaging agents.

Recent evidence suggests that a mechanism of taxol-mediated apoptosis may be limited by high levels of p26Bcl-2 or, alternatively, taxol induces inactivation or inhibition of mechanisms by which p26Bcl-2 is able to block apoptosis (Tang *et al.,* 1994). In an estrogen responsible experimental model it has been described that the increase in p26Bcl2/p21Bax induced by hormone administration limits the apoptotic effect of taxol (Huang *et al.,* 1997). Bcl-x_L is a Bcl-2-related protein potentially capable of inhibiting cell death. This protein is also phosphorylated in cell cultures, after treatment with drugs that target tubulin (Poruchynsky *et al.,* 1998). The implication of Bcl-x_L in the protection of mitochondrial perturbation observed in preapoptotic cells (Huang *et al.,* 1997) and the fact that taxol is metabolized for cytocrome *P*450 into its principal metabolite 6,α-hydroxytaxol (6HT) (Kim *et al.,* 1997; Kumar *et al.,* 1995) make of this protein an interesting subject for understanding the cytoplasmatic behavior of this drug. Bcl-x_L may limit one of the ways in which taxol induces apoptosis. Overexpression of this molecule does not affect the microtubular bundling or mitotic arrest but significantly inhibits the morphologic features associated with taxol-induced apoptosis (Kim *et al.,* 1997; Ibrado *et al.,* 1997). Analysis of taxol-treated cells in G_0/G_1 and G_2/M transitions suggest that a sustained mitotic block is not necessary to initiate cell death but cell cycle progression and *de novo* protein syntheses are prerequisites to acquiring apoptotic features. Delayed activation of $p34^{cdc2}$, coincident with DNA fragmentation, suggests that taxol is capable of activating an apoptotic pathway related to phosphoregulation, possibly involving the p34^{cdc2} kinase (Donaldson *et al.,* 1994).

Another interesting metabolic pathway activated after taxol treatment is the c-Jun N-terminal-kinase/stress activate protein kinase (JNK/SAPK), capable of phosphorylating Bcl-2. In different experimental models it has been shown that the activation of the kinases is not specific and can be initiated by different intermediate pathways after the action of several microtubule-disturbing agents (Attalla *et al.,* 1998; Wang *et al.,* 1998).

Colchicine and analogs products are examples of drugs that triggered apoptosis with an opposite effect to that described for taxol. These drugs destabilize microtubules, block or disrupt mitosis, and produce cell death with morphologic features of apoptosis. Apparently, the cytoskeletal damage induced by this drug is sufficient to kill the cell (Bonfoco *et al.,* 1995). The execution phase of apoptosis is preceded in colchicine-treated cells by loss of tau protein and fragmentation of alpha and beta tubulins (Ankarcrona *et al.,* 1996).

Treatment with colcemid induces delayed apoptosis by two different mechanisms. Continuous exposure to concentrations of colcemid sufficient enough to block cells in mitosis underwent apoptosis with a characteristic morphology; however, concentrations sufficient to cause mitotic arrest, or pulse exposure that induces mitotic block, led to the formation of disarranged mitosis with aberrant spindle function. Treatment with taxol to stabilize the microtubuli prevented the acquisition of the morphologic signs of apoptosis (Sherwood *et al.,* 1994).

It is important to remark that pathologic alterations or functional modifications in tubulin or tubulin-associated protein structures increase the resistance against drugs (Lesort *et al.,* 1998). In this sense, the acetylation of α-tubulin that is a marker of microtubule stability, has been described to be involved in apoptosis suppression (Geyp *et al.,* 1996).

In summary apoptosis may be induced by drugs that disturb the polymerization/depolymerization balance as well as by changes in the structural proteins' status. These alterations suffice to activate different cytoplasmic and or nuclear pathways to trigger apoptosis (Fig. 4).

Recently, it has been emphasized that taxol and analogs are capable of inducing gene expression independently of microtubule stabilization or disruption of mitosis (Moos and Fitzpatrick, 1998). This logical information is very important because it confirms the complex interpretation of isolated results obtained after induced cell drug injury.

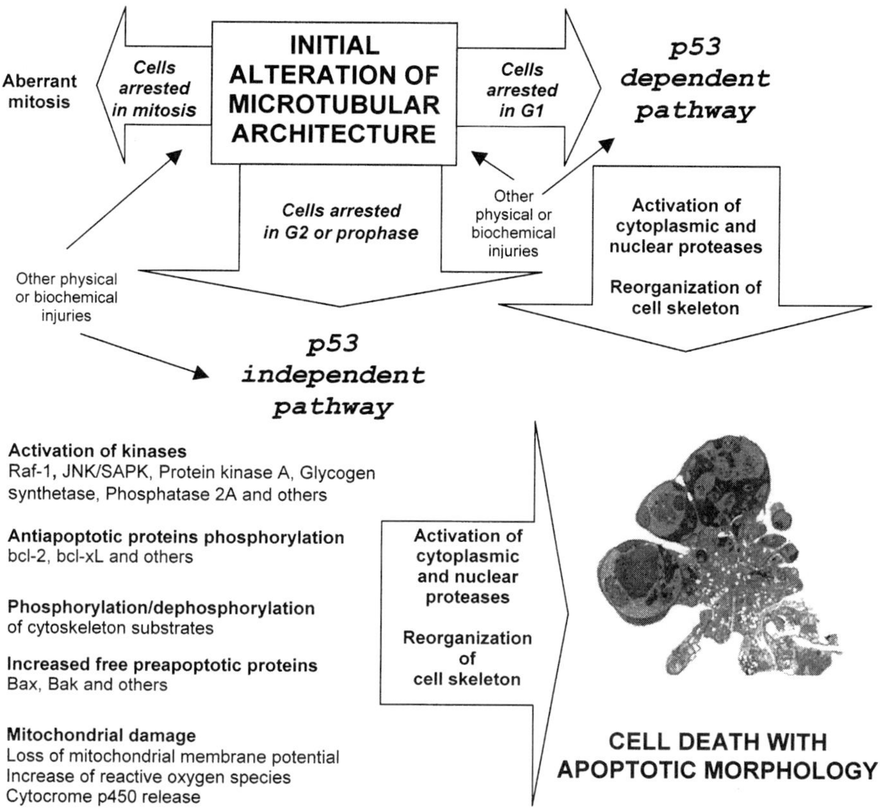

FIG. 4. Possible relation between microtubule alterations and apoptotic morphology. A puzzle with duplicated and unknown pieces.

V. Cleavage of Monomeric Cytoskeletal Structures

In several studies it has been shown that depolymerization of tubulin-based structures is associated with partial cleavage of G-actin and α- and β-tubulin. This process is observed during the execution phase of apoptosis (Hall, 1994; Ankarcrona *et al.,* 1996; Dipietrantonio *et al.,* 1996; Kayalar *et al.,* 1996). The fact that G-actin that is able to block DNase I via stoichiometric association may play a role inhibiting DNA fragmentation reveals a very interesting possibility. The cleavage of G-actin by interleukin 1β-converting enzyme (ICE) may allow endonuclease activity of Dnase I. This suggests a closed mechanism that relates nuclear and cytoplasmic events during apoptosis.

Cleavage of tubulin is apparently independent of ICE-like proteases. Its fragmentation pattern in our experimental model (unpublished results) points to the involvement of a specific proteolytic system. It is evident that limited cleavage of key cytoskeletal elements may be an important step in apoptosis, but further research to elicit these interesting questions is required.

VI. Conclusions

At a cellular level, ontogenic development and homeostasis in all-multicellular organisms include basically the events of genesis, differentiation-specialization, and cell death. Based on economy of resources, the structural machinery used in all cases must be similar and based—at a structural level—in a dramatic reorganization of cellular skeleton for mitotic apparatus formation so as to obtain a definitive cell shape and differentiated functions or die in the case of apoptosis.

The most characteristic changes that cells undergo when entering apoptosis are loss of adherence, changes in cell-surface structures, and blebbing and fragmentation into apoptotic bodies depending on rearrangement of the actin and tubulin skeleton. The phosphorylation of structural and/or enzymatic proteins and proteolytic cleavage of several components of these cytoplasmic networks contribute to the recruitment and reorganization of cell structures to form the machinery that leads the cell to convolute and finally fragment into apoptotic bodies. Similarly, chromatin condensation that precedes DNA cleavage is associated with dissolution of lamins, fundamental components of cytoskeleton in nuclear areas.

Given that apoptosis is a strictly morphologic description, it is not correct to superimpose observed pictures and biochemical pathways. Moreover, it is necessary to explain the molecular cell mechanisms that lead to obtaining the characteristic cell morphology. Delimiting these differentiated questions is difficult but essential for the understanding of cytoplasmic organization and for an effective use of drugs as therapeutic agents.

Acknowledgments

This work was supported in part by University of the Basque Country (UPV 075.327-EA 149/97). Rafael Atencia is a recipient of Fundación Gangoiti grants.

REFERENCES

Ankarcrona, M., Zhivotovsky, B., Holmstrom, T., Diana, A., Eriksson, JE., Orrenius, S., and Nicotera, P. (1996). Lamin and beta tubulin fragmentation precede chromatin degradation in glutamate, induced neuronal apoptosis. *NeuroReport* **7,** 2659–2664.

Arends, M. J., Morris, R. G., and Wyllie, A. H. (1990). Apoptosis: The role of endonuclease. *Am. J. Pathol.* **136,** 593–608.

Arendt, D., Holzer, M., Fruth, R., Bruckner, M. K., and Gartner, U. (1998). Phosphorylation of tau,Abeta-formation, and apoptosis after in vivo inhibition of PP-1 and PP-2A. *Neurobiol. Aging* **19,** 3–13.

Arpin, M., Algrain, M., and Louvard, D. (1994). Membrane-actin microfilament connections: an increasing diversity of players related to band 4.1. *Curr. Opin. Cell Biol.* **6,**136–141.

Atencia, R., García-Sanz, M., Pérez-Yarza, G., Asumendi, A., Hilario, E., and Aréchaga, J. (1997). A structural analysis of cytoskeleton components during the execution phase of apoptosis. *Protoplasma* **198,** 163–169.

Attalla, H., Westberg, J. A., Andersson, L. C., Adlercreutz, H., and Makela, T. P. (1998). 2-Methoxyestradiol-induced phosphorylation of Bcl-2: uncouling from JNK/SAPK activation. *Biochem. Biophys. Res. Commun.* **247,** 616–619.

Ayscough, K. R. (1998). In vivo functions of actin-binding proteins. *Curr. Opin. Cell Biol.* **10,** 102–111.

Blagosklonny, M. V., Giannakakou, P., Deiry, W. S., Kingston, D. G., Higgs, P. I., Neckers, L., and Fojo, T. (1997). Raf, 1/bcl, 2 phosphorylation, a step from microtubule damage to cell death. *Cancer Res.* **57,** 130–135.

Bonfoco, E., Ceccatelli, S., Manzo, L., and Nicotera, P. (1995). Colchicine induces apoptosis in cerebellar granule cells. *Exp. Cell. Res.* **218,** 189–200.

Brancolini, C., Lazarevic, D., Rodriguez, J., and Schneider, C. (1997). Dismantling cell-cell contacts during apoptosis is coupled to a caspase-dependent proteolytic cleavage of β-catenin. *J. Cell Biol.* **139,** 759–771.

Bretscher, A. (1991). Microfilament structure and function in the cortical skeleton. *Annu. Rev. Cell Biol.* **7,** 337–374.

Carlier, M. F. (1998). Control of actin dynamics. *Curr. Opin. Cell Biol.* **10,** 45–51.

Caulin, C., Salvesen, G. S., and Oshima, R. G. (1997). Caspase cleavage of keratin 18 and reorganization of intermediate filaments during epithelial cell apoptosis. *J. Cell Biol.* **138,** 1379–1394.

Chou, Y. H., Skalli, O., and Goldman, R. D. (1997). Intermediate filaments and cytoplasmic networking: New connections and more functions. *Curr. Opin. Cell Biol.* **9,** 49–53.

Clawson, G. A., Norbeck, L. L., Hatem, C. L., Rhodes, C., Amiri, P., McKerrow, J. H., Patierno, S. R., and Fiskum, G. (1992). Ca(2+)-regulated serine protease associated with the nuclear scaffold. *Cell Growth Differ.* **3,** 827–838.

Clement, S., Trejo-Skalli, A. V., Gu, L., Velasco, P.T., Lorand, L., and Goldman, R. D. (1997). A transglutaminase-related antigen associates with keratin filaments in some mouse epidermal cells. *J. Invest. Dermatol.* **109,** 778–782.

Cleveland, D. W. (1995). NuMA: A protein involved in nuclear structure, spindle assembly, and nuclear re-formation. *Trends Cell Biol.* **5,** 60–64.

Condeelis, J. (1993). Life at the leading edge: The formation of cell protrusions. *Annu. Rev. Cell Biol.* **9,** 411–444.

Cooper, J. A. (1987). Effects of cytochalasin and phalloidin on actin. *J. Cell Biol.* **105,** 1473–1478.

Cotter, T. G., Lennon, S. V., Glynn, J. M., and Green, D. R. (1992). Microfilament-disrupting agents prevent the formation of apoptotic bodies in tumor cells undergoing apoptosis. *Cancer Res.* **52,** 997–1005.

Croall, D. E., and Demartino, G. N. (1991). Calcium-activated neutral protease (calpain) system: Structure, function, and regulation. *Phys. Rev.* **71,** 813–847.

Cryns, V. L., Bergeron, L., Zhu, H., Li, H. and Yuan, J. (1996). Specific cleavage of alpha-fodrin during Fas- and tumor necrosis factor-induced apoptosis is mediated by an interleukin-1beta-converting enzyme/Ced-3 protease distinct from the poly(ADP-ribose) polymerase protease. *J. Biol. Chem.* **271,** 31277–31282.

Csermely, P., and Kahn, R. C. (1992). Insulin induces the phosphorylation of DNA-binding nuclear proteins including lamins in 3T3-F442A. *Biochemistry* **31,** 9940–9946.

Dale, B. A., Presland, R. B., Lewis, S. P., Underwood, R. A, and Fleckman, P. (1997). Transient expression of epidermal filaggrin in cultured cells causes collapse of intermediate filament networks with alteration of clel shape and nuclear integrity. *J. Invest. Dermatol.* **108,** 179–187.

Delacourte, A., and Buée, L. (1997). Normal and pathological Tau proteins as factor for microtubule assembly. *Int. Rev. Cytol.* **7,** 167–183.

Dessev, G. N. (1992). Nuclear envelope structure. *Curr. Opin. Cell Biol.* **4,** 430–335.

De Vos, K., Goossens, V., Boone, E., Vercammen, D., Vancompernolle, K., Vandenabeele, P., Haegman, G., Fiers, W., and Grooten, J. (1998). The 55-kDa tumor necrosis factor receptor induces clustering of mitochondria through its membrane-proximal region. *J. Biol. Chem.* **273,** 9673–9680.

Dipietrantonio, A., Hsieh, T., and Wu, J.M. (1996). Differential effects of retinoic acid (RA) and N-(4-Hydroxiphenyl)retinamide (4-HPR) on cell growth, induction of differentiation, and changes in $p34^{cdc2}$, Bcl-2, and actin expression in the human promyelocytic HL-60 leukemic cells. *Biochem. Biophys. Res. Commun.* **224,** 837–842.

Donaldson, K.L., Goolsby, G. L., Kiener, P. A., and Wahl, A. F. (1994). Activation of p34cdc2 coincident with taxol, induced apoptosis. *Cell Growth Differ.* **5,** 1041–1450.

Endressen, P. C., Fandrem, J., Eide, T. J., and Aarbakke, J. (1995). Morphological modifications of apoptosis in HL-60 cells: Effects of homocysteine and cytochalasins on apoptosis initiated by 3-deazaadenosine. *Virchows Arch. B* **426,** 257–266.

Esclaire, F., Terro, F., Yardin, C., and Hugon, J. (1998). Neuronal apoptosis is associated with a decrease in tau mRNA expression. *NeuroReport* **9,** 1173–1177.

Fechheimer, M., and Zigmond, S. H. (1993). Focusing on unpolimerized actin. *J. Cell Biol.* **123,** 1–5.

Fesus, L., Thomazy, V., and Falus, A. (1987). Induction and activation of tissue transglutaminase during programmed cell death. *FEBS Lett.* **224,** 104–108.

Fuchs, E. (1994). Intermediate filaments and disease: Mutations that cripple cell strength. *J. Cell Biol.* **125,** 511–516.

Fuchs, E., and Cleveland, D. W. (1998). A structural scaffolding of intermediate filaments in health and disease. *Science* **79,** 514–519.

Fuchs, E., and Weber, K. (1994). Intermediate filaments: Structure, dynamics, function, and disease. *Annu. Rev. Biochem.* **63,** 345–382.

Furukawa, K., and Hotta, Y. (1993). CDNA cloning of a germ cell specific lamin B3 from mouse spermatocytes and analysis of its function by ectopic expression in somatic cells. *EMBO J.* **12,** 97–106.

Galou, M., Gao, J., Humbert, J., Mericskay, M., Li, Z., Paulin, D., and Vicart, P. (1997). The importance of intermediate filaments in the adaptation of tissues to mechanical stress: Evidence from gene knockout studies. *Biol. Cell* **89,** 85–97.

Gangemi, R. M., Tiso, M., Marchetti, C., Severi, A. B., and Fabbi, M. (1995). Taxol cytotoxicity on human leukemia cell lines is a function of their susceptibility to programmed cell death. *Cancer Chemother.* **36,** 385–392.

García-Sanz, M., Atencia, R., Pérez-Yarza, G., Asumendi, A., Hilario, E. and Aréchaga, J. (1996). Cytoarchitectural changes during retinoic acid-induced apoptosis in F9 embryonal carcinoma cells. *Int. J. Dev. Biol., Suppl.* **1,** 195–196.

Georgatos, S. D., Meier, J., and Simos, G. (1994). Lamins and lamin-associated proteins. *Curr. Opin. Cell Biol.* **6,** 347–353.

Gerace, L., and Burke, B. (1988). Functional organization of the nuclear envelope. *Annu. Rev. Cell Biol.* **4,** 335–374.

Geyp, M., Ireland, C. M., and Pittman, S. M. (1996). Resistance to apoptotic cell death in a drug resistant T cell leukaemia cell line. *Leukemia* **103,** 447–455.

Giese, G., Schmidt, J., Gilbert, M., Albrecht, R., and Traub, P. (1997). TPA induces apoptosis in MPC-11 mouse plasmacytoma cells grown in serum-free medium. *Biol. Cell.* **89,** 99–111.

Glass, C. A., Glass, J. R., Taniura, H., Hasel, K. W., Blevitt, J., and Gerace, L. (1993). The alpha-helical rod domain of human lamins A and C contains a chromatin binding site. *EMBO J.* **12,** 4413–4424.

Greidinger, E. L., Miller, D. K., Yamin, T. T., Casciola-Rosen, L., and Rosen, A. (1996). Sequential activation of three distinct ICE-like activities in Fas-ligated Jurkat cells. *FEBS Lett.* **390,** 299–303.

Hall, A. K. (1994). Molecular interactions between G-actin, DNase I and the beta-thymosins in apoptosis: A hypothesis. *Med. Hypotheses* **43,** 125–131.

Huang, Y., Ray, S., Reed, J. C., Ibrado, A. M., Tang, C., Nawabi, A., and Bhalla, K. (1997). Estrogen increases intracellular p26Bcl,2 to p21Bax ratios and inhibits taxol induced apoptosis of human breast cancer MCF,7 cells. *Breast Cancer Res. Treat.* **42,** 73–81.

Ibrado, A. M., Liu, L., and Bhalla, K. (1997). Bcl,xL overexpression inhibits progression of molecular events leading to paclitaxel induced apoptosis of human acute myeloid leukemia HL,60 cells. *Cancer Res.* **57,** 1109–1015.

Ireland, C. M., and Pittman, S. M. (1995). Tubulin alterations in taxol, induced apoptosis parallel those observed with other drugs. *Biochem. Pharmacol.* **49,** 1491–1499.

Jenkins, S. M., and Johnson, G. V. (1998). Tau complexes with phospholipase C, gamma in situ. *NeuroReport* **9,** 67–71.

Joshi, H. J. (1997). Microtubule dynamics in living cells. *Curr. Opin. Cell Biol.* **10,** 35–44.

Kaufmann, S. H. (1989). Induction of endonucleolytic DNA cleavage in human acute myelogenous leukemia cells by etoposide, camptothecin, and other cytotoxic anticancer drugs: A cautionary note. *Cancer Res.* **49,** 5870–5878.

Kayalar, C., Örd, T., Testa, M. P., Zhong, L. T., and Bredesen, D. (1996). Cleavago of actin by interleukin 1β converting enzime to reverse DNaseI inhibition. *Proc. Natl. Acad. Sci. U.S.A.* **93,** 2234–2238.

Kerr, J. F. (1971). Shrinkage necrosis: A distinct mode of cellular death. *J. Pathol.* **105,** 13–20.

Kerr, J. F., Wyllie, A. H., and Currie, A.R. (1972). Apoptosis: A basic biological phenomenon with wide-ranging implications in tissue kinetics. *Br. J. Cancer* **26,** 239–257.

Kim, C. N., Wang, X., Huang, Y., Ibrado, A. M., Liu, L., Fang, G., and Bhalla, K. (1997). Overexpression of Bcl,X(L) inhibits Ara,C, induced mitochondrial loss of cytochrome c and other perturbations that activate the molecular cascade of apoptosis. *Cancer Res.* **57,** 3115–3120.

Klymkowsky, M. W. (1997). Intermediate filaments as dynamic structures. *Cancer Metastasis Rev.* **15,** 417–428.

Kondo, T., Takeuchi, K., Doi, Y., Yonemura, S., Nagata, S., Tsukita, Y., and Tsukita, S. (1997). ERM (Ezrin/Radixin/moesin)-based molecular mechanism of microvillar breakdown at an early stage of apoptosis. *J. Cell Biol.* **139,** 749–758.

Kouchi, Z., Saido, T. C., Ohyama, H., Maruta, H., Suzuki, K., and Tanuma, S. (1997). The restrictive proteolysis of -fodrin to a 120 kDa fragment is not catalyzed by calpains during thymic apoptosis. *Apoptosis* **2,** 84–90.

Ku, N. O., Liao, J., Chou, C. F., and Omary, M. B. (1997a). Implications of intermediate filament protein phosphorylation. *Cancer Metastasis Rev.* **15,** 429–444.

Ku, N. O., Liao, J., and Omary, M. B. (1997b). Apoptosis generates stable fragments of human type I keratins. *J. Biol. Chem.* **272,** 33197–33203.

Kumar, G., Ray, S., Walle, T., Huang, Y., Willingham, M., Self, S., and Bhalla, K. (1995). Comparative in vitro cytotoxic effects of taxol and its major human metabolite 6 alpha, hydroxytaxol. *Cancer Chemother. Pharmacol.* **36,** 129–1335.

Kundsen, K. A., Soler A. P., Johnson, K. R., and Wheelock, M. J. (1995). Interaction of α-actinin with the cadherin-catenin cell-cell adhesion complex via α-catenin. *J. Cell Biol.* **130,** 60–77.

Laster, S. M., and McKenzie, J. M. (1996). Bleb formation and F-actin distribution during mitosis and tumor necrosis factor-induced apoptosis. *Microsc. Res. Tech.* **34,** 272–280.

Lazebnik, Y. A., Cole, S., Cooke, A., Nelson, W. G., and Earnshaw, C. W. (1993). Nuclear events of apoptosis in vitro in cell-free mitotic extracts: A model system for analysis of the active phase of apoptosis. *J. Cell Biol.* **123,** 7–22.

Lazebnik, Y. A., Takahashi, A., Moir, R. D., Goldman, R. D., Poirier, G. G., Kaufmann, S. H., and Earnshaw, W. C. (1995a). Studies of the lamin proteinase reveal multiple parallel biochemical pathways during apoptotic execution. *Proc. Natl. Acad. Sci. U.S.A.* **92,** 9042–9046.

Lazebnik, Y. A., Takahashi, A., Poirier, G. G., Kaufmann, S. H., and Earnshaw, W. C. (1995b). Characterization of the execution phase of apoptosis in vitro using extracts from condemned-phase cells. *J. Cell Sci.* **19,** 41–49.

Lee, M. K., and Cleveland, D. W. (1996). Neuronal intermediate filaments. *Annu. Rev. Neurosci.* **19,** 187–217.

Lesort, M., Blanchard, C., Yardin, C., Esclaire, F., and Hugon, J. (1998). Cultured neurons expressing phosphorylated tau are more resistant to apoptosis induced by NMDA or serum deprivation. *Brain Res. Mol. Brain Res.* **45,** 1127–1132.

Levee, M. G., Dabrowska, M. I., Lelli, J. L., and Hinshaw, D. B. (1996). Actin polymerization and depolymerization during apoptosis in HL-60 cells. *Am. J. Physiol.* **271,** C1981–C1992.

Liao, J., Ku, N. O., and Omary, M. B. (1997). Stress, apoptosis, and mitosis induce phosphorylation of human keratin 8 at Ser-73 in tissues and cultured cells. *J. Biol. Chem.* **272,** 17565–17573.

Liedtke, W., Edelmann, W., Bieri, P. L., Chiu, F. C., Cowan, N. J., Kucherlapati, R., and Raine, C. S. (1996). GFAP is necessary for the integrity of CNS white matter architecture and long-term maintenance of myelination. *Neuron* **17,** 607–615.

Lovric, J., Dammeier, S., Kieser, A., Mischak, H., and Kolch, W. (1998). Activated raf induces the hyperphosphorylation od stathmin and the reorganization of the microtubular network. *J. Biol. Chem.* **273,** 22849–22855.

Luderus, M. E., de Graaf, E., Mattia, E., den Blaauwen, J. L., Grande, M. A., de Jong, L., and van Driel, R. (1992). Binding of matrix attachment regions to lamin B1. *Cell (Cambridge, Mass.)* **70,** 949–959.

Maciver, S. K. (1996). Myosin II function in nonmuscle cells. *BioEssays* **18,** 179–182.

Mandal, M., Maggirwar, S. B., Sharma, N., and Kaufmann, S. H. (1996). Bcl-2 prevents CD95 (Fas/APO-1)-induced degradation of lamin B and poly(ADP-ribose) polimerase and restores the NF-kb signaling pathway. *J. Biol. Chem.* **271,** 30354–30359.

Mandelkow, E., and Mandelkow, E. M. (1995). Microtubules and microtubule-associated proteins. *Curr. Opin. Cell Biol.* **7,** 72–81.

Martin, J. M., O'Brien, G. A., Nishioka, W. K., McGahon, A. J., Mahboubi, A., Saido, T. C., and Green, D. R. (1995). Proteolysis of fodrin (non-erythroid spectrin) during apoptosis. *J. Biol. Chem.* **270,** 6425–6428.

Martins, L. M., and Earnshaw, W. C. (1997). Apoptosis: Alive and kicking in 1997. *Trends Cell Biol.* **7,** 111–114.

McKeon, F. (1991). Nuclear lamin proteins: Domains required for nuclear targettin, assembly, and cell-cycle-regulated dynamics. *Curr. Opin. Cell Biol.* **3,** 82–86.

Mills, J. C., Stone, N. L., Erhardt, J., and Pittman, R. N. (1998). Apoptotic membrane blebbing is regulated by myosin light chain phosphorylation. *J. Cell Biol.* **140,** 627–636.

Moir, R. D., Spann, T. P., and Goldman, R. D. (1995). The dynamic properties and possible functions of nuclear lamins. *Int. Rev. Cytol.* **162,** 141–182.

Moos, P. J., and Fitzpatrick, F. A. (1998). Taxol mediated gene induction is independent of microtubule stabilization, induction of transcription regulators and enzimes that modulate inflammation and apoptosis. *Proc. Natl. Acad. Sci. U.S.A.* **95,** 3896–3901.

Nagafuchi, A., and Takeichi, A. (1988). Cell binding function of E-cadherin is regulated by the cytoplasmic domain. *EMBO J.* **7,** 3679–3684.

Neamati, N., Fernandez, A., Wright, S., Kiefer, J., and McConkey, D. J. (1995). Degradation of lamin B1 precedes oligonucleosomal DNA fragmentation in apoptotic thymocytes and isolated thymocyte nuclei. *J. Immunol.* **154,** 3788–3795.

Nigg, E. A. (1992). Assembly and cell cycle dynamics of the nuclear lamina. *Semin. Cell Biol.* **3,** 245–253.

Nuydens, R., Dispersyn, G., de Jong, M., Van den Kieboon, G., Borgers, M., and Geerts, H. (1997). Aberrant tau phosphorylation and neurite retaction during NGF deprivation in P12 cells. *Biochem. Biophys. Res. Commun.***240,** 687–691.

Oberhammer, F. A., Hochegger, K., Froschl, G., Tiefenbacher, R., and Pavelka, M. (1994). Chromatin condensation during apoptosis is accompanied by degradation of lamin A+B, without enhanced activation of cdc2 kinase. *J. Cell Biol.* **4,** 827–837.

Orth, K., Chinnaiyan, A. M., Garg, M., Froelich, C. J., and Dixit, V. M. (1996). The CED-3/ICE is activated during apoptosis and cleaves the death substrate lamin A. *J. Biol. Chem.* **28,** 16443–16446.

Paddy, M. R., Belmont, A. S., Saumweber, H., Agard, D. A., and Sedat, J. W. (1990). Interphase nuclear envelope lamins form a discontinuous network that interacts with only a fraction of the chromatin in the nuclear periphery. *Cell (Cambridge, Mass.)* **62,** 89–106.

Peter, M., Nakagawa, J., Doree, M., Labbe, J. C., and Nigg, E. A. (1990). In vitro disassembly of the nuclear lamina and M phase-specific phosphorylation of lamins by cdc2 kinase. *Cell (Cambridge, Mass.)* **61,** 591–602.

Pittman, S., Geyp, M., Fraser, M., Ellem, K., Peaston, A., and Ireland, C. (1997). Multiple centrosomal microtubule organising centres and increased microtubule stability are early features of VP, 16, induced apoptosis in CCRF, CEM cells. *Leuk. Res.* **21,** 491–499.

Pittman, S. M., Strickland, D., and Ireland, C. M. (1994). Polymerization of tubulin in apoptotic cells is not cell cycle dependent. *Exp. Cell Res.* **215,** 263–272.

Pitzer, F., Dantes, A., Fuchs, T., Baumeister, W., and Amsterdam, A. (1996). Removal of proteasomes from the nucleus and their accumulation in apoptotic blebs during programmed cell death. *FEBS Lett.* **394,** 47–50.

Porter, A., Ng, P., and Jänicke, R. U. (1997). Death substrates come alive. *BioEssays* **19,** 501–507.

Poruchynsky, M. S., Wang, E. E., Rudin, C. M., Blagosklonny, M. V., and Fojo, T. (1998). Bcl,xL is phosphorylated in malignant cells following microtubule disruption. *Cancer Res.* **58,** 3331–3338.

Rao, L., Perez, D., and White, E. (1996). Lamin proteolysis facilitates nuclear events during apoptosis. *J. Cell Biol.* **135,** 1441–1455.

Red, J. C. (1997). Bcl-2 family proteins and the hormonal control of cell life and death in normalcy and neoplasia. *Vitam. Horm. (N.Y.)* **53,** 99–138.

Rimm, D. L., Koslov, E. R., Kebriaei, P., Cianci, C. D., and Morrow, J. S. (1995). α-catenin is an actin-binding and bundling protein mediating the attachment of F-actin to membrane adhesion complex. *Proc. Natl. Acad. Sci. U.S.A.* **92,** 8813–8817.

Schliwa, M. (1982). Action of cytochalsin D on cytoskeletal networks. *J. Cell Biol.* **92,** 79–91.

Schmidt, R., Michel, S., Shroot, B., and Reichert, U. (1988). Plasma membrane transglutaminase and cytosolic transglutaminase form distinct envelope-like structures in transformed human keratinocytes. *FEBS Lett.* **229,** 193–196.

Selden, L. A., Gershman, L. C., and Estes, J.E. (1980). A proposed mechanism of action of cytochalasin D on muscle actin. *Biochem. Biophys. Res. Commun.* **95,** 1854–1860.

Senior, A., and Gerace, L. (1988). Integral membrane proteins specific to the inner nuclear membrane and associated with the nuclear lamina. *J. Cell Biol.* **107,** 2029–2036.

Sherwood, S. W., Sheridan, J. P., and Schimke, R. T. (1994). Induction of apoptosis by the anti, tubulin drug colcemid, relationship of mitotic checkpoint control to the induction of apoptosis in HeLa S3 cells. *Exp. Cell Res.* **215,** 373–379.

Sirivastava, R. K., Sirivastava, A. R., Korsmeyer, S. J., Nesterova, M., Cho-Chung, Y. S., and Longo, D. L. (1998). Involvement of microtubules in the regulation of Bcl2 phosphorylation and apoptosis through cyclic AMP-dependent protein kinase. *Mol. Cell. Biol.* **18,** 3509–3517.

Takahashi, A., Alnemri, E. S., Lazebnik, Y. A., Fernandes-Alnemri, T., Litwack, G., Moir, R. D., Goldman, R. D., Poirier, G. G., Kaufmann, S. H., and Earnshaw, W. C. (1996). Cleavage of lamin A by Mch2? but not CPP32: Multiple interleukin 1beta-converting enzyme-related proteases with distinct substrate recognition properties are active in apoptosis. *Proc. Natl. Acad. Sci. U.S.A.* **93,** 8395–8400.

Takeichi, M. (1995). Morphogenetic role of classical cadherins. *Curr. Opin. Cell Biol.* **7,** 619–627.

Takeuchi, K., Sato, N., Kasahara, N., Funayama, A., Nagafuchi, A., Yonemura, S., Tsukita, Sa., and Tsukita, Sh. (1994). Perturbation of cell adhesion and microvilli formation by antisense oligonucleotides to ERM family members. *J. Cell Biol.* **125,** 1371–1384.

Tanaka, K., and Takai, Y. (1998). Control of reorganization of the actin skeleton by Rho family small GTP-binding proteins in yeast. *Curr. Opin. Cell Biol.* **10,** 112–116.

Tang, C., Willingham, M. C., Reed, J. C., Miyashita, T., Ray, S., Ponnathpur, V., Huang, Y., Mahoney, M. E., Bullock, G., and Bhalla, K. (1994). High levels of p26BCL 2 oncoprotein retard taxol, induced apoptosis in human pre, B leukemia cells. *Leukemia* **8,**1960–1969.

Tinnemans, M. F. J., Lenders, M. J. H., ten Velde, G. P. M., Ramaekers, F. C. S., and Schutte, B. (1995). Alterations in cytoskeletal and nuclear matrix-associated proteins during apoptosis. *Eur. J. Cell Biol.* **68,** 35–46.

Trejo-Skalli, A. V., Velasco, P. T., Prasanna Murthy, S. N., Lorand, L., and Goldman, R. D. (1995). Association of a transglutaminase-related antigen with intermediate filaments. *Proc. Natl. Acad. Sci. U.S.A.* **92,** 8940–8944.

Tsukita, S., and Yonemura, S. (1997). ERM (ezrin/ radixin/ moesin) family: From cytoskeleton to signal transduction. *Curr. Opin. Cell Biol.* **9,** 70–75.

Ucker, D. S., Obermiller, P. S., Eckhart, W., Apgar, J. R., Berger, N. A., and Meyers, J. (1992a). Genome digestion is a dispensable consequence of physiological cell death mediated by cytotoxic T-lymphocytes. *Mol. Cell. Biol.* **12,** 3060–3069.

Ucker, D. S., Wilson, J. D., and Hebshi, L. D. (1992b). Target cell death triggered by cytotoxic T lymphocytes: A target cell mutant distinguishes passive pore formation and active cell suicide mechanisms. *Mol. Cell. Biol.* **14,** 427–436.

Vanags, D. M., Pörn-Ares, M. I., Coppola, S., Burgess, D. H., and Orrenius, S. (1996). Protease involvement in fodrin cleavage and phosphatidylserine exposure in apoptosis. *J. Biol. Chem.* **271,** 31075–31085.

van Engeland, M., Ramaekers, F. C., Schutte, B., and Reutelingsperger, C. P. (1996). A novel assay to measure loss of plasma membrane asymmetry during apoptosis of adherent cells in culture. *Cytometry* **24,** 131–139.

van Engeland, M., Kuijpers, H. J. H., Ramaekers, F. C. S., Reutelingsperger, C. P. M., and Schutte, M. (1997). Plasma membrane alterations and cytoskeletal changes in apoptosis. *Exp. Cell Res.* **235,** 421–430.

Wade, R. H., and Hyman, A. (1997). Microtubule structure and dynamics. *Curr. Opin. Cell Biol.* **9,**12–17.

Wang, T. H., Wang, H. S., Ichijo, H., Giannakakou, P., Foster, F. J., Fojo, T., and Wimalasena, J. (1998). Microtubule-interfering agents activate c-Jun N-Terminal Kinase/ Stress-activate protein kinase through both Ras and apoptosis signal-regulatin kinase pathways. *J. Biol. Chem.* **273,** 4928–4936.

Ward, E. G., and Kirschner, W. M. (1990). Identification of cell cycle-regulated phosphorylation sites on nuclear lamin C. *Cell (Cambridge, Mass.)* **61,** 561–577.

Weaver, V., Carson, C. E., Walker, P. R., Chaly, N., Lach, B., Raymond, Y., Brown, D. L., and Sikorska, M. (1996). Degradation of nuclear matrix and DNA cleavage in apoptotic thymocytes. *J. Cell Sci.* **109,** 45–56.

Weber, K., Plessmann, U., and Ulrich, W. (1989). Cytoplasmic intermediate filament proteins of invertebrates are closer to nuclear lamins than are vertebrates intermediate filament proteins; sequence characterization of two muscle proteins of a nematode. *Annu. Rev. Biochem.* **57,** 593–625.

Woods, C. M., Zhu, J., McQueney, P. A., Bollag, D., and Lazarides, E. (1995). Taxol, induced mitotic block triggers rapid onset of a p53, independent apoptotic pathway. *Mol. Med.* **1,** 506–526.

Wyllie, A. H. (1980). Glucocorticoid-induced thymocyte apoptosis is associated with endogenous endonuclease activation. *Nature (London)* **284,** 555–556.

Wyllie, A. H., Kerr, J. F., and Currie, A. R. (1980). Cell death: The significance of apoptosis. *Int. Rev. Cytol.* **68,** 251–306.

Yaneda, K., McBridge, O. W., Korge, B. P., Kim, I. G., and Steinert, P. M. (1992). The cornified cell envelope: loricrin and transglutaminases. *J. Dermatol.* **19,** 761–764.

Zhang, J., Lesort, M., Guttmann, R. P., Johnson, G. V., Arnold, C. S., Johnson, G. V., Cole, R. N., Dong, D. L. Y., Lee, M., and Hart, G. W. (1998). The microtubule associated protein tau is extensively modified with O,linked N,acetylglucosamine. *J. Biol. Chem.* **271,** 28741–2874.

Zhivotovsky, B., Gahm, A., and Orrenius, S. (1997). Two different proteases are involved in the proteolysis of lamin during apoptosis. *Biochem. Biophys. Res. Commun.* **233,** 96–101.

Dietary Antioxidants and Cardiovascular Disease

J. MICHAEL GAZIANO

Massachusetts Veterans Epidemiology Research and Information Center and Department of Medicine, Veterans Affairs Medical Center, Brockton / West Roxbury, Massachusetts, 02132; Division of Preventive Medicine and Cardiovascular Division, Department of Medicine, Brigham and Women's Hospital and Harvard Medical School, Boston, Massachusetts, 02215.

I. Introduction

In the United States, cardiovascular disease (CVD) and cancer account for nearly two-thirds of all deaths. Although the role of diet in the development or prevention of CVD and cancer has been explored for the past half century, many issues complicate the study of these relationships. First, the risks and benefits of various aspects of diet are likely to accrue over years or even decades, and eating habits change over time. Studies of diet and chronic disease may have to continue for many years and be updated periodically. Second, quantitative assessment of diet is usually based on self-reports, which can result in misclassification. Third, until recently, quantifying micronutrient data has been problematic. Finally, dietary habits tend to be imbedded in cultural practices and associated with lifestyle factors, making it difficult to sep-

0083-6729/00 $30.00

arate the effects of diet from other related factors on a given disease.

One of the most consistent findings in dietary research is that those who consume higher amounts of fruits and vegetables have lower rates of cancer, heart disease, and stroke. Possible explanations for these apparent protective effects include higher dietary fiber intake, the replacement of fats and cholesterol, or specific micronutrients in fruits and vegetables, particularly those with antioxidant properties. In this review, I briefly summarize the basic science behind the hypothesis that antioxidants might prevent CVD, review the observational epidemiology that relates to this hypothesis, and, finally, discuss the results of randomized trials examining primary and secondary prevention of CVD through dietary supplementation.

II. Basic Science—Oxidative Damage in Biological Systems

Oxidative metabolism, the backbone of aerobic biologic systems, involves several highly reactive species known as *free radicals* that can damage cellular structures. Although many highly effective defense mechanisms protect organisms from oxidative damage induced by free radicals, such damage occurs at low level continuously, and may have a role in the pathogenesis of many diseases, including cancer, arthritis, reperfusion injury during myocardial infarction, and atherogenesis as well as aging (Southorn and Powis, 1988a).

Antioxidant vitamins can inhibit oxidation of low-density lipoprotein (LDL) and other free-radical reactions and thereby may prevent or retard atherogenesis. Vitamin C, α-tocopherol (a major component of vitamin E), and beta carotene (a provitamin A) are among the most abundant dietary antioxidants.

A. Free-Radical Generation and Reactive Oxygen Species

Normal oxidation–reduction reactions are continuously producing trace amounts of free radicals. These highly unstable compounds contain one or more unpaired electrons in their outer shells and react with any nearby susceptible compounds including DNA, proteins, lipids, and carbohydrates (Southorn and Powis, 1988b; Slater, 1984). Although free radicals last only a few milliseconds, they can initiate a chain reaction resulting in the oxidation of thousands of particles over a great distance as the unpaired electron is transferred from one molecule to another in a game of "hot potato."

Lipid peroxidation is a chain reaction initiated when a free radical removes an atom from the lipid, leaving an unpaired electron behind. This newly formed lipid free radical then binds molecular oxygen, forming a lipid peroxide radical, which in turn steals a hydrogen atom from a nearby liquid, forming a new lipid radical in a step known as *propagation.* Reactive oxygen species that can generate such a chain reaction include the superoxide anion radical (O_2^-), hydrogen peroxide (H_2O_2), the hydroxyl radical ($OH\cdot$), and singlet oxygen ($O_{2\cdot}$).

B. Antioxidant Defense Mechanisms

Several mechanisms minimize unintended free-radical formation. First, oxidative metabolism is largely compartmentalized in the mitochondria. The reactive oxygen species formed during oxidative phosphorylation are tightly bound to mitochondrial enzymes. Second, various proteins bind heavy metal ions such as copper and iron to prevent them from reacting with oxygen species that could generate formation of free radicals. The binding capacity of these proteins greatly exceeds the amount of heavy metals in the body, ensuring that few free ions exist in body fluids.

Once free radicals have formed, specific enzymes remove and/or destroy them. Several forms of superoxide dismutase (SOD) form H_2O_2 from two superoxide radical anions. Two additional enzyme systems, glutathione peroxidase and catalase, catalyze hydrogen peroxide before it can become a hydroxyl radical. Certain nonenzymatic compounds also have important roles in antioxidant defense by scavenging free radicals, terminating lipid peroxidation, and quenching reactive oxygen species. These include the antioxidant vitamins E and C, and beta carotene, a vitamin A precursor.

Vitamin E is a lipid-soluble compound found in vegetable oils, cereal grains, egg yolk, liver, milk fat, nuts, and green vegetables. α-Tocopherol, its most potent antioxidant component, accumulates in lipid membranes and circulating lipids, where it terminates lipid peroxidation by donating a hydrogen atom to the fatty acid free radical (Ingold *et al.,* 1987), forming a new, nonreactive α-tocopherol radical that may be regenerated by vitamin C (Packer *et al.,* 1979). An antioxidant role for vitamin E in humans is supported by data showing that dietary supplementation with vitamin E reduces production of pentane (a by-product of oxidation of omega-6 fatty acids measured in the breath) (Lemoyne *et al.,* 1987). Vitamin C (ascorbic acid) is a water-soluble antioxidant found in many fruits and vegetables that can scavenge peroxyl radicals (Niki *et al.,* 1983) and quench superoxide radicals (Som *et al.,* 1983). It also may regenerate oxidized vitamin E *in vivo.*

Beta carotene is the most abundant of the carotenoids, which are found in high concentrations in fresh fruits and vegetables such as carrots, squash, melons, spinach, and broccoli. Beta carotene is a lipid-soluble antioxidant that can neutralize singlet oxygen (Foote *et al.,* 1970) by absorbing the excess energy and releasing it as heat, without damaging itself in the process. Beta carotene can also prevent the oxidation of lipids *in vitro* (Burton and Ingold, 1984) and under certain conditions is also a chain-breaking antioxidant like vitamin E (Foote *et al.,* 1970). These abilities rely on a series of conjugated double bonds that provide a stable environment for the lone electron donated by the free radical (Som *et al.,* 1983). Beta carotene is concentrated in circulating lipids and in atherosclerotic plaque (Prince *et al.,* 1988).

Other compounds with antioxidant properties include urate and bilirubin; however, their physiologic significance remains unclear. Exogenous substances with demonstrated antioxidant abilities include butylated hydroxytoluene (BHT), a food preservative, and Probucol, a cholesterol-lowering agent.

III. Role of Uncontrolled Oxidation in Atherogenesis

Although the precise mechanisms and sequence of events that lead to atherogenesis are still not clearly defined, several key players and events have been identified. Basic scientific research has supported a role of oxidation—particularly of LDL—in many of these events. *In vitro* and *in vivo* studies suggest that free-radical-induced oxidative damage to LDL cholesterol significantly increases its atherogenicity (Steinberg *et al.,* 1989). Oxidized LDL may promote several steps in atherogenesis including endothelial cell damage (Hessler *et al.,* 1983; Yagi, 1984), monocyte/macrophage accumulation, foam cell accumulation (Quinn *et al.,* 1985; Schaffner *et al.,* 1980; Gerrity, 1981), and growth (Fogelman *et al.,* 1980; Goldstein *et al.,* 1979) and synthesis of autoantibodies (Salonen *et al.,* 1992). In addition, free radicals may directly damage arterial endothelium (Beckman *et al.,* 1990), promote thrombosis (Marcus *et al.,* 1977), and interfere with normal vasomotor regulation (Saran *et al.,* 1990).

A. Endothelial Damage

Vascular endothelial damage is an important initial step in the process of atherosclerosis. Several investigators have reported that oxidatively modified LDL but not native LDL is highly toxic to cultured cells (Cathcart *et al.,* 1985; Hessler *et al.,* 1983; Yagi *et al.,* 1981). This cytotoxicity is attributed to the lipid peroxides and short-chain aldehydes that can cross link membrane proteins. Once within the cell, oxidized LDL particles may damage internal structures as well. The endothelial barrier may be disrupted, allowing lipids to migrate into the intima, and the adhesiveness of the endothelium may be altered, allowing more monocytes or platelets to adhere to the endothelial walls (Berliner, 1990). Such changes could alter vascular tone, which is largely mediated by the intact arterial endothelium. Production of EDRF may also be diminished, preventing appropriate relaxation of the artery. Production of growth factors may also be disrupted. Oxidized LDL increases endothelial cell tissue factor and decreases protein C activation, and thus may increase procoagulation at the endothelial surface.

Animal studies provide further evidence that lipid oxidation can alter endothelial structure and function. When linoleate hydroperoxide but not unoxidized linoleate is given subcutaneously to rats, aortic lesions result (Cutler and Schneider). Similar studies in rabbits resulted in aortic lesions associated with increased peroxides. Intravenous administration of linoleic hydroperoxide to rabbits resulted in ulcerated endothelial lining with adherent platelets as well as increased lipid peroxide levels in the aortic intima (Yagi, Sedar).

B. Monocyte/Macrophage Accumulation

Early in fatty streak formation, most of the foam cells that accumulate in the subendothelium are derived from circulating monocytes (Fowler *et al.,* 1979; Gerrity, 1981), but later in atherogenesis, they may also come from smooth muscle cells. How these cells are recruited is not entirely clear, but the initial step appears to be their adhesion to the endothelial lining followed by migration into the intima of the arterial wall (Gerrity, 1981). Adhesion of monocytes to endothelial cells in culture is enhanced by minimally modified LDL (Berliner, 1990). Once these monocytes have attached to the arterial wall, their migration into the subendothelium is presumably influenced by various chemotactic factors.

Oxidized LDL but not native LDL is a potent *in vitro* chemoattractant for monocytes (Quinn *et al.,* 1985). This chemotactic activity resides in the lipid peroxide component, and is mostly attributed to lysolecithin produced during oxidation (Quinn). It can also immobilize the macrophage in residence (Quinn *et al.,* 1987). Cellular motility is inhibited by oxidized LDL, and again the active component is localized to the lipid component. Thus, oxidized LDL may be involved in the accumulation of monocytes in the subendothelium.

C. Foam Cell Uptake

One of the early events in atherogenesis is the accumulation of cholesterol esters in foam cells within a developing atheroma. Most of the cholesterol in foam cells is derived from LDL. LDL transports cholesterol to tissues and is essential to normal lipid metabolism. Elevated levels of LDL are clearly an important independent cardiovascular risk factor. However, massive accumulation of cholesterol in foam cells is not mediated by the native LDL receptor, which is down-regulated by a feedback mechanism. Furthermore, patients who are deficient in native LDL receptors undergo accelerated atherogenesis. Many lines of evidence have implicated oxidatively modified LDL and an alternate receptor pathway in cholesterol accumulation in foam cells (Goldstein *et al.,* 1979) and recruitment of foam cells in the early atheroma.

An alternative receptor pathway for LDL was delineated (Brown and Goldstein, 1983) more than two decades ago. This receptor, now known as the scavenger receptor, is not down-regulated by intracellular lipid accumulation and binds oxidized LDL much more avidly than native LDL. Later studies established that incubation of LDL with cultured cells (endothelial cells, smooth muscle cells, monocyte/macrophages) caused it to become oxidized, which markedly increased its uptake by monocyte/macrophages via the scavenger receptor (Fogelman *et al.,* 1980; Henriksen *et al.,* 1981).

All cultured cells clearly enhance the oxidative environment and result in the formation of fatty acid peroxides through mechanisms that may vary with cell type. Once these lipid peroxides have formed, rapid propagation causes fragmentation of fatty acid chains and apoprotein B and formation of shorter chain aldehydes. The process appears to be biphasic, with a lag phase of several hours followed by a rapid second phase. The lag is apparently due to antioxidants within the LDL particle, which are sequentially depleted (Esterbauer), starting with α-tocopherol, followed by γ-tocopherol, lycopene, and beta carotene. Treating patients with vitamin E or Probucol renders their LDL resistant to oxidation, suggesting that vitamin E and beta carotene may protect LDL from oxidation *in vivo.*

D. *In Vivo* Evidence That Oxidized LDL Is Involved in Atherogenesis

Several lines of evidence support the existence of oxidized LDL *in vivo* (Warso and Lands, 1985; Yagi, 1976). First, ceroid deposits from atherosclerotic plaques contain high levels of lipid peroxides. Second, antibodies to oxidized LDL are present in circulation and bind to atherosclerotic lesions. Second, antibodies to oxidized LDL are present in circulation and bind to atherosclerotic lesions. Third, oxidized LDL itself has been detected in the circulation, although in only trace amounts, perhaps because it is cleared rapidly by the many antioxidants in plasma (Steinberg).

The role of lipid oxidation in atherogenesis is further supported by elevated plasma levels of lipid peroxides in association with various cardiac risk factors including smoking, diabetes, and advancing age (Blackman *et al.,* 1984; Nishigaka *et al.,* 1981; Suematsu *et al.,* 1977). Among a group of patients referred for cardiac catheterization, those with angiographically proven coronary artery stenoses had higher levels of lipid peroxides than those without (Stringer *et al.,* 1989).

The susceptibility of circulating LDL to oxidation varies considerably from subject to subject. Extensive oxidation of LDL may occur in the subendothelium at the site of ongoing atherogenesis. The localized environment in the plaque may be the critical point of oxidation, and predisposition to atherosclerosis may be partially determined by individual susceptibility of LDL to oxidation.

IV. Observational Epidemiology

Many observational epidemiologic studies have supported the notion that consumption of dietary antioxidants may decrease risk of CVD. These studies are summarized next, beginning with descriptive studies and ending with ongoing randomized clinical trials.

A. Descriptive Studies

Several descriptive or ecologic studies have shown that per capita consumption of fresh fruits and vegetables is inversely correlated with risks of CVD. Two studies in the United Kingdom found an inverse association between the consumption of fresh fruits and vegetables and risk of CVD (Acheson and Williams, 1983; Armstrong *et al.,* 1975). In the United States, the declining CVD mortality has been attributed, in part, to the increased year-round availability and consumption of fresh fruits and vegetables (Verlangieri *et al,.* 1985). In addition, American per capita consumption of vitamin C intake has been inversely associated with CVD mortality rates (Ginter, 1979).

In a study of 16 European countries, lipid-standardized α-tocopherol levels were inversely associated with mortality rates due to ischemic heart disease ($r^2 = 0.44$, $P = 0.02$), while correlations for vitamin C and carotene were not significant (Gey *et al.,* 1987a,b; Gey and Puska, 1989). A similar study of four European populations found an insignificant inverse association between vitamin E intake and cardiovascular mortality (Riemersma *et al,.* 1990).

Descriptive studies cannot control for potential confounding factors such as other dietary or lifestyle characteristics, genetics, or even differing availability of health care resources. However, they are useful for formulating hypotheses, which can be tested with more analytical studies.

B. Case-Control Studies

Two case-control studies reported a significant inverse association between plasma antioxidant levels and CVD. Plasma antioxidant levels in angina patients were compared with those of healthy controls (Riemersma *et al.,* 1989, 1991). Angina patients had significantly lower levels of vitamin C, lipid-standardized vitamin E, and carotene than controls. After multivariate adjustment, the relative risk of angina between the lowest and highest quintiles of lipid-standardized vitamin E level was 2.98 (95% CI = 1.07–6.70). Vitamin C and beta carotene followed a similar trend, but adjustment for cigarette smoking significantly attenuated these relationships [relative risk = 1.63 (0.76–3.49) for vitamin C and 1.41 (0.63–3.13) for beta carotene]. Vitamin A had no apparent relationship with angina. In another study, leukocyte ascorbic acid levels were significantly lower among those with angiographically documented coronary disease compared with controls ($P < 0.001$) (Ramirez and Flowers, 1980).

In a recent case-control study (EURAMIC Study), adipose tissue levels of α-tocopherol and beta carotene were obtained in 683 cases of myocardial infarction and 727 hospital-based controls (Kardinaal *et al.,* 1993). Mean levels of beta carotene were 0.35 μg/g in cases and 0.42 μg/g in controls, while α-tocopherol levels were 193 and 192 μg/g for cases and controls, respectively. The multivariate odds ratio in the lowest quintile beta carotene level compared to the highest was 1.78 (95% CI = 1.17–2.71) after controlling for a number of potential confounders, with a significant trend across quintiles (P for trend = 0.001). The associations were strongest among current and former smokers. In contrast, low α-tocopherol levels were not associated with apparent increased risk, with an odds ratio of myocardial infarction in the lowest compared to the highest category of 0.83 (95% CI = 0.57–1.21) and no significant trend across quintiles of α-tocopherol tissue levels (*P* for trend = 0.27).

C. Prospective Cohort Studies of Dietary Intake

Although case-control studies are often efficient and less costly than cohort studies, selection and recall bias may affect risk estimates. Prospective cohort studies are less subject to these types of bias, because information is collected before disease develops. Most but not all of the prospective cohort studies examining the role of dietary intake of antioxidants and cardiovascular disease are consistent with an inverse association.

1. *The Nurses Health Study (NHS)*

This prospective cohort study of 121,000 U.S. female nurses aged 30–55 is the largest to examine the relationship of antioxidants with cardiovascular disease (Manson *et al.,* 1991, 1992; Stampfer *et al.,* 1993). The study began in 1976, and participants were followed for 8 years. Biennial questionnaires elicited information about a wide variety of coronary risk factors. In 1980 and 1984, questions were included about the usual frequency of consuming certain foods over the previous year, with nine options ranging from "never" to "six or more times per day." The dietary questionnaires themselves, with documentation of their reproducibility and validity, have been published (Willett *et al.,* 1985; Salvini *et al.,* 1989). Intake scores were computed by multiplying how frequently each food was consumed by the nutrient content of the specified portions. Composition values for vitamins were obtained from the U.S. Department of Agriculture and other published sources (Adams, 1975; Consumer and Food Economics Institute, 1982; Willett and Stampfer, 1986).

During 8 years of follow-up (671,185 person-years), there were 552 cases of coronary disease, including 150 deaths and 437 nonfatal myocardial infarctions. Women in the highest quintile of beta carotene consumption had a 22% risk reduction (relative risk = 0.78; 95% CI = 0.59–1.03; *P* for trend across quintiles = 0.02) when compared to those in the lowest. For vitamin E, the relative risk was 0.66 (95% CI = 0.50–0.87) in the highest intake quintile (*P* for trend = 0.001), an effect attributable almost entirely to supplements, rather than diet. The relative risk for vitamin C was 0.80 (95% CI = 0.58–1.10), and across quintiles there was no significant trend after controlling for vitamin E and multivitamin intake, which were highly correlated with vitamin C consumption (P for trend = 0.15). When the intake of beta carotene, vitamin E, and vitamin C were combined into a total antioxidant score, the relative risk for coronary disease was 0.54 (95% CI = 0.40–0.73) among those in the highest quintile compared to the lowest (P for trend = 0.001).

Ischemic strokes were documented in 183 cases. When women in the highest quintile of intake (foods plus supplements) were compared to those in the lowest, the relative risk of ischemic stroke after adjustment for age, cigarette smoking, and other vascular risk factors was 0.75 (95% CI = 0.48–1.17) for vitamin C, 0.76 (95% CI = 0.49=1.20) for vitamin E, and 0.61 (95% CI, 0.39–0.98) for beta carotene. For increasing quintiles of a total antioxidant vitamin score the relative risks were 1.0 (referent), 0.93, 0.71, 0.62, and 0.46 (P for trend = 0.01) after adjustment for other risk factors. Intake of these micronutrients from foods accounted for most of the association; multivariate relative tasks for increasing quintiles of a score derived from foods alone (excluding supplements) were 1.0, 0.94, 0.59, 0.63, and 0.59 (P for trend = 0.007).

2. *The Health Professional Follow-Up Study (HPFS)*

The HPFS, a more recent prospective cohort study using the same methodology as the NHS, examined dietary antioxidants in men over a 4-year period beginning in 1986 (Rimm *et al.*, 1993). Of 39,000 men who began the study with no history of vascular disease or other condition necessitating dietary changes, there were 667 major coronary events (360 revascularizations, 209 nonfatal myocardial infarctions, and 106 fatal myocardial infarctions) in 4 years. When men in the highest quintile of beta carotene intake were compared with those in the lowest quintile, the relative risk was 0.75 (95% CI = 0.57–0.99; P for trend = 0.04). Men in the highest quintile of vitamin E consumption had a relative risk of 0.68 (95% CI = 0.51–0.90; P for trend = 0.01) when compared with those in the lowest, and the effect was largely confined to those who consumed more than 100 IU of vitamin E supplements daily for 2 or more years. The relative risk for vitamin C was 1.29 (P for trend = 0.10).

3. *The First National Health and Nutrition Examination Survey (NHANSE-1)*

The 11,349 participants in the NHANSE-1 study, aged 25–74, were flowed for a median of 10 years. The standardized cardiovascular mortality rate among those with the highest vitamin C intake was 34% lower (relative risk = 0.66, 95% CI = 0.53–0.82) than expected (Enstrom *et al.*, 1992). Vitamin C supplements explained most of the association.

This study did not examine the correlation of vitamin C supplementation with either vitamin E or multivitamins, which eliminated a crude association in both the NHS and HPFS.

4. *The Massachusetts Elderly Cohort Study*

The Massachusetts Elderly Cohort Study, which related lifestyle factors to health and functional status in the elderly, examined dietary information obtained through annual mailings and personal interviews in 1976 and 1980 with an average follow-up of 4.75 years (Gaziano *et al.*, 1992). Of the 1299 participants, 151 died of cardiovascular causes, including 47 fatal myocardial infarctions (MI). The relative risks of cardiovascular death from lowest to highest quartile of beta carotene were 1.00 (referent), 0.75, 0.65, and 0.57, respectively (P for trend = 0.016), after controlling for confounders that included age, gender, smoking, alcohol consumption, cholesterol intake, and functional status. The corresponding relative risks for fatal MI were 1.00 (referent), 0.77, 0.59, and 0.32 (P for trend = 0.02).

5. *The Swedish Cohort Study*

A prospective study of 1462 Swedish women found no correlation between vitamin C intake as estimated from a 24-hour recall dietary history and cardiovascular mortality after 12 years of follow-up after controlling for age (Lapidus *et al.*, 1986). The relatively small number of deaths on this cohort may have limited the power to detect small to moderate benefits.

6. *Prospective Blood-Based Studies*

Three studies measured plasma antioxidants levels at baseline for the entire cohort. The Basel Prospective Study measured baseline antioxidants in 2974 middle-aged men, who were followed for several years. The risk of death from coronary heart disease was increased among those in the lowest quartile of carotene (relative risk = 1.53; 95% CI = 1.07–2.20) and vitamin C (relative risk = 1.25; 95% CI = 0.77–2.01) level compared with those in the highest (Gey *et al.*, 1993; Eichholzer *et al.*, 1992). In addition, low plasma concentrations of carotene accompanied by low plasma concentrations of vitamin C were associated with an elevated risk of ischemic heart disease mortality (relative risk = 1.96, 95% CI = 1.10–3.50). Risk of death from stroke was increased among those with both low beta carotene and low vitamin C levels (relative risk = 4.17; 95% CI = 1.68–10.33). There was no apparent relationship of lipid-standardized vitamin E levels with death from coronary heart disease or stroke.

In the Lipid Research Clinic Coronary Primary Prevention Trial, baseline carotenoid level after 14 years of follow-up was inversely correlated with risk of MI after adjustment for age, smoking, HDL, and LDL (Morris *et al.*, 1993). In the Kupio Ischemic Heart Disease Study plasma beta carotene and vitamin E levels were inversely related to progression of carotid artery wall thickness (Salonen *et al.*, 1993).

An alternative to measuring plasma antioxidant levels for an entire cohort is to use a nested case-control design in which baseline blood samples are collected and frozen at the beginning of the trial. Subjects who later develop cardiovascular disease are matched with healthy controls, and their baseline blood samples are compared. One such study found a significant inverse association between baseline beta carotene levels and subsequent MI (Street *et al.*, 1991). Two other nested case-control studies found no association between vitamin A and vitamin E levels and vascular mortality (Kok *et al.*, 1987; Salonen *et al.*, 1985); however, the blood samples in these two studies were stored at $-20°C$, and the stability of antioxidants at this temperature is questionable. In addition, the vitamin A level may not accurately reflect beta carotene status.

V. Randomized Trials

For most hypotheses, randomized trials are neither necessary nor desirable (Hennekens *et al.*, 1994); however, when searching for small to moderate effects, the amount of uncontrolled confounding interest in observational studies may be as large as the likely risk reduction. For these reasons, reliable data on the relationship of antioxidants and cardiovascular disease can only emerge from large-scale randomized trials of adequate dose and duration, in which investigators allocate subjects at random to either active treatment or placebo. When the sample size is sufficient, randomized trials can avoid some limitations of observational studies by distributing the known and unknown confounding variables among treatment groups. Results of completed and ongoing randomized trials on dietary antioxidants in the primary and secondary prevention of cardiovascular disease are summarized next.

A. Primary Prevention Trials

1. *The Chinese Cancer Prevention Trial*

The first large-scale randomized primary prevention trial of vitamin supplements in the prevention of cancer was conducted among a poorly nourished population in Linxian, China (Blot *et al.*, 1993). Overall,

29,584 men and women were randomized to one of eight treatment arms consisting of various combinations of nine vitamins and minerals. Overall mortality was significantly reduced among those assigned a cocktail of beta carotene (15 mg daily), α-tocopherol (30 mg daily), and selenium (50 μg daily), largely due to a reduction in stomach cancer mortality. Although cardiovascular mortality was also assessed, heart disease rates were relatively low. Fewer than 9% of the total deaths were due to ischemic heart disease. Cerebrovascular disease accounted for 26% of the total deaths, and there was an insignificant reduction in the risk of cerebrovascular disease mortality (relative risk = 0.90; 95% CI = 0.76–1.07). However, in this population, most strokes are likely to be hemorrhagic, which are not usually due to atherosclerotic disease.

2. *The Alpha-Tocopherol Beta Carotene (ATBC) Cancer Prevention Trial*

The ATBC Trial was the first large-scale randomized trial of antioxidant vitamins in a well-nourished population. This 2 × 2 factorial trial tested the effect of synthetic α-tocopherol (50 mg daily) and synthetic beta carotene (20 mg daily) in the prevention of lung cancer among 29,133 Finnish male smokers (Alpha-Tocopherol, Beta Carotene Cancer Prevention Study Group, 1994). After adjustment for testing multiple hypotheses, there were no increases or decreases in risk that could not be explained plausibly by chance.

Nevertheless, some findings were unexpected. There was no reduction in the rate of lung cancer, risk of ischemic heart disease (relative risk = 0.95; 95% CI = 0.85–1.05), or ischemic stroke mortality (relative risk = 0.84; 95% CI = 0.59–1.19) among those assigned to vitamin E. In a subsequent report, the risk of developing angina was lower among those receiving vitamin E. However, the dose of vitamin E was not much higher than the U.S. recommended daily allowance (relative risk = 0.91; 95% CI = 0.83–0.99) (Rapola *et al.,* 1996). Some observational research suggests that higher dose supplementation may be required to reduce the risk of heart disease. The apparent benefits among those who took vitamin E supplements in the NHS (Stampfer *et al.,* 1993) and the HPFS (Rimm *et al.,* 1993) were largely confined to those who used an average daily dose of 100 IU or more.

There was an apparent increase in the risk of hemorrhagic stroke in the α-tocopherol treatment group compared to placebo (relative risk = 1.50; 95% CI = 1.03–2.20), a finding at variance with the lower stroke rates among those assigned antioxidant vitamins in the Linxian, China, trial. Hemorrhagic stroke was not a prespecified end point, and while this finding could be attributed to an antiplatelet effect of vitamin E, it also could be the result of the play of chance.

Unexpectedly, the incidence of lung cancer was increased among those assigned to receive beta carotene. Supplementation had no apparent protective effect with respect to deaths from ischemic heart disease and stroke; in fact, there were slightly more ischemic heart disease deaths (relative risk = 1.12; 95% CI = 1.00–1.25) among those assigned beta carotene. There was no reduction in risk of angina among the beta carotene group (relative risk = 1.06; 95% CI = 0.97–1.16).

3. *The Physicians' Health Study (PHS)*

The PHS is a randomized, double-blind, placebo-controlled trial of beta carotene (50 mg on alternate days) among 22,071 U.S. male physicians, aged 40–84, of whom 11% were current and 39% past smokers at baseline in 1982 (Hennekens *et al.,* 1996). By December 31, 1995, the scheduled end of the trial, fewer than 1% were lost to morbidity and mortality follow-up and compliance was 78% among those assigned to active pills. After more than 12 years of treatment and follow-up, among 11,036 participants randomized to beta carotene and 11,035 to placebo, there were virtually no early or late differences for cardiovascular disease deaths (relative risk = 1.09; 95% CI = 0.93–1.27); myocardial infarction (relative risk = 0.96; 95% CI = 0.84–1.09); stroke (relative risk = 0.96; 95% CI = 0.83–1.11); or a composite of the previous three end points (relative risk = 1.00; 95% CI = 0.91–1.09) between treatment groups. There was also no significant benefit or harm for total malignant neoplasms (relative risk = 0.98; 95% CI = 0.91–1.06), cancer mortality (relative risk = 1.02; 95% CI = 0.89–1.18), or lung cancer. Among current or past smokers, there were likewise no significant early or late effects of beta carotene on any of these end points. This large-scale randomized trial among apparently healthy well-nourished men provides substantial evidence that 12 years of beta carotene supplementations confers neither benefit nor harm with respect to CVD disease mortality or malignant neoplasms.

4. *The Beta Carotene and Retinol Efficacy Trial (CARET)*

In this trial of 18,314 men and women at high risk of lung cancer due to cigarette smoking and/or occupational exposure to asbestos, a combined daily treatment of beta carotene (30 mg) and retinol (25,000 IU) was evaluated (Omenn *et al.,* 1996). Data were collected at annual clinic visits. The study was stopped prematurely when no benefit was detected over the projected funding period and there was a trend toward increased lung cancer in the treatment group. After four years of treatment and follow-up there was an excess of total deaths (relative risk = 1.17; 95% CI = 1.03–1.33), a trend toward excess cardiovascular deaths (relative risk = 1.26; 95% CI = 0.99–1.61), and excess cases of lung cancer among those assigned beta carotene and vitamin A. Data on specific other cardiovascular outcomes including nonfatal events are not yet available.

5. *The Skin Cancer Prevention Study*

In this multicenter, double-blind, placebo-controlled trial, investigators randomized 1805 men and women with history of skin cancer to 50 mg of beta carotene daily or placebo (Greenberg *et al.,* 1996). More than 80% of participants reported taking more than 50% of their study pills after a median follow-up of 8.2 years by questionnaire and visits to dermatologists. Death certificates were obtained for all deaths that occurred during follow-up and cause of death was coded by trained nosologists. There was no significant risk reduction in terms of total (relative risk = 1.05; 95% CI = 0.83–1.32) or cardiovascular death (relative risk = 1.15; 95% CI = 0.81–1.63).

B. Secondary Prevention Trials

1. *Small-Scale Trials*

Beginning in the 1950s, several small-scale trials have tested the effects of antioxidants among individuals with various forms of atherosclerotic disease including claudication and angina as well as following angioplasty. Benefits of supplemental vitamin E were observed in each of three studies among patients with claudication (Livingston and Jones, 1958; Williams *et al.,* 1971; Haeger, 1974). However, the utility of these data is limited by the small sample sizes, high dropout rates, and lack of blinding. Two more recent trials tested the effect of vitamin E in the treatment of angina pectoris with equivocal results. There was an insignificant trend toward improved angina pain score in a 9-week placebo-controlled trial among stable angina patients who consumed 3200 IU of vitamin E daily (Anderson and Reid, 1974). A double-blind crossover trial of 52 angina pectoris patients (Gillian *et al.,* 1977) receiving 1600 IU of vitamin E daily for 6 months found no apparent benefit of vitamin E treatment as measured by exercise tolerance, symptoms of angina pectoris, or left ventricular function. These studies provide no clear evidence that short-term treatment with vitamin E benefits angina patients; however, both the small sample size and short duration of treatment may have limited their statistical power to detect small to moderate benefits.

One recent small-scale trial tested the effect of vitamin E supplementation on restenosis among 100 subjects after percutaneous transluminal coronary angioplasty (DeMaio *et al.,* 1992). Restenosis is likely the result of an accelerated atherosclerotic process. Subjects were treated with 400 IU of vitamin E daily after angioplasty. There was an apparent, though not statistically significant, 30% reduction in the risk of restenosis as measured by later catheterization or exercise test.

A subgroup of the PHS consisting of more than 300 doctors who had chronic stable angina or a prior coronary revascularization procedure but no history of myocardial infarction or stroke was analyzed (Gaziano *et al.,* 1990). Subjects who received beta carotene had a 51% reduction (relative risk = 0.49, 95% CI = 0.29–0.88) in risk of major coronary events, and a 54% reduction (relative risk = 0.46, 95% CI = 0.24–0.85) in risk of major vascular events after an average of 5 years of treatment. These findings were considerably attenuated with longer follow-up. In a similar subgroup of the ATBC trial, among those with angina at baseline, risk of cardiovascular mortality or events was not reduced by consumption of beta carotene or vitamin E (Rapola *et al.,* 1996).

2. *Cambridge Heart Antioxidant Study (CHAOS)*

In the CHAOS, a double-blind, placebo-controlled trial, 2002 patients with angiographically proven coronary artery disease were assigned randomly to supplemental vitamin E (546 patients were treated with 800 IU and 489 with 400 IU daily after a protocol change) or placebo ($n = 967$) (Stephens *et al.,* 1996). Median follow-up was 510 days. Compared to those receiving placebo, those assigned to vitamin E had a significantly lower risk of subsequent nonfatal MI (relative risk = 0.53; 95% CI = 0.11–0.47) and a combined end point of nonfatal MI and cardiovascular death (relative risk = 0.53; 95% CI = 0.34–0.83) than those receiving placebo. However, there was a nonsignificant excess of cardiovascular deaths (relative risk = 1.18; 95% CI = 0.62–2.27). Due to the relatively small study sample size various baseline characteristics were not equivalent in the two treatment groups; the placebo group had fewer women, lower cholesterol levels, and lower systolic blood pressure levels.

C. Ongoing Large-Scale Trials

More reliable data should soon be forthcoming that will further define the role of antioxidants in atherosclerotic disease. Currently, several large-scale randomized trials are examining the role of dietary antioxidants in both primary and secondary prevention of cardiovascular disease and cancer. Three large-scale trials are testing antioxidant supplements among those without known atherosclerotic disease. The Women's Health Study (WHS) is testing vitamin E and low-dose aspirin in the primary prevention of cardiovascular disease and cancer in

40,000 healthy, U.S. health professionals. The Heart Protection Study is testing a cocktail of vitamin E, beta carotene, and vitamin C in a factorial design with a cholesterol-lowering medicine among 20,000 higher risk individuals with coronary risk factors but no known cardiovascular disease. The PHS will continue treatment with beta carotene for an additional 5 years with willing participants and vitamin E, vitamin C, and a multivitamin will be added in a factorial design.

Three large-scale secondary prevention trials are currently under way. The Women's Antioxidant Cardiovascular Study (WACS) is a secondary prevention trial of vitamin C, vitamin E, and beta carotene in a factorial design that has randomized approximately 8000 female health professionals with reported cardiovascular disease or with several coronary risk factors. The Heart Outcomes Prevention Evaluation (HOPE) Study is testing vitamin E among 9000 men and women with prior MI or stroke or known peripheral vascular disease. The Gruppo Italiano per lo Studio della Sopravvivenza nell'Infarcto Miocardico (GISSI) is conducting an unblinded trial of vitamin E among those with a recent MI. In addition, several small-scale angiographic trials are testing antioxidant supplements alone or in various combinations among individuals with coronary artery disease.

VI. Conclusions

Antioxidants are a possible but as yet unproven means to reduce risks of cardiovascular disease. Results of recent randomized trials raise the possibility that some of the benefits from observational epidemiology may have been overestimated (Hennekens and Buring, 1994). Beta carotene supplementation appears to confer no overall benefit in the primary prevention of cardiovascular disease among well-nourished individuals. However, whether risk can be reduced among those with disease or lower baseline levels of beta carotene remains unclear. For vitamin E, there have only been two primary prevention trials, both of which had methodologic limitations. Recent trial data are compatible with the possibility that vitamin E reduces reinfarction among those with coronary disease; however, this requires confirmation in the current ongoing large-scale randomized trials. The ongoing randomized trials will provide valuable information on which to base both general health policy and clinical decision making for individuals. Although whether antioxidant supplementation reduces the risks of cardiovascular disease remains unclear, consumption of fruits and vegetables high in these micronutrients is an important part of a healthy diet.

Note added in proof: The GISSI study has recently been published (GISSI-Prevenzione Investigators, 1999). There were nonsignificant trends toward modest reductions in most cardiovascular endpoints among those assigned to vitamin E. The HOPE trial was terminated early due to a benefit of the ACE inhibitor ramapril in reducing CVD events. There was no reduction in CVD events among those randomized to vitamin E compared with placebo (S. Yusef, HOPE Study Investigator, unpublished data presented at the European Cardiology Society annual meeting, 1999).

REFERENCES

Acheson, R. M., and Williams, D. R. R. (1983). Does consumption of fruit and vegetables protect against stroke? *Lancet* **1,** 1191–1193.

Adams, C. F. (1975). "Nutritive Values of American Foods," No. 456. U.S. Department of Agriculture, Washington, DC.

Alpha-Tocopherol, Beta Carotene Cancer Prevention Study Group (1994). The effect of vitamin E and beta carotene on the incidence of lung cancer and other cancers in male smokers. *N. Engl. J. Med.* **330,** 1029–1035.

Anderson, T. W., and Reid, W. (1974). A double-blind trial of vitamin E in the treatment of angina pectoris. *Am. Heart J.* **93,** 444–449.

Armstrong, B. K., Mann, J. L., Adelstein, A. M., and Eskin, F. (1975). Commodity consumption and ischemic heart disease mortality, with special reference to dietary practices. *J. Chronic Dis.* **36,** 673–677.

Beckman, J. S., Beckman, T. W., Chen, J., Marshall, P. A., and Freeman, B. A. (1990). Apparent hydroxyl radical production by peroxynitrite: Implications for endothelial injury from nitric oxide and superoxide. *Proc. Natl. Acad. Sci. U.S.A.* **87,** 1620–1624.

Berliner, J. A., Territo, M. C., Sevanian, A., Ramin, S., Kim, J. A., Bamshad, B., Esterson, M., and Fogelman, A. (1990). Minimally modified LDL stimulates monocyte endothelial interactions. *J. Clin. Invest.* **85,** 1260–1266.

Blackman, B. C., White, P., Tsou, W., and Finkel, D. (1984). Peroxidation of plasma and platelet lipids in chronic cigarette smokers and insulin-dependent diabetics. *Ann. N.Y. Acad. Sci.* **435,** 385.

Blot, W. J., Li, J. Y., Taylor, P. R., Guo, W., Dawsey, S., Wang, G. Q., Yang, C. S., Zheng, S. F., Gail, M., Li, G. Y., Yu, Y., Liu, B. Q., Tangrea, J., Sun, Y. H., Liu, F., Fraumeni, J. F., Jr., Zhang, Y. H., and Li, B. (1993). Nutrition intervention trials in Linxian, China: Supplementation with specific vitamin/mineral combinations, cancer incidence, and disease-specific mortality in the general population. *J. Natl. Cancer Inst.* **85,** 1483–1492.

Brown, M. S., and Goldstein, J. L. (1983). Lipoprotein metabolism in the macrophage: Implications for cholesterol deposition in atherosclerosis. *Annu. Rev. Biochem.* **52,** 223–261.

Burton, G. W., and Ingold, K. U. (1984). Beta-carotene: An unusual type of lipid antioxidant. *Science* **224,** 569–573.

Cathcart, M. K., Morel, D. W., and Chisolm, G. M. (1985). Monocytes and neutrophils oxidize low density lipoprotein making it cytotoxic. *J. Leukocyte Biol.* **38,** 341–350.

Consumer and Food Economics Institute (1982). "Composition of Foods: Fruits and Fruit Juices, Raw, Processed, Prepared," Handb. 8-9. U.S. Department of Agriculture, Washington, DC.

DeMaio, S. J., King, S. B., III, Lembo, N. J., Roubin, G. S., Hearn, J. A., Bhagavan, H. N., and Sgoutas, D. S. (1992). Vitamin E supplementation, plasma lipids and incidence of restenosis after percutaneous transluminal coronary angioplasty (PTCA). *J. Am. Coll. Nutr.* **11,** 131–138.

Eichholzer, M., Stahelin, H. B., and Gey, K. F. (1992). Inverse correlation between essential antioxidants in plasma and subsequent risk to develop cancer, ischemic heart disease and stroke, respectively: 12-year follow-up of the Prospective Basel Study. *In* "Free Radicals and Aging" (I. Emerit and B. Chance, eds.), pp. 398–410. Birkhaeuser, Basel.

Enstrom, J. E., Kanim, L. E., and Klein, M. A. (1992). Vitamin C intake and mortality among a sample of the United States population. *Epidemiology* **3,** 194–202.

Fogelman, A. M., Schechter, I., Seager, J., Hokom, M., Child, J. S., and Edwards, P. A. (1980). Malondialdehyde alteration of low density lipoproteins leads to cholesterol ester accumulation in human monocytemacrophages. *Proc. Natl. Acad. Sci. U.S.A.* **77,** 2214–2218.

Foote, C. S., Denny, R. W., Weaver, L., Chang, Y., and Peters, J. (1970). Quenching of singlet oxygen. *Ann. N.Y. Acad. Sci.* **171,** 139–148.

Fowler, S., Shio, H., and Haley, N. J. (1979). Characteristics of lipid laden aortic cells from cholesterol-fed rabbits. IV. Investigation of macrophage properties of aortic cell populations. *Lab. Invest.* **41,** 372–378.

Gaziano, J. M., Manson, J. E., Ridker, P. M., Buring, J. E., and Hennekens, C. H. (1990). Beta carotene therapy for chronic stable angina. *Circulation* **82** (4, Suppl. III), III.

Gaziano, J. M., Manson, J. E., Branch, L. G., LaMott, F., Colditz, G. A., Buring, J. E., and Hennekens, C. H. (1992). Dietary beta carotene and decreased cardiovascular mortality in an elderly cohort. *J. Am. Coll. Cardiol.* **19,** 377.

Gerrity, R. G. (1981). The role of monocyte in atherogenesis. I. Transition of blood-borne monocytes into foam cells in fatty lesions. *Am. J. Pathol.* **103,** 181–190.

Gey, K. F., and Puska, P. (1989). Plasma vitamins E and A inversely correlated to mortality from ischemic heart disease in cross-cultural epidemiology. *Ann. N.Y. Acad. Sci.* **570,** 254–282.

Gey, K. F., Brubache, G. B., and Stahelin, H. B. (1987a). Plasma levels of antioxidant vitamins in relation to ischemic heart disease and cancer. *Am. J. Clin. Nutr.* **45,** 1368–1377.

Gey, K. F., Stahelin, H. B., Puska, P., and Evans, A. (1987b). Relationship of plasma vitamin C to mortality from ischemic heart disease. *Ann. N.Y. Acad. Sci.* **498,** 110–123.

Gey, K. F., Stahelin, H. B., and Eichholzer, M. (1993). Poor plasma status of carotene and vitamin C is associated with higher mortality from ischemic heart disease and stroke: Prospective Basel Study. *Clin. Invest.* **71,** 3–6.

Gillian, R. E., Mandell, B., and Warbasse, J. R. (1977). Quantitative evaluation of vitamin E in the treatment of angina pectoris. *Am. Heart J.* **93,** 444–449.

Ginter, E. (1979). Decline of coronary mortality in the United States and vitamin C. *Am. J. Clin. Nutr.* **32,** 511–512.

GISSI-Prevenzione Investigators. (1999). Dietary supplementation with n-3 polyunsaturated fatty acids and vitamin E after myocardial infarction: Results of the GISSI-Prevenzione trial. *Lancet,* **354,** 447–445.

Goldstein, J. L., Ho, Y. K., Basu, S. K., and Brown, M. S. (1979). Binding site on macrophages that mediates uptake and degradation of acetylated low density lipoprotein, producing massive cholesterol deposition. *Proc. Natl. Acad. Sci. U.S.A.* **76,** 333–337.

Greenberg, E. R., Baron, J. A., Karagas, M. R., Stukel, T. A., Nierenberg, D. W., Stevens, M. M., Mandel, J. S., and Haile, R. W. (1996). Mortality associated with low plasma concentration of beta carotene and the effect of oral supplementation. *JAMA, J. Am. Med. Assoc.* **275,** 660–703.

Haeger, K. (1974). Long-time treatment of intermittent claudication with Vitamin E. *Am. J. Clin. Nutr.* **27,** 1179–1181.

Hennekens, C. H., and Buring, J. E. (1993). Observational evidence. *Ann. N.Y. Acad. Sci.* **703,** 18–24.

Hennekens, C. H., Buring, J. E., and Peto, R. (1994). Antioxidant vitamins—benefits not yet proven. *N. Engl. J. Med.* **330,** 1080–1081.

Hennekens, C. H., Buring, J. E., Manson, J. E., Stampfer, M., Rosner, B., Cook, N. R., Belanger, C., LaMotte, F., Gaziano, J. M., Ridker, P. M., Willett, W., and Peto, R. (1996). Lack of effect of long term supplementation with beta carotene on the incidence of malignant neoplasms and cardiovascular disease. *N. Engl. J. Med.* **334,** 1145–1149.

Henriksen, T., Mahoney, E. M., and Steinberg, D. (1981). Enhanced macrophage degradation of low density lipoprotein previously incubated with endothelial cells: Recognition by receptors for acetylated low density lipoprotein. *Proc. Natl. Acad. Sci. U.S.A.* **78,** 6499–6503.

Hessler, J. R., Morel, D. W., James, L. J., and Chisolm, G. M. (1983). Lipoprotein oxidation and lipoprotein-induced cytotoxicity. *Arteriosclerosis* **3,** 215–222.

Ingold, K. U., Webb, A. C., Witter, D., Burton, G. W., Metcalf, T. A., and Muller, D. P. R. (1987). Vitamin E remains the major lipid-soluble, chain-breaking antioxidant in human plasma even in individuals suffering severe vitamin E deficiency. *Arch. Biochem. Biophys.* **259,** 224–225.

Kardinaal, A. F. M., Kok, F. J., Ringstad, J., Gomez-Aracena, J., Mazaev, V. P., Kohlmeier, L., Martin, B. C., Aro, A., Kark, J. D., Delgado-Rodriguez, M., Riemersma, R. A., Huttunen, J. K., and Martin-Moreno, J. M. (1993). Antioxidants in adipose tissue and risk of myocardial infarction: The EURAMIC study. *Lancet* **342,** 1379–1384.

Kok, F. J., de Bruijn, A. M., Vermeeren, R., Hofman, A., VanLaar, A., de Bruin, M., Hermus, R. T. J., and Valkenberg, H. A. (1987). Serum selenium, vitamin antioxidants and cardiovascular mortality: A 9 year follow-up study in the Netherlands. *Am. J. Clin. Nutr.* **45,** 462–468.

Lapidus, L., Anderson, H., Bengtson, C., and Bosceus, I. (1986). Dietary habits in relation to incidence of cardiovascular disease and death in women: A 12 year follow-up of participants in the study of women in Gothenberg, Sweden. *Am. J. Clin. Nutr.* **44,** 444–448.

Lemoyne, M., Van Gossum, A., Kurian, R. T., Ostro, M., Axler, J., and Jeejeebhoy, K. N. (1987). Breath pentane analysis as an index of lipid peroxidation: A functional test of vitamin E status. *Am. J. Clin. Nutr.* **46,** 267–272.

Livingston, P. D., and Jones, C. (1958). Treatment of intermittent claudication with vitamin E. *Lancet* **2,** 602–604.

Manson, J. E., Stampfer, M. J., Willett, W. C., Colditz, G. A., Rosner, B., Speizer, F. E., and Hennekens, C. H. (1991). A prospective study of antioxidant vitamins and incidence of coronary heart disease in women. *Circulation* **84** (Suppl. II).

Manson, J. E., Stampfer, M. J., Willett, W. C., Colditz, G. A., Rosner, B., Speizer, F. E., and Hennekens, C. H. (1992). A prospective study of vitamin C and incidence of coronary heart disease in women. *Circulation* **85,** 865.

Marcus, A. J., Silk, S. T., Safier, L. B., and Ullman, H. L. (1977). Superoxide production and reducing activity in human platelets. *J. Clin. Invest.* **59,** 149–158.

Morris, D. L., Kritchevsky, S. B., and Davis, C. E. (1993). Serum carotenoids and coronary heart disease in the Lipid Research Clinics Coronary Primary Prevention Trial. *Circulation* **87,** 2 (abstr.).

Niki, E., Saito, T., and Kamiya, Y. (1983). The role of vitamin C as an antioxidant. *Chem. Lett.,* pp. 631–632.

Nishigaka, I., Hagihara, M., Tsunekawa, H., Maseki, M., and Yagik, K. (1981). Lipid peroxide levels of serum lipoprotein fractions of diabetic patients. *Biochem. Med.* **25,** 373–378.

Omenn, G. S., Goodman, G. E., Thornquist, M. D., Blames, J., Cullen, M. R., Glass, A., Keogh, J. P., Meyskens, F. L., Valanis, B., Williams, J. H., Barnhart, S., and Hammer, S. (1996). Effects of a combination of beta carotene and vitamin A on lung cancer and cardiovascular disease. *N. Engl. J. Med.* **334,** 1150–1155.

Packer, J. E., Slater, T. F., and Wilson, R. L. (1979). Direct observation of a free radical interaction between vitamin E and vitamin C. *Nature (London)* **278,** 737–738.

Prince, M. R., LaMuraglia, G. M., and MacNichol, E. F. (1988). Increased preferential absorption in human atherosclerotic plaque with oral beta carotene: Implications for laser endarterectomy. *Circulation* **78,** 338–344.

Quinn, M. T., Parthasarathy, S., and Steinberg, D. (1985). Endothelial cell-derived chemotactic activity for mouse peritoneal macrophages and the effects of modified forms of low density lipoprotein. *Proc. Natl. Acad. Sci. U.S.A.* **82,** 5949–5953.

Quinn, M. T., Parthasarathy, S., and Steinberg, D. (1987). Oxidatively modified low density lipoproteins: A potential role in recruitment and retention of monocyte/macrophages during atherogenesis. *Proc. Natl. Acad. Sci. U.S.A.* **84,** 2995–2998.

Ramirez, J., and Flowers, N. C. (1980). Leukocyte ascorbic acid and its relationship to coronary heart disease in man. *Am. J. Clin. Nutr.* **33,** 2079–2087.

Rapola, J. M., Virtamo, J., Haukka, J. K., Heinonen, O. P., Albanes, D., Taylor, P. R., and Huttunen, J. K. (1996). Effect of vitamin E and beta carotene on the incidence of angina pectoris: A randomized, double-blind, controlled trial. *JAMA, J. Am. Med. Assoc.* **275,** 693–698.

Riemersma, R. A., Wood, D. A., Macintyre, C. C. H., Elton, R. A., Gey, K. F., and Oliver, M. F. (1989). Low plasma vitamin E and C increased risk of angina in Scottish men. *Ann. N.Y. Acad. Sci.* **570,** 291–295.

Riemersma, R. A., Oliver, M., Elton, M. A., Alfthan, G., Vartiainen, E., Salo, M., Rubba, P., Mancici, M., Georgi, H., Vuilleumier, J., and Gey, K. F. (1990). Plasma antioxidants and coronary heart disease: Vitamins C and E and selenium. *Eur. J. Clin. Nutr.* **44,** 143–150.

Riemersma, R. A., Wood, D. A., Macintyre, C. H. H., Elton, R. A,. Gey, K. F., and Oliver, M. F. (1991). Risk of angina pectoris and plasma concentrations of vitamins A, C, E, and carotene. *Lancet* **337,** 1–5.

Rimm, E. B., Stampfer, M. J., Ascherio, A., Giovannucci, E., Colditz, G. A., and Willett, W. C. (1993). Dietary intake and risk of coronary heart disease among men. *N. Engl. J. Med.* **328,** 1450–1456.

Salonen, J. T., Salonen, R., Penttila, I., Herranen, J., Jauhiainen, M., Kantola, M., Lappetelainen, R., Maenpaa, P., Alfthan, G., and Puska, P. (1985). Serum fatty acids, apolipoproteins, selenium and vitamin antioxidants and risk of death from coronary artery disease. *Am. J. Cardiol.* **56,** 226–231.

Salonen, J. T., Yla-Herttuala, S., Yamamoto, R., Butler, S., Korpela, H., Salonen, R,. Nyyssonen, K., Palinski, W., and Witztum, J. L. (1992). Autoantibody against oxidized LDL and progression of carotid atherosclerosis. *Lancet* **339,** 883–887.

Salonen, J. T., Nyyssonen, K., Parviainen, M. T., Kantola, M., Korpela, H., and Salonen, R. (1993). Low plasma beta carotene, vitamin E and selenium levels associated with accelerated carotid atherogenesis in hypercholesterolemic eastern Finnish men. *Circulation* **87,** 1.

Salvini, S., Hunter, D. J., Sampson, L., Stampfer, M. J., Colditz, G. A., Rosner, B., Willett, W. C. (1989). Food-based validation of a dietary questionnaire: The effects of week-to-week variation in food consumption. *Int. J. Epidemiol.* **18,** 858–867.

Saran, M., Michael, C., and Bors, W. (1990). Reaction of NO with O_2. Implications for the action of endothelium-derived relaxing factor (EDRF). *Free Radical Res. Commun.* **10,** 221–226.

Schaffner, T., Taylor, K., Bartucci, E. J., Fisher-Dzoga, K., Beeson, J. H., Glagov, S., and Wissler, R. W. (1980). Arterial foam cells with distinctive immunomorphologic and histochemical features of macrophages. *Am. J. Pathol.* **100,** 57–73.

Slater, T. F. (1984). Free radical mechanisms in tissue injury. *Biochem. J.* **222,** 1–15.

Som, S., Raha, C., and Chatterje, I. B. (1983). Ascorbic acid: A scavenger of superoxide radical. *Acta Amino Enzymol.* **5,** 243–250.

Southorn, P. A., and Powis, G. (1988a). Free radicals in medicine. I. Chemical nature and biologic reactions. *Mayo Clin. Proc.* **63,** 381–389.

Southorn, P. A., and Powis, G. (1988b). Free radicals in medicine. II. Involvement in human disease. *Mayo Clin. Proc.* **63,** 390–408.

Stampfer, M. J., Hennekens, C. H., Manson, J. E., Colditz, G. A., Rosner, B., and Willett, W. C. (1993). A prospective study of vitamin E consumption and risk of coronary disease in women. *N. Engl. J. Med.* **328,** 1444–1449.

Steinberg, D., Parthasarathy, S., Carew, T., Khoo, J., and Witztum, J. (1989). Beyond cholesterol-modifications of low-density lipoprotein that increase its atherogenicity. *N. Engl. J. Med.* **320,** 915–924.

Stephens, N. G., Parsons, A., Schofield, P. M., Kelly, F., Cheeseman, Mitchinson, M. J., and Brown, M. J. (1996). Randomized controlled trial of vitamin E in patients with coronary disease: Cambridge Heart Antioxidant Study (CHAOS). *Lancet* **347,** 781–786.

Street, D. A., Comstock, G. W., Salkeld, R. M., Schuep, W., and Klag, M. (1991). A population based case-control study of serum antioxidants and myocardial infarction. *Am. J. Epidemiol.* **134,** 719–720.

Stringer, M. D., Gorog, P. G., Freeman, A., and Kakkar, V. V. (1989). Lipid peroxides and atherosclerosis. *Br. Med. J.* **298,** 281–284.

Suematsu, T., Kamada, T., Abe, H., Kikuchi, S., and Yagi, K. (1977). Serum lipoperoxide level in patients suffering from liver disease. *Clin. Chim. Acta* **79,** 267.

Verlangieri, A. J., Kapeghian, J. C., El-Dean, S., and Bush, M. (1985). Fruit and vegetable consumption and cardiovascular disease mortality. *Med. Hypotheses* **16,** 7–15.

Warso, M. A., and Lands, W. E. (1985). Presence of lipid hydroperoxide in human plasma. *J. Clin. Invest.* **75,** 667–671.

Willett, W. C., and Stampfer, M. J. (1986). Total energy intake: Implications for epidemiologic analyses. *Am. J. Epidemiol.* **124,** 17–27.

Willett, W. C., Sampson, L., Stampfer, M. J., Rosner, B., Bain, C., Wistch, J., Hennekens, C. H., Speizer, F. E. (1985). Reproducibility and validity of a semiquantitative food frequency questionnaire. *Am. J. Epidemiol.* **122,** 51–65.

Williams, H. T. G., Fenna, D., and MacBeth, R. A. (1971). Alpha tocopherol in the treatment of intermittent claudication. *Surg., Gynocol. Obstet.* **132,** 662–666.

Yagi, K. (1976). A simple fluorometric assay for lipoperoxide in blood plasma. *Biochem. Med.* **15,** 212–216.

Yagi, K. (1984). Increased serum lipid peroxides initiate atherogenesis. *BioEssays* **1,** 58–60.

Yagi, K., Ohkawa, H., Onishi, N., Yamashita, M., and Nakashima, T. (1981). Lesion of aortic intima caused by intravenous administration of linoleic acid hydroperoxide. *J. Appl. Biochem.* **3,** 58–65.

VITAMINS AND HORMONES, VOL. 58

Melatonin Receptors and Ligands

PAULA A. WITT-ENDERBY* AND PUI-KAI LI†

Department of Pharmacology and Toxicology and †Department of Medicinal Chemistry, Graduate School of the Pharmaceutical Sciences, School of Pharmacy, Duquesne University, Pittsburgh, Pennsylvania 15282

The goal of the article is to provide a clearer understanding of how melatonin and its related analogs interact with melatonin receptors with the hope of developing important tools and agents of significant clinical and scientific importance. The review provides a compilation of the currently published melatonergic ligands and their relative affinities for melatonin receptors and discusses the importance of developing reversible, high-affinity, and subtype selective melatonin receptor antagonists. In addition, the review discusses the utility of developing high-affinity *charged* melatonergic ligands and *irreversible* ligands. Finally, the review discusses some of the problems associated with the current models used to study receptor pharmacology and function. As the availability of tools increases in the melatonin receptor field, a great body of knowledge is also gained about the structure of the melatonin receptor and the role that specific melatonin receptor subtypes have in physiologic processes. Further design, synthesis, and application of melatonergic ligands will lead us to a clearer understanding of the role that melatonin and its receptors play in humans.

I. Introduction

The hormone melatonin is unique for many reasons: (1) Its synthesis and release is under the control of both circadian and diurnal cues (Re-

0083-6729/00 $30.00

iter, 1991; Sack *et al.,* 1998); (2) the hormone is very potent, producing physiologic effects in the picomolar range (Morgan *et al.,* 1994); (3), its physiologic effects are vast (Brzezinski, 1997; Dawson and van den Heuvel, 1998); and (4) the molecule can act through receptors (Morgan *et al.,* 1994; Reppert *et al.,* 1994, 1995a; Witt-Enderby and Dubocovich, 1996; Dubocovich *et al.,* 1998a) and independent of receptors (Benítez-King and Anton-Tay, 1993). The goal of this review is to provide a clearer understanding of how melatonin and its related analogs interact with melatonin receptors with the hope of developing important tools and agents of significant clinical and scientific importance.

A. Synthesis of Melatonin

N-acetyl-5-methoxytryptamine, better known as melatonin, is a hormone produced mainly in the pineal gland but also in the retina (Reiter, 1991, 1998; Brzezinski, 1997). The pathway by which melatonin is synthesized is similar for both the pineal gland and the retina in that trytophan is ultimately converted to melatonin. Tryptophan, taken in from the diet, is converted via an intermediate step into serotonin, which is then converted to *N*-acetyl-serotonin and ultimately to melatonin by the enzymes *N*-acetyl-transferase and 5-hydroxyindole-*O*-methyl transferase, respectively (Reiter, 1998; Li *et al.,* 1998). The synthesis and secretion of melatonin is stimulated by darkness as detected in the retina and inhibited by light. The onset of darkness stimulates the retina to directly synthesize and secrete melatonin locally and/or send signals to the suprachiasmatic nucleus of the hypothalamus (Li *et al.,* 1998; Reiter, 1998). Through a neural pathway, norepinephrine is released at the pinealocytes of the pineal gland, which contains α_1- and β_1-adrenergic receptors. Binding of norepinephrine to these receptors increases intracellular concentrations of cAMP within the pinealocyte, which then increases the production of *N*-acetyl-transferase, the rate limiting enzyme in melatonin synthesis (Reiter, 1991; Li *et al.,* 1998). As a result, the increase in *N*-acetyl-transferase allows for an increase in the synthesis of melatonin, which then passively diffuses into the bloodstream (Reiter, 1998).

B. Physiologic Effects of Melatonin

Melatonin can affect a number of processes once released into the bloodstream. Studies have indicated a hypnotic effect caused by melatonin when given orally that is beneficial for sleep disorders (Wurtman and Zhdanova, 1995; Shochat *et al.,* 1998). In addition, melatonin entrains circadian rhythms (Cassone *et al.*, 1993; Cassone and Natesan, 1997; Benloucif *et al.*, 1997; Sack *et al.*, 1998) and, therefore, helps people with jet lag or affective disorders (Sack *et al.*, 1998; Brzezinski, 1997). Melatonin influences reproduction and sexual maturation (Silman, 1991; Morgan *et al.*, 1994; Maywood and Hastings, 1995; Reiter, 1998; Kripke *et al.*, 1998) both centrally and locally. The central effects of melatonin on reproduction have been widely studied (Reiter, 1998); a majority of the studies suggest an inhibitory effect of melatonin on the GnRH pulse generator (Silman, 1991; Messager *et al.*, 1996; Reiter, 1998; Kripke *et al.*, 1998). However, more recent evidence suggests that melatonin may have an effect at the level of the gonads (Ayre *et al.*, 1992; Yie *et al.*, 1995; Murayama *et al.*, 1997).

Melatonin has effects on cardiovasculature (Viswanathan *et al.*, 1990; Capsoni *et al.*, 1994; Krause *et al.*, 1995; Geary *et al.*, 1997; Doolen *et al.*, 1998) and may regulate renal function (Song *et al.*, 1996, 1997). In addition, melatonin may play a protective role in cancer (Ying *et al.*, 1993; Molis *et al.*, 1994; Cos and Sanchez-Barcelo, 1995; Cos *et al.*, 1996) by lengthening cell cycle times (Cos *et al.*, 1996) or decreasing the transcription of the estrogen receptor gene (Molis *et al.*, 1994). Melatonin may also play a role in brain function (Avery *et al.*, 1998); however, the mechanisms underlying such functions are not known. Presently, it is known that melatonin exerts a strong influence on cellular morphology similar to that of neurons (Benítez-King and Anton-Tay, 1993), which may have important clinical implications for the treatment of diseases associated with the brain.

C. Melatonin and Its Function within the Cell

Melatonin is highly lipophilic and, therefore, can enter fluids, tissues, and cells throughout the body (Reiter, 1991; Menendez-Pelaez and Reiter, 1993; Finocchiaro and Gliken, 1998) and accumulate up to 30 times its concentration within the blood (Benítez-King *et al.*, 1990; Reiter, 1991). Also, it has been proposed that cells can actively uptake and accumulate melatonin (Menendez-Pelaez and Reiter, 1993; Benítez-King and Anton-Tay, 1993). Therefore, it is possible if not probable that melatonin can achieve pharmacologic levels at the level of the cell. This has important physiologic implications because higher concentrations of melatonin may be able to activate signal transduction paths other than the cAMP-dependent cascade. In addition, melatonin may directly bind to intracellular proteins located in either the cytosol (Benítez-King and Anton-Tay, 1993; Starkey, 1996; Anton-Tay *et al.*, 1998a; Betthai *et al.*, 1998) or nucleus (Wiesenberg *et al.*, 1995; Missbach *et al.*, 1996) trig-

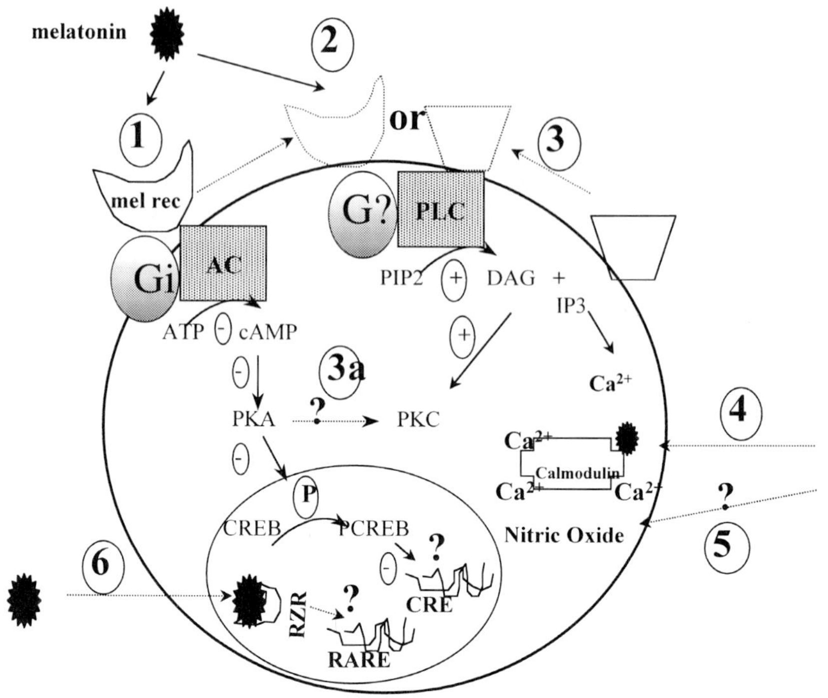

FIG. 1. Proposed mechanisms of melatonin receptor signal transduction at the level of the cell. As shown, melatonin could achieve diversity of action within the body by utilizing multiple paths for transducing its signal. In mechanism (1) is established through many reports that activation of melatonin receptors by melatonin results in a decrease in adenylyl cyclase activity, resulting in a decrease in cAMP levels. The decrease in cAMP decreases PKA activity and consequently decreases the activation or phosphorylation of cyclic AMP response element binding (CREB) protein. The consequences of CREB protein inactivation have not been established; however, it may decrease the transcription of genes containing cyclic AMP response elements (CREs). As shown in mechanisms 2, 3, and 3a, melatonin can also stimulate the PLC-dependent pathway. However, how melatonin activates this pathway is unclear. As shown in mechanism (2), the melatonin receptor (shown as the dotted receptor of same shape as the melatonin receptor) may couple directly to PLC and increase the production of phosphoinositides. Also, in other models, melatonin does not stimulate directly this pathway but merely enhances the activity of this pathway stimulated by another receptor class [dotted receptor of a different shape; mechanism (3)]. The ability of melatonin to "cross-talk" to this signal transduction cascade is not yet known but may occur through kinases. PKA and PKC [mechanism (3a)]. Besides acting through the receptor, melatonin can also act independently of receptors by direct binding to cytosolic proteins [mechanism (4) or (5)] or nuclear proteins [mechanism (6)].

gering yet other signal transduction components (Fig. 1, mechanisms 4, 5, and 6, respectively). The ability of melatonin to act through receptors or independent of receptors may contribute to the diversity of melatonin action *in vivo.*

D. Melatonin Receptors

As mentioned previously, melatonin may produce its effects via receptors. Melatonin receptors are G-protein-coupled receptors characterized by a seven transmembrane-spanning domain. Melatonin receptors denoted as the mt_1 subtype [or formerly known as the Mel_{1a} receptor (Reppert *et al.,* 1994)] are so named because this receptor has a known molecular structure but does not have well-defined functional characteristics in native tissue. The ML_2 melatonin receptor [formerly known as the Mel_{1b} receptor (Reppert *et al.,* 1995a)] is so named because this receptor has a defined function in native tissue. The MT_3 melatonin receptor (formerly known as ML_2) is so named because this receptor has been defined functionally and pharmacologically but has an unknown molecular structure (Dubocovich *et al.,* 1998b). The mt_1 melatonin receptor is expressed in the suprachiasmatic nucleus of the hypothalamus (SCN) (Reppert *et al.,* 1994; Kokkola and Laitinen, 1998), and in the basolateral membranes of the proximal tubules in the kidneys (Song *et al.,* 1997). The MT_2 melatonin receptor is expressed in both the brain and retina (Reppert *et al.,* 1995a; Dubocovich *et al.,* 1997; Dubocovich *et al.,* 1998a) and possibly in vascular smooth muscle of rat caudal artery (Doolen *et al.,* 1998).

Little is known about the signal transduction cascades involved with melatonin receptor signaling though research is progressing. Studies indicate that melatonin receptors are coupled to an inhibition of adenylyl cyclase (Niles and Hashemi, 1990) resulting in decreases in intracellular cAMP levels via pertussis toxin-sensitive G proteins (Reppert *et al.,* 1994, 1995a; Morgan *et al.,* 1994; Capsoni *et al.,* 1994; Witt-Enderby and Dubocovich, 1996; Vanecek and Watanabe, 1998). The attenuation of cAMP results in a decrease in the activation of PKA (Hazlerigg *et al.,* 1991; Witt-Enderby *et al.,* 1998) and consequently a decrease in the phosphorylation of CREB (Ginty *et al.,* 1993; A. McNulty *et al.,* 1996; S. McNulty *et al.,* 1994; Witt-Enderby *et al.,* 1998; von Gall *et al.,* 1998; Kopp *et al.,* 1999) (Fig. 1, mechanism 1).

Besides the cAMP-dependent cascade, melatonin also stimulates phosphoinositide hydrolysis (Eison and Mullins, 1993; Popova and Dubocovich, 1995; Fig. 1, mechanism 2) and activates PKC-dependent pathways (Godson and Reppert, 1997; McArthur *et al.,* 1997; Anton-Tay *et al.,* 1998b; Betthai *et al.,* 1998; Barrett *et al.,* 1998; Ross *et al.,* 1998; Fig. 1, mechanism 3). However, the mechanisms underlying these effects on PKC are not clear. Recently, reports have demonstrated an effect of melatonin on nitric oxide synthesis (Starkey, 1996; Betthai *et al.,* 1998); however, these mechanisms are not yet established.

As previously mentioned, various subtypes of melatonin receptors have been cloned and studies are beginning to demonstrate their unique tissue distribution and functional roles throughout the body (Song *et al.,* 1997; Dubocovich *et al.,* 1997; Doolen *et al.,* 1998; Dubocovich *et al.,* 1998a; Clemens *et al.,* unpublished results). Thus, it is becoming increasingly more important to develop high-affinity and subtype-selective melatonin receptor antagonists to determine whether certain physiologic effects of melatonin are mediated through specific receptor subtypes. Presently, there are reports of putative MT_2 melatonin receptor antagonists (Dubocovich *et al.,* 1997; Teh and Sugden, 1998) and their ability to antagonize functional effects in various tissues has also been demonstrated (Dubocovich *et al.,* 1997, 1998a; Doolen *et al.,* 1998).

E. Regulation of Melatonin Receptors

Even though the half-life of melatonin within the body is relatively short (10–40 min), the levels of melatonin peak and persist for about 8 h each night (Reiter, 1991). Therefore, each night melatonin receptors are exposed to melatonin for prolonged periods of time. Following prolonged exposure to an agonist, regulation of receptors can occur. This seems likely for melatonin receptors as studies have shown that melatonin receptors are highly regulated and display different periods of "sensitivity" throughout the course of a 24-hour day and following melatonin exposure. For example, the density and affinity of melatonin receptors change throughout the 24-hour day and following melatonin exposure (Tenn and Niles, 1993; Gauer *et al.,* 1994; Morgan *et al.,* 1994; Song *et al.,* 1996; Witt-Enderby *et al.,* 1998). Perhaps, critical to the normal functioning of melatonin within the body is its ability to alter the function of its own receptors. One form of regulation of receptors, known as *receptor desensitization,* is a process by which a receptor becomes refractory to an agonist over time. This "turning off" of receptors may be the result of receptor/G-protein uncoupling, or receptor internalization and/or receptor down-regulation. Receptors may use one form of the aforementioned modes of regulation, two, or all three (Kenakin, 1997; Carman and Benovic, 1998).

Currently, it is unclear as to the mode of regulation utilized by melatonin receptors.

It has been difficult to conclude with certainty which form of regulation melatonin receptors use due to a lack of high-affinity, subtype selective, and reversible antagonists. The use of the *agonist* radioligand 2-[^{125}I]-iodomelatonin has been beneficial for elucidating melatonin receptor location; however, due to its agonist nature its use in studying melatonin receptor regulation has been difficult. Agonists bind to receptors coupled to their respective G proteins. Thus, if a receptor becomes uncoupled from its G protein, then an agonist will bind with low affinity to its receptor (Carman and Benovic, 1998). This is an important concept to understand in the melatonin receptor field. For example, a decrease in total 2-[^{125}I]-iodomelatonin binding could be misinterpreted as melatonin receptor down-regulation when, in fact, the decrease in 2-[^{125}I]-iodomelatonin binding may actually be due to a loss of affinity of the radioligand for its receptor due to receptor/G-protein uncoupling. As for assessing whether or not melatonin receptors internalize following agonist exposure, this form of regulation cannot be studied because of the lipophilic nature of 2-[^{125}I]-iodomelatonin (see Fig. 2). Thus, until a ligand is developed for assessing this form of receptor desensitization, whether or not melatonin receptors internalize remains a mystery.

1. *Importance of Developing Tools to Study Melatonin Receptor Regulation*

Internalization is the process by which receptors are sequestered and removed from the membrane surface following agonist exposure. This form of desensitization is utilized by many other G-protein-coupled receptors including muscarinic cholinoceptors (Galper *et al.*, 1982) and β-adrenoceptors (von Zastrow and Kobilka, 1994). Whether or not melatonin receptors internalize following melatonin exposure has not been determined due to the unavailability of specific probes, that is, charged melatonin receptor ligands that are unable to penetrate membranes and bind only to surface melatonin receptors (Fig. 2). Recently, however, a charged melatonin receptor ligand has been developed (Li *et al.*, 1997b) and radiolabeled (Chu and Li, 1998). Even though the affinity of this ligand for melatonin receptors is low, it is a step in the right direction for the future development of higher affinity and charged melatonergic ligands.

Therefore, critical to the advancement of the melatonin receptor field is the development of high-affinity and subtype selective *antagonists,*

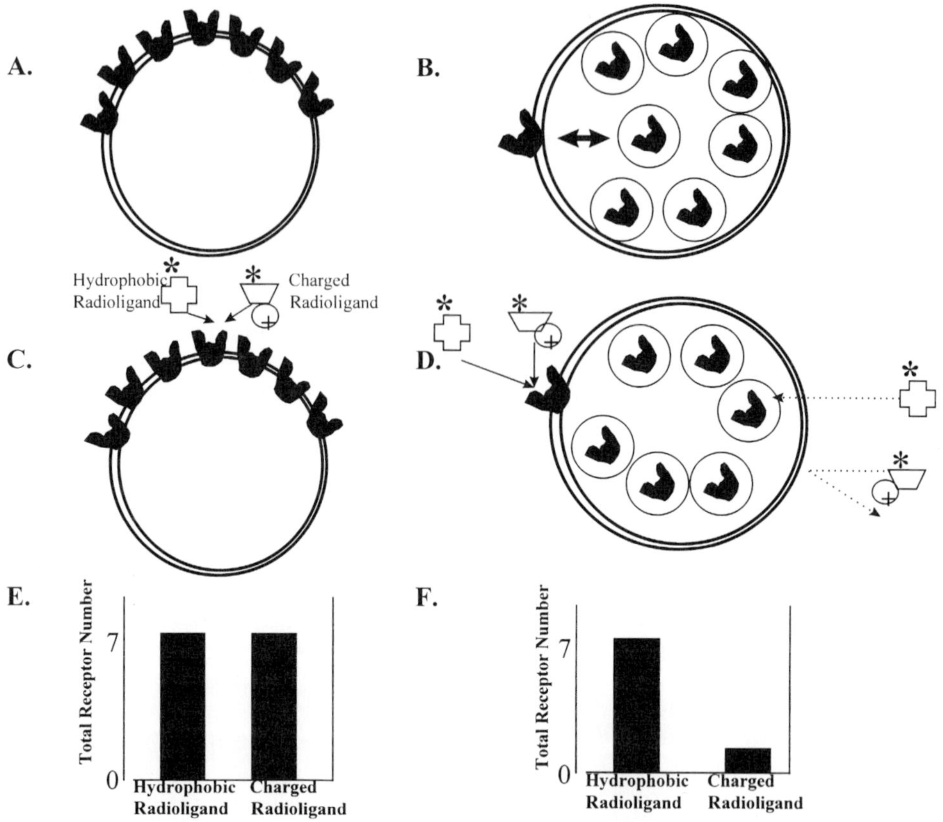

FIG. 2. Rationale behind the design of charged melatonergic ligands. (A) Cell with all of its plasma membrane-bound receptors at the surface. (B) Following agonist exposure, internalization of receptors has occurred where a majority of the receptors are contained within vesicles inside the cell. The double-sided arrow indicates that the receptors could be "recycled" back to the membrane surface, demonstrating that internalization is a "reversible" process. (C) Both hydrophobic and charged radioligands can bind to receptors located at the surface. (D) However, following internalization, the charged radioligand will not be able to bind to the receptors within the cell because the charge prevents the radioligand from penetrating cellular membranes. In contrast, the hydrophobic radioligand will be able to bind to receptors both at the surface and within the cell. (E) The determination of total receptor number on cells presented in part A using either radioligand would result in a similar total receptor number. (F) However, if internalization and no down-regulation of receptors occurs as depicted in part B, then a decrease in charged radioligand binding with no change in hydrophobic radioligand binding would result.

that is, antagonists for assessing melatonin receptor down-regulation and charged melatonin receptor antagonists for assessing melatonin receptor internalization.

2. *Importance of Developing Tools by Which to Study Melatonin Receptor Subtype Function and Regulation*

The preceding discussion focused on the development of *reversible* melatonin receptor antagonists; however, the development of high-affinity and *irreversible* melatonin ligands also has an important place in the study of ligand–receptor interactions. Besides providing qualitative information about the receptor (i.e., molecular size, structure, glycosylation states, receptor reserves) (Birdsall *et al.,* 1979; Burgen *et al.,* 1974; Curtis *et al.,* 1989; Lohse *et al.,* 1986), irreversible melatonergic ligands could also aid in the determination of the domains within the receptor important for ligand binding.

In the melatonin receptor field, affinity probes are in an early yet promising stage of development. Presently, two melatonergic affinity ligands have been developed: *N*-bromoacetyl-2-iodo-5-methoxytryptamine (Laudon and Zisapel, 1991; Nelson *et al.,* 1993) and *N*-[2-(2-bromoacetyl-7-methoxynaphthyl)ethyl]propionamide (BMNEP) (Witt-Enderby *et al.,* 1997). These affinity probes have provided valuable qualitative information about melatonin receptors. In addition, the more recently synthesized affinity ligand, BMNEP, may be an important development for the study of MT_2 melatonin receptors as BMNEP alkylates MT_2 melatonin receptors with high affinity (Witt-Enderby *et al.,* 1997). Recently, it has been determined that BMNEP acts as a full agonist at MT_2 melatonin receptors resulting in their permanent activation while having (at best) partial agonist activity at the mt_1 receptor (Fig. 3). In addition, the radiolabeled version of BMNEP, that is, $[^{14}C]$BMNEP (synthesis to be reported elsewhere) results in the visualization of the MT_2 melatonin receptor via affinity gel electrophoresis (Fig. 4). The development of BMNEP could lead to important future developments because (1) BMNEP binds irreversibly and permanently "activates" the MT_2 melatonin receptor, which could lead to elucidation of the domains within the MT_2 melatonin receptor involved in its activation by agonist; (2) BMNEP discriminates between the mt_1 and MT_2 melatonin receptors *functionally,* which could lead to the development of selective melatonin receptor ligands; and (3) $[^{14}C]$BMNEP allows for the "first time" visualization of the human MT_2 melatonin receptor. Thus, use of this compound or development of future analogs of BMNEP could lead to great advances in understanding MT_2 melatonin receptor function within the body.

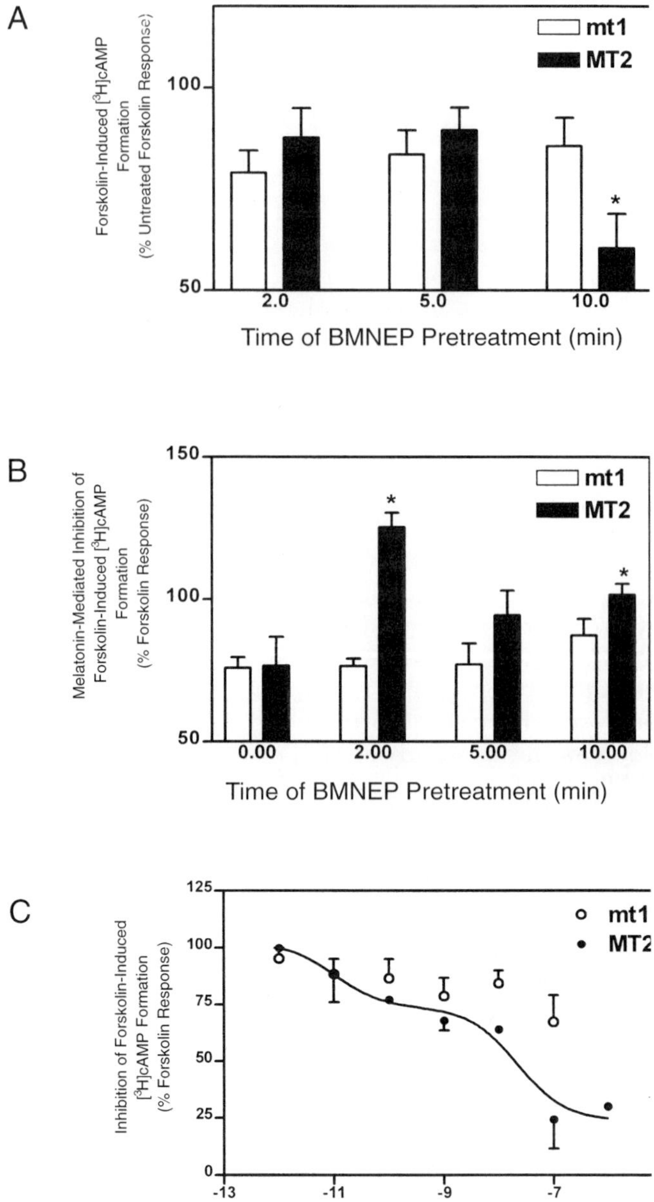

FIG. 3. Permanent activation of MT_2 melatonin receptors by BMNEP. CHO cells expressing either the mt_1 or MT_2 cells were incubated with 1 n*M* BMNEP for 0–10 min at 25°C. Following the incubation with BMNEP, the ability of forskolin to stimulate cAMP accumulation or melatonin to inhibit forskolin-induced cAMP formation was assessed in

II. Structure–Affinity Relationships between Melatonergic Ligands with Melatonin Receptors

With the continual development and synthesis of melatonergic analogs, a vast amount of knowledge about the structure–affinity relationships between these ligands and melatonin receptors is emerging. This section provides a compilation of the currently published melatonergic ligands and their relative affinities for melatonin receptors. In an attempt to summarize the conclusions from these published reports, generic structures containing different nuclear cores have been created. Much of the reported data on the synthesis of melatonergic ligands have centered around these basic structures where modifications at the various R groups have been made. However, it is important to mention that these structures are not all inclusive of all of the reported melatonergic ligands synthesized and the binding affinities of these analogs are expressed relative to the affinity of melatonin generated for the receptor in that particular model system.

Analogs containing the indole nuclear core (compounds 1–37) are described in Tables I through V.

Table I Conclusion: The methoxy group (R_1) and the ethyl amide group (R_3) are critical for melatonin receptor affinity. Changing the methoxy group (R_1) to a H, OH, alkyl, or halogen or other alkoxy groups results in a decrease in receptor affinity.

---

these cells. (A) In BMNEP-treated cells, the forskolin response when compared to vehicle-treated cells was significantly ($P < 0.05$) reduced in MT_2 cells, suggesting that the MT_2 melatonin receptor is activated and causing an inhibition of adenylyl cyclase. (B) In vehicle-treated cells, melatonin (1 n*M*) inhibited forskolin-induced cAMP accumulation. However, in cells pretreated with BMNEP for 2 and 10 min, the ability of melatonin to inhibit forskolin-induced cAMP accumulation was significantly ($P < 0.05$) attenuated in MT_2 cells compared to untreated cells. At first glance, these data could suggest that BMNEP has no activity at the MT_2 receptor. However, as shown in part A, the forskolin response in BMNEP-treated cells is already attenuated. Thus, it could be that the MT_2 receptor is already maximally activated by BMNEP. To determine this possibility, the ability of BMNEP (1 p*M* to 1 μ*M*) to inhibit forskolin-induced cAMP accumulation was assessed in mt_1-CHO and MT_2-CHO cells using methods described previously (Witt-Enderby and Dubocovich, 1996). (C) As shown, BMNEP inhibited forskolin-induced cAMP accumulation in MT_2-CHO cells with a maximal inhibition (~70%) occurring at 100 n*M* and with high potency of [IC_{50} = 1.3 n*M* (0.35–4.8 n*M*), $n = 3$]. In contrast, the ability of BMNEP to inhibit forskolin-induced cAMP accumulation in mt_1-CHO cells was variable and thus no potency could be calculated. Thus, BMNEP, at mt_1 melatonin receptors, acted as either an antagonist or a partial agonist. The results from these experiments indicate that BMNEP is a full agonist at MT_2 receptors and may possibly be an antagonist/partial agonist at the mt_1 receptor.

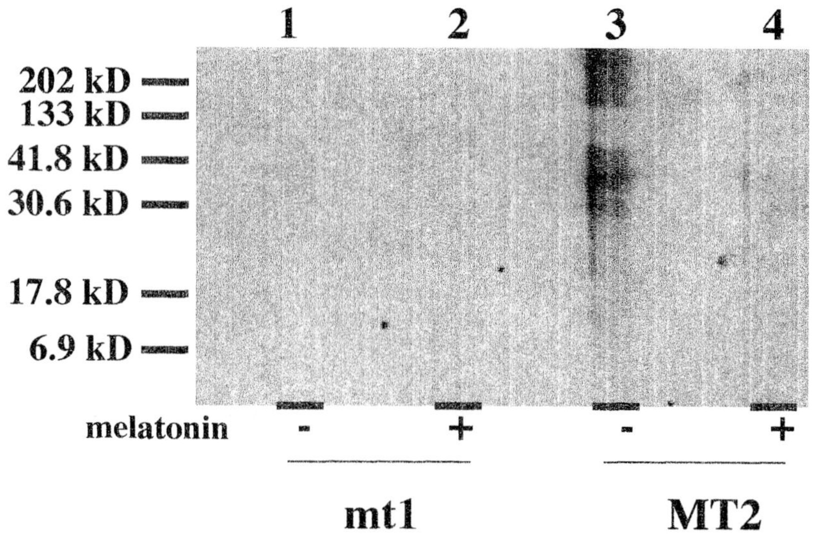

FIG. 4. Affinity gel electrophoresis of melatonin receptors following incubation with [^{14}C]BMNEP. To further characterize BMNEP at MT_2 melatonin receptors, BMNEP was radiolabeled with [^{14}C] (procedure for radiolabeling will be published elsewhere). Cell membranes were prepared from CHO cells expressing the neomycin-resistant plasmid alone (neo-CHO)or from cells expressing either the mt_1 melatonin receptor (mt_1-CHO) or the MT_2 melatonin receptor (MT_2-CHO) as already described (Witt-Enderby and Dubocovich, 1996). Cell membranes (~50 μg) were added to reaction tubes containing [^{14}C]BMNEP (1 n*M*) in the absence or presence (nonspecific binding) of melatonin (1 μ*M*). Cells were incubated for 10 min at 25°C followed by repeated washes. Membrane proteins were then separated by SDS–polyacrylamide gel electrophoresis using kaleidoscope prestained standards (BioRad, Hercules, CA) as already described (Witt-Enderby *et al.*, 1998) and proteins were visualized by exposure to film for 3 months. As shown in lane 3, a radioactive signal is apparent between molecular weight sizes of 35–40 kDa. This signal is absent in the other lanes and in the lanes containing cold excess melatonin. These data suggest that [^{14}C]BMNEP is able to bind to MT_2 melatonin receptors in a specific manner. This compound could be used as a basis for the development of future MT_2 selective irreversible ligands.

Table II Conclusion: Melatonin receptors can accommodate certain increases in the size of the alkyl group (up to C_3H_7) attached to the carbonyl group of the ethyl amide side chain (R_3). Any increase in the size larger than C_3H_7 or branching will result in a decrease in affinity of these ligands for melatonin receptors. In addition, conversion of the ethyl amide side chain to an ethyl ester decreases the affinity of this ligand (compound 20) for melatonin receptors.

Table III Conclusions: Compounds 21–28 show that substitutions at the 2-position (R_4) of melatonin with a halogen, CH_3, or a phenyl will

TABLE I

STRUCTURES AND AFFINITIES OF MELATONIN ANALOGS FOR MELATONIN RECEPTORS

Compound #	R_1	R_3	RA[a]	Species	Ref.
Melatonin	OCH_3	$CH_2CH_2NHCOCH_3$	1		
1	H	$CH_2CH_2NHCOCH_3$	1621	Chicken brain	Nelson *et al.* (1993)
1	H	$CH_2CH_2NHCOCH_3$	3041	Chicken brain	Sugden *et al.* (1995)
1	H	$CH_2CH_2NHCOCH_3$	739	Chicken brain	Sicsic *et al.* (1997)
2	OCH_3	$CH_2CH_2NH_2$	10533	Chicken brain	Sugden *et al.* (1995)
3	OH	$CH_2CH_2NHCOCH_3$	2033	Chicken brain	Sugden *et al.* (1995)
4	CH_3	$CH_2CH_2NHCOCH_3$	612	Chicken brain	Sugden *et al.* (1995)
5	CH_2CH_3	$CH_2CH_2NHCOCH_3$	407	Quail optic tecta	Mor *et al.* (1998)
6	F	$CH_2CH_2NHCOCH_3$	135	Quail optic tecta	Mor *et al.* (1998)
7	Cl	$CH_2CH_2NHCOCH_3$	50	Quail optic tecta	Mor *et al.* (1998)
8	Br	$CH_2CH_2NHCOCH_3$	28	Quail optic tecta	Mor *et al.* (1998)
9	OCH_2Ph	$CH_2CH_2NHCOCH_3$	708	Chicken brain	Sugden *et al.* (1995)
10	OCH_2CH_2OH	$CH_2CH_2NHCOCH_3$	490	Chicken brain	Mor *et al.* (1998)

[a]RA = relative affinity of analog compared to affinity of melatonin (i.e., K_i or IC_{50} analog/K_i or IC_{50} melatonin) for melatonin receptors.

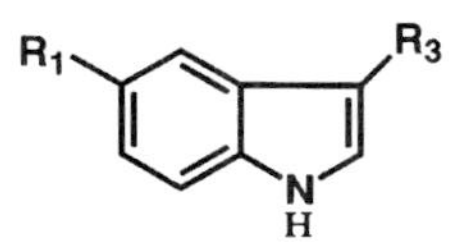

TABLE II

STRUCTURES AND AFFINITIES OF MELATONIN ANALOGS FOR MELATONIN RECEPTORS

Compound #	R_1	R_3	RA[a]	Species	Ref.
Melatonin	OCH_3	$CH_2CH_2NHCOCH_3$	1		
11	OCH_3	$CH_2CH_2NHCOC_2H_5$	0.225	Chicken brain	Nelson *et al.* (1993)
12	OCH_3	$CH_2CH_2NHCOC_2H_5$	0.47	Chicken retina	Copinga *et al.* (1993)
13	OCH_3	$CH_2CH_2NHCOC_3H_7$	0.107	Chicken brain	Nelson *et al.* (1993)
14	OCH_3	$CH_2CH_2NHCOC_3H_7$	1.28	Chicken retina	Copinga *et al.* (1993)
15	OCH_3	$CH_2CH_2NHCOC_4H_9$	52	Chicken brain	Sugden *et al.* (1995)
16	OCH_3	$CH_2CH_2NHCOCF_3$	0.8	Ovine pars tuberalis	Marot *et al.* (1998)
17	OCH_3	$CH_2CH_2NHCOC_3H_5$	0.0003	Ovine pars tuberalis	Depreux *et al.* (1994)
18	OCH_3	$CH_2CH_2NHCOCH_2Br$	0.238	Chicken brain	Nelson *et al.* (1993)
19	OCH_3	$CH_2CH_2NHCO\text{-}iC_3H_7$	25	Chicken brain	Sugden *et al.* (1995)
20	OCH_3	$CH_2CH_2OCOCH_3$	1009	Chicken brain	Sugden *et al.* (1995)

[a]RA = relative affinity of analog compared to affinity of melatonin (i.e., K_i or IC_{50} analog/K_i or IC_{50} melatonin) for melatonin receptors.

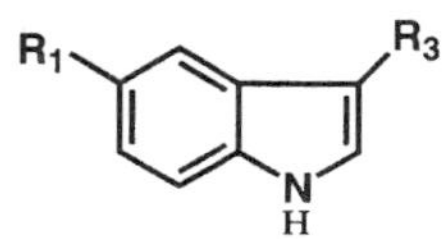

TABLE III

STRUCTURES AND AFFINITIES OF MELATONIN ANALOGS FOR MELATONIN RECEPTORS

Compound #	R_2	R_4	R_5	R_6	R_7	RA[a]	Species	Ref.
21	H	CH_3	H	H	H	0.39	Quail brain	Spadoni *et al.* (1993)
22	H	i-C_3H_7	H	H	H	3.91	Quail brain	Spadoni *et al.* (1993)
23	H	cyclohexyl	H	H	H	4.82	Quail brain	Spadoni *et al.* (1993)
24	H	phenyl	H	H	H	0.052	Quail brain	Spadoni *et al.* (1993)
25	H	Cl	H	H	H	0.12	Quail brain	Spadoni *et al.* (1993)
26	H	Br	H	H	H	0.053	Quail brain	Spadoni *et al.* (1993)
27	H	I	H	H	H	0.019	Quail brain	Spadoni *et al.* (1993)
27	H	I	H	H	H	0.04	Chicken brain	Sicsic *et al.* (1997)
28	H	CH_2Ph	H	H	H	37	mt_1 in CHO cells	Dubocovich *et al.* (1997)
28	H	CH_2Ph	H	H	H	1.39	MT_2 in CHO cells	Dubocovich *et al.* (1997)
29	H	H	H	H	F	1.5	Chicken brain	Sugden *et al.* (1995)
30	H	H	H	H	Cl	2.42	Chicken brain	Sugden *et al.* (1995)
31	H	H	H	H	OH	26.3	Chicken brain	Sugden *et al.* (1995)
32	H	H	H	H	OCH_3	132.1	Chicken brain	Sugden *et al.* (1995)
33	H	H	H	OCH_3	H	198.4	Chicken brain	Mor *et al.* (1998)
34	Br	Br	H	H	H	9.42	Quail optic tecta	Mor *et al.* (1998)
35	H	H	CH_3	H	H	11.0	Quail optic tecta	Mor *et al.* (1998)
36	H	H	Ph	H	H	623.0	Quail optic tecta	Mor *et al.* (1998)
37	H	H	CH_2Ph	H	H	470.5	Quail optic tecta	Mor *et al.* (1998)

[a] RA = relative affinity of analog compared to affinity of melatonin (i.e., K_i or IC_{50} analog/K_i or IC_{50} melatonin) for melatonin receptors.

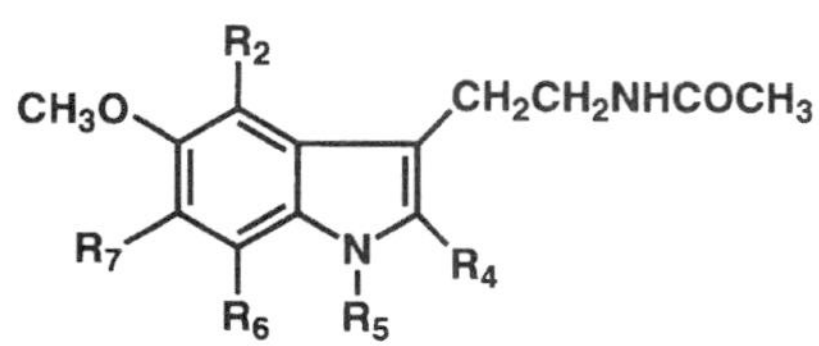

TABLE IV
STRUCTURES AND AFFINITIES OF MELATONIN ANALOGS FOR MELATONIN RECEPTORS

Compound #	Z	Y	RA[a]	Species	Ref.
38	O	C	1.58	Ovine pars tuberalis	Marot *et al.* (1998)
38	O	C	10.8	Ovine pars tuberalis	Depreux *et al.* (1994)
39	S	C	1.26–7.64	Ovine pars tuberalis	Depreux *et al.* (1994); Marot *et al.* (1998)
40	N	N	3200	Ovine pars tuberalis	Marot *et al.* (1998)

[a]RA = relative affinity of analog compared to affinity of melatonin (i.e., K_i or IC_{50} analog/K_i or IC_{50} melatonin) for melatonin receptors.

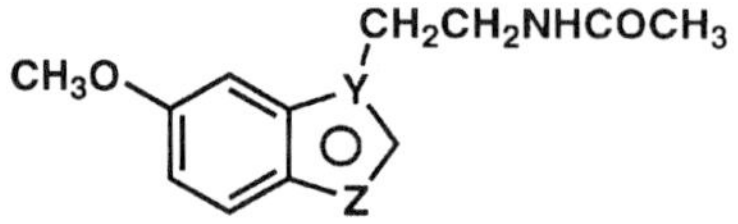

enhance ligand affinity for melatonin receptors. The conclusions drawn from compounds 29–33 are that any substitution at the 6-position (R_7) results in equal or a decrease in affinity of these analogs for melatonin receptors. As shown by compound 34, simultaneous substitutions at both the R_4 (compound 26) and R_2 positions results in a decrease in affinity of the ligand for melatonin receptors. Thus, this implies that a substitution at the R_2 position will negate the increase in affinity generated by an R_4 substitution only (compound 26). Substitution at the 1-position (R_5) or the 7-position (R_6) of melatonin results in a decrease in affinity as revealed by compounds 35–37.

Table IV Conclusion: The indole ring of melatonin can be replaced with a benzofuran (compound 38) or a benzothiophene (compound 39) and result in a slight decrease in affinity. However, replacement of the indole ring with benzimidazole will dramatically lower the affinity (compound 40).

Table V Conclusion: Incorporation of a double bond at the ethylene group in the ethylamido side chain (R_3; compound 41) or shortening the ethylamido side chain (R_3) to a methylamido side chain (compound 42) results in a decrease in the affinity of the analog for the receptor. In addition, altering the distance between the methoxy group (R_1) and the ethyl amido group (R_3) of melatonin (compounds 43, 44, and 46) results in a decrease in affinity. Switching the methoxy group from R_1 to R_7 and concurrently switching the ethylamido group from R_3 to R_5 (compound 45) does not dramatically alter the affinity because the substitutions on

TABLE V

STRUCTURES AND AFFINITIES OF MELATONIN ANALOGS FOR MELATONIN RECEPTORS

Compound #	R_1	R_2	R_3	R_4	R_5	R_6	R_7	RA[a]	Species	Ref.
Melatonin	OCH_3	H	$CH_2CH_2NHCOCH_3$	H	H	H	H	1		
41	OCH_3	H	$CH{=}CHNHCOCH_3$ (trans)	H	H	H	H	13.2	Chicken brain	Spadoni *et al.* (1997)
42	OCH_3	H	$CH_2NHCOCH_2CH_3$	H	H	H	H	>10,000	Chicken brain	Spadoni *et al.* (1997)
43	OCH_3	$CH_2CH_2NHCOCH_3$	H	H	H	H	H	590	Chicken brain	Spadoni *et al.* (1997)
44	H	OCH_3	H	$CH_2CH_2NHCOCH_3$	H	H	H	2600	mt_1 in NIH 3T3 cells	Spadoni *et al.* (1998)
45	H	H	H	H	$CH_2CH_2NHCOCH_3$	H	OCH_3	5.45	Quail optic tecta	Tarzia *et al.* (1997)
46	H	H	$CH_2CH_2NHCOCH_3$	H	H	H	OCH_3	2511	Quail optic tecta	Mor *et al.* (1998)

[a]RA = relative affinity of analog compared to affinity of melatonin (i.e., K_i or IC_{50} analog/K_i or IC_{50} melatonin) for melatonin receptors.

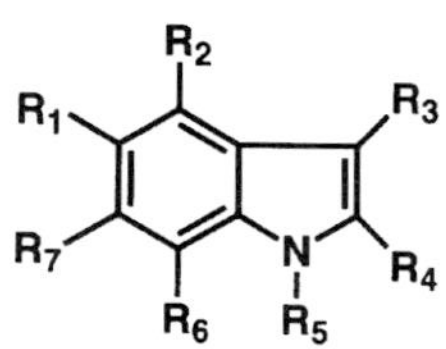

R_5 and R_7 are symmetrical and, therefore, the distance between them remains relatively unaltered.

Analogs containing either the naphthalene or quinoline nuclear core (compounds 47–84) are described in Tables VI and VII.

Table VI Conclusion: Substitution of the indole ring of melatonin with a naphthalene nucleus will result in a ligand (compound 48) with no change in affinity compared to melatonin. Thus, the naphthalene ring completely substitutes for the indole ring. Substitutions made on this naphthalene derivative demonstrate similar binding affinity trends as for melatonin (Table II). That is, the receptor exhibits the same bulk tolerance on the alkyl group attached to the amide carbonyl (R_3) on the naphthalene derivative (compounds 49–56).

Table VII Conclusion(s): Substitutions at the 2-position (R_4) and the 3-position (R_5) of compound 48 also has a certain bulk tolerance (compounds 57–82). In addition, replacement of the naphthalene ring with a quinoline nucleus (compounds 83 and 84) results in ligands with a binding affinity comparable to melatonin.

Melatonin receptor ligands with the simplest structural features are derivatives of phenylalkyl amides (compounds 85–109), as described in Tabe VIII.

Table VIII Conclusion: Both 2 or 3 methoxy substituted phentlalkyl amides have moderate to high affinity for melatonin receptors. Compound 97 has a relative affinity (RA) of 1.68, the highest affinity ligand in the series. Its high affinity for melatonin receptors and its structural simplicity indicate that it possesses the minimal structural requirement for receptor recognition.

Conformationally restricted ligands (compounds 110–129) are described in Table IX.

Table IX Conclusion: It has been shown that both the methoxy and the ethylamido side chain are necessary for high binding affinity to the receptor and that the distance between the group is also a critical factor (Garratt *et al.,* 1994a). Because both the methoxy group and the ethylamido side chain have a different degree of flexibility and can adopt different conformations, conformationally restricted ligands could be used to probe the melatonin receptor binding site for optimum positions of these groups. Compounds 110–129 are conformationally restricted melatonin receptor ligands that contain either an indole, naphthalene, or phenylamide nuclear core. The ethylamido side chains are in differ-

TABLE VI
STRUCTURES AND AFFINITIES OF MELATONIN ANALOGS FOR MELATONIN RECEPTORS

Compound #	R_1	R_3	RA[a]	Species	Ref.
Melatonin			1		
47	OCH_3	CH_2CH_2NHCOH	3.71	Chicken brain	Sicsic *et al.* (1997)
48	OCH_3	$CH_2CH_2NHCOCH_3$	0.79–1.09	Chicken brain	Yous *et al.* (1992); Sicsic *et al.* (1997)
49	OCH_3	$CH_2CH_2NHCOC_2H_5$	0.24–0.27	Chicken brain	Yous *et al.* (1992); Sicsic *et al.* (1997)
50	OCH_3	$CH_2CH_2NHCOC_3H_7$	0.24–0.27	Chicken brain	Yous *et al.* (1992); Sicsic *et al.* (1997)
51	OCH_3	$CH_2CH_2NHCO\text{-}iC_3H_7$	25.0	Chicken brain	Yous *et al.* (1992)
52	OCH_3	$CH_2CH_2NHCOC_3H_5$	0.005	Chicken brain	Yous *et al.* (1992)
53	OCH_3	$CH_2CH_2NHCOC_4H_7$	264.5	Chicken brain	Yous *et al.* (1992)
54	OCH_3	$CH_2CH_2NHCOC_4H_9$	0.037	Chicken brain	Yous *et al.* (1992)
55	OCH_3	$CH_2CH_2NHCOC_5H_{11}$	3092	Chicken brain	Yous *et al.* (1992)
56	OCH_3	$CH_2CH_2NHCOC_6H_{13}$	>20,000	Chicken brain	Yous *et al.* (1992)

[a]RA = relative affinity of analog compared to affinity of melatonin (i.e., K_i or IC_{50} analog/K_i or IC_{50} melatonin) for melatonin receptors.

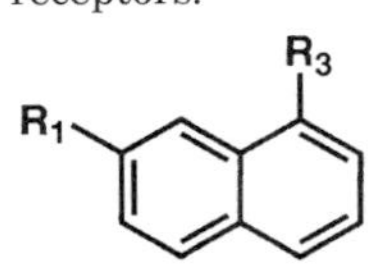

TABLE VII
STRUCTURES AND AFFINITIES OF MELATONIN ANALOGS FOR MELATONIN RECEPTORS

Compound #	R_1	R_3	R_4	R_5	X	RA^a	Species	Ref.
57	OCH_3	$CH_2CH_2NHCOCH_3$	H	H	CH	0.79–1.09	Chicken brain	Yous *et al.* (1992); Sicsic *et al.* (1997)
58	H	$CH_2CH_2NHCOCH_3$	OCH_3	H	CH	4.0	Chicken brain	Sicsic *et al.* (1997)
59	OCH_3	$CH_2CH_2NHCOCH_3$	OCH_3	H	CH	0.35	Chicken brain	Sicsic *et al.* (1997)
60	OCH_3	$CH_2CH_2NHCOC_2H_5$	OCH_3	H	CH	0.107	Chicken brain	Sicsic *et al.* (1997)
61	OCH_3	$CH_2CH_2NHCOC_2H_5$	OCH_3	H	CH	0.495	mt_1 in CHO cells	Li *et al.* (1997b)
62	OCH_3	$CH_2CH_2NHCOC_2H_5$	OCH_3	H	CH	1.31	MT_2 in CHO cells	Li *et al.* (1997b)
63	OCH_3	$CH_2CH_2NHCOC_2H_5$	OC_2H_5	H	CH	1.95	mt_1 in CHO cells	Li *et al.* (1997b)
64	OCH_3	$CH_2CH_2NHCOC_2H_5$	OC_2H_5	H	CH	4.13	MT_2 in CHO cells	Li *et al.* (1997b)
65	OCH_3	$CH_2CH_2NHCOC_2H_5$	OC_3H_7	H	CH	40.5	mt_1 in CHO cells	Li *et al.* (1997b)
66	OCH_3	$CH_2CH_2NHCOC_2H_5$	OC_3H_7	H	CH	55.1	MT_2 in CHO cells	Li *et al.* (1997b)
67	OCH_3	$CH_2CH_2NHCOC_2H_5$	OC_4H_9	H	CH	35.5	mt_1 in CHO cells	Li *et al.* (1997b)
68	OCH_3	$CH_2CH_2NHCOC_2H_5$	OC_4H_9	H	CH	23.1	MT_2 in CHO cells	Li *et al.* (1997b)
69	OCH_3	$CH_2CH_2NHCOC_2H_5$	OCH_2CH_2OH	H	CH	33.1	mt_1 in CHO cells	Li *et al.* (1997b)
70	OCH_3	$CH_2CH_2NHCOC_2H_5$	OCH_2CH_2OH	H	CH	5.65	MT_2 in CHO cells	Li *et al.* (1997b)
71	OCH_3	$CH_2CH_2NHCOC_2H_5$	OCH_2CH_2I	H	CH	500	mt_1 in CHO cells	Li *et al.* (1997b)
72	OCH_3	$CH_2CH_2NHCOC_2H_5$	OCH_2CH_2I	H	CH	251	MT_2 in CHO cells	Li *et al.* (1997b)
73	OCH_3	$CH_2CH_2NHCOC_2H_5$	$OCH_2CH_2N(CH_3)_2$	H	CH	210	mt_1 in CHO cells	Li *et al.* (1997b)

74	OCH_3	$CH_2CH_2NHCOC_2H_5$	$OCH_2CH_2N(CH_3)_2$	H	CH	34.5	MT_2 in CHO cells	Li *et al.* (1997b)
75	OCH_3	$CH_2CH_2NHCOCH_3$	H	CH_2Ph	CH	16	Ovine pars tuberalis	Marot *et al.* (1998)
76	OCH_3	$CH_2CH_2NHCOCH_3$	H	$(CH_3)_2CH_3$	CH	6.3	Ovine pars tuberalis	Marot *et al.* (1998)
77	OCH_3	$CH_2CH_2NHCOC_3H_5$	H	$COCH_3$	CH	50	Ovine pars tuberalis	Marot *et al.* (1998)
78	OCH_3	$CH_2CH_2NHCOCH_3$	H	$COCH_2CH_3$	CH	50	Ovine pars tuberalis	Marot *et al.* (1998)
79	OCH_3	$CH_2CH_2NHCOCH_3$	H	$CO\text{-}C_3H_5$	CH	40	Ovine pars tuberalis	Marot *et al.* (1998)
80	OCH_3	$CH_2CH_2NHCOC_3H_5$	H	CH_2CH_3	CH	16	Ovine pars tuberalis	Marot *et al.* (1998)
81	OCH_3	$CH_2CH_2NHCOCH_3$	H	$CH_2C_3H_5$	CH	7.94	Ovine pars tuberalis	Marot *et al.* (1998)
82	OCH_3	$CH_2CH_2NHCOCH_3$	H	COPh	CH	0.32	Ovine pars tuberalis	Marot *et al.* (1998)
83	OCH_3	$CH_2CH_2NHCOCH_3$	H	H	N	45	mt_1 in CHO cells	Li *et al.* (1997a)
84	OCH_3	$CH_2CH_2NHCOCH_3$	H	H	N	6	MT_2 in CHO cells	Li *et al.* (1997a)

[a]RA = relative affinity of analog compared to affinity of melatonin (i.e., K_i or IC_{50} analog/K_i or IC_{50} melatonin) for melatonin receptors.

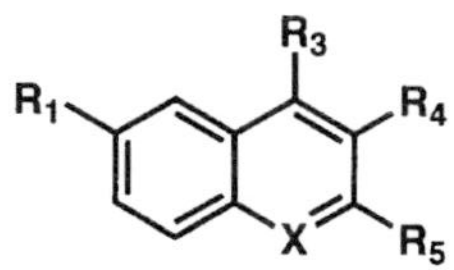

TABLE VIII
STRUCTURES AND AFFINITIES OF MELATONIN ANALOGS FOR MELATONIN RECEPTORS

Compound #	R_1	R_2	R_3	R_4	n	R	RA[a]	Species	Ref.
85	OCH_3	H	H	H	2	CH_3	295–971	Chicken brain	Copinga *et al.* (1993); Langlois *et al.* (1995); Garratt *et al.* (1996); Sicsic *et al.* (1997)
86	OCH_3	H	H	H	2	C_2H_5	228–1384	Chicken brain	Langlois *et al.* (1995); Garratt *et al.* (1996); Sicsic *et al.* (1997)
87	OCH_3	H	H	H	2	C_3H_7	116–1267	Chicken brain	Langlois *et al.* (1995); Garratt *et al.* (1996)
88	OCH_3	H	H	OCH_3	2	CH_3	104	Chicken brain	Sicsic *et al.* (1997)
89	OCH_3	H	H	OCH_3	2	CF_3	148	Chicken brain	Langlois *et al.* (1995)
90	OCH_3	H	H	OCH_3	2	C_2H_5	26	Chicken brain	Langlois *et al.* (1995)
91	OCH_3	H	H	OCH_3	2	C_3H_5	576	Chicken brain	Langlois *et al.* (1995)
92	OCH_3	H	H	OCH_3	2	C_4H_9	>1500	Chicken brain	Langlois *et al.* (1995)
93	OCH_3	H	H	CH_3	2	CH_3	287	Chicken brain	Langlois *et al.* (1995)
94	OCH_3	H	H	CH_3	2	C_2H_5	95.6	Chicken brain	Sicsic *et al.* (1997)
95	OCH_3	H	H	C_2H_5	2	CH_3	295–324	Chicken brain	Sicsic *et al.* (1997); Langlois *et al.* (1995)
96	OCH_3	H	H	Br	2	CH_3	61.6–67.7	Chicken brain	Sicsic *et al.* (1997); Langlois *et al.* (1995)

97	OCH_3	H	H	Br	2	C_2H_5	1.68–11.9	Chicken brain	Sicsic *et al.* (1997); Langlois *et al.* (1995)
98	OCH_3	H	H	Ph	2	CH_3	443	Chicken brain	Sicsic et al. (1997)
99	H	OCH_3	H	H	2	CH_3	377–1623.7	Chicken brain	Langlois *et al.* (1995); Garratt *et al.* (1996); Sicsic *et al.* (1997)
100	H	OCH_3	H	H	2	C_2H_5	105.1	Chicken brain	Garratt *et al.* (1996)
101	H	OCH_3	H	H	2	C_3H_7	67.8	Chicken brain	Garratt *et al.* (1996)
102	H	H	OCH_3	H	2	CH_3	>10,000	Chicken brain	Garratt *et al.* (1996)
103	OCH_3	H	H	H	3	CH_3	2423.7	Chicken brain	Garratt *et al.* (1996)
104	H	OCH_3	H	H	3	CH_3	106.8	Chicken brain	Garratt *et al.* (1996)
105	H	OCH_3	H	H	3	C_2H_5	9.49	Chicken brain	Garratt *et al.* (1996)
106	H	OCH_3	H	H	3	C_3H_7	9.32	Chicken brain	Garratt *et al.* (1996)
107	H	F	H	H	3	CH_3	3254	Chicken brain	Garratt *et al.* (1996)
108	H	Cl	H	H	3	CH_3	1078	Chicken brain	Garratt *et al.* (1996)
109	H	Br	H	H	3	CH_3	1444	Chicken brain	Garratt *et al.* (1996)

[a]RA = relative affinity of analog compared to affinity of melatonin (i.e., K_i or IC_{50} analog/K_i or IC_{50} melatonin) for melatonin receptors.

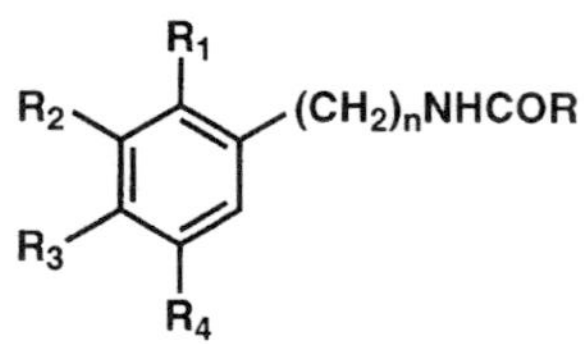

TABLE IX

STRUCTURES AND AFFINITIES OF MELATONIN ANALOGS FOR MELATONIN RECEPTORS

	Compound #	n	R	RA[a]	Species	Ref.
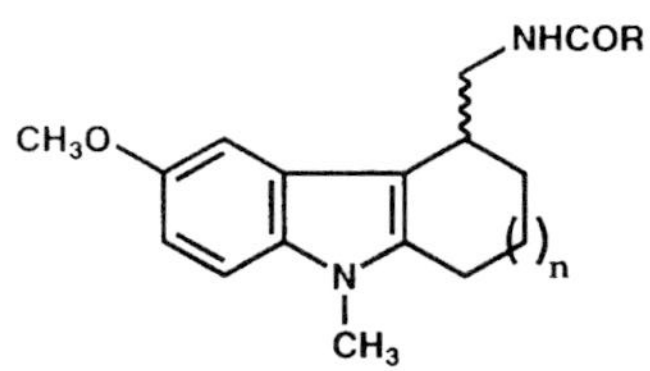 Compounds 110-117	110	0	CH_3	273	Chicken brain	Davies *et al.* (1998)
	111 S (−) isomer	0	C_2H_5	2.88	Chicken brain	Davies *et al.* (1998)
	112 R (+) isomer	0	C_2H_5	412	Chicken brain	Davies *et al.* (1998)
	113 S (−) isomer	1	CH_3	0.63	Chicken brain	Davies *et al.* (1998)
	114 R (+) isomer	1	CH_3	82	Chicken brain	Davies *et al.* (1998)
	115	2	CH_3	40.7	Chicken brain	Davies *et al.* (1998)
	116 S (−) isomer	2	C_2H_5	11.0	Chicken brain	Davies *et al.* (1998)
	117 R (+) isomer	2	C_2H_5	341.5	Chicken brain	Davies *et al.* (1998)
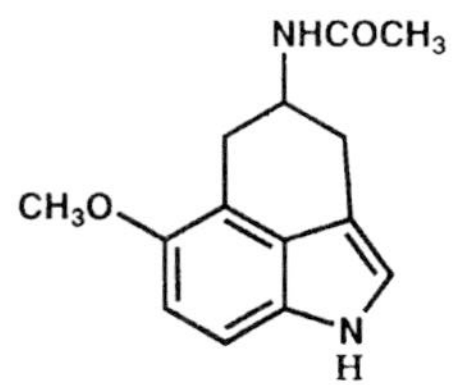 Compounds 118	118			4.4	Quail optic tecta	Spadoni *et al.* (1997)
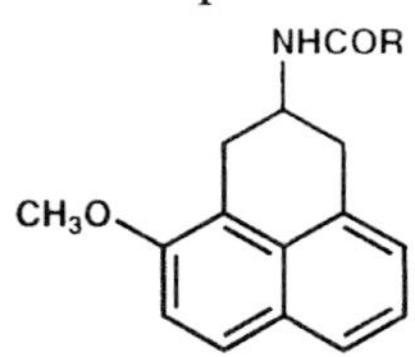 Compounds 119-120	119		CH_3	62.4	Chicken brain	Spadoni *et al.* (1997)
	120		C_2H_5	13.0	Chicken brain	Spadoni *et al.* (1997)

Structure	Compound	n	R	RA[a]	Tissue	Reference
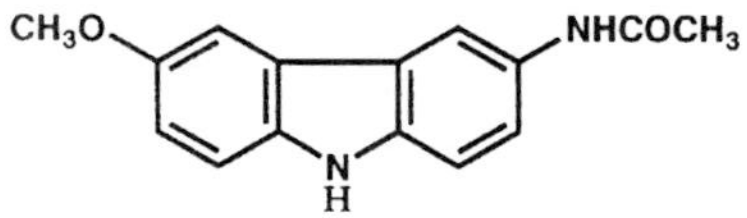 Compounds 121	121			1344	Quail optic tecta	Spadoni *et al.* (1997)
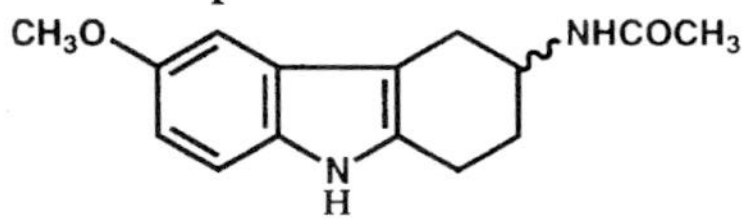 Compounds 122	122			371	Chicken brain	Garratt *et al.* 1994b)
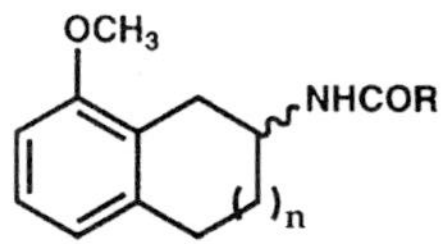 Compounds 123-128	123 S (−) isomer	0	C_2H_5	>10,000	Chicken retina	Jansen *et al.* (1996)
	124 R (+) isomer	0	C_2H_5	1359	Chicken retina	Jansen *et al.* (1996)
	125 S (−) isomer	1	CH_3	44.4	Chicken retina	Jansen *et al.* (1996)
	126 R (+) isomer	1	CH_3	9951	Chicken retina	Jansen *et al.* (1996)
	127 S (−) isomer	2	C_2H_5	44.6	Chicken retina	Jansen *et al.* (1996)
	128 R (+) isomer	2	C_2H_5	461.5	Chicken retina	Jansen *et al.* (1996)
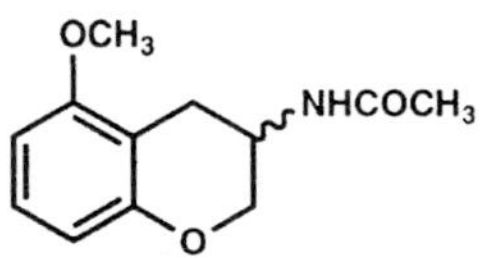 Compounds 129	129			162	Chicken brain	Sugden (1994)

[a]RA = relative affinity of analog compared to melatonin (i.e., K_i or IC_{50} analog/K_i or IC_{50} melatonin) for melatonin receptors.

ent degrees of restricted rotational freedom. In general, analogs with the folded ethylamido side chain (e.g., compound 113—S(−) isomer) have higher affinity than the analogs with the side chain in the extended conformation (compounds 121 and 122).

III. Complications Involved with Generalizing about Structure–Function Relationships of the Current Melatonin Receptor Ligands

Currently, it is very difficult to generalize about the structure–activity relationships between existing melatonergic ligands and melatonin receptors because assessment of the affinities of these analogs (in many instances) was performed in one model system, whereas assessment of the function of these analogs was performed in a different model system. Because each model system may contain a heterogeneous population of melatonin receptors (Table X), then the conclusions drawn about the stucture–*function* relationships of many of these analogs for a particular melatonin receptor *subtype* cannot be made without hesitation. Thus it is important to assess both the pharmacology and function of melatonergic analogs in the same model, but the question remains, which model system?

Though many different models exist to study the effects of melatonin, there are benefits and limitations to all (Kenakin, 1997). Tissue models are beneficial because tissue is taken directly from the body and may be more reflective of an *in vivo* state. However, limitations to using tissues are that (1) extracted tissue may contain a heterogeneous population of receptors (Table X) and (2) melatonin receptor expression levels are very low, making it difficult to study ligand-dependent activation of melatonin receptors. Another model used to study ligand–receptor interactions is primary cultures, that is, cells taken from the animal and cultured *in vitro.* This model, like the tissue, has its benefits and limitations as well. The benefit to using such a model is that, like the tissue, this tissue comes from the animal and contains all of the necessary components for proper receptor expression and signaling. However, limitations to using this model, like tissue, is that these cells may contain a heterogeneous population of receptors. A third type of *in vitro* model, that is, transfected cells is also widely used and it too has its benefits and limitations. The biggest benefit to using this model is that it is possible to determine the affinities and potencies of various analogs for a *defined* population of receptors (Witt-Enderby and Dubocovich, 1996). Therefore, more direct determinations of melatonergic ligand affinity and potency could be made on specific receptors (i.e., the mt_1 or

TABLE X
COMMONLY USED MODEL SYSTEMS FOR RADIOLIGAND BINDING AND FUNCTIONAL ANALYSIS OF MELATONERGIC LIGANDS

Model	Melatonin receptor subtype(s)	Reference(s)
Chicken retina	Unknown	Jansen *et al.* (1996)
Chicken brain	mt_1, MT_2 Mel_{1c}	Reppert *et al.* (1995b); Pang *et al.* (1997); Kokkola and Laitinen, (1998)
Hamster brain	Unknown (MT_3??) (Djungarian hamster); mt_1 (Siberian hamster)	Dubocovich, (1995); Kokkola and Laitinen, (1998)
Rabbit retina	Unknown (MT_2?)	Dubocovich *et al.* (1997)
Quail optic tecta	Unknown	Cozzi et al., (1993)
Quail retina	Unknown	Underwood *et al.* (1988); Manglapus *et al.* (1999)
Xenopus melanophores	Mel_{1c}	Kokkola and Laitinen, (1998)
Pars tuberalis	mt_1, MT_2 (rat); mt_1 (ovine)	Masson-Pevet *et al.* (1998); Barrett *et al.* (1998)
mt_1 in NIH 3T3 cells	mt_1	LeClerc *et al.* (1998); Spadoni *et al.* (1998)
mt_1 in CHO cells	mt_1	Witt-Enderby and Dubocovich, (1996)
MT_2 in CHO cells	MT_2	Witt-Enderby *et al.* (1998)

MT_2 receptors) without the confounding influence of other melatonin receptor subtypes. The limitation to this model, however, is that the receptor is being overexpressed and is placed in an artificial environment. However, these limitations could be overcome by establishing that the pharmacology (i.e., affinity of melatonin for the receptor) and function (i.e., potency of melatonin for the receptor) of the receptor expressed in these surrogate cells is similar to those expressed in endogenous tissue.

IV. CONCLUSIONS

In conclusion, melatonin produces diverse actions within the body. These variable effects of melatonin *in vivo* may be due to (1) its ability to act at receptors located at the plasma membrane, cytosol, or nucleus; (2) its ability to act independent of a receptor; (3) its ability to activate multiple signal transduction pathways; (4) its ability to regulate melatonin receptor function; and, finally, (5) its ability to interact with multiple receptor subtypes throughout the body. Thus, the need for high-affinity and selective melatonin receptor antagonists is becoming very evident. The field is in an exciting state of development with the synthesis and application of many melatonergic ligands (reversible antagonists, irreversible and subtype-selective agonists, charged melatonergic ligands, and MT_2-selective antagonists). As the availability of tools increases in the melatonin receptor field, a great body of knowledge is also gained about the structure of the melatonin receptor and the role that specific melatonin receptor subtypes have in physiologic processes. Further design, synthesis, and application of melatonergic ligands will lead us to a clearer understanding of the role that melatonin and its receptors play in man.

Acknowledgment

The authors are indebted to Joerg Brockmann and Michael Jarzynka, R.Ph., for performing the experiments involving BMNEP. In addition, the authors would like to thank Dr. Robert W. Brueggemeier and Dr. Michael V. Darby (The Ohio State University, Department of Medicinal Chemistry) for the synthesis of [^{14}C]BMNEP.

REFERENCES

Anton-Tay, F., Martínez, I., Tovar R., and Benítez-King, G. (1998a). Modulation of the subcellular distribution of calmodulin by melatonin in MDCK cells. *J. Pineal Res.* **24,** 35–42.

Anton-Tay, F., Ramírez, G., Martínez, I., and Benítez-King, G. (1998b). *In vitro* stimulation of protein kinase C by melatonin. *Neurochem. Res.* **23**(5), 601–606.

Avery, D., Lenz, M., and Landis, C. (1998). Guidelines for prescribing melatonin. *Ann. Med.* **30,** 122–130.

Ayre, E. A., Yuan, H., and Pang, S. F. (1992). The identification of I-125-labeled iodomelatonin binding sites in the testes and ovaries of the chicken (*Gallus domesticus*). *J. Endocrinol.* **133,** 5–11.

Barrett, P., Davidson, G., Hazlerigg, D. G., Morris, M. A., Ross, A. W., and Morgan, P. J. (1998). Mel1a melatonin receptor expression is regulated by protein kinase C and an additional pathway addressed by the protein kinase C inhibitor RO 31–8220 in ovine pars tuberalis cells. *Endocrinology (Baltimore)* **139,** 163–171.

Benítez-King, G., and Anton-Tay, F. (1993). Calmodulin mediates melatonin cytoskeletal effects. *Experientia* **49,** 635–641.

Benítez-King, G., Huerto-Delgadillo, L., and Anton-Tay, F. (1990). Melatonin effects on the cytoskeletal organization of MDCK and neuroblastoma N1E-115 cells. *J. Pineal Res.* **9,** 209–220.

Benloucif, S., Masana, M.I., and Dubocovich, M.L. (1997). Responsiveness to melatonin and its receptor expression in the suprachiasmatic nucleus of young and old C3H/HeN mice. *Am. J. Physiol.* **42,** R1855-R1860.

Betthai, I., Guerrero, J. M., Reiter, R. J., and Osuna, C. (1998). Physiological concentrations of melatonin inhibit the norepinephrine-induced activation of prostaglandin E2 and cyclic AMP production in rat hypothalamus: A mechanism involving inhibition of nitric oxide synthase. *J. Pineal Res.* **25,** 34–40.

Birdsall, N. J. M., Burgen, A. S. V., and Hulme, E. C. (1979). A study of the muscarinic receptor by gel electrophoresis. *Br. J. Pharmacol.* **66,** 337–342.

Brzezinski, A. (1997). Melatonin in humans. *N. Engl. J. Med.* **336**(3), 186–195.

Burgen, A. S. V., Hiley, C. R., and Young, J. M. (1974). The properties of muscarinic receptors in mammalian cerebral cortex. *Br. J. Pharmacol.* **51,** 279–285.

Capsoni, S., Viswanathan, M., Oliveira, A. M., and Saavedra, J. M. (1994). Characterization of melatonin receptors and signal transduction system in rat arteries forming the circle of Willis. *Endocrinology (Baltimore)* **135,** 373–378.

Carman, C. V., and Benovic, J. L. (1998). G-protein-coupled receptors: Turn-ons and turn-offs. *Signaling mechanisms. Curr. Opin. Neurobiol.* **8,** 335–344.

Cassone, V. M., and Natesan, A. K. (1997). Time and time again: The phylogeny of melatonin as a transducer of biological time. *J. Biol. Rhythms* **12**(6), 489–497.

Cassone, V. M., Warren, W. S., Brooks, D. S., and Lu, J. (1993). Melatonin, the pineal gland, and circadian rhythms. *J. Biol. Rhythms* **8,** S73-S81.

Chu, G.-H., and Li, P.-K. (1998). Synthesis of a radiolabeled charged melatonin receptor ligand. *J. Labelled Compd. Radiopharm.* **41,** 435–441.

Copinga, S., Tepper, P. G., Grol, C. J., Horn, A. S., and Dubocovich, M. L. (1993). 2-Amido-8-methoxytetralins: A series of nonindolic melatonin-like agents. *J. Med. Chem.* **36,** 2891–2898.

Cos, S., and Sanchez-Barcelo, E. J. (1995). Melatonin inhibition of MCF-7 human breast-cancer cells growth: Influence of cell proliferation rate. *Cancer Lett.* **93,** 207–212.

Cos, S., Recio, J., and Sanchez-Barcelo, E. J. (1996). Modulation of the length of the cell cycle time of MCF-7 human breast cancer cells by melatonin. *Life Sci.* **58,** 811–816.

Cozzi, B., Stankov, B., Viglietti-Panzica, C., Capsoni, S., Aste, N., Lucini, V., Fraschini, F., and Panzica, G. (1993). Distribution and characterization of melatonin receptors in the brain of the Japanese quail, *Coturnix japonica. Neurosci. Lett.* **150,** 149–152.

Curtis, C. A. M., Wheatley, M., Bansal, S., Birdsall, N. J. M., Eveleigh, P., Pedder, E. K., Poyner, D., and Hulme, E.C. (1989). Propylbenzilylcholine mustard labels an acidic residue in transmembrane helix 3 of the muscarinic receptor. *J. Biol. Chem.* **264,** 489–495.

Davies, D. J., Garratt, P. J., Tocher, D. A., Vonhoff, S., Davies, J., Teh, M. T., and Sugden, D. (1998). Mapping the melatonin receptor. 5. Melatonin agonists and antagonists derived from tetrahydrocyclopent[*b*]indoles, tetrahydocarbozoles and hexahydrocyclohept[*b*]indoles. *J. Med. Chem.* **41,** 451–467.

Dawson, D., and van den Heuvel, C. J. (1998). Integrating the actions of melatonin on human physiology. *Ann. Med.* **30,** 95–102.

Depreux, P., Lesieur, D., Mansour, H. A., Morgan, P., Howell, H. E., Renard, P., Caignard, D. H., Pfeiffer, B., Delagrange, P., Guardiola, B., Yous, S., Demarque, A., Adam, G., and Andrieux, J. (1994). Synthesis and structure-activity relationships of novel naphthalenic and bioisosteric related amidic derivatives as melatonin receptor ligands. *J. Med. Chem.* **37,** 3231–3239.

Doolen, S., Krause, D., Dubocovich, M. L., and Duckles, S. P. (1998). Melatonin mediates two distinct responses in vascular smooth muscle. *Eur. J. Pharmacol.* **345,** 67–68.

Dubocovich, M. L. (1995). Melatonin receptors: Are there multiple subtypes? *Trends Pharmacol. Sci.* **16,** 50–56.

Dubocovich, M. L., Masana, M. I., Iacob, S., and Sauri, D. M. (1997). Melatonin receptor antagonists that differentiate between the human Mel1a and Mel1b recombinant subtypes are used to assess the pharmacological profile of the rabbit retina ML1 presynaptic heteroreceptor. *Naunyn-Schmiedeberg's Arch Pharmacol.* **355,** 355–365.

Dubocovich, M. L., Yun, K., Al-Ghoul, W. M., Benloucif, S., and Masana, M. I. (1998a). Selective MT2 melatonin receptor antagonists block melatonin-mediated phase advances of circadian rhythms. *FASEB J.* **12,** 1211–120.

Dubocovich, M. L, Cardinali, D. P., Guardiola-Lemaitre, B., Hagan, R. M., Krause, D. N., Sugden, D., Yocca, F. D., and Vanhoutte, P. M. (1998b). Melatonin receptors. "The IUPHAR Compendium of Receptor Characterization and Classification." IUPHAR Media, London.

Eison, A. S., and Mullins, U. L. (1993). Melatonin binding sites are functionally coupled to phosphoinositide hydrolysis in Syrian hamster RPMI 1846 melanoma cells. *Life Sci.* **53,** 393–398.

Finocchiaro, L. M. E., and Gliken, G. C. (1998). Intracellular melatonin distribution in cultured cell lines. *J. Pineal Res.* **24,** 22–34.

Galper, J. B., Dziekan, L. C., O'Hara, D. S., and Smith, T. W. (1982). The biphasic response of muscarinic cholinergic receptors in cultured heart cells to agonists. *J. Biol. Chem.* **257,** 10344–10356.

Garratt, P. J., Jones, R., Rowe, S. J., and Sugden, D. (1994a). Mapping the melatonin re ceptor. 1. The 5-methoxyl group of melatonin is not an essential requirement for biological activity. *Bioorg. Med. Chem. Lett.* **4,** 1555–2558.

Garratt, P. J., Vonhoff, S., Rowe, S. J., and Sugden, D. (1994b). Mapping the melatonin receptor. 2. Synthesis and biological activity of indole derived melatonin analogues with restricted conformations of the C-3 amidoethane side chain. *Bioorg. Med. Chem. Lett.* **4,** 1559–1564.

Garratt, P. J., Travard, S., Vonhoff, S., Tsotinis, A., and Sugden, D. (1996). Mapping the melatonin receptor. 4. Comparison of the binding affinities of a series of substituted phenylalkyl amides. *J. Med. Chem.* **39,** 1797–1805.

Gauer, F., Masson-Pevet, M., Stehle, J., and Pevet, P. (1994). Daily variations of melatonin receptor density of rat pers tuberalis and suprachiasmatic nuclei are distinctly regulated. *Brain Res.* **641,** 92–98.

Geary, G. G., Krause, D. N., and Duckles, S. P. (1997). Melatonin directly constricts rat cerebral arteries through modulation of potassium channels. *Am. J. Physiol.* **273,** H1530–H1536.

Ginty, D. D., Kornhauser, J. M., Thompson, M. A., Bading, H., Mayo, K. E., Takahashi, J. S., and Greenberg, M. E. (1993). Regulation of CREB phosphorylation in the suprachiasmatic nucleus by light and a circadian clock. *Science* **260,** 238–241.

Godson, C., and Reppert, S. M. (1997). The mel1a melatonin receptor is coupled to parallel signal transduction pathways. *Endocrinology (Baltimore)* **138,** 397–404.

Hazlerigg, D. G., Morgan, P. J., Lawson, W., and Hastings, M. H. (1991). M.H. melatonin inhibits the activation of cyclic AMP-dependent protein kinase in cultured pars tuberalis cells from ovine pituitary. *J. Neuroendocrinol.* **3,** 597–603.

Jansen, J. M., Copinga, S., Gruppen, G., Molinari, E. J., Dubocovich, M. L., and Grol, C. J. (1996). The high-affinity melatonin binding site probed with conformationally restricted ligands-I. Pharmacophore and minireceptor models. *Bioorg. Med. Chem.* **4,** 1321–1332.

Kenakin, T. (1997). "Human Recombinant Receptor Systems in Pharmacologic Analysis of Drug-receptor Interaction," 3rd ed., pp.57–77. Lippincott-Raven, Philadelphia.

Kokkola, T., and Laitinen, J. T. (1998). Melatonin receptor genes. *Ann Med.* **30,** 88–94.

Kopp, M. D. A., Schomerus, C., Dehghani, F., Korf, H.-W., and Meissl, H. (1999). Pituitary adenylate cyclase-activating polypeptide and melatonin in the suprachiasmatic nucleus: Effects on the calcium signal transduction cascade. *J. Neurosci.* **19,** 206–219.

Krause, D. N., Barrios, V. E., and Duckles, S. P. (1995). Melatonin receptors mediate potentiation of contractile responses to adrenergic nerve stimulation in rat caudal artery. *Eur. J. Pharmacol.* **276,** 207–213.

Kripke, D. F., Elliott, J. A., Youngstedt, S. D., and Smith, J. S. (1998). Melatonin: Marvel or marker? *Ann. Med.* **30,** 81–87.

Langlois, M., Brémont, B., Shen, S., Poncet, A., Andrieux, J., Sicsic, S., Serraz, I., Mathé-Allainmat, M., Renard, P., and Delagrange, P. (1995). Design and synthesis of new naphthalenic derivatives as ligands for 2-[^{125}I]Iodomelatonin binding sites. *J. Med. Chem.* **38,** 2050–2060.

Laudon, M., and Zisapel, N. (1991). Melatonin binding proteins identified in the rat brain by affinity labeling. *FEBS Lett.* **288,** 105–108.

Leclerc, V., Fourmaintraux, E., Depreux, P., Lesieur, D., Morgan, M., Howell, H. E., Renard, P., Caignard, D. H., Pfeiffer, B., Delagrange, P., Guardiola-Lemaitre, B., and Andrieux, J. (1998). Synthesis and structure-activity relationships of novel naphthalenic and bioisoteric related amidic derivatives as melatonin receptor ligands. *Bioorg. Med. Chem.* **6,** 1875–1887.

Li, P. K., Chu, G. H., Gillen, M. L., and Witt-Enderby, P. A. (1997a). Synthesis and receptor binding studies of quinolinic derivatives as melatonin receptor ligands. *Bioorg. Med. Chem. Lett.* **7,** 2177–2180.

Li, P. K, Chu, G.-H., Gillen, M. L., Parekh, T., and Witt-Enderby, P. A. (1997b). The development of a charged melatonin receptor ligand. *Bioorg. Med. Chem. Lett.* **7,** 2409–2414.

Li, X., Borjigin, J., and Snyder, S. H. (1998). Molecular rhythms in the pineal gland. *Curr. Opin. Neurobiol.* **8,** 648–651.

Lohse, M. J., Klotz, K.-N., and Schwabe, U. (1986). Agonist photoaffinity labeling of A1 adenosine receptors: Persistent activation reveals spare receptors. *Mol. Pharmacol.* **30,** 403–409.

Manglapus, M. K., Iuvone, P. M., Underwood, H., Pierce, M. E., and Barlow, R. B. (1999). Dopamine mediates circadian rhythms of rod-cone dominance in the japanese quail retina. *J. Neurosci.* **19,** 4132–4141.

Marot, C., Chavatte, P., Morin-Allory, L., Viaud, M. C., Guillaumet, G., Renard, P., Lesieur, D., and Michel, A. (1998). Pharmacophoric search and 3D-QSAR comparative molecular field analysis studies on agonists of melatonin sheep receptors. *J. Med. Chem.* **41,** 4453–4465.

Masson-Pevet, M., Recio, J., Guerrero, H. Y., Mocaer, E., Delagrange, P., Guardiola-Lemaitre, B., and Pevet, P. (1998). Effects of two melatonin analogues, S-20098 and S-20928, on melatonin receptors in the pars tuberalis of the rat. J. *Pineal Res.* **25,** 172–176.

Maywood, E. S., and Hastings, M. H. (1995). Lesions of the iodomelatonin-binding sites of the mediobasal hypothalamus spare the lactotropic, but block the gonadotropic response of male syrian hamsters to short photoperiod and to melatonin. *Endocrinology (Baltimore)* **136,** 144–153.

McArthur, J., Hunt, A. E., and Gillette, M. U. (1997). Melatonin action and signal transduction in the rat suprachiasmatic circadian clock: Activation of protein kinase C at dusk and dawn. *Endocrinology (Baltimore)* **138,** 627–634.

McNulty, A., Ross, A. W., Shiu K. Y., Morgan, P. J., and Hastings, M. H. (1996). Phosphorylation of CREB in ovine pars tuberalis is regulated both by cyclic AMP-dependent and cyclic AMP-independent mechanisms. *J. Neuroendocrinol.* **8,** 635–645.

McNulty, S., Ross, A. W., Barret, P., Hastings, M. H., and Morgan, P. J. (1994). Melatonin regulates the phosphorylation of CREB in ovine pars tuberalis. *J. Neuroendocrinol.* **6,** 523–532.

Menendez-Pelaez, A., and Reiter, R. J. (1993). Distribution of melatonin in mammalian tissues: The relative importance of nuclear versus cytosolic localization. *J. Pineal Res.* **15,** 59–69.

Messager, S., Caillol, M., Rossano, B., and Martinet, L. (1996). Effect of melatonin on release of gonadotropin-releasing hormone and cyclic AMP from the rat hypothalamus: An *in vitro* study. *J. Neuroendocrinol.* **8,** 801–807.

Missbach, M., Jagher, B., Sigg, I., Nayeri, S., Carlberg, C., and Wiesenberg, I. (1996). Thiazolidine diones, specific ligands of the nuclear receptor retinoid Z receptor/retinoid acid receptor-related orphan receptor a with potent antiarthritic activity. *J. Biol. Chem.* **271,** 13515–13522.

Molis, T. M., Spriggs, L. L., and Hill, S. M. (1994). Modulation of estrogen receptor mRNA expression by melatonin in MCF-7 human breast cancer cells. *Mol. Endocrinol.* **8,** 1681–1690.

Mor, M., Rivara, S., Silva, C., Bordi, F., Plazzi, P. V., Spadoni, G., Diamantini, G., Balsamini, C., Tarzia, G., Frachini, F., Lucini, V., Nonno, R., and Stankov, B. M. (1998). *J. Med. Chem.* **41,** 3831–3844.

Morgan, P. J., Barrett, P., Howel, H. E., and Helliwell, R. (1994). Melatonin receptors: Localization, molecular pharmacology and physiological significance. *Neurochem. Int.* **24,** 101–146.

Murayama, T., Kawashima, M., Takahashi, T., Yasuoka, T., Kuwayama, T., and Tanaka, K. (1997). Direct action of melatonin on hen ovarian granulosa cells to lower responsiveness to luteinizing hormone. *Proc. Soc. Exp. Biol. Med.* **215,** 386–392.

Nelson, W., Chong, S., Evans, J. E., and Sugden, D. (1993). N-bromoacetyl 5-methoxytryptamine: An irreversible melatonin ligand? *Biochem. Biophys. Res. Commun.* **193,** 1355–1361.

Niles, L. P., and Hashemi, F. (1990). Picomolar-affinity binding and inhibition of adenylate cyclase activity by melatonin in Syrian hamster hypothalamus. *Cell. Mol. Neurobiol.* **10,** 553–557.

Pang, C. S., Tang, P. L., Song, Y., Pang, S. F., Ng, K. W., Guardiola-Lemaitre, B., Delagrange, P., and Brown, G. M. (1997). Differential inhibitory effects of melatonin analogs and three naphthalenic ligands on 2-[125]iodomelatonin binding to chicken tissues. *J. Pineal Res.* **23,** 148–155.

Popova, J. S., and Dubocovich, M. L. (1995). Melatonin receptor-mediated stimulation of phosphoinositide breakdown in chick brain slices. *J. Neurochem.* **64,** 130–139.

Reiter, R. J. (1991). Pineal melatonin: Cell biology of its synthesis and of its physiological interactions. *Endocr. Rev.* **12,** 151–180.

Reiter, R. J. (1998). Melatonin and human reproduction. *Ann Med.* **30,** 103–108.

Reppert, S. M., Weaver, D. R., and Ebisawa, T. (1994). Cloning and characterization of a mammalian melatonin receptor that mediates reproductive and circadian responses. *Neuron.* **13,** 1177–1185.

Reppert, S. M., Godson, C., Mahle, C. D., Weaver, D. R., Slaugenhaupt, S. A., and Gusella, J. F. (1995a). Molecular characterization of a second melatonin receptor expressed in human retina and brain: The Mel1b melatonin receptor. *Proc. Natl. Acad. Sci. U.S.A.* **92,** 8734–8738.

Reppert, S. M., Weaver, D. R., Cassone, V. M., Godson, C., and Kolakowski, L. F., Jr. (1995b). Melatonin receptors are for the birds: Molecular analysis of two receptor subtypes differentially expressed in chick brain. *Neuron* **15,** 1003–1015.

Ross, A. W., Webster, C. A., Thompson, M., Barrett, P., and Morgan, P. J. (1998). A novel interaction between inhibitory melatonin receptors and protein kinase C-dependent signal transduction in ovine pars tuberalis cells. *Endocrinology (Baltimore)* **139,** 1723–1730.

Sack, R. L., Lewy, A. J., and Hughes, R.J. (1998). Use of melatonin for sleep and circadian rhythm disorders. *Ann Med.* **30,** 115–121.

Shochat, T., Haimov, I., and Lavie, P. (1998). Melatonin—the key to the gate of sleep. *Ann Med.* **30,** 109–114.

Shui-Wang, Y., Niles, L. P., and Crocker, C. (1993). Human malignant melanoma cells express high-affinity receptors for melatonin: Antiproliferative effects of melatonin and 6-chloromelatonin. *Eur. J. Pharmacol.* **246,** 89–96.

Sicsic, S., Serraz, I., Andrieux, J., Bremont, B., Mathe-Allainmat, M., Poncet, A., Shen, S., and Langlois, M. (1997). Three-dimensional quantitative structure-activity relationship of melatonin receptor ligands: A comparative molecular field analysis study. *J. Med. Chem.* **40,** 739–748.

Silman, R. (1991). Melatonin and the human gonadotrophin-releasing hormone pulse generator. *J. Endocrinol.* **128,** 7–11.

Song, Y., Pang, C. S., Ayre, E. A., Brown, G. M., and Pang, S. F. (1996). Melatonin receptors in the chicken kidney are up-regulated by pinealectomy and linked to adenylate cyclase. *Eur. J. Endocrinol.* **135,** 128–133.

Song, Y., Chan, C. W. Y., Brown, G. M., Pang, S. F., and Silverman, M. (1997). Studies of the renal action of melatonin: Evidence that the effects are mediated by 37 kDa receptors of the Mel1a subtype localized primarily to the basolateral membrane of the proximal tubule. *FASEB J.* **11,** 93–100.

Spadoni, G., Stankov, B., Durnati, A., Biella, G., Lucini, V., Salvatori, A., and Fraschini, F. (1993). 2-substituted 5-methoxy-N-acyltryptamines: Synthesis, binding affinity for the melatonin receptor and evaluation of the biological activity. *J. Med. Chem.* **36,** 4069–4074.

Spadoni, G., Balsamini, C., Diamantini, G., Di Giacomo, B., Tarzia, G., Mor, M., Plazzi, P. V., Rivara, S., Lucini, V., Nonno, R., Pannacci, M., Fraschini, F., and Stankov, B. M. (1997). Conformationally restrained melatonin analogues: Synthesis, binding affinity for the melatonin receptor, evaluation of the biological activity and molecular modeling study. *J. Med. Chem.* **40,** 1990–2002.

Spadoni, G., Balsamini, C., Bedini, A., Diamantini, G., Di Giacomo, B., Tontini, A, Tarzia, G., Mor, M., Plazzi, P. V., Rivara, S., Nonno, R., Pannacci, M., Lucini, V., Frachini, F., and Stankov, B. M. (1998). 2-[N-Acylamino(C1-C3)alkyl]indoles as MT1 melatonin receptor partial agonists, antagonists and putative inverse agonists. *J. Med. Chem.* **41,** 3624–3634.

Starkey, S. (1996). Melatonin and 5-hydroxytryptamine phase-advance the rat circadian clock by activation of nitric oxide synthesis. *Neurosci. Lett.* **211,** 199–202.

Sugden, D. (1994). N-acyl-3-amino-5-methoxychromans: A new series of non-indolic melatonin analogues. *Eur. J. Pharmacol.* **254,** 271–275.

Sugden, D., Chong, N. W. S., and Lewis, D. F. V. (1995). Structural requirements at the melatonin receptor. *Br. J. Pharmacol.* **114,** 618–623.

Tarzia, G., Diamantini, G., Di Giacomo, B., Spadoni, G., Esposti, D., Nonno, R., Lucini, V., Pannacci, M., Fraschini, F., and Stankov, B. M. (1997). 1-(2-Alkanamidoethyl)-6-methoxyindole derivatives: A new class of potent indole melatonin analogues. *J. Med. Chem.* **40,** 2003–2010.

Teh, M. T., and Sugden, D. (1998) Comparison of the structure-activity relationships of melatonin receptor agonists and antagonists: Lengthening the N-acyl side-chain has differing effects on potency on Xenopus melanophores. *Naunyn Schmiedeberg's Arch. Pharmacol.* **358,** 522–528.

Tenn, C., and Niles, L. P. (1993). Physiological regulation of melatonin receptors in rat suprachiasmatic nuclei: Diurnal thythmicity and effects of stress. *Mol. Cell. Endocrinol.* **98,** 43–48.

Underwood, H., Siopes, T., and Barrett, R. K. (1988). Does a biological clock reside in the eye of the quail? *J. Biol. Rhythms* **3,** 323–331.

Vanecek, J., and Watanabe, K. (1998). Melatonin inhibits the increase of cyclic AMP in rat suprachiasmatic neurons induced by vasoacitve intestinal peptide. *Neurosci. Lett.* **252,** 21–24.

Viswanathan, M., Laitinen, J. T., and Saavedra, J. (1990). Expression of melatonin receptors in arteries involved in thermoregulation. *Proc. Natl. Acad. Sci. U.S.A.* **87,** 6200–6203.

von Gall, C., Duffield, G. E., Hastings, M. H., Kopp, M. D. A., Dehghani, F., Korf, H.-W., and Stehle, J. H. (1998). CREB in the mouse SCN: A molecular interface coding the phase-adjusting stimuli light, glutamate, PACAP, and melatonin for clockwork access. *J. Neurosci.* **18,** 10389–10397.

von Zastrow, M., and Kobilka, B. L. (1994). Antagonist-dependent and -independent steps in the mechanism of adrenergic receptor internalization. *J. Biol.Chem.* **269,** 18448–18452.

Wiesenberg, I., Missbach, M., Kahlen, J.-P., Schrader, M., and Carlberg, C. (1995). Transcriptional activation of the nuclear receptor RZRα by the pineal gland hormone melatonin and identification of CGP 52608 as a synthetic ligand. *Nucleic Acids Res.* **23,** 327–333.

Witt-Enderby, P. A., and Dubocovich, M. L. (1996). Characterization and regulation of the human ML1A melatonin receptor stably expressed in Chinese hamster ovary cells. *Mol. Pharmacol.* **50,** 166–174.

Witt-Enderby, P. A., Chu, G. H., Gillen, M. L., and Li, P.-K. (1997). The development of a high-affinity ligand that binds irreversibly to mel1b melatonin receptors. *J. Med. Chem. Lett.* **40,** 4195–4198.

Witt-Enderby, P. A., Masana, M. I., and Dubocovich, M. L. (1998). Physiological exposure to melatonin supersensitizes the cyclic AMP-dependent signal transduction cascade in CHO cells expressing the human mt1 melatonin receptor. *Endocrinology (Baltimore)* **139,** 3064–3071.

Wurtman, R. J., and Zhdanova, I. (1995). Improvement of sleep quality by melatonin. *Lancet* **346,** 1491.

Yie, S.-M., Niles, L. P., and Younglai, E. V. (1995). Melatonin receptors on human granulosa cell membranes. *J. Clin. Endocrinol. Metab.* **80,** 1747–1749.

Ying, S., Niles, L. P., and Crocker, C. (1993). Human malignant melanoma cells express high-affinity receptors for melatonin: Antiproliferative effects of melatonin and 6-chloromelatonin. *Eur. J. Pharmacol.* **246,** 89–96.

Yous, S., Andrieux, J., Howell, H. E., Morgan, P. J., Renard, P., Pfeiffer, B., Lesieur, D., and Guardiola-Lemaitre, B. (1992). Novel naphthalenic ligands with high affinity for the melatonin receptor. *J. Med. Chem.* **35,** 1484–1486.

VITAMINS AND HORMONES, VOL. 58

Vitamin K-Dependent Proteins

GARY L. NELSESTUEN, AMIT M. SHAH, AND STEPHEN B. HARVEY

Department of Biochemistry, University of Minnesota, St. Paul, Minnesota 55108

Vitamin K is required for the synthesis of γ-carboxyglutamate (Gla) during postribosomal protein modification. Substrates include blood clotting proteins, bone proteins, cell signaling, and receptor proteins. In addition, Gla is a component of short toxin peptides from the marine snail *Conus*. Studies of structure–function relationships are the most advanced for the blood coagulation proteins. Reviews of vitamin K action and blood coagulation are presented. Special focus is on the structure–function role of Gla in blood coagulation and the impact of this amino acid on enzyme reaction kinetics. This amino acid forms calcium and membrane binding sites for these proteins. Two proposed mechanisms of protein–membrane attachment are reviewed. One involves membrane attachment by protein insertion into the hydrocarbon region of the membrane, while another considers attachment by specific interactions with phospholipid head groups. Membrane attachment generates the potential for several forms of nonclassical enzyme kinetic behaviors, all of which have been observed *in vitro*. For example, the reaction may be limited by properties of the enzyme active site, a condition that allows use of classic steady-state enzyme kinetic parameters. However, the reaction may be limited by substrate binding to the membrane, by substrate flux through solution, and/or by solvent flow rates across the membrane surface. These states provide special mechanisms that are not anticipated by classical steady-state kinetic derivations. They may

0083-6729/00 $30.00

be used to regulate coagulation *in vivo*. Overall, vitamin K research spans the spectrum of biological research and experience. Exciting new ideas and findings continue to emanate from vitamin K-related research.

I. Introduction

Vitamin K, discovered by Henrik Dam and his colleagues in 1929 (Dam, 1929, 1930, 1934, 1935), is essential for several important physiologic processes (for reviews of vitamin K, see Dowd *et al.,* 1995; Suttie, 1985). The biochemical role of vitamin K is the posttranslational conversion of a number of glutamic acids in precursor proteins to γ-carboxyglutamic acids (Gla) in the mature proteins. This novel amino acid was initially identified in bovine prothrombin (Magnusson *et al.,* 1974; Nelsestuen *et al.,* 1974; Stenflo, 1974) and was subsequently identified in other vitamin K-dependent plasma proteins, including factor VII, factor IX, factor X, protein C, protein S, and protein Z. Gla residues are critical for calcium binding to these proteins in order to create a membrane contact site. Without Gla, the proteins lack physiologic function.

Six of the seven plasma proteins just listed function in blood coagulation and anticoagulation. Recently, a number of other proteins have been discovered. Some have known function while others have only been identified by sequence homology to the blood clotting proteins. Two bone proteins, osteocalcin and matrix Gla protein (Gallop *et al.,* 1980; Price, 1989), are vitamin K dependent. Although early studies of function were inconclusive, recent experiments with gene knockouts have suggested that osteocalcin is involved in the limitation of bone growth (Ducy *et al.,* 1996) and that bone matrix protein is needed for prevention of calcification of soft tissues such as arteries (Luo *et al.,* 1997). The mechanisms of these functions are not well understood. The newly discovered growth arrest-specific protein, Gas 6, is also dependent on vitamin K (Manfioletti *et al.,* 1993). Gas 6 may be a ligand for members of a family of tyrosine kinases (Stitt *et al.,* 1995; Varnum *et al.,* 1995). Recent sequence analysis suggests vitamin K-dependent peptide sequences in abundant transmembrane proteins of the spinal cord and thyroid (Kulman *et al.,* 1997). Other regions of these proteins suggest a possible role in cell signaling. Finally, the conotoxins include vitamin K-dependent peptides found in the marine snail *Conus.* These are peptides of about 20 residues that include neurotoxin functions. Recent structural studies suggest the role of Gla in metal ion binding to these materials (Chen *et al.,* 1998; Rigby *et al.,* 1997). Thus, vitamin K or vitamin K inhibitors, like dicoumarol or Warfarin, can impact on many biologic processes in mammals, other vertebrates, and even some invertebrate species. This review concentrates on the proteins of the blood coagulation cascade, where knowledge of structure–function is most advanced.

Understanding of the role of vitamin K is not complete until the function of the resulting structure, γ-carboxyglutamic acid, is known. Function includes knowledge of protein structure, the physical interactions created by this structure, and the impact of these interactions on the relevant biological processes. Advances in these latter topics are highlighted in the latter parts of this review. However, to present a more complete picture of vitamin K, we begin with a summary of Gla synthesis and a description of the principal coagulation systems in which this amino acid functions.

II. Synthesis of Gla

The synthesis of Gla from glutamic acid by γ-glutamyl carboxylase occurs in the endoplasmic reticulum of liver cells (Carlisle and Suttie, 1980). Although seldom discussed, one role of the vitamin seems noncontroversial. That is, transformation of the reduced vitamin (Fig. 1) to vitamin K epoxide, like virtually all direct oxidations by molecular oxygen, is a thermodynamically favorable process. In contrast, the fixation of CO_2 is energetically unfavorable. For example, carboxylation of pyruvate to oxaloacetate has a standard free-energy change of about +6 kcal/mol. At intracellular metabolite concentrations, condensation of two substrates (CO_2 plus the target of carboxylation) to one product (oxaloacetate) creates a further free-energy change of at least +4 kcal/mol (Nelsestuen, 1989). Thus, all carboxylation reactions must be linked to an energetically favorable event in order to generate product. Direct oxidation of reduced vitamin K to vitamin K epoxide provides the energy for this carboxylation.

The manner in which carboxylation is coupled to vitamin K oxidation is not as clear. The vitamin appears to add oxygen to generate a hydroperoxide intermediate (Dowd *et al.,* 1995). The hydroperoxide anion is suggested to abstract the γ-hydrogen from a glutamic acid (Ham and Dowd, 1990), generating the necessary intermediate for carboxylation. This role for the vitamin does not appear unique, since many enzymes abstract a proton alpha to a carboxyl in the course of an enzyme-catalyzed reaction. Examples include the enolase enzyme family (Babbitt

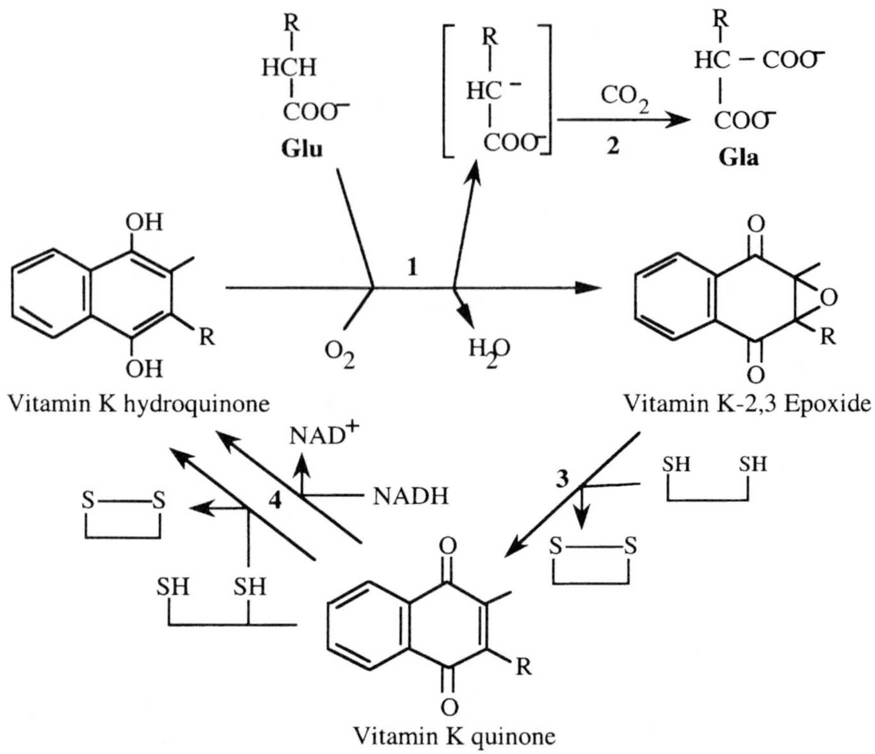

FIG. 1. Vitamin K reaction cycle.

et al., 1996) where lysine or glutamate are used to abstract the alpha proton. Although much is known, full understanding of the carboxylation mechanism remains a future goal.

Regeneration of vitamin K hydroquinone is catalyzed by vitamin K–2,3-epoxide reductase, which can utilize thiols such as lipoic acid. The 2,3-epoxide of vitamin K is first converted to vitamin K quinone by an enzyme that uses free thiols as the reducing agent. The subsequent reduction of vitamin K quinone to the hydroquinone may be catalyzed by the same enzyme (Gardill and Suttie, 1990) or by a related enzyme sharing many of the same properties as vitamin K–2,3-epoxide reductase. NADH-linked dehydrogenases are also able to reduce vitamin K quinone (Fasco and Principe, 1980). Reactions 3 and 4 are severely inhibited by coumarin anticoagulants (Fasco and Principe, 1982).

These mechanisms have been supported by observation of significant levels of vitamin K hydroquinone in liver microsomes from Warfarin-resistant rats (Fasco *et al.,* 1983). In addition, recent reports suggest that the enzyme is highly processive, catalyzing multiple carboxylations during a single association between γ-glutamyl carboxylase and its substrate (Morris *et al.,* 1995). Recent efforts have also focused on substrate specificity, including the role of specific amino acids of the signal peptide in recognition by the carboxylase (Furie *et al.,* 1997; Li *et al.,* 1997; Sugiura *et al.,* 1997). Another recent review outlines some of these results (Furie and Furie, 1997).

III. Blood Coagulation Cascade

The vascular system provided a major evolutionary advantage by providing transport of essential nutrients and oxygen to tissues. However, this required a hemostatic system that would rapidly recognize wounds and would generate an insoluble plug to stop blood loss. The main constituents of the clot include platelets and arrays of cross-linked fibrin that form an insoluble fibrous network. The physiologic response requires a series of reactions that behave in a localized, amplified, and highly modulated manner. Recent reviews on blood coagulation can be consulted for more detail (Davie *et al.,* 1991; Furie and Furie, 1988, 1992; Kalafatis *et al.,* 1994b; Mann *et al.,* 1990).

Figure 2 presents an abbreviated version of the coagulation cascade that highlights the roles of vitamin K-dependent proteins and emphasizes their homology. The major participants include the vitamin K-dependent substrates, prothrombin, factor X, factor IX, and factor VII.

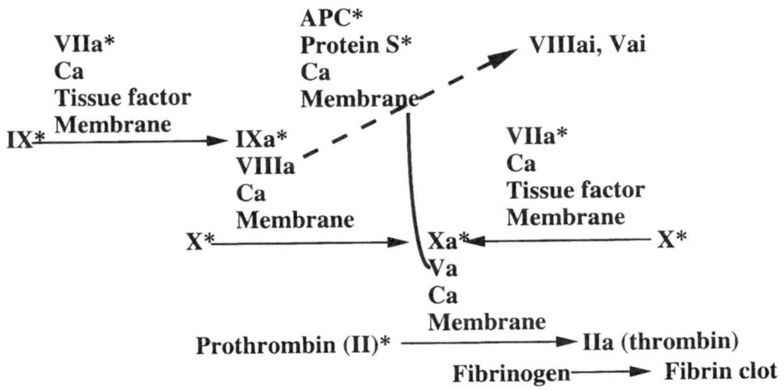

Fig. 2. Steps involving vitamin K proteins in the blood coagulation cascade. Procoagulation reactions are shown in solid lines, whereas inhibition reactions are shown by dashed lines. Vitamin K-dependent proteins have an asterisk (*).

These are zymogens that are activated to serine proteases by other proteases further up the cascade. Other proteins, including factor Va, factor VIIIa, and tissue factor (TF), serve as cofactors to enhance the rate of zymogen activation. Factors V and VIII are converted to their active forms by limited proteolysis, often by thrombin, the product of the cascade. Factors V and VIII are also membrane-binding proteins that utilize non-Gla-dependent membrane contact sites. Tissue factor is an integral membrane protein that is not subject to activation by proteolysis. However, it may also require an activation step, as suggested by "cryptic" expression in normal cell membranes (Bach and Moldow, 1997).

Figure 3 provides a representation of a typical enzyme complex, showing very general relationships of enzyme, cofactor, and substrate. The drawing shows the common theme of a membrane that serves as the staging site for assembly of the enzyme, which in all cases is a serine protease plus a cofactor protein. Membrane attachment of the substrate will contribute to substrate–enzyme assembly. Few details are depicted in this drawing. For example, the cofactor may participate in substrate as well as enzyme binding.

The coagulation cascade has two arms, described as the intrinsic and extrinsic pathways. The intrinsic pathway is initiated by factors XII and XI, proteases that are not vitamin K dependent and are therefore not included in Fig. 2. The extrinsic pathway, so called because one of the components, tissue factor, is located in tissue adventitia, comes in contact with blood only after vascular damage. TF is an integral membrane glycoprotein consisting of a 219-amino-acid extracellular domain, a 23-residue transmembrane region, and a 21-residue cytoplasmic domain (Morrissey *et al.,* 1987; Scarpati *et al.,* 1987; Spicer *et al.,* 1987). Upon vascular damage and TF exposure, TF forms a one-to-one complex with factor VII. This assembly potentiates the conversion of

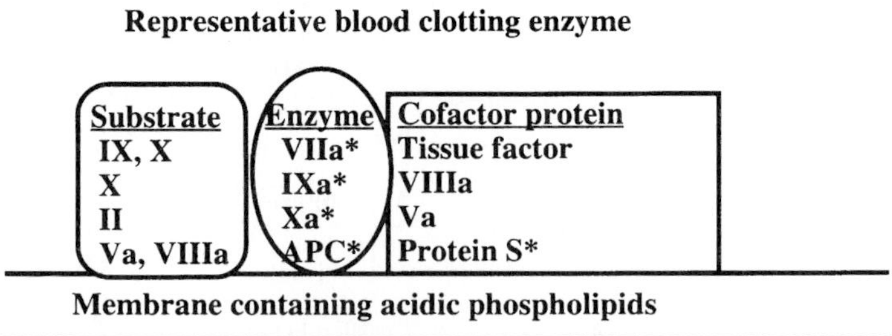

FIG. 3. Typical enzyme–substrate complexes involving vitamin K-dependent proteins.

factor VII to VIIa by autolysis (Bach *et al.,* 1986). Activation of factor VII is also accomplished by proteolysis of the Arg_{152}-Ile peptide bond (Hagen *et al.,* 1986) by a number of proteases such as factors Xa and IXa.

Once formed, the VIIa–TF complex converts factors X and IX to serine proteases (factor Xa and factor IXa). TF is a positively enhancing cofactor for factor VIIa, causing a conformational change in factor VIIa (Nemerson and Gentry, 1986). It enhances the rate of factor X activation by 4000-fold over the rate with factor VIIa alone (Bach *et al.,* 1986). Thus, TF serves as a cell binding protein, as an activator to factor VII, and as a cofactor for factor VIIa.

The generation of factor Xa allows the formation of prothrombinase, which produces thrombin from prothrombin. Prothrombinase is composed of factor Xa and factor Va, which assemble on a procoagulant membrane surface in the presence of calcium (Tracy *et al.,* 1981). Factor V, a non-vitamin-K-dependent protein, circulates in plasma as a 330,000-Da single chain glycoprotein (Nesheim *et al.,* 1979a). It is also located in the platelet α-granules (Tracy *et al.,* 1982). Proteolysis by thrombin converts factor V to factor Va (Esmon, 1979; Nesheim and Mann, 1979). It also can be activated by factor Xa (Foster *et al.,* 1983; Monkovic and Tracy, 1990). Factor Va is composed of a heavy chain (M_r = 105,000) containing the amino terminal part of factor V and a light chain (M_r = 74,000) containing the carboxyl-terminal part of factor V (Esmon, 1979). Recent studies have found two forms of human factor Va in plasma with different procoagulant activities (Hoekema *et al.,* 1997; Rosing *et al.,* 1993). Although understanding of membrane contact by factors V and VIII is incomplete, they may involve ionic as well as nonionic forces (Cutsforth *et al.,* 1996; Gilbert and Arena, 1995; Pusey *et al.,* 1982).

As with other cofactors, factor Va enhances prothrombinase activity by several thousand-fold (Nesheim *et al.,* 1979b). After thrombin is formed, it catalyzes the conversion of fibrinogen to fibrin by limited proteolysis, which leads to fibrin self-association to a polymer and the insoluble blood clot.

Activation of factor IX by factor VIIa–TF occurs in a manner that is condition dependent (Baur *et al.,* 1990; Østerud and Rapaport, 1977). At low TF concentrations, factor IX may be a more suitable substrate than factor X (Komiyama *et al.,* 1990; Zur and Nemerson, 1980). In mixtures, factor IX also appears to be the preferred substrate (Lawson and Mann, 1991). The newly generated factor IXa forms a complex with factor VIIIa in the presence of calcium and phospholipid to generate the other Xase enzyme. Like factor Va in the prothrombinase, factor VIIIa participates as a positively enhancing cofactor by increasing the V_{max}

of the reaction by several thousand-fold (van Dieijen *et al.,* 1981). Factor VIII is activated by thrombin (Eaton *et al.,* 1986; Lollar and Parker, 1989) in a manner analogous to factor V activation.

A popular model for coagulation involves use of both the intrinsic and extrinsic pathways of blood coagulation. First, exposure of TF by vascular injury and the assembly of the factor VIIa–TF complex triggers the hemostatic response through activations of small quantities of factor X and factor IX. The continued formation of factor Xa is then accomplished by the factor VIIIa–factor IXa complex. Factor Xa forms the terminal procoagulant complex, prothrombinase, with factor Va. Prothrombinase then produces thrombin, which leads to the formation of the fibrin network. This model demonstrates the involvement of the extrinsic pathway during the initiation of blood clotting, while the intrinsic pathway is the primary route for the continued growth of the clotting response.

Genetic deficiency in the coagulation proteins leads to bleeding disorders known as hemophilia. These disorders establish the importance of these proteins. For example, deficiency of factor VIII results in the most common bleeding disorder, hemophilia A (Sadler and Davie, 1987). Another common disorder, hemophilia B, is a consequence of factor IX deficiency (Hedner and Davie, 1987). Molecular defects in other proteins, such as factor X and factor VII, also cause bleeding tendencies. For example, substitution of Gla-7 by Gly in factor X (Rudolph *et al.,* 1996) results in reduced clotting activity.

Vitamin K-dependent proteins can be effective in treatment of hemophilia. While a role for factor IX in treatment of hemophilia B is apparent, high doses of factor VIIa are also effective (Hedner, 1996).

IV. Inhibition of Blood Coagulation

Hemostasis requires systems that inactivate the coagulation response and prevent overcoagulation. A number of serine protease inhibitors act directly on the vitamin K-dependent protease family. Examples include antithrombin III (AT-III), a 58,000-Da protein that is a member of the serpin inhibitor family and circulates in plasma at a concentration of 2–3 μM. The primary targets of AT-III are thrombin and factor Xa. Inhibition is greatly accelerated by heparin, a normal constituent of mast cells and a clinical anticoagulant (reviewed by Becker and Alpert, 1990). The importance of AT-III was confirmed by the association of thrombosis tendency with heterozygous AT-III deficiency (for a review on AT-III, see Olson *et al.,* 1993). HC-II is another thrombin inhibitor that is present at a relatively high concentration. This inhibition is also accelerated by heparin as well as by dermatan sulfate (for a review on HC-II, see Tollefsen, 1995). HC-II inhibition may be of lesser importance, since heterozygous deficiency of HC-II is found in about 1% of the population and does not appear to be a risk factor for thrombosis.

Tissue factor pathway inhibitor (TFPI) is present at relatively low concentrations. It binds at or near the active site of factor Xa in a calcium-dependent manner (for a review on TFPI, see Rapaport, 1991). The TFPI–factor Xa–calcium complex binds to the factor VIIa–TF complex, resulting in the formation of a TF–factor VIIa–TFPI–factor Xa quaternary complex. Bivalent inhibition of factor Xa and the factor VIIa–TF complex provides an effective mechanism to shut down the activation of factors IX and X.

Protein C, a 60,000-Da vitamin K-dependent protein, is also a critical anticoagulant. Several recent reviews of the protein C pathway have been presented (Dahlbäck, 1991; Esmon, 1989; Walker and Fay, 1992) and a brief summary is included here. Like the vitamin K-dependent proteins of the procoagulant pathway, protein C is the zymogen of a serine protease. It is converted to activated protein C (APC) by proteolysis. The enzyme is thrombin in complex with a membrane-associated cofactor protein, thrombomodulin. APC then inactivates two procoagulant proteins, factors Va and VIIIa, by limited proteolysis (Fig. 2).

A second vitamin K-dependent protein, protein S, serves as a nonenzymatic cofactor for APC. Protein S circulates in human plasma as a mixture of free protein and in a reversible complex with C4b binding protein, a component of the classical complement cascade (Dahlbäck and Stenflo, 1981). When bound to C4b binding protein, protein S does not function as a cofactor for APC (Dahlbäck, 1986). Although current ideas suggest a 1:1 complex of APC and protein S on the membrane, as shown in Fig. 3, details are not clear. Protein S has been reported to increase APC activity by 10-fold (Walker, 1980) or by as little as 2-fold (Bakker *et al.,* 1992). Human protein S may act under select conditions. For example, it was shown to enhance the APC-dependent cleavage of membrane-bound human factor Va at Arg^{306} (Rosing *et al.,* 1995). It may function to alleviate the protective effect of factor Xa on factor Va (Rosing *et al.,* 1995; Solymoss *et al.,* 1988). Finally, protein S may inhibit prothrombinase through direct interactions with factors Va and Xa (Hackeng *et al.,* 1994; Heeb *et al.,* 1993, 1994; Mitchell *et al.,* 1988). Regardless of the mechanism of action of protein S, a role in anticoagulation is suggested by thrombotic complications in patients deficient in protein S (Griffin *et al.,* 1981).

The biological significance of protein C has been established by severe thrombotic episodes in individuals who present homozygous protein C deficiency (Clouse and Comp, 1986). Another hereditary defect is characterized by a poor anticoagulant response to APC, referred to as APC resistance. This disorder is at least 10 times more common than all other known genetic thrombosis risk factors combined and is correlated with a mutation in factor V (for a review on APC resistance, see Dahlbäck, 1995). The heavy chain of plasma human factor Va contains three peptide bonds that can be cleaved by APC. These are at Arg^{306}, Arg^{506}, and Arg^{679} (Kalafatis *et al.,* 1994a; Nicolaes *et al.,* 1995). Factor V of APC-resistant patients has a single substitution of Arg^{506} to glutamine (Bertina *et al.,* 1994).

Overall, vitamin K-dependent proteins are critical components of anticoagulation pathways. This function may find clinical uses. For example, thrombosis during sepsis may arise from a decrease in protein C levels. Supplementation with protein C may be an effective treatment for this condition (Fourrier *et al.,* 1995; Rintala *et al.,* 1998).

V. Structural Design of Vitamin K-Dependent Plasma Proteins

The vitamin K-dependent plasma proteins of blood coagulation and anticoagulation cascades have common, elegant structural strategies (reviewed by Sunnerhagen *et al.,* 1996; Tulinsky, 1991). The general picture is summarized by examination of the factor VIIa–TF complex shown in Fig. 4. Approximately the amino terminal 45 residues comprise the "Gla domain," which contains 9–13 γ-carboxyglutamic acid residues, depending on the vitamin K-dependent protein. This domain is discussed in a separate section later. The Gla domain is followed by other recognizable protein domains such as the kringle (prothrombin) or regions homologous to the epidermal growth factor (EGF domains, all other plasma proteins; see Fig. 4, factor VII). Finally, the proteins contain a large, carboxy-terminal domain that contains the major enzymatic function of blood coagulation factors II, VII, IX, X, and protein C.

The number of domains intermediate between the Gla and enzymatic domains is variable. For example, prothrombin contains a short disulfide loop plus two kringles. Factor VII contains two EGF domains (Fig. 4). Protein S contains a disulfide loop that is homologous to the prothrombin structure, plus four EGF domains. These intermediate domains are involved in protein–protein interactions and/or may provide "spacers" that allow proper positioning of other parts of the molecule for optimum biological function (Sunnerhagen *et al.,* 1996). Both of these functions are illustrated by the factor VIIa–TF structure (Fig. 4).

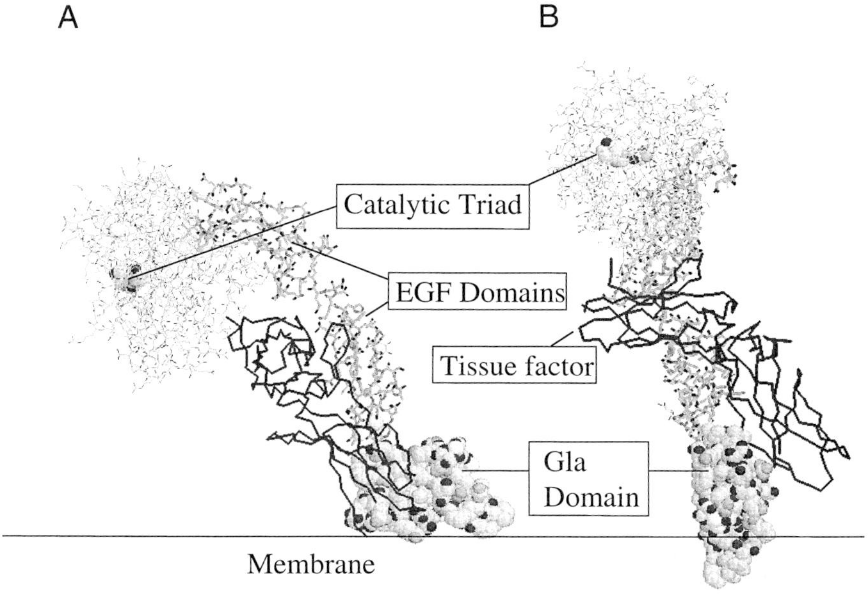

FIG. 4. Factor VIIa–TF complex with two possible membrane contact sites (A and B). The structures are shown from the RasMol program, using PDB file 1DAN. Tissue factor is given in a peptide backbone trace. Residues of the Gla domain are space filled, those of the EGF domains are shown in "sticks," and the catalytic domain of factor VIIa is in wire-frame, except for active site residues Ser-195 and His-57, which are space filled.

The catalytic domains of factors II, VII, IX, X, and protein C are homologous to the pancreatic serine proteases, trypsin and chymotrypsin. The corresponding, nonserine protease domain of protein S is homologous to sex-hormone-binding globulin (Baker *et al.,* 1987; Gershagen *et al.,* 1991). The carboxyl terminal of the seventh plasma protein, protein Z, is not a serine protease precursor and its function is not known (Höjrup *et al.,* 1985).

Overall, it is quite evident that the vitamin K-dependent plasma proteins were formed by a series of gene duplications to produce a family of proteins with similarities in some domains but large differences in others. The differences provide for divergent functions.

VI. STRUCTURE–FUNCTION OF THE GLA DOMAIN

A major achievement in vitamin K research was the solution of the three-dimensional structure of a Gla domain, prothrombin fragment

1 (Fig. 5) (Seshadri *et al.,* 1994; Soriano-Garcia *et al.,* 1992). Very similar structures have been reported for the Gla domains of factor VIIa (Fig. 4, Banner *et al.,* 1996)) and factor IX (Freedman *et al.,* 1995a,b). The structure revealed many unusual features. For example, most Gla residues point inward toward calcium ions bound in the interior of the domain. This was unexpected. Many early ideas anticipated that anionic Gla residues would reside on the protein surface where they could interact with calcium ions, which, in turn, interact with anionic phospholipids of the membrane. This mechanism of membrane association is often referred to as a *calcium-bridging model.* This concept may suit other proteins such as annexins (Swairjo *et al.,* 1995) and serum amyloid P (Emsley *et al.,* 1994), which also show calcium-dependent membrane binding. In keeping with the function of a novel amino acid, however, many new structure–function aspects have been discovered, or remain to be fully explained, for the vitamin K-dependent proteins.

A. Calcium Binding Sites

Despite two carboxyl groups per side chain and the presence of side-by-side pairs of Gla residues in the protein sequence, the affinity of Gla for calcium is quite low. Individual or paired Gla residues do not create functional calcium binding sites, even at extracellular calcium concentrations. Proper protein folding is necessary to generate adequate affinity (Nelsestuen *et al.,* 1976). The Gla domain of prothrombin appears to

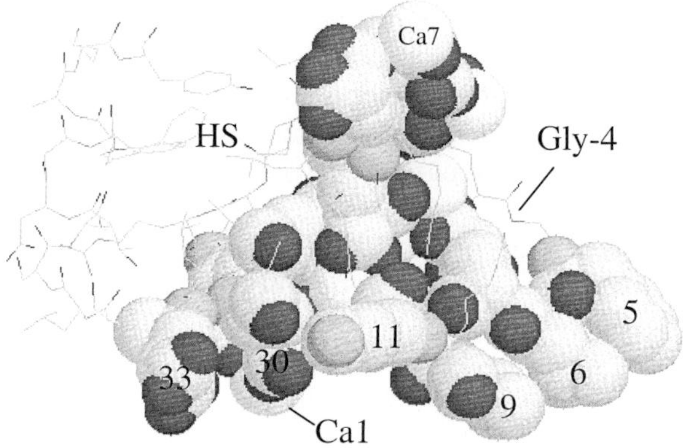

Fig. 5. The Gla domain of prothrombin (Protein Data Bank reference 2PF2). Residues 1–45 are shown with the program RasMol. Gla residues are depicted in space-filled form, as are residues 5, 6, 9, and 11. Calcium ions 1 and 7 are labeled. Other calcium ions are located between these two ions.

be disordered in the absence of calcium (Park and Tulinsky, 1986; Seshadri *et al.,* 1991; Tulinsky *et al.,* 1988) and becomes highly ordered (Soriano-Garcia *et al.,* 1992) upon calcium binding. This differs from factor X, which is found in a distinct, but very different, conformation in the absence of calcium (Sunnerhagen *et al.,* 1995). The calcium-dependent peptide folding transition is evident from changes in CD spectra and intrinsic protein fluorescence.

Bovine prothrombin is unique and shows a calcium-dependent transition with kinetics that resemble *trans-* to *cis*-proline isomerization (Marsh *et al.,* 1979) with a half-time of several minutes at room temperature and activation energy of about 20 kcal/mol (Nelsestuen, 1976). Few other protein conformational changes have these kinetic properties. The conformational change is essential for membrane interaction. This conformational change was apparent in residues 1–45 of bovine prothrombin, which contains a single proline at residue 22 (Evans and Nelsestuen, 1996b). Thus, it was unexpected that proline-22 appeared to be in the *trans* configuration in the X-ray structure of bovine prothrombin fragment 1 (Soriano-Garcia *et al.,* 1992). Furthermore, recent modeling studies have suggested little impact of the *cis* or *trans* conformations of proline-22 on other parts of the Gla domain (Perera *et al.,* 1998). How proline isomerization impacts membrane association is therefore unclear. Some adjustments in the structure may be needed (Evans and Nelsestuen, 1996b) or other explanations for the conformational change determined.

The crystal structure shows seven calcium ions in the Gla domain of prothrombin (Fig. 5 Soriano-Garcia *et al.,* 1992) and factor VIIa (Figure 4) (Banner *et al.,* 1996). Placement of these metal ions is nearly identical in the two proteins. A stoichiometry of seven was found for direct calcium binding in solution (Deerfield *et al.,* 1987). Four of the seven calcium ions (Ca-2, -3, -4 and -5) are buried. These four plus one more form a linear array that passes directly through the center of the Gla domain. Only two calcium ions are well exposed to solvent. In the presence of acidic phospholipid membranes, prothrombin binds additional calcium ions (Evans and Nelsestuen, 1994; Nelsestuen and Lim, 1977; Sommerville *et al.,* 1986), suggesting that the protein crystal structure is insufficient to explain membrane contact. A number of multivalent cations, such as magnesium and strontium, cause a protein conformational change but only a few metals are able to support protein association with membranes (Nelsestuen *et al.,* 1976). The ability of calcium to accept variable coordination numbers and geometries may be critical to forming the membrane contact site.

The Gla domain has several other notable features (Soriano-Garcia *et al.,* 1992). The amino terminus forms three hydrogen bonding interactions with Gla-17, Gla-21, and Gla-27, which makes the amino group inaccessible to solvent or to water-soluble chemical reagents (Welsch and Nelsestuen, 1988). Derivatization of the amino terminal, even with retention of charge, causes loss of membrane binding function. Arg-55 of the second disulfide loop forms a salt bridge with Gla-15 and Gla-20. This interaction may be important to high calcium binding affinity. For example, truncated proteins such as prothrombin residues 1–45 have lowered affinity for calcium (Evans and Nelsestuen, 1996b). Some contribution of other parts of the protein is also suggested by factor IX (Freedman *et al.,* 1995b) and protein C (Colpitts and Castellino, 1994) where residues 1–45 are insufficient to bind calcium with the affinity of the intact protein. The Cys-18–Cys-23 disulfide is buried near the aromatic cluster of Phe-41, Trp-42, and Tyr-45 (Fig. 5). This interaction is the basis for fluorescence quenching of tryptophan upon calcium binding. The Sr-prothrombin fragment 1 structure contains an additional metal ion bound to Gla-30 (Seshadri *et al.,* 1994) (Fig. 5). This metal ion may constitute a part of the membrane contact site (McDonald *et al.,* 1997b).

The helix between residues 36 to 47 is often referred to as the *helical stack* (HS, Fig. 5) and may also contribute to protein folding stability. That it is not essential for membrane contact is suggested by the ability of prothrombin residues 1–38 to bind to membranes (Evans and Nelsestuen, 1996a). Lowered protein folding stability of the 1–38 peptide was evident from its requirement for higher calcium concentrations. Preceding the HS region, there are two shorter helical segments and a reverse turn, which are antiparallel to the HS. The amino-terminal 12 residues present the shape of an ω-loop. This contains a cluster of three solvent-exposed hydrophobic residues, Phe-5, Leu-6, and Val-9.

B. Membrane Association

The nature of the membrane contact site is still under investigation. There are few precedents for peripheral calcium-dependent membrane-binding protein structures so that calcium and/or membrane contact sites cannot be identified by homology to known structures such as a transmembrane helix or EF-hand. Most of the experimental evidence can be interpreted in a manner that argues against earlier, simple models for membrane contact. In fact, it is possible that no function of the Gla domain has been obvious from initial inspections or has fulfilled initial expectations.

In a search for the membrane contact site, a concern is the necessary size of the site, or "How many attachments are needed to create the observed contact affinity?" Enzyme reaction with low molecular weight substrates provides a useful comparison. The K_M of aldolase for Fru-1,6-bisphosphate and that of phosphoglycerate kinase for phosphoglycerate are each about 10 μM. A K_D of this magnitude represents about 75% of the free energy of prothrombin–membrane association. Thus, a small region of the Gla domain and/or the membrane would seem sufficient for membrane contact. In fact, a large contact interface involving many Gla residues and/or calcium ions may be precluded. For example, phospholipase A shows a membrane-binding site consisting of several surface hydrophobic residues plus ionic contact between one phospholipid head group and a calcium ion bound to the protein (Scott *et al.*, 1991). The affinity of phospholipase A for phospholipid vesicles is so high that dissociation rates are measured in hours (Jain *et al.*, 1995). Annexin proteins may utilize 10–15 bridging calcium ions plus some hydrophobic contacts between the protein and membrane (Evans and Nelsestuen, 1994; Swairjo *et al.*, 1995). Again, these proteins bind to membranes with high affinity, giving dissociation times of hours or even days (Bazzi and Nelsestuen, 1991; Evans and Nelsestuen, 1994). In contrast, the vitamin K-dependent proteins show dissociation rates from PS/PC (20/80) that are measured in milliseconds to seconds (Lu and Nelsestuen, 1996a; McDonald *et al.*, 1997b). Thus, the membrane contact region of the vitamin K-dependent proteins needs to, or even must, involve a small area of the Gla domain and/or the membrane.

An important definition for data interpretation is the magnitude of a ΔK_D that signifies a "significant" change in the binding site. Regardless of impact on biological function, a protein modification that generates a 10-fold change in membrane binding affinity has altered a small portion of the interaction energy (a 10-fold change in K_D corresponds to a $\Delta\Delta G$ of 1.4 kcal/mol versus a total ΔG of -10 kcal/mol for membrane binding with $K_D = 10^{-7}$ M) and has left the binding site nearly unchanged. With this interpretation, it is surprising to learn the extent to which the Gla domain can be modified with a small change in the membrane-binding function.

Evidence of critical residues of the Gla domain are rather limited. Only three of the nine conserved Gla residues appear to require both carboxyls for high function. Replacement of each Gla residue of human prothrombin (Ratcliffe *et al.*, 1993) and factor X (Larson *et al.*, 1998) by Asp showed high loss of function for only three sites: Gla-17, -27, and -30 (numbered with homology to bovine prothrombin, Fig. 5). Similar results were obtained for human protein C, except that requirement for

Gla-30 was not as apparent (Christiansen *et al.,* 1994). Nonconservative changes include Gla-6 mutation to Val in human protein C (Christiansen *et al.*, 1994). Although enzyme activity was greatly reduced, membrane affinity was reduced to a small degree, by the criterion given earlier. Nonconservative replacement of the only conserved basic residue of the Gla domain, Arg-15 of human protein C, resulted in decreased enzyme function, but membrane binding was reduced by only a few fold (Thariath and Castellino, 1997).

Based on comparison of sequence and membrane affinity of the naturally occurring vitamin K-dependent proteins, a membrane contact site centered at Gla-30 of fragment 1 was proposed ((McDonald *et al.,* 1997a,b); Fig. 5). Various physical forces, including a single ion pair in an isolated environment (McDonald *et al.*, 1997a), may be involved. This binding site seems consistent with the essential nature of Gla-30, -26, and of nearby Gla-17. In contrast to Gla-30, the Gla residues that bind Ca-7 in human protein C were not highly important so that Ca-7 did not appear to be important to membrane contact. A high importance of adjacent Gla residues indicated that calcium 2 was important in membrane association (Zhang and Castellino, 1993).

Further evidence supporting this membrane contact site was obtained by site-directed mutations of nearby residues in factors VII and protein C. Replacement of Pro-10 and Lys-32 (homologous to 11 and 33 of prothrombin, Fig. 5) of human factor VII enhanced membrane affinity and protein function in many circumstances (Shah *et al.,* 1998). Although improvement of binding affinity was important, the magnitude of change (about 20-fold) suggested a peripheral role for these residues. Site-directed mutation of Pro-10 of bovine protein C to His resulted in enhanced membrane affinity (Shen *et al.,* 1997). Change of Ser-11 to Gly in human protein C also provided enhanced membrane affinity (Shen *et al.,* 1998). Ser at this position (homologous to residue 12 of prothrombin, Fig. 5) is unique to human protein C. All other proteins of this family contain Gly. Other site-directed mutations in this region were able to enhance calcium binding to the protein without improving membrane contact affinity (Shen *et al.*, 1998). All of these changes occurred in the vicinity of the newly proposed membrane contact site (Fig. 4) (McDonald *et al.*, 1997b).

The membrane contact site must provide consistency with biophysical results. It was surprising to learn that, if calcium is maintained at saturating levels, membrane affinity of the Gla domain is relatively insensitive to ionic strength of the solution (McDonald *et al.*, 1997a; Resnick and Nelsestuen, 1980). This property greatly limited possible models of ionic interaction between the protein and membrane but did not rule out the involvement of ionic forces. For example, single charge interactions are quite insensitive to ionic strength, but could support membrane interaction if the ion pair were isolated from the aqueous environment (McDonald *et al.*, 1997a). Alternatively, binding may occur through charges that are closely spaced on each component. The protein and membrane components may have neutral charge due to multiple cation and anionic groups on each. Ionic attraction may only occur at short distances, where cations on one site attract anions on the other component and vice versa. Since attraction only occurs at short distance, shielding by ions in solution would not impact on binding affinity. This was referred to as a chelation model (Resnick and Nelsestuen, 1980). Thus, specific forms of ionic contact may be compatible with the experimental result.

The membrane contact site must explain other experimental evidence. For example, vitamin K-dependent proteins project far from the membrane surface (Lim *et al.*, 1977) and the active sites of several enzymes are estimated to be 60–80 Å above the membrane surface (Armstrong *et al.*, 1990; McCallum *et al.*, 1996, 1997; Mutucumarana *et al.*, 1992; Yegneswaran *et al.*, 1997). In all cases tested, the cofactor protein decreased the distance from the active site to the membrane surface. This can be rationalized from the structure in Fig. 4A. When bound to the membrane by itself, free rotation about the EGF domains should create a large average distance between the active site and membrane surface. Tissue factor would decrease VIIa flexibility and place the active site closer to the membrane. However, like many attempts to describe the ultimate function of Gla in forming a membrane contact site, this argument is circumstantial.

An alternative membrane contact mechanism is shown in Fig. 4B. An interesting feature of the prothrombin Gla domain is a cluster of hydrophobic residues (Phe-5, Leu-6 and Val-9, Fig. 5) on the protein surface and a popular model for membrane contact has suggested insertion of this group into the hydrocarbon region of the membrane (Figure 4B) (Christiansen *et al.*, 1995; Freedman *et al.*, 1996; Seshadri *et al.*, 1994; Zhang and Castellino, 1994). We feel that this model is less likely and summarize the reasons for this suggestion.

The naturally occurring proteins (McDonald *et al.*, 1997b) show a tremendous variation of amino acids on various surfaces of the Gla domain. Variation includes the hydrophobic residues. Factor IX contains Lys at position 5 (sequences summarized in McDonald *et al.* (1997b) while protein S contains Thr-8 (homologous to Val-9 of prothrombin). Biophysical properties provide little support for protein penetration into the hydrocarbon region of the membrane. Affinity of prothrombin for phospholipid monolayers was influenced to a small degree by packing density of the phospholipid head groups (Mayer *et al.,* 1983). Surface pressure changes appeared to be phospholipid specific, rather than protein specific (Ellison and Castellino, 1998). These proteins also showed little selectivity for binding to small phospholipid vesicles, with highly curved surfaces and substantially exposed hydrocarbon regions, versus large, planar vesicles that have tightly packed headgroups and little hydrophobic exposure (Lu and Nelsestuen, 1996a). Proteins that are known to penetrate into the hydrocarbon region show high selectivity for small vesicles over large ones (Abbott and Nelsestuen, 1987; Greenhut *et al.,* 1986; Silversmith and Nelsestuen, 1986).

A striking change is the removal of the amino terminal three residues of prothrombin fragment 1 with only small loss of membrane affinity at saturating calcium (Nelsestuen, 1988; Weber *et al.,* 1992). As outlined earlier, a free amino terminal group is essential for membrane association. This group is buried in the Gla domain and forms multiple hydrogen bonds to Gla residues. It was therefore surprising to find that prothrombin residues 4–156 retained an excellent membrane contact site (a modest fivefold reduction of membrane affinity, Weber *et al.,* 1992). It is likely that Gly-4 of the 4–156 peptide fulfills the role of Ala-1. The amino group of Gly-4 is identified in Fig. 5. Its relocation to the position of Ala-1 requires a movement of about 10 Å and severe relocation of Phe-5 and Leu-6. That this major change in the integrity of the hydrophobic cluster was accompanied by a small change in free energy of protein–membrane affinity, at saturating calcium, suggested little role for Phe-5 and Leu-6 in actual membrane contact. That the truncated protein showed reduced affinity for calcium (Weber *et al.,* 1992) indicated that many amino acid residues were important for native calcium binding affinity, a process that can be separated from the membrane contact event (Nelsestuen and Lim, 1977).

Site-directed mutagenesis shows many changes with small impact. Replacement of Phe-4 of human protein C (homologous to Phe-5 of prothrombin) with Gln had no detected impact on membrane contact affinity (Christiansen *et al.,* 1995). Transfer of a phenylalanine side chain from water to organic solvent is accompanied by a free-energy change of about −3.6 kcal/mol (Fersht, 1985), equal to a 400-fold change in K_D. This mutant argued against the simple hydrophobic contact mechanism shown in Fig. 4B. A modified hydrophobic interaction was proposed that used positions 6 and 9 (Christiansen *et al.,* 1995). However, replacement of Leu-6 or Val-9 of factor IX by a large hydrophobic photoaffinity probe, para-benzoyl-L-phenylalanine, had no detected impact on membrane affinity (Freedman *et al.,* 1996). Upon irradiation, the photoaffinity probe reacted with phospholipids, indicating that these residues were close to the membrane. However, low impact of the initial derivative would argue against insertion of these sites into the hydrocarbon region of the membrane. Replacement of Leu-5 by Trp also had little impact on membrane affinity (Jalbert *et al.,* 1996). Thus, increases in the hydrophobic character of the cluster have little impact on membrane affinity, suggesting that its insertion into the membrane is not an important aspect of membrane contact.

Overall, the search for the membrane contact mechanism is still incomplete. Few initial expectations have been confirmed and the function of Gla in this critical process is still not obvious. Other functions of Gla may prove more obvious. This amino acid forms low-affinity calcium-binding sites and may interact well with calcium crystals. Thus, use of Gla-containing proteins to limit or modify growth of calcium-phosphate crystals may be an early function, and the development of a membrane-binding site may be a recent recruitment of Gla during evolutionary developments.

VII. Role of Membrane Binding in Coagulation

A. Qualitative Descriptions of Membrane Function

Features of membrane-bound blood coagulation reactions are depicted in Fig. 6. Some qualitative membrane functions seem to be widely accepted. For example, the membrane is a regulation mechanism that localizes blood coagulation to the site of injury. This is shown in Fig. 6 by the fact that the membrane-bound enzymes remain at the location of the immobile membrane component, despite solvent flow over the surface. Under normal circumstances, vitamin K-dependent proteins exist as soluble proteins in the circulation. The exteriors of normal cell membranes lack anionic phospholipids such as phosphatidylserine (PS), the key to membrane binding by these proteins. Phosphatidylserine may appear on a cell surface by virtue of proteins, labeled scramblases (Basse *et al.,* 1996; Zhao *et al.,* 1998), which flip the PS from the inner to outer membrane surface of cells on proper stimulation. Alternatively, PS may be exposed by cell rupture or loss of the energy required to maintain phospholipid asymmetry of the cell membrane. In any event, the immobile cell serves to localize this essential component of the coagulation cascade.

A second, widely acknowledged property is that the membrane serves to "lower the substrate requirement" of the enzyme. While this is often

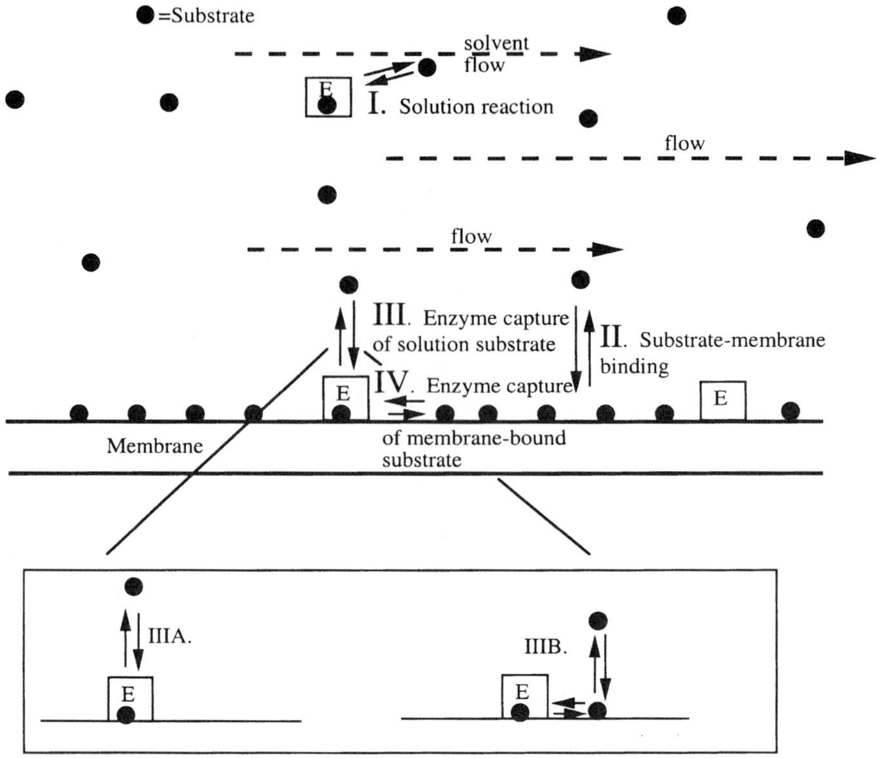

FIG. 6. Possible events in blood clotting reactions. Reaction I shows enzyme–substrate reaction in solution. Reaction II shows substrate binding to the membrane surface. Reactions IIIA and IIIB show two varieties of enzyme capture of substrate from solution. Reaction IV shows enzyme capture of substrate from a population that is bound to the membrane. Flow of solvent is also depicted.

referred to as a lowering of K_M, that designation is not correct for all situations, as outlined later. In solution, the prothrombin concentration at half-maximum reaction velocity for prothrombinase is nearly 50 μM, far above the 1.5-μM concentration of prothrombin in plasma (Rosing *et al.,* 1980). On the membrane surface, the prothrombin concentration at half-maximum reaction has been reported to range from 400 nM (Pusey and Nelsestuen, 1983) to as low as 3 nM (Billy *et al.,* 1995; Giesen *et al.,* 1991; Lu and Nelsestuen, 1996b), depending on the membrane used in the assay.

Two mechanisms for lowering the substrate requirement have been considered extensively. Both of these have assumed diffusional equilibrium, a property that may not be correct for all circumstances (see later discussion). One mechanism consists of the accumulative effect of multiple contacts. The interaction of substrate with soluble enzyme (reaction I, Fig. 6) involves protein–protein contact only. On the membrane surface (reaction III, Fig. 6), substrate binding includes protein–phospholipid contact as well. Due to multiple contacts, the membrane-bound enzyme will have higher affinity for substrate than the soluble enzyme (Nelsestuen, 1978; Pusey and Nelsestuen, 1983). Mechanism III includes two forms (IIIA and IIIB, Fig. 6). The important feature is enzyme–phospholipid capture of substrate from solution, either by initial contact with the enzyme or membrane. An intermediate pool of substrate at the membrane surface is not important to reaction velocity.

The second mechanism was described in early expectations. It was suggested that the free proteins were present at low concentration in plasma and that association with a membrane would increase local concentrations, thereby enhancing reaction velocity at that site (reaction IV, Fig. 6) (Nesheim *et al.,* 1984; Rosing *et al.,* 1980). Recent studies indicate that both of these mechanisms may occur, depending on the membrane composition used in the study (Lu and Nelsestuen, 1996b).

B. Quantitative Descriptions of Membrane Function

A major biochemical goal is the reduction of a biological process to mathematical description. A number of studies have utilized classic kinetic modeling approaches to study individual steps of coagulation (Rosing *et al.,* 1980) or to model the entire cascade (Jones and Mann, 1994). Other efforts, as well as this review, have concentrated on more elementary questions about the reaction on a membrane. For example, "Is the use of a steady-state parameter such as (apparent) K_M appropriate?" or "Should nonclassical mechanisms be considered?"

Two mechanisms for lowering substrate titration curves are shown in Fig. 6. For membranes of low PS content, binding affinity of the substrate, prothrombin, is insufficient to create a local high substrate density (Nelsestuen, 1978; Pusey and Nelsestuen, 1983; van Rijn *et al.,* 1984). Direct capture by mechanism III (A or B) must occur. Prothrombin concentrations at the midpoint of the titration (half of V_{max}) are typically 100–400 n*M* (Nelsestuen, 1978; Pusey and Nelsestuen, 1983; van Rijn *et al.,* 1984), well below plasma concentrations of prothrombin. Alternatively, membranes with high affinity for substrate produce a high local substrate density and substrate concentrations at the midpoint of the titration can be as low as 5 n*M* (Lu and Nelsestuen, 1996b). It is not clear that the latter situations will be important *in vivo,* where substrate concentrations are much higher than 5 n*M*.

Nevertheless, the ability to use either solution substrate or membrane-bound substrate creates complex behaviors. For example, membranes with low surface area (small vesicles of 30 nm in diameter) may not bind enough substrate to provide a constant supply to the enzyme. The reaction can be limited by product dissociation from the membrane particle (Lu and Nelsestuen, 1996b). In this case, maximum velocity is lower than that provided by enzyme catalysis (Fig. 7A, dashed line). The enzyme site is never rate limiting. Rather, titrations follow equilibrium II, substrate–membrane binding. A different result is obtained with large vesicles. In this case, substrate binding to the surface (step II, Fig. 6) is rate limiting at low substrate, but the enzyme became saturated with substrate before the vesicle surface is saturated with substrate. This is analogous to a coupled enzyme assay where the reaction is limited by one enzyme at low substrate but another at high substrate. The result is nonlinearity of Lineweaver-Burk or Eadie-Hofstee plots (Fig. 7A, solid line). No single steady-state parameter such as K_M can describe the entire substrate titration curve. A high density of enzymes will decrease the surface area per enzyme and the behavior of large membrane particles may mimic that of small membrane particles. Overall, enzyme kinetic behavior can depend on particle size, rates of substrate association and dissociation (governed by membrane composition), and the number of enzymes per membrane particle. No simple set of constants can describe all of these situations.

It is possible that some of these complications are not relevant *in vivo*. Although mechanism IV applied to some *in vitro* situations, it required membranes of high, nonphysiologic PS content (about 50%) (Lu and Nelsestuen, 1996b). Consequently, it is not clear that mechanism IV will be important *in vivo*.

Membrane-bound systems offer still other forms of kinetic behavior that may be significant *in vivo*. Most *in vitro* reactions utilize small membrane vesicles; so-called "large" vesicles have diameters of only about 100 nm. In contrast, typical cell diameters are 7000 nm and the entire clotting surface may include many cells. A general description of the rate constant is required to describe the potential impacts of these changes. Classical biochemical and chemical education presents the law of the rate constant in an oversimplified manner. Velocities of bimolecular reactions are described by a rate constant, k [Eq. (1)], with units of $M^{-1}s^{-1}$.

$$E + S \rightarrow ES \qquad v = k_1[E][S]. \tag{1}$$

In reality, the rate constant is the product of two terms. One describes collision of the particles in solution (Smoluchowski theory

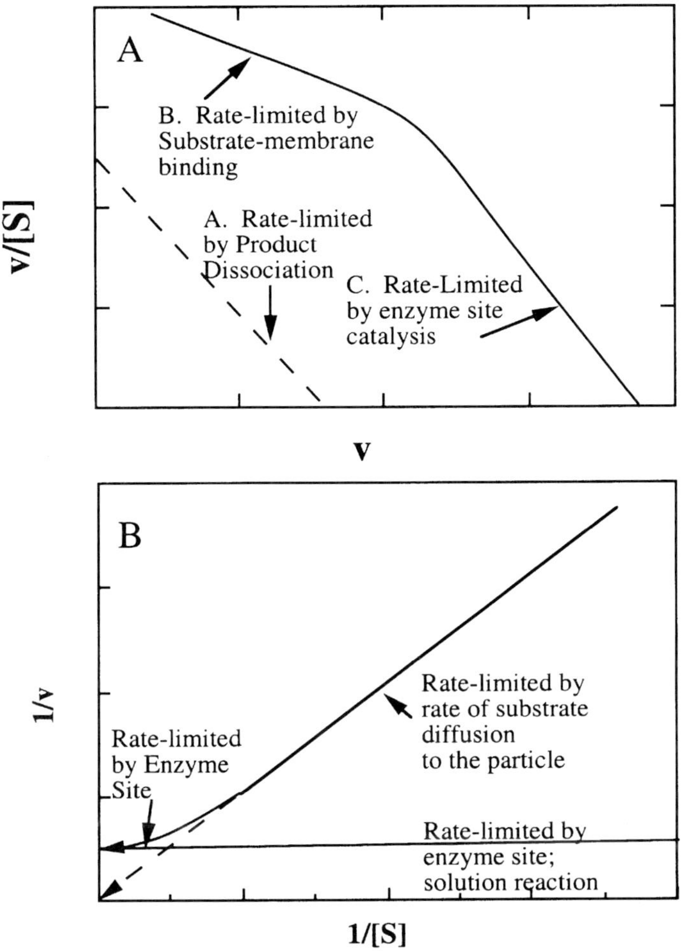

FIG. 7. Possible kinetic properties of membrane-bound enzymes. (A) General behavior of prothrombinase that was assembled on large vesicles of 50% phosphatidylserine (solid line) and on small vesicles of the same composition (dashed line) (Lu and Nelsestuen, 1996b). Data are plotted by the Eadie-Hofstee method so that the classical K_M would equal -1/slope. Plot curvature is caused by a change in the rate-limiting process (see text). (B) Behavior of an enzyme reaction limited by substrate flux to the enzyme-containing particle. At low substrate, the plot extrapolates to zero on the Lineweaver-Burk plot, indicating the lack of "site" behavior. At high substrate concentration, the reaction is limited by enzyme catalysis and intercepts the vertical axis at $1/V_{max}$ for the enzyme site. The straight-line plot shows the behavior of the same amount of soluble enzyme. This behavior is from a bacterial enzyme system (Martinez *et al.,* 1996).

$k_{coll}=4\pi N_{av}Da/1000$, where N_{av} is Avogadro's number, D is the sum of particle diffusion constants, and a is the sum of their radii). The second is the probability of binding once collision has occurred [$Ns/(Ns+\pi a)$, where N is the number of vacant binding sites in a particle and s is the capture radius of the binding site]. The probability term was developed by Berg and Purcell (1977) to describe the kinetics of phage binding to receptors on a bacterial surface. A more complete expression for the actual rate constant (k_{obs}) is given in Eq. (2):

$$k_{obs} = (4\pi N_{Av}Da/1000)[Ns/(Ns+\pi a)]. \tag{2}$$

The critical feature of classical enzyme kinetics, which allows the use of a simple k for each step of even the most complex enzymes, is that $Ns<<\pi a$ in all cases so that k_{obs} [Eq. (2)] is reduced to the combination of constants given in Eq. (3):

$$k_{obs}=4N_{Av}Ds/1000. \tag{3}$$

This represents a "site" rate constant that describes properties of the site, s. The size of the enzyme-containing particle (a) is irrelevant. Kinetic parameters such as K_M or apparent K_M are ratios of such "site" rate constants. This simplification works very well until one encounters multisite particles such as membrane-bound enzymes.

A major change occurs when N of Eq. (2) becomes very large. The rate constant may approach a second extreme where $Ns>>\pi a$. Under these conditions, the "site" term of Eq. (2) disappears and the rate constant describes the behavior of the "particle", a:

$$k_{obs} = (4\pi N_{Av}Da/1000)(1) \qquad v=k_{obs}[\text{particle-1}][\text{particle-2}]. \tag{4}$$

Under these conditions, the reaction is determined by substrate flux. The reaction solution is heterogeneous since the enzyme-containing surface depletes substrate in its vicinity. Steady-state plots such as Lineweaver-Burk or Eadie-Hofstee become nonlinear. As substrate is increased, N, the number of vacant sites, declines. Finally, at high substrate concentration, the value of N becomes low and the rate constant switches from that in Eq. (4) to that in Eq. (3). The reaction switches from one type of kinetic behavior to another. This type of behavior can apply to local high enzyme concentrations on a membrane, as well as to the entire particle. In either case, high enzyme density can create a perfectly reactive region or particle, one that captures every substrate molecule that collides with its surface. An interesting feature of this state is that substrate concentration at half-maximum velocity is higher than that for the same amount of soluble enzyme. This property is illustrated by the comparison to the behavior of free enzyme in Fig. 7B. Flux-limited kinetics of factor Xase enzyme on cell surfaces has been reported (McGee *et al.,* 1992). A modified analysis of this study would suggest that apparent K_M is not applicable to a reaction with these characteristics.

Substrate flow over the membrane surface is still another concern (Fig. 6). If the surface has a high density of enzyme, the reaction can be limited by arrival of substrate at the surface due to diffusion plus solvent flow. This is described by Eq. (5). For solvent flow (V_f) through a tube of length L and radius r, with substrate concentration [S], the product concentration at the end of the tube will be Pe(D) (Laidler and Bunting, 1980):

$$\mathrm{Pe}(D) = [DL/R^2V_f]^{2/3}[\mathrm{S}]. \tag{5}$$

As for flux-limited reactions, the lack of a "site" term in the rate expression results in nonclassical kinetics. Such flow/diffusion-limited kinetics have been demonstrated for enzyme-lined tubes that contain both prothrombinase (Billy *et al.,* 1995) and factor Xase (Andree *et al.,* 1994) enzymes. As flow or substrate is increased, the rate of substrate arrival at the surface will exceed that which can be captured by the enzyme. The reaction becomes limited by the enzyme site. This is analogous to the change in the rate-limiting step described for diffusional reactions (Fig. 7B). The result will be nonlinear Lineweaver-Burk or Eadie-Hofstee plots, and the inability to predict reaction properties by classical application of constants such as $K_{M(apparent)}$.

Given this tremendous diversity of possible kinetic reaction mechanisms that occur *in vitro,* it seems clear that a definitive description of blood clotting kinetics *in vivo* will not be possible in the near future. It is also clear that vitamin K research spans the scientific research spectrum from enzyme reaction mechanism involving vitamin K to cells and complex enzyme kinetic problems of the *in vivo* state. Research into the vitamin K-dependent proteins promises a rich future. New findings may add to the potential use of these proteins for study of topics from enzymology to biological cell design. A growing area of interest in vitamin K-dependent proteins is their use to treat disease.

ACKNOWLEDGMENT

Supported in part by grant HL60859 from the National Institutes of Health. A. M. S. was supported in part by the Arnold H. Johnson Fellowship and S. B. H. by NIH Predoctoral Fellowship ES05688.

REFERENCES

Abbott, A. J., and Nelsestuen, G. L. (1987). Association of a protein with a membrane vesicles at the collisional limit: Studies with blood coagulation factor Va-light chain also suggest major differences between small and large unilamellar vesicles. *Biochemistry* ***26,*** 7994–8003.

Andree, H. A. M., Contino, P. B., Repke, D., Gentry, R., and Nemerson, Y. (1994). Transport rate limited catalysis on macroscopic surfaces. The activation of factor X in a continuous flow enzyme reactor. *Biochemistry* **33,** 4368–4374.

Armstrong, S. A., Husten, E. J., Esmon, C. T., and Johnson, A. E. (1990). The active site of membrane-bound meizothrombin. A fluorescence determination of its distance from the phospholipid surface and its conformational sensitivity to calcium and factor Va. *J. Biol. Chem.* **265,** 6210–6218.

Babbitt, P. C., Hasson, M. S., Wedekind, J. E., Palmer, D. R. J., Barrett, W. C., Reed, G. H., Rayment, I., Ringe, D., Kenyon, G. L., and Gerlt, J. A. (1996). The enolase superfamily: A general strategy for enzyme-catalyzed abstraction of the alpha-protons of carboxylic acids. *Biochemistry* **36,** 16489–16501.

Bach, R., Gentry, R., and Nemerson, Y. (1986). Factor VII binding to tissue factor in reconstituted phospholipid vesicles: Induction of cooperativity by phosphatidylserine. *Biochemstry* **25,** 4007–4020.

Bach, R. R., and Moldow, C. F. (1997). Mechanism of tissue factor activation on HL-60 cells. *Blood* **89,** 3270–3276.

Baker, M. E., French, F. S., and Joseph, D. R. (1987). Vitamin K-dependent protein S is similar to rat androgen-binding protein. *Biochem. J.* **243,** 293–296.

Bakker, H. M., Tans, G., Janssen-Claessen, T., Thomassen, C. L. G. D., Hemker, H. C., Griffin, J. H., and Rosing, J. (1992). The effect of phospholipids, calcium ions and protein S on rate constants of human factor Va inactivation by activated human protein C. *Eur. J. Biochem.* **208,** 171–178.

Banner, D. W., D'Arcy, A., Chene, C., Winkler, F. K., Guha, A., Konigsberg, W. H., Nemerson, Y., and Kirchhofer, D. (1996). The crystal structure of the complex of blood coagulation factor VIIa with soluble tissue factor. *Nature (London)* **380,** 41–46.

Basse, F., Stout, J. G., Sims, P. J., and Wiedmer, T. (1996). Isolation of an erythrocyte membrane protein that mediates Ca2+-dependent transbilayer movement of phospholipid. *J. Biol. Chem.* **271,** 17205–17210.

Baur, K. A., Kass, B. L., ten Cate, H., Hawiger, J. J., and Rosenberg, R. D. (1990). Factor IX is activated in vivo by the tissue factor mechanism. *Blood* **76,** 731–736.

Bazzi, M. D., and Nelsestuen, G. L. (1991). Highly sequential binding of protein kinase C and related proteins to membranes. *Biochemistry* **30,** 7970–7977.

Becker, R. C., and Alpert, J. S. (1990). The impact of medical therapy on hemorrhagic complications following coronary artery bypass grafting. *Arch. Intern. Med.* **150,** 2016–2021.

Berg, H. C., and Purcell, E. M. (1977). Physics of chemoreception. *Biophys. J.* **20,** 193–219.

Bertina, R. M., Koeleman, B. P. C., Koster, T., Rosendaal, F. R., Dirven, R. J., de Ronde, H., van der Velden, P. A., and Reitsma, P. H. (1994). Mutation in blood coagulation factor V associated with resistance of activated protein C. *Nature (London)* **369,** 64–67.

Billy, D., Speijer, H., Willems, G., Hemker, H. C., and Lindhout, T. (1995). Prothrombin activation by prothrombinase in a tubular flow reactor. *J. Biol. Chem.* **270,** 1029–1034.

Carlisle, T. L., and Suttie, J. W. (1980). Vitamin K dependent carboxylase: Subcellular location of the carboxylase and enzymes involved in vitamin K metablosim in rat liver. *Biochemistry* **19,** 1161–1167.

Chen, Z., Blandl, T., Prorok, M., Warder, S. E., Li, L., Zhu, Y., Pedersen, L. G., Ni, F., and Castellino, F. J. (1998). Conformational changes in conantokin-G induced upon binding of calcium and magnesium as revealed by NMR structural analysis. *J. Biol. Chem.* **273,** 16248–16258.

Christiansen, W. T., Tulinsky, A., and Castellino, F. J. (1994). Functions of individual gamma-carboxyglutamic acid (Gla) residues of human protein C. Determination of functional nonessential Gla residues and correlations with their mode of binding to calcium. *Biochemistry* **33,** 14993–15000.

Christiansen, W. T., Jalbert, L. R., Robertson, R. M., Jhingan, A., Prorok, M., and Castellino, F. J. (1995). Hydrophobic amino acid residues of human anticoagulation protein C that contribute to its functional binding to phospholipid vesicles. *Biochemistry* **34,** 10376–10382.

Clouse, L. H., and Comp, P. C. (1986). The regulation of hemostasis: The protein C system. *N. Engl. J. Med.* **314,** 1298–1304.

Colpitts, T. L., and Castellino, F. J. (1994). Calcium and phospholipid binding properties of synthetic gamma-carboxyglutamic acid-containing peptides with sequence counterparts in human protein C. *Biochemistry* **33,** 3501–3508.

Cutsforth, G. A., Koppaka, V., Krishnaswamy, S., Wu, J. R., Mann, K. G., and Lentz, B. R. (1996). Insights into the complex association of bovine factor Va with acidic-lipid-containing synthetic membranes. *Biophys. J.* **70,** 2938–2949.

Dahlbäck, B. (1986). Inhibition of protien Ca cofactor function of hman and bovine protiens by C4b-binding protein. *J. Biol. Chem.* **261,** 12022–12027.

Dahlbäck, B. (1991). Protein S and C4b-binding protein: Components involved in the regulation of the protein C anticoagulant system. *Thromb. Haemostasis* **66,** 49–61.

Dahlbäck, B. (1995). The protein C anticoagulant system: Inherited defects as basis for venous thrombosis. *Thromb. Res.* **77,** 1–43.

Dahlbäck, B., and Stenflo, J. (1981). High molecular weight complex in human plasma between vitamin K-dependent protein S and complement component C4b-binding protein. *Proc. Natl. Acad. Sci. U.S.A.* **78,** 2512–2516.

Dam, H. (1929). Cholesterinstoffwechsel in Huhnereiern und Huhnchen. *Biochem. Z.* **215,** 475–492.

Dam, H. (1930). Cholesterol synthesis in the animal body. *Biochem. Z.* **220,** 158–163.

Dam, H. (1934). Hemmorrhages in chicks reared on artificial diets. A new deficiency disease. *Nature (London)* **133,** 909–910.

Dam, H. (1935). Antihaemorrhagic vitamin of chick. *Biochem. J.* **29,** 1273–1285.

Davie, E. W., Fujikawa, K., and Kisiel, W. (1991). The coagulation cascade: Initiation, maintenance, and regulation. *Biochemistry* **30,** 103663–10370.

Deerfield, D. W., II, Olson, D. L., Berkowitz, P., Byrd, P. A., Koehler, K. A., Pedersen, L. G., and Hiskey, R. G. (1987). Mg binding by bovine prothrombin fragment 1 via equilibrium dialysis and the relative roles of Mg and Ca in blood coagulation. *J. Biol. Chem.* **262,** 4017–4021.

Dowd, P. H., Naganathan, S., and Hershine, R. (1995). The mechanism of action of vitamin K. *Annu. Rev. Nutr.* **15,** 419–440.

Ducy, P., Desbois, C., Boyce, B., Pinero, G., Story, B., Dunstan, C., Smith, E., Bonadio, J., Goldstein, S., Gundberg, C., Bradley, A., and Karsenty, G. (1996). Increased bone formation in osteocalcin-deficient mice. *Nature (London)* **382,** 448–452.

Eaton, D., Rodriguez, H., and Vehar, G. A. (1986). Proteolytic processing of human factor VIII. Correlation of specific cleavages by thrombin, factor Xa, and activated protein C with activation and inactivation of factor VIII coagulant activity. *Biochemistry* **25,** 505–512.

Ellison, E. H., and Castellino, F. J. (1998). Adsorption of vitamin K-dependent blood coagulation proteins to spread phospholipid monolayers as determined from combined measurements of the surface pressure and surface protein concentration. *Biochemistry* **37,** 7997–8003.

Emsley, J., White, H. E., O'Hara, B. P., Oliva, G., Srinivasan, N., Tickle, I. J., Blundell, T. L., Pepys, M. B., and Wood, S. P. (1994). Structure of pentameric human serum amyloid P component. *Nature (London)* **367,** 338–345.

Esmon, C. T. (1979). The subunit structure of thrombin-activated factor V. *J. Biol. Chem.* **254,** 964–973.

Esmon, C. T. (1989). The roles of protein C and thrombomodulin in the regulation of blood coagulation. *J. Biol. Chem.* **264,** 4743–4746.

Evans, T. C., Jr., and Nelsestuen, G. L. (1994). Calcium and membrane-binding properties of monomeric and multimeric annexin II. *Biochemistry* **33,** 13231–13238.

Evans, T. E., Jr., and Nelsestuen, G. L. (1996a). Importance of Cis-proline 22 and the aromatic stack (residues 41–45) for prothrombin-membrane binding. *Protein Sci.* (Suppl 1) **5,** 163 abs.

Evans, T. E., Jr., and Nelsestuen, G. L. (1996b). The importance of cis-proline 22 in the membrane-binding conformation of bovine prothrombin. *Biochemistry* **35,** 8210–8215.

Fasco, M. J., and Principe, L. M. (1980). Vitamin K1 hydroquinone formation catalyzed by a microsomal reductase system. *Biochem. Biophys. Res. Commun.* **97,** 1487–1492.

Fasco, M. J., and Principe, L. M. (1982). R- and S-Warfarin inhibition of vitamin K and vitamin K 2,3-epoxide reductase activities in the rat. *J. Biol. Chem.* **257,** 4894–4901.

Fasco, M. J., Preusch, P. C., Hildebrandt, E., and Suttie, J. W. (1983). Formation of hydroxyvitamin K by vitamin K epoxide reductase of warfarin-resistant rats. *J. Biol. Chem.* **258,** 4372–4380.

Fersht, A. R. (1985). In "Enzyme Structure and Mechanism," 2nd ed., pp. 293–310. Freeman, New York.

Foster, W. B., Nesheim, M. E., and Mann, K. G. (1983). The factor Xa-catalyzed activation of factor V. *J. Biol. Chem.* **258,** 13970–13977.

Fourrier, F., Jourdain, M., Tournois, A., Caron, C., Goudemand, J., and Chopin, C. (1995). Coagulation inhibitor substitution during sepsis. *Intensive Care Med.* **21** (Suppl. 2), S264–S268.

Freedman, S. J., Furie, B. C., Furie, B., and Baleja, J. D. (1995a). Structure of the metal-free Gla membrane binding region of factor IX by two-dimensional NMR spectroscopy. *J. Biol. Chem.* **270,** 7980–7987.

Freedman, S. J., Furie, B. C., Furie, B., and Baleja, J. D. (1995b). Structure of the calcium ion-bound gamma-carboxyglutamic acid-rich domain of factor IX. *Biochemistry* **34,** 12126–12137.

Freedman, S. J., Blostein, M. D., Baleja, J. D., Jacobs, M., Furie, B. C., and Furie, B. (1996). Identification of the phospholipid binding site in the vitamin K-dependent blood coagulation protein factor IX. *J. Biol. Chem.* **271,** 16227–16236.

Furie, B., and Furie, B. C. (1988). The molecular basis of blood coagulation. *Cell (Cambridge, Mass.)* **53,** 505–518.

Furie, B., and Furie, B. C. (1992). Molecular and cellular biology of blood coagulation. *N. Engl. J. Med.* **326,** 800–806.

Furie, B. C., and Furie, B. (1997). Structure and mechanism of action of the vitamin K-dependent gamma-glutamyl carboxylase: Recent advances from mutagenesis studies. *Thromb. Haemostasis* **78,** 595–598.

Furie, B. C., Ratcliffe, J. V., Tward, J., Jorgensen, M. J., Blaszkowsky, L. S., DiMichele, D., and Furie, B. (1997). The gamma-carboxylation recognition site is sufficient to direct vitamin K-dependent carboxylation on an adjacent glutamate-rich region of thrombin in a propeptide-thrombin chimera. *J. Biol. Chem.* **272,** 28258–28262.

Gallop, P. M., Lian, J. B., and Hauschka, P. V. (1980). Carboxylated calicum-binding proteins and vitamin K. *N. Engl. J. Med.* **302,** 1460–1466.

Gardill, S. L., and Suttie, J. W. (1990). Vitamin K epoxide and quinone reductase activites. Evidence for reduction by a common enzyme. *Biochem. Pharmacol.* **40,** 1055–1061.

Gershagen, S., Fernlund, P., and Edenbrandt, C. M. (1991). The genes for SHBG/ABP and the SHBG-like region of vitamin K-dependent protein S have evolved from a common ancestral gene. *J. Steroid Biochem. Mol. Biol.* **40,** 763–769.

Giesen, P. L. A., Willems, G. M., and Hermens, W. T. (1991). Production of thrombin by the prothrombinase complex is regulated by membrane-mediated transport of prothrombin. *J. Biol. Chem.* **266,** 1379–1382.

Gilbert, G. E., and Arena, A. A. (1995). Phosphatidylethanolamine induces high affinity binding sites for factor VIII on membranes containing Phosphatidylserine. *J. Biol. Chem.* **270,** 18500–18505.

Greenhut, S. F., Bourgeois, V. R., and Roseman, M. A. (1986). Distribution of cytochrome b5 between small and large unilamellar phospholipid vesicles. *J. Biol. Chem.* **261,** 3670–3675.

Griffin, J. H., Evatt, B., Zimmerman, T. S., Kleiss, A. J., and Wideman, C. (1981). Deficiency of protein C in congenital thrombotic disease. *J. Clin. Invest.* **68,** 1370–1373.

Hackeng, T. M., van't Veer, C., Meijers, J. C. M., and Bouma, B. N. (1994). Human protein S inhibits prothrombinase complex activity on endothelial cells and platelets via direct interactions with factors Va and Xa. *J. Biol. Chem.* **269,** 21051–21058.

Hagen, F. S., Gray, C. L., O'Hara, P., Grant, F. J., Saari, G. C., Woodbury, R. G., Hart, C. E., Insley, M., Kisiel, W., Kurachi, K., and Davie, E. W. (1986). Characterization of a cDNA coding for human factor VII. *Proc. Natl. Acad. Sci. U.S.A.* **83,** 2412–2416.

Ham, S. W., and Dowd, P. (1990). On the mechanism of action of vitamin K. A new nonenzymatic model. *J. Am. Chem. Soc.* **112,** 1660–1661.

Hedner, U. (1996). Dosing and monitoring NovoSeven treatment. *Haemostasis* **26** (Suppl. 1), 102–108.

Hedner, U., and Davie, E. W. (1987). In "Hemostasis and Thrombosis" (R. W. Colman, J. Hirsh, V. J. Marder, and E. W. Salzman, eds.), pp. 39–47. Lippincott, Philadelphia.

Heeb, M. J., Mesters, R. M., Tans, G., Rosing, J., and Griffin, J. H. (1993). Binding of protein S to factor Xa associated with inhibition of prothrombinase that is independent of activated protein C. *J. Biol. Chem.* **269,** 2872–2877.

Heeb, M. J., Rosing, J., Bakker, H. M., Fernandez, J. A., Tans, G., and Griffin, J. H. (1994). Protein S binds to and inhibits factor Xa. *Proc. Natl. Acad. Sci. U.S.A.* **91,** 2728–2732.

Hoekema, L., Nicolaes, G. A. F., Hemker, C., Tans, G., and Rosing, J. (1997). Human factor Va1 and factor Va2: Properties in the procoagulant and anticoagulant pathways. *Biochemistry* **36,** 3331–3335.

Höjrup, P., Jensen, M. S., and Petersen, T. E. (1985). Amino acid sequence of bovine protein Z: A vitamin K-dependent serine protease homolog. *FEBS Lett.* **184,** 333–338.

Jain, M. K., Gelb, M. H., Rogers, J., and Berg, O. G. (1995). Kinetic basis for interfacial catalysis by phospholipase A2. In "Methods in Enzymology" (D. L. Purich, ed.), Vol. 249, pp. 567–614. Academic Press, San Diego, CA.

Jalbert, L. R., Chan, J. C. Y., Christiansen, W. T., and Castellino, F. J. (1996). The hydrophobic nature of residue-5 of human protein C is a major determinant of its functional interactions with acidic phospholipid vesicles. *Biochemistry* **35,** 7093–7099.

Jones, K. C., and Mann, K. G. (1994). A model for the tissue factor pathway to thrombin. II. A mathematical simulation. *J. Biol. Chem.* **269,** 23367–23373.

Kalafatis, M., Rand, M. D., and Mann, K. G. (1994a). The mechanism of inactivation of human factor V and human factor Va by activated protein C. *J. Biol. Chem.* **269,** 31869–31880.

Kalafatis, M., Swords, N. A., Rand, M. D., and Mann, K. G. (1994b). Membrane-dependent reactions in blood coagulation: Role of the vitamin K-dependents enzyme complexes. *Biochim. Biophys. Acta* **1227,** 113–129.

Komiyama, Y., Pedersen, A. H., and Kisiel, W. (1990). Proteolytic activation of human factors IX and X by recombinant human factor VIIa: Effects of calcium, phospholipids, and tissue factor. *Biochemistry* **29,** 9418–9425.

Kulman, J. D., Harris, J. E., Haldeman, B. A., and Davie, E. W. (1997). Primary structure and tissue distribution of two novel proline-rich gamma-carboxyglutamic acid proteins. *Proc. Natl. Acad. Sci. U.S.A.* **94,** 9058–9062.

Laidler, K. J., and Bunting, P. S. (1980). The kinetics of immobilized enzyme systems. *In* "Methods in Enzymology" (D. L. Purich, ed.), Vol. 64, 397–427. Academic Press, New York.

Larson, P. J., Camire, R. M., Wong, D., Fasano, N. C., Monroe, D. M., Tracy, P. B., and High, K. A. (1998). Structure/function analyses of recombinant variants of human factor Xa: Factor Xa incorporation into prothrombinase on the thrombin-activated platelet surface is not mimicked by synthetic phospholipid vesciles. *Biochemistry* **37,** 5029–5038.

Lawson, J. H., and Mann, K. G. (1991). Cooperative activation of human factor IX by the human extrinsic pathway of blood coagulation. *J. Biol. Chem.* **266,** 11317–11327.

Li, S., Furie, B. C., Furie, B., and Walsh, C. T. (1997). The propeptide of the vitamin K-dependent carboxylase substrate accelerates formation of the gamma-glutamyl carbanion intermediate. *Biochemistry* **36,** 6384–6390.

Lim, T. K., Bloomfield, V. A., and Nelsestuen, G. L. (1977). Stucture of the prothrombin and blood clotting factor X-membrane complexes. *Biochemistry* **16,** 4177–4181.

Lollar, P., and Parker, C. G. (1989). Subunit structure of thrombin-activated porcine factor VIII. *Biochemistry* **28,** 666–674.

Lu, Y., and Nelsestuen, G. L. (1996a). Dynamic features of prothrombin interaction with phospholipid vesicles of different size and composition: Implications for protein-membrane contact. *Biochemistry* **35,** 8193–8200.

Lu, Y., and Nelsestuen, G. L. (1996b). The prothrombinase reaction: Mechanism switching between Michaelis Menten and non-Michaelis-Menten behaviors. *Biochemistry* **35,** 8201–8209.

Luo, G., Ducy, P., McKee, M. D., Pinero, G. J., Loyer, E., Behringer, R. R., and Karsenty, G. (1997). Spontaneous calcification of arteries and cartilage in mice lacking matrix GLA protein. *Nature (London)* **386,** 78–81.

Magnusson, S., Sottrup-Jensen, L., Petersen, T. E., Morris, H. R., and Dell, A. (1974). Primary structure of the vitamin K-dependent part of prothrombin. *FEBS Lett.* **44,** 189–193.

Manfioletti, G., Brancolini, C., Avanzi, G., and Schneider, C. (1993). The protein encoded by a growth arrest-specific gene (gas6) is a new member of the vitamin K-dependent proteins related to protein S, a negative coregulator in the blood coagulation cascade. *Mol. Cell. Biol.* **13,** 4976–4985.

Mann, K. G., Nesheim, M. E., Church, W. R., Haley, C. P., and Krishnaswamy, S. (1990). Surface-dependent reactions of the vitamin K-dependent enzyme complexes. *Blood* **76,** 1–16.

Marsh, H. C., Scott, M. E., Hiskey, R. G., and Koehler, K. A. (1979). The nature of the slow metal ion-dependent conformational transition in bovine prothrombin. *Biochem. J.* **183,** 513–517.

Martinez, M. B., Flickinger, M., and Nelsestuen, G. L. (1996). Accurate kinetic modeling of alkaline phosphatate in the *Eschericia coli* periplasm: Implications for enzyme-properties and substrate diffusion. *Biochemistry* **35,** 1179–1186.

Mayer, L. D., Nelsestuen, G. L., and Brockman, H. L. (1983). Prothrombin association with phospholipid monolayers. *Biochemistry* **22,** 316–321.

McCallum, C. D., Hapak, R. C., Neuenschwander, P. F., Morrissey, J. H., and Johnson, A. E. (1996). The location of the active site of blood coagulation factor VIIa above the membrane surface and its reorientation upon association with tissue factor. A fluorescence energy transfer study. *J. Biol. Chem.* **271,** 28168–28175.

McCallum, C. D., Su, B., Neuenschwander, P. F., Morrissey, J. H., and Johnson, A. E. (1997). Tissue factor positions and maintains the factor VIIa active site far above the membrane surface even in the absence of the factor VIIa Gla domain. A fluorescence resonance energy transfer study. *J. Biol. Chem.* **272,** 30160–30166.

McDonald, J. F., Evans, T. C., Jr., Emeagwali, D. B., Hariharan, M., Allewell, N. M., Pusey, M. L., Shah, A. M., and Nelsestuen, G. L. (1997a). Ionic properties of membrane association by vitamin K-dependent proteins: The case for univalency. *Biochemistry* **36,** 15589–15598.

McDonald, J. F., Shah, A. M., Schwalbe, R. A., Kisiel, W., Dahlback, B., and Nelsestuen, G. L. (1997b). Comparison of naturally occurring vitamin K-dependent proteins: Correlation of amino acid sequences and membrane binding properties suggests a membrane contact site. *Biochemistry* **36,** 5120–5127.

McGee, M. P., Li, L. C., and Xiong, H. (1992). Diffusion control in blood coagulation. Activation of factor X by factors IXa/VIIIa assembled on human monocyte membranes. *J. Biol. Chem.* **267,** 24333–24339.

Mitchell, C. A., Kelemen, S. M., and Salem, H. H. (1988). The anticoagulant properties of a modified form of protein S. *Thromb. Haemostasis* **60,** 298–304.

Monkovic, D. D., and Tracy, P. B. (1990). Activation of human factor V by factor Xa and thrombin. *Biochemistry* **29,** 1118–1128.

Morris, D. P., Stevens, R. D., Wright, D. J., and Stafford, D. W. (1995). Processive post-translational modification. *J. Biol. Chem.* **270,** 30491–30498.

Morrissey, J. H., Fakhrai, H., and Edgington, T. S. (1987). Molecular cloning of the cDNA for tissue factor, the cellular receptor for the initiation of the coagulation protease cascade. *Cell (Cambridge, Mass.)* **50,** 129–135.

Mutucumarana, V. P., Duffy, E. J., Lollar, P., and Johnson, A. E. (1992). The active site of factor IXa is located far above the membrane surface and its conformation is altered upon association with factor VIIIa. A fluorescence study. *J. Biol. Chem.* **267,** 17012–17021.

Nelsestuen, G. L. (1976). Role of gamma-carboxyglutamic acid. An unusual protein transition required for the calcium-dependent binding of prothrombin to phospholipid. *J. Biol. Chem.* **251,** 5648–5656.

Nelsestuen, G. L. (1978). Interaction of vitamin K-dependent proteins with calcium ions and phospholipid membranes. *Fed. Proc., Fed. Am. Soc. Exp. Biol.* **37,** 2621–2625.

Nelsestuen, G. L. (1988). Basis for prothrombin-membrane binding. In "Current Advances in Vitamin K Research" (J. W. Suttie, ed.), pp. 335–339. Elsevier, New York.

Nelsestuen, G. L. (1989). A partial remedy for the non-relationship between reversiblitiy of a reaction in the cell and the value of ΔGo'. *Biochem. Educ.* **4,** 190–192.

Nelsestuen, G. L., and Lim, T. K. (1977). Equilibria involved in prothrombin- and blood clotting factor X-membrane binding. *Biochemistry* **16,** 4164–4171.

Nelsestuen, G. L., Zytokovicz, T. H., and Howard, J. B. (1974). The mode of action of vitamin K. Identification of gamma-carboyglutamic acid as a component of prothrombin. *J. Biol. Chem.* **249,** 6347–6350.

Nelsestuen, G. L., Broderius, M., and Martin, G. (1976). Role of gamma-carboxyglutamic acid. Cation specificity of prothrombin and factor X-phospholipid binding. *J. Biol. Chem.* **251,** 6886–6893.

Nemerson, Y., and Gentry, R. (1986). An ordered addition, essential activation model of the tissue factor pathway of coagulation: Evidence for a conformational cage. *Biochemistry* **25,** 4020–4033.

Nesheim, M. E., and Mann, K. G. (1979). Thrombin-catalyzed activation of single chain bovine factor V. *J. Biol. Chem.* **254,** 1326–1334.

Nesheim, M. E., Myrmel, K. H., Hibbard, L., and Mann, K. G. (1979a). Isolation and characterization of single chain bovine factor V. *J. Biol. Chem.* **254,** 508–517.

Nesheim, M. E., Taswell, J. B., and Mann, K. G. (1979b). The contribution of bovine Factor V and Factor Va to the activity of prothrombinase. *J. Biol. Chem.* **254,** 10952–10962.

Nesheim, M. E., Tracy, R. P., and Mann, K. G. (1984). "Clotspeed," a mathematical simulation of the functional properties of prothrombinase. *J. Biol. Chem.* **259,** 1447–1453.

Nicolaes, G., Tans, G., Thomassen, C., Hemker, H. C., Pabinger, I., Varadi, K., Schwarz, H., and Rosing, J. (1995). Peptide bond clevages and loss of functional activity during inactivation of factor Va and factor Va (R506Q) by APC. *J. Biol. Chem.* **270,** 21158–21166.

Olson, S. T., Bjork, I., and Shore, J. D. (1993). Kinetic characterization of heparin-catalyzed and uncatalyzed inhibition of blood coagulation proteinases by antithrombin. *In* "Methods in Enzymology" (L. Lorand and K. G. Mann, eds.), vol. 222, 525–559. Academic Press, San Diego, CA.

Østerud, B., and Rapaport, S. I. (1977). Activation of factor IX by the reaction product of tissue factor and factor VII: Additional pathway for initiating blood coagulation. *Proc. Natl. Acad. Sci. U.S.A.* **74,** 5260–5264.

Park, C. H., and Tulinsky, A. (1986). Three-dimensional structure of the kringle sequence: Structure of prothrombin fragment 1. *Biochemistry* **25,** 3977–3982.

Perera, L., Darden, T. A., and Pedersen, L. G. (1998). Trans-cis isomerization of proline 22 in bovine prothrombin fragment 1: A surprising result of structural characterization. *Biochemistry* **37,** 10920–10927.

Price, P. A. (1989). Gla-containg proteins of bone. *Connect. Tissue Res.* **21,** 51–60.

Pusey, M. L., and Nelsestuen, G. L. (1983). Physical significance of Km in the prothrombinase reaction. *Biochem. Biophys. Res. Commun.* **114,** 526–532.

Pusey, M. L., Mayer, L. D., Wei, G. J., Bloomfield, V. A., and Nelsestuen, G. L. (1982). Kinetic and hydrodynamic analysis of blood clotting factor V-membrane binding. *Biochemistry* ***21,*** 5262–5269.

Rapaport, S. I. (1991). The extrinsic pathway inhibitor: A regulator of tissue factor-dependent blood coagulation. *Thromb. Haemostasis* **66,** 6–15.

Ratcliffe, J. V., Furie, B., and Furie, B. C. (1993). The importance of specific gamma-carboxyglutamic acid residues in prothrombin. Evaluation by site-specific mutagenesis. *J. Biol. Chem.* **268,** 24339–24345.

Resnick, R. M., and Nelsestuen, G. L. (1980). Prothrombin-membrane interaction. Effects of ionic strength, pH and temperature. *Biochemistry* **19,** 3028–3033.

Rigby, A. C., Baleja, J. D., Li, L., Pedersen, L. G., Furie, B. C., and Furie, B. (1997). Role of gamma-carboxyglutamic acid in the calcium-induced structural transition of conantokin G, a conotoxin from the marine snail *Conus geographus. Biochemistry* **36,** 15677–15684.

Rintala, E., Seppala, O. P., Kotilainen, P., Pettila, V., and Rasi, V. (1998). Protein C in the treatment of coagulopathy in meningococcal disease. *Crit. Care Med.* **26,** 965–968.

Rosing, J., Tans, G., Govers-Riemslag, J. W. P., Zwaal, R. F. A., and Hemker, H. C. (1980). The role of phosholipids and factor Va in the prothrombinase complex. *J. Biol. Chem.* **255,** 274–283.

Rosing, J., Bakker, H. M., Thomassen, M. C., Hemker, H. C., and Tans, G. (1993). Characterization of two forms of human factor Va with different cofactor activities. *J. Biol. Chem.* **268,** 21130–21136.

Rosing, J., Hoekema, L., Nicolaes, G. A. F., Thomassen, C. G. D., Hemker, H. C., Varadi, K., Schwarz, H. P., and Tanz, G. (1995). Effects of protein S and factor Xa on peptide bond cleavages during inactivation of factor Va and factor VaR506Q by activated protein C. *J. Biol. Chem.* **270,** 27852–27858.

Rudolph, A. E., Mullane, M. P., Porche-Sorbet, R., Tsuda, S., and Miletich, J. P. (1996). Factor X St. Louis II. Identification of a glycine substitution at residue 7 and characterization of the recombinant protein. *J. Biol. Chem.* **271,** 28601–28606.

Sadler, J. E., and Davie, E. W. (1987). Hemophilia. In "The Molecular Basis of Blood Diseases" (G. Stamatoyannopoulos, A. W. Nienhuis, P. Leder, and P. W. Majerus, eds.), pp. 575–630. Saunders, Philadelphia.

Scarpati, E. M., Wen, D., Broze G. J., J., Miletich, J. P., Fandermeyer, R. R., Siegel, N. R., and Sadler, J. E. (1987). Human tissue factor: cDNA sequence and chromosome localization of the gene. *Biochemistry* **26,** 5234–5238.

Scott, D. L., White, S. P., Browning, J. L., Rosa, J. J., Gelb, M. H., and Sigler, P. B. (1991). Structures of free and inhibited human secretory phospholipase A2 from inflammatory exudate. *Science* **254,** 1007.

Seshadri, T. P., Tulinsky, A., Skrzypczak-Jankun, E., and Park, C. H. (1991). Structure of bovine prothrombin fragment 1 refined at 2.25 Å resolution. *J. Mol. Biol.* **220,** 481–494.

Seshadri, T. P., Skrzypczak-Jankun, E., Yin, M., and Tulinsky, A. (1994). Differences in the metal ion structure between Sr and Ca-prothrombin fragment 1. *Biochemistry* **33,** 1087–1092.

Shah, A. M., Kisiel, W., Foster, D. C., and Nelsestuen, G. L. (1998). Manipulation of the membrane binding site of vitamin K-dependent eProteins: Enhanced biological function of human factor VII. *Proc. Natl. Acad. Sci. U.S.A.* **95,** 4229–4234.

Shen, L., Shah, A. M., Dahlback, B., and Nelsestuen, G. L. (1997). Enhancing the activity of protein C by mutagenesis to improve the membrane-binding site: Studies related to proline-10. *Biochemistry* **36,** 16025–16031.

Shen, L., Shah, A. M., Dahlback, B., and Nelsestuen, G. L. (1998). Enhancement of human protein C function by site-directed mutagenesis of the gamma-carboxyglutamic acid domain. *J. Biol. Chem.* **273,** 31086–31091.

Silversmith, R. E., and Nelsestuen, G. L. (1986). Interaction of complement proteins C5b-6 and C5b-7 with phospholipid vesicles: Effects of phospholipid structural features. *Biochemistry* **25,** 7717–7725.

Solymoss, S., Tucker, M. M., and Tracy, P. B. (1988). Kinetics of inactivation of membrane-bound factor Va by activated protein C. *J. Biol. Chem* **263,** 14884–14890.

Sommerville, L. E., Resnick, R. M., Thomas, D. D., and Nelsestuen, G. L. (1986). Terbium probe of calcium-binding sites on the prothrombin-membrane complex. *J. Biol. Chem.* **261,** 6222–6229.

Soriano-Garcia, M., Padmanabhan, K., de Vos, A. M., and Tulinsky, A. (1992). The Ca2+ ion and membrane binding structure of the Gla domain of Ca-prothrombin fragment 1. *Biochemistry* **31,** 2554–2566.

Spicer, E. K., Horton, R., Bloem, L., Bach, R., Williams, K. R., Guha, A., Kraus, J., Lin, T. C., Nemerson, Y., and Konigsberg, W. H. (1987). Isolation of cDNa clones coding for human tissue factor: Primary structure of the protein and cDNA. *Proc. Natl. Acad. Sci. U.S.A.* **84,** 5148–5152.

Stenflo, J. (1974). Vitamin K and the biosynthesis of prothrombin. IV. Isolation of peptides containing prosthetic groups from normal prothrombin and the corresponding peptides from dicumarol-induced prothrombin. *J. Biol. Chem.* **249,** 5527–5535.

Stitt, T. N., Conn, G., Gore, M., Lai, C., Bruno, J., Radziejewski, C., Mattson, K., Fisher, J., Gies, D. R., Jones, P. F., Masiakowski, P., Ryan, T. E., Tobkes, N. J., Chen, D. H., DiStefano, P. S., Long, G. L., Basilico, C., Goldfarb, M. P., Lemke, G., Glass, D. J., and Yacopoulos, G. D. (1995). The anticoagulation factor protein S and its relative, Gas6, are ligands for the Tyro 3/Axl family of receptor tyrosine kinases. *Cell (Cambridge, Mass.)* **80,** 661–670.

Sugiura, I., Furie, B., Walsh, C. T., and Furie, B. C. (1997). Propeptide and glutamate-containing substrates bound to the vitamin K-dependent carboxylase convert its vitamin K epoxidase function from an inactive to an active state. *Proc. Natl. Acad. Sci. U.S.A.* **94,** 9069–9074.

Sunnerhagen, M., Forsén, S., Hoffren, A.-M., Drakenberh, T., Teleman, O., and Stenflo, J. (1995). Structure of the Ca-free GLA domain sheds light on membrane binding of blood coagulation proteins. *Nat. Struct. Biol.* **6,** 504–509.

Sunnerhagen, M., Drakenberg, T., Forsén, S., and Stenflo, J. (1996). Effect of Ca on the structure of vitaminK-dependent coagulation factors. *Haemostasis* **26,** 45–53.

Suttie, J. W. (1985). Vitamin K-dependent carboxylase. *Annu. Rev. Biochem.* **54,** 459–477.

Swairjo, M. A., Concha, N. O., Kaetzel, M. A., Dedman, J. R., and Seaton, B. A. (1995). Ca2+-bridging mechanism and phospholipid head group recognition in the membrane-binding protein annexin V. *Nat. Struct. Biol.* **2,** 968–974.

Thariath, A., and Castellino, F. J. (1997). Highly conserved residue Arginine-15 is required for th Ca2+-dependent properties of the gamma carbosyglutaminc acid domain of human anticoagulation protein C and activated protein C. *Biochem. J.* **322,** 309–315.

Tollefsen, D. M. (1995). Insight into the mechanism of action of heparin cofactor II. *Thromb. Haemostasis* **74,** 1209–1214.

Tracy, P. B., Nesheim, M. E., and Mann, K. G. (1981). Coordinate binding of factor Va and factor Xa to the unstimulated platelet. *J. Biol. Chem.* **256,** 743–751.

Tracy, P. B., Eide, L. L., Bowie, E. J., and Mann, K. G. (1982). Radioimmunoassay of factor V in human plasma and platelets. *Blood* **60,** 59–63.

Tulinsky, A. (1991). The structures of domains of blood proteins. *Thromb. Haemostasis* **66,** 16–31.

Tulinsky, A., Park, C. H., and Skrzypczak-Jankun, E. (1988). Structure of prothrombin fragment 1 refined at 2.8 Å resolution. *J. Mol. Biol.* **202,** 885–901.

van Dieijen, G., Tans, G., Rosing, J., and Hemker, H. C. (1981). The role of phospholipid and factor VIIIa in the activation of bovine factor X. *J. Biol. Chem.* **256,** 3433–3442.

van Rijn, J. L., Grovers-Riemslag, J. W., Zwaal, R. F., and Rosing, J. (1984). Kinetic studies of prothrombin activation: Effect of factor Va and phopholipids on the formation of the enzyme substrate complex. *Biochemistry* **23,** 4557–4564.

Varnum, B. C., Young, C., Elliott, G., Garcia, A., Bartley, T. D., Fridell, Y. W., Hunt, R. W., Trail, G., Clogston, C., Toso, R. J., Yanagihara, D., Bennett, L., Sylber, M., Merewether, L. A., Tseng, A., Escobar, E., Liu, E. T., and Yamane, H. K. (1995). Axl receptor tyrosine kinase stimulated by the vitamin K-dependent protein encoded by growth-arrest-specific gene 6. *Nature (London)* **373,** 623–626.

Walker, F., and Fay, D. (1992). Regulation of blood coagulation by the protein C system. *FASEB J.* **6,** 2561–2567.

Walker, F. J. (1980). Regulation of activated protein C by a new protein. A possible function for bovine protein S. *J. Biol. Chem.* **255,** 5521–5524.

Weber, D. J., Berkowitz, P., Panek, M. G., Huh, N. W., Pedersen, L. G., and Hiskey, R. G. (1992). Modifications of bovine prothrombin fragment 1 in the presence and absence of Ca (II) ions. Loss of positive cooperativity in Ca(II) ion binding for the modified proteins. *J. Biol. Chem.* **267,** 4564–4569.

Welsch, D. J., and Nelsestuen, G. L. (1988). Amino terminal alanine functions in a calcium-specific process essential for membrane binding by prothrombin fragment 1. *Biochemistry* **27,** 4939–4945.

Yegneswaran, S., Wood, G. M., Esmon, C. T., and Johnson, A. E. (1997). Protein S alters the active site location of activated protein C above the membrane surface. A fluorescence resonance energy transfer study of topography. *J. Biol. Chem.* **272,** 25013–25021.

Zhang, L., and Castellino, F. J. (1993). The contributions of individual gamma-carbosyglutamic acid residues in the calcium-dependent binding of recombinant human protein C to acidic phospholipid vesicles. *J. Biol. Chem.* **268,** 12040–12045.

Zhang, L., and Castellino, F. J. (1994). The binding energy of human coagulation protein C to acidic phospholipid vesicles contains a major contribution from leucine 5 in the gamma-carboxyglutamic acid domain. *J. Biol. Chem.* **269,** 3590–3595.

Zhao, J., Zhou, Q., Wiedmer, T., and Sims, P. J. (1998). Level of expression of phospholipid scramblase regulates induced movement of phosphatidylserine to the cell surface. *J. Biol. Chem.* **273,** 6603–6606.

Zur, M., and Nemerson, Y. (1980). Kinetics of factor IX activation via the extrinsic pathway. Dependence of Km on tissue factor. *J. Biol. Chem.* **255,** 5703–5707.

VITAMINS AND HORMONES, VOL. 58

Steroid/Nuclear Receptor Coactivators

J. DON CHEN

Department of Pharmacology and Molecular Toxicology, University of Massachusetts Medical School, Worcester, Massachusetts 01655–0126

In higher eukaryotes, steroids/thyroid hormones and many lipophilic compounds regulate cellular physiology through binding to the steroid/nuclear receptor proteins. Steroid/nuclear receptors are ligand-dependent transcriptional activators that can stimulate gene expression. This transcriptional activation plays a pivotal role in hormone-regulated physiological and pharmacological responses. In recent years, several steroid/nuclear receptor cofactors have been identified and found to interact with the receptor and modulate

0083-6729/00 $30.00

its transcriptional activity. Among these cofactors, a family of three coactivators has been the focus of intense studies. Although gaps remain, progress has been made in understanding how a given coactivator interacts with the receptor and promotes transcriptional activation. We are beginning to understand coactivator action; for instance, several studies have established the molecular basis of antagonism by anti-hormones and the connection of coactivators with human cancers.

I. Introduction

Lipophilic steroids, including estrogen, progesterone, androgens, glucocorticoid and minerocorticoid, thyroid hormones, retinoids, vitamin D_3, and peroxisome proliferators regulate diverse biological activities including cell proliferation, differentiation, development, and homeostasis. The activities of these compounds are thought to be mediated by members of the steroid/nuclear receptor superfamily, most of which are ligand-regulated transcriptional activators (Mangelsdorf *et al.,* 1995; Mangelsdorf and Evans, 1995; Kastner *et al.,* 1995; Thummel, 1995; Beato *et al.,* 1995). A distinct domain structure, including an N-terminal DNA-binding domain (DBD) and a C-terminal ligand-binding domain (LBD) characterizes the steroid/nuclear receptors. The DBD binds to specific DNA sequences located within promoter regions of target genes. The LBD binds to specific ligand, which in turn controls the receptor's transcriptional activity by triggering conformational changes in the receptors that affect protein-protein interaction and transcriptional activation. Thus, lipophilic nuclear hormones display diverse biological effects owing to transcriptional activity driven by steroid/nuclear receptors.

The steroid/nuclear receptor assumes both active and inactive conformations depending on ligand binding to the receptor's LBD. The LBD mediates not only ligand-binding, but also protein–protein interaction, transcriptional activation, and transcriptional repression. Located near the C terminus of the receptor, the conserved AF-2 helix domain (also known as AF2-AD, τC, or τ4) plays a crucial role in regulating ligand-dependent transcriptional activity. Several lines of investigation have helped elucidate receptor-driven transcriptional activation (Beato *et al.,* 1995). Notably, squelching effects between different receptors provide evidence that cofactors regulate receptor-mediated transcriptional activity (Meyer *et al.,* 1989; Barettino *et al.,* 1994). The identification of receptor-associated proteins involves both genetic and biochemical approaches such as the yeast two-hybrid screen and far-Western-based expression screening (Seol *et al.,* 1995; Cavaillès *et al.,* 1994; Halachmi *et al.,* 1994; Kurokawa *et al.,* 1995; Eggert *et al.,* 1995; Chen and Evans, 1995; Horlein *et al.,* 1995).

The two important classes of steroid/nuclear receptor cofactors are the transcriptional corepressors and coactivators. Corepressors interact with unliganded receptors to inhibit target gene expression. The silencing mediator for retinoid and thyroid hormone receptors (SMRT) and the nuclear receptor corepressor (N-CoR) are examples of the corepressors (Chen and Evans, 1995; Horlein *et al.,* 1995). In contrast, coactivators interact with liganded receptors to enhance transcriptional activation. Many potential coactivators have been identified (Horwitz *et al.,* 1996; Glass *et al.,* 1997; Chen and Li, 1998), including transcriptional intermediate factor 1 (TIF1) (Le Douarin *et al.,* 1995), receptor interacting protein 140 (RIP140) (Cavaillès *et al.,* 1995), androgen receptor activator 70 (ARA70) (Yeh and Chang, 1996), and steroid receptor coactivators (SRCs) (Oñate *et al.,* 1995; Takeshita *et al.,* 1996; Zhu *et al.,* 1996; Li *et al.,* 1997; Anzick *et al.,* 1997; Yao *et al.,* 1996; Chen *et al.,* 1997; Kamei *et al.,* 1996; Torchia *et al.,* 1997). In addition, the general transcriptional coactivators SWI/SNF (Chiba *et al.,* 1994; Khavari *et al.,* 1993; Muchardt and Yaniv, 1993) and CREB/E1A-binding protein (CBP/p300) (Arany *et al.,* 1994; Chrivia *et al.,* 1993; Kwon *et al.,* 1994) enhance transcriptional activation by steroid/nuclear receptors. The role of ligand is to induce corepressor dissociation and coactivator recruitment. Such exchange of corepressors and coactivators on DNA-bound receptors is thought to underscore the mechanism of ligand-dependent transcriptional activation (Fig. 1).

Among the steroid/nuclear receptor coactivators, the SRC family has been the focus of recent intense studies. Compelling evidence suggests that SRC coactivators regulate the transcriptional activity of many steroid/nuclear hormone receptors. Extensive investigations have detailed SRC-receptor interactions at the molecular level and described the mechanism of SRC-regulated transcription. Inactivation of one SRC coactivator in mouse demonstrated that this coactivator is required for maximal hormone responses. In addition, another SRC coactivator is amplified in breast, ovarian, and pancreatic cancers, suggesting an important role for these coregulators in cell growth and differentiation. Understanding SRC coactivators may provide a model system and new insights for therapeutic intervention of hormone-related human diseases. This chapter is intended to summarize recent findings about the function and mechanism of action of the SRC coactivators.

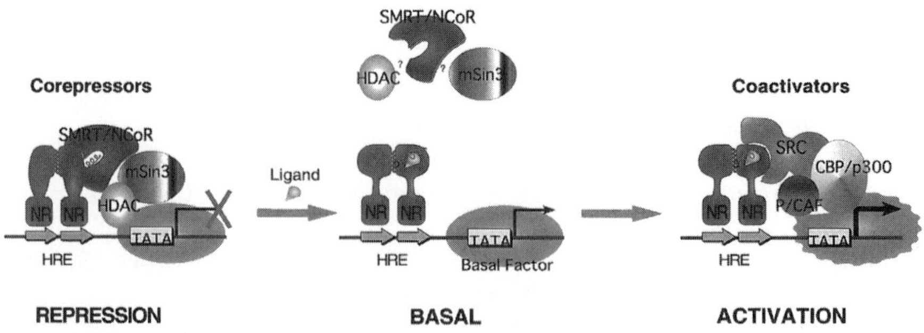

FIG. 1. Model of transcriptional regulation by steroid/nuclear hormone receptors. In the absence of hormone, DNA-bound unliganded receptors recruit nuclear receptor corepressors such as SMRT/NCoR, which target mSin3 and histone deacetylases (HDAC) to catalyze deacetylation of histones, leading to chromatin condensation and transcriptional repression. Hormone binding induces conformational changes of the receptors, resulting in the dissociation of corepressors and recruitment of coactivators, which catalyze acetylation of histones, leading to opening of chromatin and transcriptional activation.

II. Identification of SRC Coactivators

A. SRC-1

Steroid receptor coactivator 1 (SRC-1), the first member of the SRC family, was isolated in a yeast two-hybrid screen using PR_B-LBD as bait (Oñate *et al.,* 1995). A C-terminal 197-amino-acid (aa) fragment of human SRC-1 was identified that interacts with PR in an agonist-dependent manner (Oñate *et al.,* 1995). Later work identified several isoforms of SRC-1 from both human and mouse libraries as TR, PPAR, or CBP/p300 interacting protein (Takeshita *et al.,* 1996; Kalkhoven *et al.,* 1998; Zhu *et al.,* 1996; Kamei *et al.,* 1996; Yao *et al.,* 1996). RT-PCR analysis confirms two SRC-1 isoforms, SRC-1a and SRC-1e (Kalkhoven *et al.,* 1998). SRC-1e differs from SRC-1a at the C termini; the 1441-aa-long SRC-1a contains 56 unique residues and lacks the most C-terminal 14 amino acids present in SRC-1e (1399-aa) (Kalkhoven *et al.,* 1998), suggesting a potential functional difference between these two isoforms. For simplicity, the full-length SRC-1a will be referred to as SRC-1 unless specified otherwise.

B. SRC-2 and SRC-3

Following the identification of SRC-1, glucocorticoid receptor interacting protein 1 (GRIP1) was isolated in yeast two-hybrid screen (Hong *et al.*, 1996, 1997). At about the same time, transcriptional intermediate factor 2 (TIF2) was identified as a 160-kDa human protein that interacts with liganded ER and RAR in a far-Western-based assay (Voegel *et al.*, 1996). Human TIF2 contains 1464 amino acids that are similar to the mouse GRIP1 (over 94% identity), suggesting that TIF2 and GRIP1 are the mouse and human orthologs. In addition, NCoA-2 was reported as a mouse variant of GRIP1 (Torchia *et al.*, 1997). The 1463-aa-long NCoA-2 is nearly identical to GRIP1, except for several single amino acid substitutions and two unrelated gaps at residues 251–320 and 959–982 of GRIP1. In addition, a rat homolog of TIF2 was recently identified as a PPARα-interacting protein in a yeast two-hybrid screen (Leers *et al.*, 1998). Sequence comparison between GRIP1/TIF2/NCoA-2 and SRC-1 reveals high similarity (Fig. 2), especially in the N-terminal domain, which is related to the bHLH (basic-helix-loop-helix)-PAS (Per-Arnt-Sim) domains in many transcriptional regulators (Swanson *et al.*, 1995; Lindebro *et al.*, 1995; Zelzer *et al.*, 1997). While bHLH-PAS is the most conserved domain among the SRC family members, its function remains undetermined in this coactivator family. Because GRIP1/TIF2/NCoA-2 is an SRC-1-related gene, it will be referred to as SRC-2 unless specified otherwise.

The SRC family was established when p/CIP, RAC3, ACTR, AIB1, TRAM-1, and SRC-3 were cloned and found as the third member of the

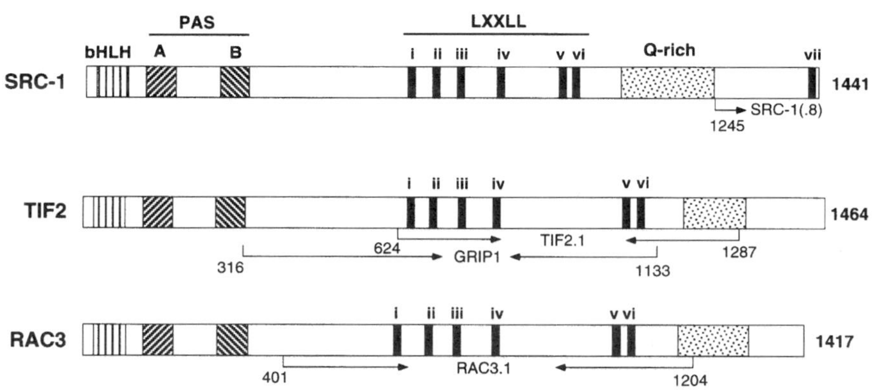

FIG. 2. Schematic representation of the structural domains of SRC family proteins. The N-terminal region contains highly conserved bHLH, PAS A and B domains. The central regions contain six LXXLL motifs (i to vi). SRC-1 contains an additional LXXLL motif at its C terminus. In addition, the C-terminal region contains a glutamine-rich domain, where consecutive glutamine track is present in RAC3 but not in TIF2 or SRC-1. SRC-1 is also known as NCoA-1 or NRC-1. TIF2 is also known as GRIP1 and NCoA-2. RAC3 is also known as p/CIP, ACTR, AIB1, SRC-3, and TRAM-1. The regions encoded by the original clones RAC3.1, TIF2.1, GRIP1, and SRC-1(.8) are indicated with arrows.

family (Torchia *et al.,* 1997; Chen *et al.,* 1997; Anzick *et al.,* 1997; Li *et al.,* 1997; Takeshita *et al.,* 1997; Suen *et al.,* 1998). p/CIP was identified as a CBP-interacting protein (Torchia *et al.,* 1997). RAC3 was found as an RAR-interacting protein (Li *et al.,* 1997). ACTR was identified as an hRARβ-stimulatory protein (Chen *et al.,* 1997). AIB1 was isolated as a gene amplified from the long arm of chromosome 20 (Anzick *et al.,* 1997). TRAM-1 was isolated as a TR-interacting protein (Takeshita *et al.,* 1997), while SRC-3 was reported as an ER-interacting protein (Suen *et al.,* 1998). Since p/CIP/RAC3/ACTR/AIB1/TRAM-1/SRC-3 is highly related to SRC-1 and SRC-2, it will be referred to as SRC-3 unless otherwise specified. SRC-3 shared about 46% identity with SRC-2, and about 36% identity with SRC-1. Recently, SRC-3 was also identified in *Xenopus* as an RXR-interacting protein (Kim *et al.,* 1998). The xSRC-3 shares 72% identity with SRC-3, 45% with SRC-2, and 38% with SRC-1, suggesting that xSRC-3 may be the homolog of human SRC-3. Analyses of the chromosomal locations of the three SRC coactivator genes mapped AIB1 to chromosome 20q12, a region amplified in breast cancer (Anzick *et al.,* 1997). The SRC-1 gene was mapped to chro-

TABLE I
SYNONYMS OF SRC COACTIVATORS

	Species	Genbank	References
SRC-1			
SRC-1	Human	U90661/U40396	Oñate *et al.* (1995)
F-SRC-1	Human	U59302	Takeshita *et al.* (1996)
hSRC-1a	Human	AJ000881	Kalkhoven *et al.* (1998)
SRC-1e	Human	AJ000882	Kalkhoven *et al.* (1998)
mSRC-1	Mouse	U64828	Yao *et al.* (1996)
mSRC-1a	Mouse	U56920	Kamei *et al.* (1996)
mNRC-1	Mouse	U64606	Zhu *et al.* (1996)
SRC-2			
TIF2	Human	X97674	Voegel *et al.* (1996)
rTIF2	Rat	AF000582	Leers *et al.* (1998)
GRIP1	Mouse	U39060	Hong *et al.* (1997)
NCoA-2	Mouse	AF000582	Kamei *et al.* (1996)
SRC-3			
RAC3	Human	AF010227	Li *et al.* (1997)
SRC-3	Human		Suen *et al.* (1998)
AIB1	Human	AF012108	Anzick *et al.* (1997)
ACTR	Human	AF036892	Chen *et al.* (1997)
TRAM-1	Human	AF016031	Takeshita *et al.* (1997)
p/CIP	Mouse	AF000581	Torchia *et al.* (1997)
xSRC-3	*Xenopus*	AF044080	Kim *et al.* (1998)

mosome band 2p23 and TIF2 was mapped to 8q21.1 (Kalkhoven *et al.*, 1998), where no amplification in breast cancer was found. Table I lists SRC coactivators synonyms.

C. Structural Domains of SRC Coactivators

Each SRC coactivator contains an N-terminal bHLH domain found in many transcriptional regulators (Fig. 3). The bHLH domain can act as a DNA-binding and/or dimerization interface in several transcription factors (Murre *et al.*, 1989a,b). Immediately adjacent to the bHLH motif is a region similar to the PAS domain found in Period (Per), Aryl hydrocarbon receptor (AhR), the AhR nuclear translocator protein (Arnt), and single-minded (Sim). The PAS domain is present in a class of proteins involved in regulation of *Drosophila* development and cellular signaling induced by hypoxia and dioxin treatment. Interestingly, the bHLH-PAS domain is the most conserved region within the SRC coactivators, indicating the SRC coactivators belong to a large family of bHLH-PAS-containing proteins. The PAS domains in AhR, Arnt, and Sim play important roles in protein–protein interaction, heterodimeric partner selection, and target gene specificity (Swanson *et al.*, 1995; Lindebro *et al.*, 1995; Zelzer *et al.*, 1997). Although the function of the bHLH-PAS domain in SRC coactivators remains unknown, this region could possibly mediate intra- or intermolecular interaction.

Apart from SRC-1 and SRC-2, the structure of SRC-3 is unique in that it contains consecutive poly-glutamine (poly-Q) tracks that result from expansion of CAG repeats. Expansion of poly-Q track is associated with several human diseases (Koshy and Zoghbi, 1997; Reddy and Housman, 1997; Butler *et al.*, 1998). In SRC-3, two poly-Q tracks are found at both ends of the glutamine-rich domain. At the C-terminal location, three of the five cloned human SRC-3 alleles contain 26 consecutive glutamines, while two of them contain 29 consecutive residues. Consistently, polymerase chain reaction products of the SRC-3 CAG repeats at this location revealed size polymorphism (Shirazi *et al.*, 1998). Interestingly, *Xenopus* SRC-3 contains only four consecutive glutamines at this position. Likewise, the mouse SRC-3 (p/CIP) contains no significant poly-Q track at this site. The corresponding region of SRC-2 contains three and four consecutive glutamines in the mouse and human SRC-2, respectively, while no obvious glutamine repeats are present in SRC-1 at this location. The second poly-Q track is more evident within the mouse SRC-3 (p/CIP), which contains a track of 23 consecutive glutamines at a position around residue 1000 near the N-terminal end of the glutamine-rich domain. This location contains five consecu-

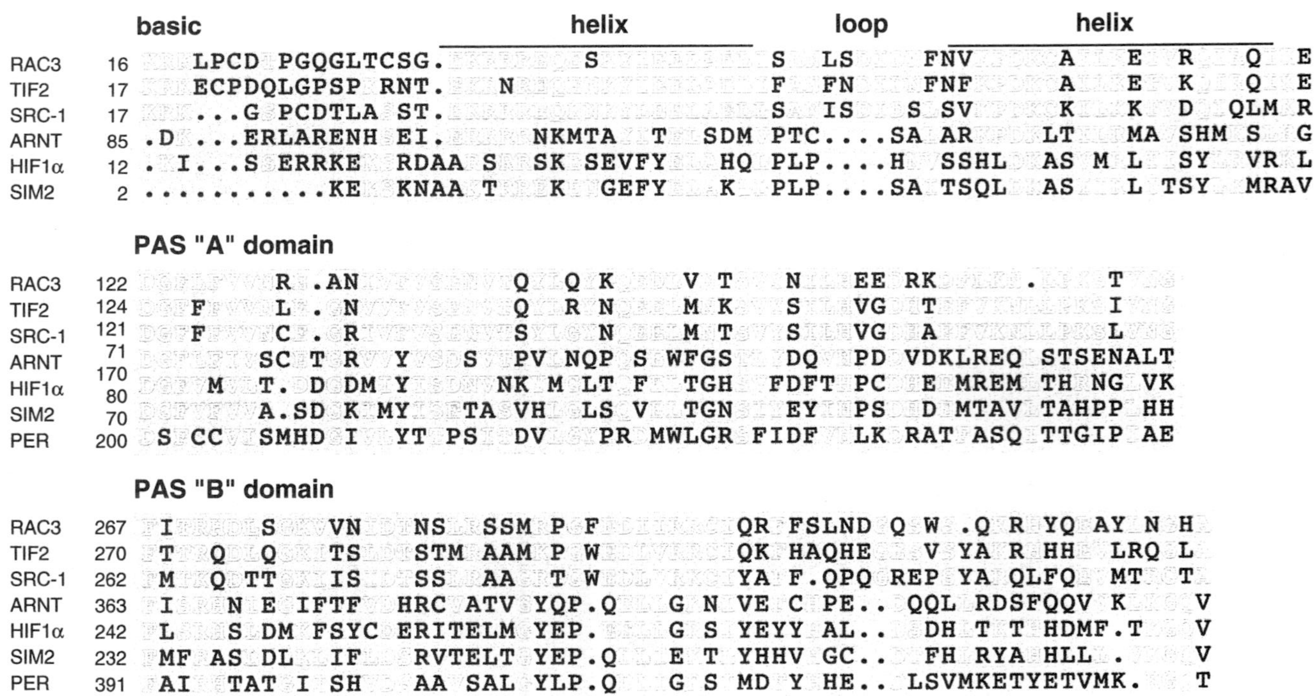

FIG. 3. Comparison of the bHLH-PAS domains of SRC coactivators. The bHLH and PAS domains of RAC3, TIF2, and SRC-1 are compared with other bHLH-PAS proteins. The conserved residues are colored white. The dots are gaps introduced to maximize the alignment.

tive glutamines in all five human SRC-3 proteins and four glutamines in the *Xenopus* protein. Conceivably, the relative length and position of these poly-Q tracks may distinguish functional differences among SRC coactivator members and alleles.

III. Interaction with Nuclear Receptors

A. Ligand-Dependent Interaction

Implication of SRCs as transcriptional coactivators for steroid/nuclear receptors first came from the observation that an SRC interacts with a receptor in a ligand-dependent manner, suggesting a role in transcriptional activation. Ligand-dependent interactions of SRCs with steroid/nuclear receptors have been demonstrated with multiple assay systems. Using the yeast two-hybrid assay, all three SRC coactivators have been shown to interact with multiple members of the steroid/nuclear receptor family in a ligand-dependent manner (Oñate *et al.,* 1995; Chen *et al.,* 1997; Li *et al.,* 1997; Voegel *et al.,* 1998; Hong *et al.,* 1997). In addition, coimmunoprecipitation and subcellular colocalization also detect *in vivo,* ligand-dependent interactions of SRC with steroid/nuclear receptors. Specifically, subcellular colocalization has been used to analyze ligand-dependent interaction of TIF2 with RAR, ER, and PR (Voegel *et al.,* 1996). Because a truncated TIF2 mutant (TIF2.1) does not contain a nuclear localization signal and remains in the cytoplasm, it demonstrates ligand-dependent translocation from the cytoplasm into the nucleus via interaction with liganded receptors targeted to the nucleus (Voegel *et al.,* 1996).

The association of two proteins *in vivo* involves the formation of a complex containing other proteins. Therefore, *in vivo* interaction observed in the two-hybrid and coimmunoprecipitation assays is usually insufficient to conclude direct interaction between two proteins. GST pull-down and far-Western assays are commonly utilized for detecting protein–protein interaction *in vitro.* In addition, the far-Western assay is also used for screening interacting clones; in fact, this method identified two mouse SRC-1s and the TRAM-1 clone (Kamei *et al.,* 1996; Yao *et al.,* 1996; Takeshita *et al.,* 1997). Figure 4 shows an example of ligand-dependent interaction between RAC3 and VDR analyzed by far-Western analysis. The ligand-dependent interaction with steroid/nuclear receptors suggests that SRCs are components of a transcriptionally active complex. Consistently, the SRC coactivator does not interact with steroid/nuclear receptors bound to antagonist.

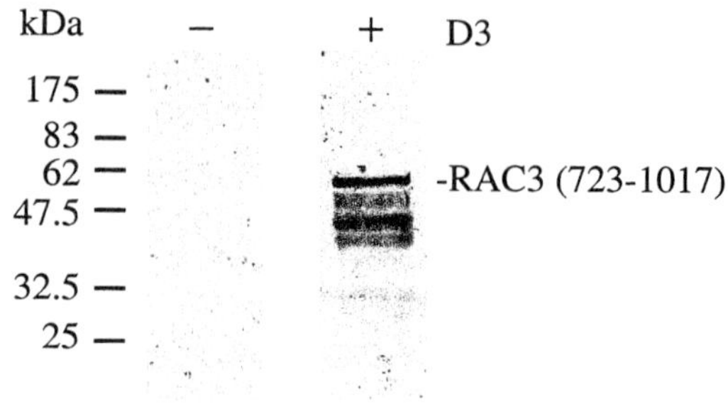

FIG. 4. Ligand-dependent interaction of RAC3 with nuclear receptor. The purified GST-RAC3 (723–1017) fusion protein was separated by SDS-PAGE and analyzed by far-Western analyses for interaction with ^{35}S-labeled hVDR in the absence (−) or presence (+) of 1 μM 1α,25-dihydroxyvitamin D_3. The position of the intact GST-RAC3 (723–1017) fusion protein is as indicated. The smaller peptides that also interact with VDR represent degradation products of GST-RAC3 (723–1017).

B. Interacting Interfaces

The mechanism by which SRC coactivators interact with steroid/nuclear receptors has been extensively analyzed by both biochemical and X-ray crystallography studies. These studies led to the identification of several conserved LXXLL motifs (where L is leucine and X is any amino acid) that are responsible for interaction with liganded receptors and transcriptional activation (Fig. 5). Six LXXLL motifs are located at the central region of all three SRC coactivators and one SRC-1a-unique motif is located at the C terminus (Heery *et al.,* 1997; Li *et al.,* 1997; Torchia *et al.,* 1997). A sequence resembling LXXLL motif has also been identified as receptor-interacting box (NR-box) in TIF1β (Le Douarin *et al.,* 1996) and in other steroid/nuclear receptor interacting proteins such as RIP140, CBP/p300, and TRIPs (Heery *et al.,* 1997).

The interacting domain between SRC-1 and ER was mapped first by a series of deletion mutants, where aa 570–780 and 1241–1441 of SRC-1a bound ER in an agonist-dependent fashion (Heery *et al.,* 1997; Henttu *et al.,* 1997; Kalkhoven *et al.,* 1998). The aa 570–780 fragment contains three conserved LXXLL motifs (i, ii, iii) and the C-terminal aa 1241–1441 fragment contains one motif (vii), which is present only in SRC-1a but not in SRC-1e or other SRC coactivators. Additionally, other regions of SRC-1a are also capable of binding ER and PR in an AF2- and ligand-independent manner, but the significance of these interactions remains unclear (Oñate *et al.,* 1998; Kalkhoven *et al.,* 1998).

Experiments using site-directed mutants and synthetic peptides have provided strong evidence for LXXLL motifs in mediating interaction with liganded steroid/nuclear receptors. First, a series of Gal4 DBD fusions with each LXXLL motif (i, ii, iii, or vii) interacts independently with ER in a ligand-dependent manner (Heery *et al.,* 1997), suggesting that an individual motif is sufficient for mediating the interaction. This study suggests that motif ii interacts most tightly with ER, while other unrelated sequences containing similar LXXLL core sequences failed to interact, suggesting that the LXXLL alone is insufficient and that flanking residues are also important for the interaction. Similarly, others observed strong ligand-dependent interactions of motif ii and the C-terminal motif of mouse SRC-1 (NCoA-1) with ER and RAR (Torchia *et al.,* 1997). In support of these observations, replacing the leucine doublet of the C-terminal LXXLL motif of SRC-1a with alanines disrupted the interaction with liganded receptors (Heery *et al.,* 1997).

Motifs			
i.	RAC3	(615-631)	SKGHKKLLQLLTCSSDD
	TIF2	(640-651)	SKGQTKLLQLLTTKSDD
	SRC1	(632-643)	SQTSHKLVQLLTTTAEE
ii.	RAC3	(678-695)	LQEKHRILHKLLQNGNSP
	TIF2	(683-699)	LKEKHKILHRLLQDSSSP
	SRC1	(683-699)	LTERHKILHRLLQEG.SP
iii.	RAC3	(730-749)	KKKE..NNALLRYLLDRDDPSD
	TIF2	(738-753)	KKKE...NALLRYLLDKDDTKD
	SRC1	(739-757)	KKKESKDHQLLRYLLDKDE.KD
iv.	RAC3	(1025-1041)	QNRPLLRNSLDDLVGPP
	TIF2	(1051-1067)	QNRQPFGSSPDDLLCPH
	SRC1	(904-920)	SEDQCISSQLDELLCPP
v.	RAC3	(1045-1061)	EGQSDERALLDQLHTLL
	TIF2	(1071-1087)	ESPSDEGALLDQLYLAL
	SRC1	(924-940)	EGRNDEKALLEQLVSFL
vi.	RAC3	(1069-1084)	LEEIDRALGIPELVNQ
	TIF2	(1093-1108)	LEEIDRALGIPELVSQ
	SRC1	(948-962)	LAELDRALGIDKLV.Q
vii.	SRC1	(1424-1440)	QTPQAQQKSLLQQLLTE

FIG. 5. The LXXLL motifs of SRC coactivators. The amino acid residues are shown at right in parentheses. The first six motifs are surrounded by highly charged residues and motifs ii, iv, v, and vi were predicted to form α-helical structures.

Because the three LXXLL motifs (i, ii, iii) are sufficient to interact with liganded receptors, the relative contribution and the specificity of each motif become important to understanding the mechanism of coactivator–receptor interaction. Systematic analysis of each LXXLL motif in the context of full-length SRC-1e protein suggests that loss of individual motif has little effect on the ability of SRC-1e to bind ER or to enhance its transcriptional activity (Heery *et al.,* 1997). Conversely, mutation of motif ii in combination with motif i or iii, or both, drastically reduces binding to liganded ER, and the ability to enhance ER-mediated transcription in transfected cells. However, combined mutation of motifs i and iii had less effect. When the mutation was generated in the central receptor interacting domain (635–760) of NcoA-1, it appears that mutation of motif ii is sufficient to abrogate interaction with liganded ER and RAR (Torchia *et al.,* 1997). Because disruption of motif ii blocks the function of motif i and iii within the minimal interacting domain of NCoA-1, but not in the full-length SRC-1e protein, it is conceivable that other interacting surfaces may contribute to stabilizing the interaction. These studies indicate that motif ii of SRC-1 is the preferred site for interaction with liganded ER, while motifs i and iii may contribute to optimal binding and activation of ER in intact cells.

The relative contribution of LXXLL motifs in mediating interaction with different receptors has also been analyzed by peptide competition assay (Heery *et al.,* 1997; Torchia *et al.,* 1997; Darimont *et al.,* 1998). Consistent with mutational studies, motif ii of SRC-1 is most important for interaction with RAR and TR, whereas C-terminal motif vii is most prominent for interaction ER (Heery *et al.,* 1997; Torchia *et al.,* 1997). For instance, an excess of 24-aa oligopeptide encompassing motif ii of NCoA-1 effectively blocked interaction between liganded RAR and NCoA-1 *in vitro,* but a peptide corresponding to motif i was less effective (Torchia *et al.,* 1997). Similarly, an excess 14-aa oligopeptide encompassing the C-terminal motif of SRC-1 blocks interaction between liganded ER and SRC-1a, but a peptide-containing leucine doublet mutant has no effect (Heery *et al.,* 1997). Also, a 13-aa peptide of GRIP1 motif ii inhibits the interaction between GRIP1 and TRβ LBD, while substitution of the leucine residues with alanines eliminates such inhibitory effect (Darimont *et al.,* 1998). Substitution of the leucine residues with phenylalanines also reduced the competition, suggesting that efficient interaction does not simply rely on the hydrophobicity of the LXXLL motif but rather on the stereochemical property of the side chain of leucine (Darimont *et al.,* 1998). Although motif ii of GRIP1 is the preferred sequence for ER binding, motif iii is preferred by GR

(Ding *et al.,* 1998). Another study shows that motif ii of rTIF2 is most critical for binding with PPARα and TRα, while motif i is the preferred site for RXRβ and motif iii is preferred by GR (Leers *et al.,* 1998). These studies suggest that steroid/nuclear receptors may interact with a given SRC coactivator through a precise arrangement of multiple LXXLL motifs.

C. Determinants of LXXLL Specificity

The presence of four different LXXLL motifs that can each interact independently with liganded receptors suggests a requirement for specificity, since it is not likely that all motifs interact simultaneously with a given receptor. In addition, different coactivators compete rather than cooperate for binding to a given receptor (Leers *et al.,* 1998), suggesting coactivator preference for the receptor. Thus, it is important to understand the mechanism of selectivity of LXXLL motifs for specific receptors. Using synthetic chimeric peptide, the preference of GRIP1 motif ii for TRβ appears determined by sequences adjacent to the LXXLL core residues. This was shown by a chimeric peptide containing adjacent sequences of motif ii and LXXLL of motif iii, which competes equally well as the intact motif ii for TRβ interaction (Darimont *et al.,* 1998). Conversely, a chimeric peptide containing LXXLL of motif ii flanked by sequences adjacent to a VP16 Fxxhh motif competes poorly for TRβ interaction. These studies suggest that both the LXXLL residues and the adjacent sequences of motif ii contribute to TR binding. In contrast, a chimeric peptide containing adjacent sequences of motif ii and LXXLL of motif iii competes equally well for GR binding as intact motif iii. Consistently, a chimeric peptide containing motif iii adjacent sequences and motif ii core competes inefficiently with GR binding. Therefore, the LXXLL core can dictate the selectivity of GR for its preference of motif iii over motif ii of GRIP1.

The specificity determinant of LXXLL motifs on transcriptional coactivation by NCoA-1 (SRC-1) has also been analyzed by site-directed mutagenesis and microinjection assay (McInerney *et al.,* 1998). For microinjection assay, a β-galactosidase reporter driven by specific response elements is injected into cell nuclei, along with specific antibody and a rescuing plasmid (McInerney *et al.,* 1998; Torchia *et al.,* 1997 L. Xu *et al.,* 1998; Korzus *et al.,* 1998). The requirement of specific LXXLL motifs of NCoA-1 for transactivation by different receptors was determined by antibody injection to inhibit reporter gene activation, along with a plasmid expressing wild-type or LXXLL mutant of NCoA-1. In this study, injection of anti-NCoA-1 IgG inhibits transactivation by ER,

PR, RAR, TR, and PPARγ (McInerney *et al.,* 1998), while coinjection of wild-type NCoA-1 reverses the IgG-mediated inhibition completely (Fig. 6). Coinjection of different LXXLL mutants elicits distinct levels of rescue. For instance, wild-type NCoA-1 and the motif i or iii mutants are capable of reversing IgG-mediated transcriptional inhibition. Double mutation of motifs i and iii had no effect on the coactivation function on ER-mediated transcriptional activation. In contrast, mutation of motif ii abrogated the ability to rescue IgG-mediated inhibition. These studies suggest that motif ii of NCoA-1 is sufficient for supporting ER activation, consistent with previous transient transfection and peptide competition studies (Ding *et al.,* 1998; Heery *et al.,* 1997; Kalkhoven *et al.,* 1998). Mutation of motif ii seems to play a more profound effect in the injection assay, but this might be due to a more severe mutation used in the injection assay (LXXLL→LAAAA) than in the transfection assay (LXXLL→LXXAA). It was also found that PR and PPARγ require both motifs i and ii, but not iii, while RAR and TR require motifs ii and iii, but not i, suggesting a distinct pattern of LXXLL motif requirement for different receptors. In addition, the LXXLL motif preference by PPARγ appears to be regulated by ligands. While troglitazone (TGZ; thiazolidinedione)-activated PPARγ prefers motif ii over i, prostaglandin J2 metabolites (PGJ2) promote an equivalent, partial requirement for both motif i and ii, but indomethacin alters the preference to motif i over ii (McInerney *et al.,* 1998). This specificity appears to depend on amino acids carboxy terminal to the LXXLL core. Consistently, distinct carboxy-terminal amino acids are required for PPARγ activation in response to different ligands (McInerney *et al.,* 1998). To-

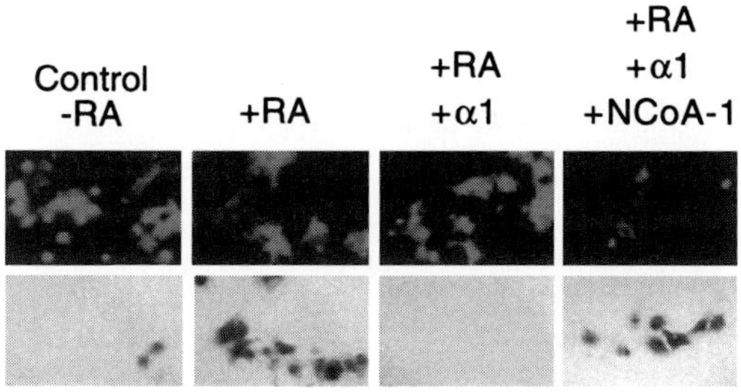

FIG. 6. Probing SRC coactivator function by microinjection assay. Microinjection of affinity-purified anti-NCoA-1 IgG blocked retinoic-acid-dependent activation of the RARE/LacZ reporter. The RA-dependent expression of reporter gene was fully rescued by coinjection of NCoA-1 expression vector. Photomicrographs of rhodomine-stained injected cells and the corresponding protein of X-Gal staining. [Adapted by permission from Fig. 4 of Torchia, J., *et al.* (1997). The transcriptional co-activator p/CIP binds CBP and mediates nuclear receptor function. *Nature* **387,** 677–684.]

gether, these studies suggest that LXXLL motifs may serve overlapping roles for both receptor-specific and ligand-specific assembly of a coactivator complex.

The contribution of each LXXLL motif residue for interaction with different receptors has also been analyzed by systemic mutation in combination with microinjection assays (McInerney *et al.,* 1998). The sequences encompassing the eight amino-terminal or carboxy-terminal flanking residues of motif ii of SRC-1 were mutated to alanines and their abilities to restore transactivation by RAR, TR, and ER were analyzed. It was shown that the flanking amino-terminal residues are not essential, while the eight carboxy-terminal residues are required for SRC-1-mediated coactivation on RAR, TR, and ER (McInerney *et al.,* 1998). Additionally, residues +12 and +13 (the first L in LXXLL is designated +1) are required for ER binding, while residues at +6, +7, +11, and +13 are important for interaction with RAR-RXR heterodimer on DNA template. Similar experiments also reveal that amino acids at positions +6, +11, and +13 of motif ii are critical for NCoA-1 binding to TGZ-activated PPARγ. Intriguingly, when PPARγ is activated by BRL49653, distinct residues at +8, +9, +10, +12, and +13 become important for NCoA-1 binding. These studies suggest a ligand-specific alteration of receptor structure, which may impose a requirement for different LXXLL residues to achieve high-affinity interactions with the SRC coactivators.

Because many nuclear receptors seem to require two functional LXXLL motifs on one SRC molecule for maximal interaction, it is conceivable that spacing between two motifs may be important for such recognition. Accordingly, deletion of 30 amino acids from the conserved spacing of 50 amino acids between motifs ii and iii severely inhibits the capability of SRC-1 to restore IgG-mediated inhibition on RAR or TR transactivation (McInerney *et al.,* 1998). This deletion does not have any effect on the ability of SRC-1 to rescue PPARγ function, consistent with observation that motif iii is not essential for SRC-1 coactivation of PPARγ. However, shortening the spacing between motifs i and ii inhibits the function of SRC-1 to support PPARγ transactivation, consistent with a requirement for both motif i and ii of SRC-1 for PPARγ activation. These studies suggest that appropriately spaced LXXLL motifs are essential for maximal SRC-1 function.

IV. X-Ray Crystal Structures

Biochemical studies suggest that interactions between SRC coactivators and steroid/nuclear receptors involve LXXLL motifs of the coactivators and the AF-2 helix of the receptors. Because steroid/nuclear receptors usually form dimers on DNA template, and SRC contains multiple LXXLL motifs, the mechanism of interaction is expected to be complex but precise to allow receptor specificity and coactivator selection. Recently, the interaction surface between SRC coactivator and receptor has been analyzed by X-ray crystallography studies and scanning mutagenesis studies. These studies led to the discovery of a hydrophobic cleft (groove) on the surface of receptor LBD, which appears to bind directly to the LXXLL motif of a given SRC coactivator. This hydrophobic cleft is induced upon agonist binding, consistent with ligand-dependent interaction. The interaction also involves the C terminus AF-2 helix (H12), which undergoes a drastic conformational change in response to ligand binding and forms part of the hydrophobic cleft. The interactions observed in the crystal structure are consistent with many biochemical data, and correlate precisely with the role of AF-2 helix in mediating both SRC interaction and ligand-dependent transcriptional activation (Kalkhoven *et al.,* 1998). This section summarizes the characteristics of the hydrophobic cleft and detailed mechanisms of the formation and composition of this coactivator-binding site.

A. Hydrophobic Cleft of TR LBD

Based on the TR LBD X-ray crystallographic structure, 37 surface residues of hTRβ1 LBD were systemically mutated and tested for interactions with GRIP1 (Feng *et al.,* 1998). As expected, mutations of surface residues in helix 12 (L454R and E457K) of TR abolished GRIP1 binding. Two mutations in helix 3 (V284R and K288A) and two in helix 5 (I302R and K306A) also impaired binding, suggesting that both helix 3 and helix 5 also contribute to the formation of a coactivator-binding site. Point mutations that diminish GRIP1 binding (V284R, K288A, I302R, L454R, and E457K) also show decreased binding to SRC-1a, suggesting that different SRC coactivators may interact with a similar set of TR surface residues. Furthermore, transient transfection assay indicates that mutations with impaired GRIP1 binding also show diminished ligand-dependent transactivation function, which in turn could be partially restored by overexpression of GRIP1. Several control experiments demonstrate that these mutants are still efficient in hormone binding, heterodimerization, DNA binding, and inhibition of AP-1 activity, suggesting a direct involvement of coactivator binding in ligand-dependent transcriptional activation. The TR surface residues required for binding to GRIP1 and SRC-1 are highly conserved among members of the steroid/nuclear receptor family, suggesting a similar coactivator-binding surface among different nuclear receptors. Consistently, the corresponding mutations (K362A, V376R, and E542K) in hERα also abolished GRIP1 binding and inhibited transcriptional activation. Similarly, the lysine 366 of mouse ERα, which aligns to the K362 residue in human ERα in the predicted helix 3, is also essential for E2-dependent transactivation and binding to coactivators SRC-1 and TIF2 (Henttu *et al.*, 1997).

The critical residues identified by the scanning surface mutagenesis for coactivator binding appear to encircle a small hydrophobic cleft on the surface of TR-LBD (Fig. 7). Ligand binding results in the formation

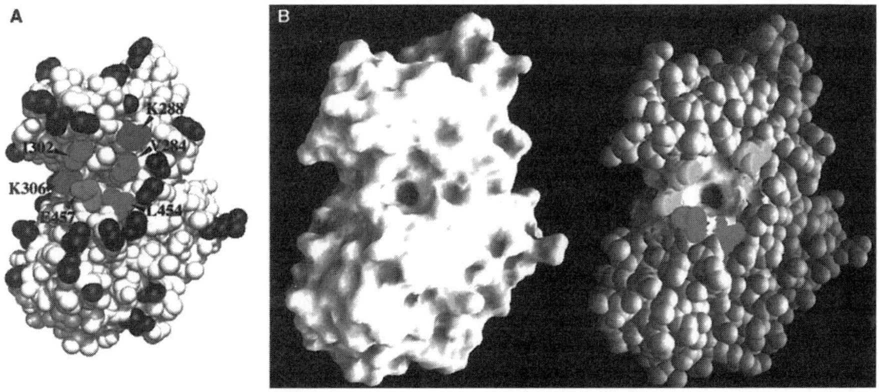

FIG. 7. A hydrophobic cleft on TR LBD involved in binding of SRC coactivators. A small cluster of effective mutations that surround a surface cleft containing central hydrophobic residues was identified by scanning surface mutagenesis. (A) A space-filling model of the TR LBD shows the LBD surface locations of mutations made in the full-length hTRβ1. Mutated residues that have no effect on GRIP1 binding or on activation in HeLa cells are shaded black. Mutated residues with diminished GRIP1 and SRC-1a binding and diminished activation in HeLa cells are shaded gray. (B) The AF-2 surface contains a cleft, one side of which is formed by conformationally hormone-responsive residues. Left, a view of the TR LBD molecular surface, showing the concave surfaces in gray; note the cavity at the center of the figure. Right, a space-filling model of the TR LBD, overlayed with a molecular surface view restricted to a 12-Å radius of the hydrophobic cavity. [Reprinted with permission from Feng W., *et al.* (1998). Hormone-dependent coactivator binding to a hydrophobic cleft on nuclear receptors. *Science* **280,** 1747–1749. Copyright © 1998 American Association for the Advancement of Science.]

of this surface by folding the carboxyl-terminal AF-2 helix against a scaffold of H3, H4, and H5. It was predicted that this small hydrophobic cleft will match a complementary surface of the LXXLL motif with the hydrophobic residues driving coactivator-binding reaction (Feng *et al.*, 1998).

B. Structure of TRβ:LXXLL Peptide Complex

The interacting interface between TR LBD and the LXXLL motif has now been revealed by X-ray crystallography. The crystal structure of hTRβ LBD complexed with T3 (3,3′,5-triiodo-L-thyronine) and a 13-aa peptide KHKILHTLLQDSS encompassing the LXXLL motif ii of GRIP1 was determined (Darimont *et al.*, 1998). The crystal contains two asymmetric monomers of the TRβ LBD with each monomer binding to one peptide. The structure of the hTRβ LBD is similar to that of the rTRα LBD (Wagner *et al.*, 1995) and consists of 12 α-helices and 4 β-strands organized in three layers. The LXXLL peptide forms an amphipathic α-helix of about three turns for the core residues. The helical structure of the peptide may be induced by complex formation since far UV-CD spectrum of the peptide indicates a random coil conformation in the absence of TRβ LBD. In the crystal structure, the hydrophobic face of the peptide helix contacts a hydrophobic groove formed by 16 residues from helices H3, H4, H5, and H12 of the hTRβ LBD. The 16 residues are I280, T281, V283, V284, A287, and K288 from H3; F293 from H4; Q301, I302, L305, K306, and C308 from H5; and L454, E457, V458, and F459 from H12. These residues are arranged in a way that the hydrophobic residues form the floor of the groove and the charged residues line the rim. The three leucines of the LXXLL core, L690, L693, and L694, are buried within the hydrophobic groove (Fig. 8). The L690 residue makes van der Waals contacts with L454 and V458 of H12, and I689 packs against L454 of H12 outside the edge of the groove. L693 contacts V284 of H3, whereas L694 contacts F293 and L305 of H4 and H5, respectively. This structure is consistent with results obtained in scanning surface mutagenesis, confirming the importance of V284 of H3 and L454 of H12 for *in vitro* binding with both GRIP1 and SRC-1a (Feng *et al.*, 1998).

C. Structure of the ERα:LXXLL Peptide Complex

The crystal structure of diethylstilbestrol (DES)-bound ERα LBD complexed with a LXXLL peptide (motif ii) of GRIP1 has also been determined (Shiau *et al.*, 1998). The overall structure of the ER-peptide

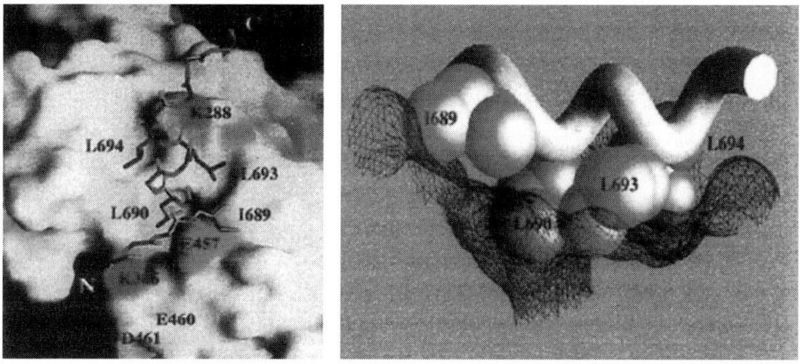

FIG. 8. The TRβ LBD:LXXLL peptide interface revealed by X-ray crystal structure. The side chains of the LXXLL motif ii of GRIP1 are shown in a CPK representation, with the main chain of the peptide drawn as a Cα worm. The three leucines fit into pockets on the molecular surface of the TRβ LBD, depicted as mesh, whereas the nonconserved isoleucine residue rests on the edge of the surface cleft. [Adapted by permission from Fig. 4 of Darimont, B. D., *et al.* (1998). Structure and specificity of nuclear receptor coactivator interactions. *Genes & Development* **12,** 3343–3356.]

complex is similar to that of the TR-peptide complex (Darimont *et al.,* 1998). The LXXLL peptide binds as a short amphipathic α-helix to a hydrophobic groove formed on the surface of the LBD (Fig. 9). This LXXLL binding surface of ER LBD is located at the same position as the hydrophobic cleft of TR LBD. The asymmetric unit of the ER complex contains a noncrystallographic dimer (Tanenbaum *et al.,* 1998; Brzozowski *et al.,* 1997; Shiau *et al.,* 1998), consistent with ER's function as a homodimer (Mangelsdorf and Evans, 1995). In the ternary complex, one LXXLL peptide is bound to each LBD in a hydrophobic groove composed of residues from helices H3, H4, H5, and H12 and the turn between H3 and H4 (Fig. 9A). The ends of this hydrophobic groove are charged, similar to the coactivator-binding pockets found in the TR-LXXLL peptide complex (Darimont *et al.,* 1998) (Fig. 9C). In the crystal structure, L690 forms van der Waals contacts with I358, V376, L379, E380, and M543, whereas L694 makes van der Waals contacts with I358, K362, L372, Q375, V376, and L379 of the ER LBD on the hydrophobic floor. In contrast, I689 and L693 of the LXXLL helix rest against the rim of the groove. The side chain of I689 lies in a depression formed by D538, L539, and E542 and the side chain of L693 makes nonpolar contacts with I358 and L539 of the ER LBD. In addition to the hydrophobic interactions, the LXXLL helix appears to be stabilized by capping interactions with E542 and K362 of ER LBD at opposite ends of the LXXLL

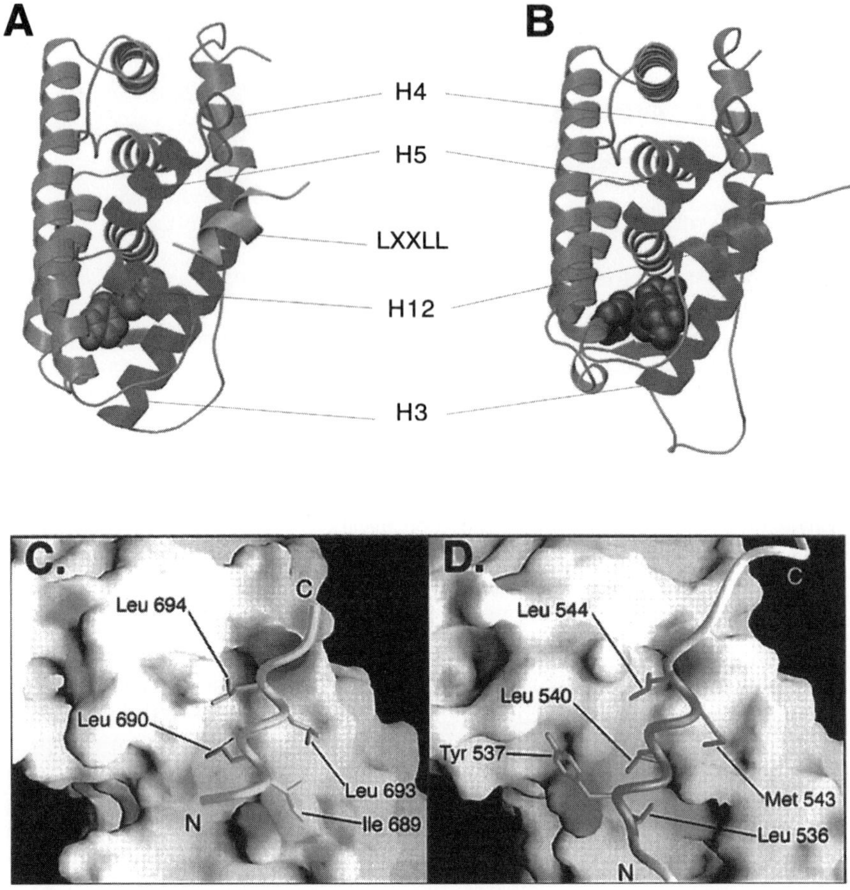

FIG. 9. Structure of ERα-LXXLL Peptide Complex. (A) Structure of the DES-ER-peptide complex. The coactivator peptide and the LBD are shown as ribbon drawings. DES is shown in space-filling representation. (B) Structure of the OHT-ER LBD complex. The LBD is depicted as a ribbon drawing. OHT is shown in space-filling representation. (C) A molecular surface representation of the LBD bound to DES. The side chains of Leu-690 and Leu-694 of the coactivator peptide are bound in a hydrophobic groove and those of Ile-689 and Leu-693 rest against the edge of this groove. (D) A molecular surface representation of the LBD bound to OHT. Whereas the side chains of Leu-540 and Leu-544 of helix 12 are embedded in the hydrophobic groove, that of Met-543 lies along the edge of this groove. [Adapted by permission from Figs. 2 and 3 of Shiau, A. K., *et al.* (1998). The structural basis of estrogen receptor/coactivator recognition and the antagonism of this interaction by tamoxifen. *Cell* **95,** 927–937.]

helix by forming hydrogen bonds. The importance of these interactions observed in the crystal structures was confirmed by a series of site-directed mutations of the ER LBD. Mutations that perturb the hydrophobic characteristic of the LXXLL-binding groove, or that prevent the formation of the capping interactions (K362A and E542K), abolish ligand-dependent interaction between ER and GRIP1 (Shiau *et al.*, 1998), indicating that both capping and hydrophobic packing interactions are important.

Importantly, the coactivator-binding hydrophobic groove of ER LBD is occluded in the LBD bound with the antagonists 4-hydroxytamoxifen (OHT) (Shiau *et al.*, 1998) or Rolaxifen (RAL) (Brzozowski *et al.*, 1997) due to misalignment of the AF-2 helix (Fig. 9B). This AF-2 helix appears to mimic the interactions of the LXXLL peptide with the LBD in the antagonist-bound complex, providing a molecular basis for mechanism of antagonism by OHT and RAL. In the crystal structures, the ER agonists DES and E2 are completely buried within a hydrophobic cavity of the LBDs (Brzozowski *et al.*, 1997; Shiau *et al.*, 1998). In contrast, the binding of antagonist OHT or RAL induces a conformation that differs from the structure driven by DES or E2 binding. Thus, the receptor antagonism by OHT and RAL is attributed to their bulky side chains that project out of the ligand-binding pocket between helices 3 and 11. Consequently, the positive-charged side chains of OHT and RAL produce steric clashes with the hydrophobic side chain of L540 in AF-2 helix, shifting this helix from over the ligand-binding pocket to the hydrophobic region of the LXXLL-binding groove (Fig. 9B). It appears that an LXXLL-like sequence (LXXML) within ER helix 12 binds intramolecularly to the coactivator-binding pocket of LBD (Fig. 9D). Therefore, the binding of antagonist to ER promotes an AF-2 helix conformation that cripples the AF-2 surface and inhibits binding of SRC coactivator by blocking the hydrophobic groove required for binding of LXXLL motif. Since the LXXLL-like motif of ER is not shared by all other nuclear receptors, other mechanisms of antagonism might be utilized by different steroid/nuclear receptors.

D. Structure of the PPARγ:SRC-1 Complex

The crystal structures of an apo-PPARγ LBD and a ternary complex containing the PPARγ LBD, the antidiabetic ligand rosiglitazone (BRL49653), and an 88-aa fragment of SRC-1 have also been described recently (Nolte *et al.*, 1998) (Fig. 10). In both the apo and the ternary complex structures, the PPARγ LBD forms a noncrystallographic dimer, similar to the RXRα and ERα crystal structures (Bourguet *et al.*, 1995;

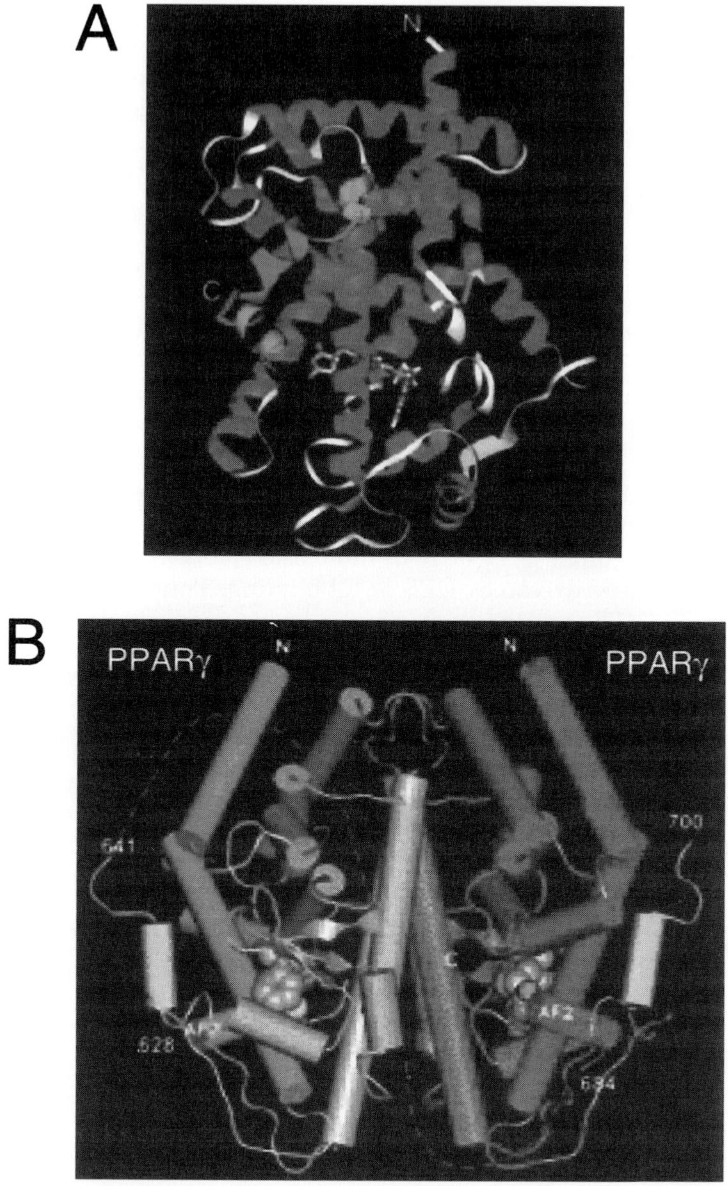

FIG. 10. Structure of the PPARγ-rosiglitazone–SRC-1 ternary complex. (A) Ribbon drawing showing the ternary complex of PPARγ LBD, BRL49653, and the LXXLL helix domain of SRC-1. Rosiglitazone (stick diagram) binds in a deep cavity of the protein and provides a network of polar interactions that include the AF-2 domain. (B) Ribbon drawing of the PPARγ LBD dimer and SRC-1, including the ligand rosiglitazone. The struc-

Brzozowski *et al.,* 1997). The structure of the PPARγ LBD is very similar to the overall fold of other steroid/nuclear receptors, except that it contains an extra helix between the first β-strand and H3. The PPARγ structure reveals a large T-shaped ligand-binding pocket. The ligand rosiglitazone occupies about 40% of this cavity in the ternary complex. The remaining cavity of the ligand-binding pocket may allow free interaction with ligands in a relatively nonspecific manner, resulting in flexibility on ligand binding by PPARγ.

The crystal structure of the PPARγ–rosiglitazone–SRC-1 ternary complex shows that SRC-1 binds to a liganded PPARγ homodimer, with one LXXLL binding to one molecule and the second LXXLL binding to the other molecule (Fig. 10B). The connecting sequences between these two LXXLL motifs of SRC-1 were not defined. In the ternary complex, E471 and K301 of PPARγ appear to define a "charge clamp" that allows the placement of LXXLL motif into the coactivator-binding site (Fig. 10A). At one end of the coactivator-binding site, the side chain of E471 forms hydrogen bonds with the backbone amides of K632 and L633 in motif i and with the backbone amides of K688, I689, and L690 in motif ii. At the other end of the binding site, the side chain of K301 forms hydrogen bonds with two backbone carbonyls of L636 and T639 in motif i and L693 and L694 in motif ii. The corresponding residues of both E471 and K301 in TR and ER also are important in coactivator binding and transactivation (Henttu *et al.,* 1997; Darimont *et al.,* 1998). Similar to the TR-LXXLL and ER-LXXLL complexes, the hydrophobic face of the LXXLL helix of SRC-1, packs into a hydrophobic pocket formed between E471 and K301 by H3, H4, H5, and H12 of PPARγ. The hydrophobic core of the LXXLL motif is buried within the binding surface and amino acids L633, L636 and L690, L693 of the two SRC-1 LXXLL motifs interact hydrophobically with L468 and L318 of the PPARγ LBD. The residues at positions −3 and −2 of the LXXLL motif do not appear to make any significant interactions with the LBD. The amino acid at position −1 fits in a shallow pocket created by P467 and L468 of the AF-2 helix H12 and the +4 residue of the LXXLL motif. The amino acids

ture of SRC-1 was determined from amino acids 628–640 and 684–703 and was crystallographically refined. Very weak electron density from residues 670 to 684 was visible but was not crystallographically refined and is shown as a dashed line. SRC-1 amino acids 642–669 were disordered and not structurally determined. The diagram shows how one SRC-1 molecule, with two interacting domains, forms a complex with a PPARγ homodimer. The dashed line connecting the two structurally determined domains of SRC-1 is the proposed connection between these two domains. [Adapted by permission from Figs. 2 and 3 of Nolte, R. T., *et al.* (1998). Ligand binding and co-activator assembly of the peroxisome proliferator-activated receptor-γ. *Nature* **395,** 137–143.]

at +2 and +3 of the LXXLL motif face out into solution and make no contacts with the LBD, consistent with the lack of sequence conservation and other mutagenesis studies (McInerney *et al.,* 1998; Darimont *et al.,* 1998). The two leucines at positions +4 and +5 lie in a hydrophobic pocket and, therefore, are most critical for stabilizing the interaction by forming hydrogen bonds with the clamping residue K301 of the LBD.

According to this model, the length and orientation of the LXXLL motif is vital for proper backbone interactions with E471 in AF-2 helix and with K301 in helix H3. The E471 and K301 are highly conserved among nuclear receptors and are important for transcriptional activation and coactivator binding. These two residues appear to define a ligand-dependent "charge clamp" that positions the LXXLL motif into a hydrophobic pocket in the receptor LBD. The observation that two LXXLL motifs of one SRC-1 molecule make simultaneous contact with a PPARγ homodimer suggests a cooperative binding of the LXXLL motifs to a receptor dimer. The existence of a third LXXLL motif within SRC coactivators may allow combinatorial regulation and optimal interaction for different receptors. In the active ternary complex, the two PPARγ LBDs have nearly identical conformations. In contrast, one AF-2 helix in the apo-PPARγ homodimer adopts an extended inactive conformation, projecting away from the LBD, whereas the other AF-2 helix is folded against the LBD, adopting an active conformation. It is possible that the unliganded receptor can assume both active and inactive conformations, with the ligand acting to lock the receptor into the active conformation as proposed by the "mouse trap" model (Renaud *et al.,* 1995). However, the "inactive" AF-2 helix appears to contact the charge clamp of the active AF-2 helix in a crystallographically related PPARγ molecule. It is believed that this arrangement of an AF-2 helix in the LXXLL binding pocket may underlie allosteric inhibition observed with specific partners of RXR.

E. Model of Allosteric Inhibition

Functional studies suggested that RXR-PPAR heterodimer could be activated by both PPAR and RXR ligands, whereas RXR-RAR heterodimer is selectively activated by RAR ligand only (Kliewer *et al.,* 1992; Kurokawa *et al.,* 1994). The differential ligand responsiveness may be due to allosteric inhibition of the binding of ligands to RXR by RAR, but not by PPAR, in the respective heterodimers. It was proposed that allosteric inhibition of RXR by RAR is a result of the placement of

RXR AF-2 helix in the LXXLL-binding pocket of RAR in the absence of ligand (Westin *et al.,* 1998). Upon binding of RAR ligands, an LXXLL motif is recruited, displacing the RXR AF-2 helix and allowing RXR ligands to bind (Fig. 11). Consequently, the second LXXLL motif will then bind to the RXR molecule.

This model is supported by several observations. First, an RXR-specific ligand LG268 can only stimulate the binding of SRC-1 to RXR-RAR heterodimers in the presence of an RAR-specific ligand TTNPB, indicating that the interaction of SRC-1 with RAR may relieve the

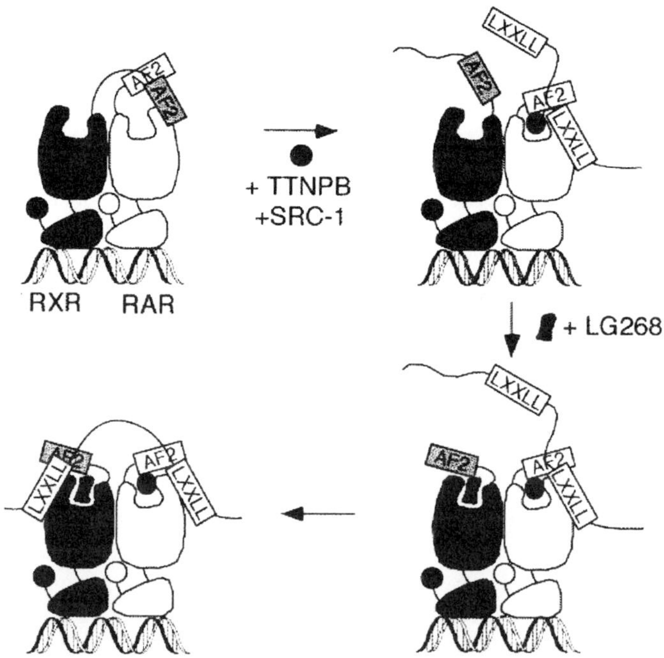

FIG. 11. Model of allosteric inhibition of RXR-RAR heterodimer. In the absence of ligand, the AF-2 helix of RXR is docked to the RAR coactivator-interaction site, preventing the binding of RXR ligands. In response to RAR-specific ligand, one of the three LXXLL motifs is recruited to RAR, resulting in displacement of the RXR AF-2 helix from RAR (step 1). The release of the RXR AF-2 domain relieves allosteric inhibition, allowing ligands to bind to RXR (step 2). The binding of an RXR ligand can then promote the interaction of a second LXXLL motif from the same SRC-1 molecule with RXR, stabilizing the complex (step 3). [Adapted by permission from Fig. 5 of Westin, S., *et al.* (1998). Interaction controlling the assembly of nuclear-receptor heterodimers and co-activators. *Nature* **395,** 199–202.]

allosteric inhibition on RXR. Accordingly, binding of the RXR-specific ligand LG69 to RXR-RAR heterodimers was induced by combination of TTNPB with SRC-1, suggesting that activation of RAR by TTNPB and SRC-1 permits binding of ligands to RXR. Consistent with this, overexpression of SRC-1 also enhances transcription induced by RXR-specific ligand LG268 in the presence of TTNPB. These studies suggest that ligand activation of RAR recruits coactivators, which may relieve allosteric inhibition on RXR, allowing RXR to bind ligands and to interact with coactivators. Second, two LXXLL motifs of SRC-1 are required for the cooperative effects of two ligands on binding of SRC-1 to a heterodimer of RXR-RAR or PPARγ-RXR. This suggests that each LXXLL motif may contact one molecule of the dimer. Third, both AF-2 domains of the RXR-RAR heterodimer are required for the cooperative effects of two ligands to recruit SRC-1. Deletion of an AF-2 helix from one receptor partially increases SRC-1 binding to the partner and completely blocks the cooperative effects of two ligands to recruit SRC-1. These data suggest an inhibitory role of the AF-2 helix on SRC-1 binding to the partner and that both AF-2 domains of the heterodimer are required for cooperative recruitment of SRC-1. Fourth, the X-ray crystal structure of apo-PPARγ reveals that the AF-2 helix of one PPARγ molecule interacts with the LXXLL binding pocket of another PPARγ in a different, crystallographically related dimer. Molecular modeling of the RXR-RAR heterodimer shows that the AF-2 helix of RXR could be rotated to contact the LXXLL-binding pocket of RAR. Presumably, such an interaction would prevent AF-2 helix-dependent closure of the ligand-binding pocket of RXR, suggesting a structural basis for allosteric inhibition by RAR on ligand binding of RXR. Also, the RXR AF-2 helix is required for binding of RXR ligands since RXRΔ443-RAR heterodimer does not bind well to RXR-specific ligand in the presence of TTNPB and SRC-1. In addition, synthetic coactivator LXXLL peptides can relieve the inhibition on RXR ligand binding. A synthetic RXR AF-2 peptide binds to the unliganded RAR with a higher affinity than the coactivator LXXLL peptide, and binding of RXR AF-2 peptide to RAR is displaced from RAR by LXXLL peptides. Finally, GST-RXR-AF2 helix fusion protein binds to RAR efficiently, and such binding is inhibited by SRC-1 in the presence of RAR ligand. In contrast, GST-RXR-AF2-helix interacts poorly with PPARγ, consistent with the observation that PPARγ does not inhibit ligand binding of RXR.

Consistent with the model, mutations in the AF-2 helix of ER that affect its AF-2 function and mutations that affect dimerization both impair SRC-1 binding (Kalkhoven *et al.,* 1998). For instance, R507A and L511A mutations in ER that do not affect hormone binding appear to inhibit binding of SRC-1 (Kalkhoven *et al.,* 1998). Conversely, the G525R mutation, which still allows dimerization but is unable to bind ligands, also inhibits binding of SRC-1. In addition, while SRC-1 interacts with an ER homodimer containing two functional AF-2 domains in a gel retardation assay, SRC-1 could not form a complex with an ER homodimer containing defective AF-2 helix (Kalkhoven *et al.,* 1998). These studies suggest that, in addition to hormone binding and AF-2 function, homodimerization of ER is also required for efficient recruitment of SRC-1. Together, these studies support a hypothesis that one coactivator molecule interacts simultaneously with two subunits of the receptor homo- or heterodimer through two LXXLL motifs of one coactivator molecule. However, another study using gel shift assay for analyzing interaction between rTIF2 and TR/RXR heterodimer concluded that two coactivator molecules bind to a heterodimeric receptor complex (Leers *et al.,* 1998). Further studies are necessary to understand the exact stoichiometry of the coactivator–receptor complex and the possible differences among different receptor–coactivator complexes.

V. Mechanism of Transactivation

A. Activation Domains

Modulation of the transcriptional activities of steroid/nuclear receptors by coactivators is a complex process involving enzymatic remodeling of chromatin as well as communication with basal transcriptional machinery at specific promoters. One common property of transcriptional coactivators is the ability to activate transcription when recruited to a promoter via protein–protein interaction with DNA binding proteins. Such a recruitment event can be mimicked by fusing coactivator with a heterologous DBD. Using Gal4-DBD fusion, all three SRC coactivators have been shown to contain intrinsic transcription activation function (Li *et al.,* 1997; Zhu *et al.,* 1996; Torchia *et al.,* 1997; Voegel *et al.,* 1998; Suen *et al.,* 1998; Hong *et al.,* 1997; Chen *et al.,* 1997; Kurokawa *et al.,* 1998; Oñate *et al.,* 1998) (Fig. 12). Fusion proteins of Gal4-DBD and full-length mSRC-1 (Zhu *et al.,* 1996), GRIP1 (Hong *et al.,* 1997), NCoA-1 and p/CIP (Torchia *et al.,* 1997) efficiently activate transcription from a Gal4-driven promoter in both mammalian and yeast cells. Comparison of the transactivation activity between Gal4-NCoA1 and Gal4-p/CIP suggests stronger activation function for NCoA-1 than p/CIP (Torchia *et al.,* 1997). This is consistent with a finding in the same study that p/CIP exhibits three- to fivefold less coacti-

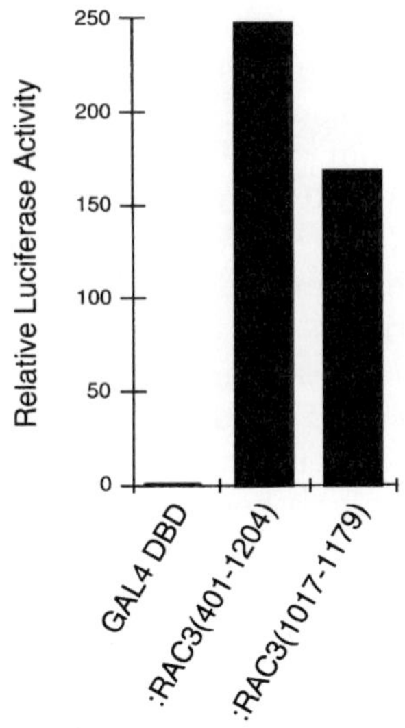

FIG. 12. Transcriptional activation by the SRC coactivator. Transcriptional activation by RAC3 in mammalian cells. The indicated RAC3 fragments were expressed as Gal4-DBD fusion proteins from the CMV promoter. The relative fold-induction is determined by comparing with activity of Gal4-DBD.

vation activity on RAR and ER than NCoA-1. However, all three SRC coactivators contain potent transcriptional activation domains, and SRC-3 and SRC-1 have comparable levels of coactivation function on RAR, PR, and TR (Li *et al.,* 1997; Takeshita *et al.,* 1997).

Within SRC-1, three independent activation domains have been described. One study found the first activation domain (AD1) at the N-terminal 93 amino acids that cover the bHLH region (Oñate *et al.,* 1998). However, another study showed no detectable transactivation function with a fusion of Gal4 DBD and amino acids 1–198 of hSRC-1 (Kalkhoven *et al.,* 1998). Therefore, a transcriptional suppressor domain may exist within the conserved PAS-A region. The second activation domain (AD2) was mapped to amino acids 781–988 or 840–948 of hSRC1 (Oñate *et al.,* 1998; Kalkhoven *et al.,* 1998), and 896–1200 or 947–1084 of NCoA-1 (mSRC-1) (McInerney *et al.,* 1998; Kurokawa *et al.,* 1998). The AD2 domains in SRC-2 and SRC-3 have been mapped to amino acids 1010–1131 of TIF2 (Voegel *et al.,* 1998), 1017–1179 of RAC3 (Li *et al.,* 1997), 1038–1088 of ACTR (Chen *et al.,* 1997), and 896–1200 of p/CIP (Kurokawa *et al.,* 1998). Therefore, the minimal AD2 of SRC-3 is located within a 50-aa fragment. Direct comparison of the AD2 activity between p/CIP and NCoA-1 shows that they can activate transcription equally well (Kurokawa *et al.,* 1998). In addition to AD1 and AD2 domains, a third activation domain (AD3) was observed in all three SRC coactivators. The AD3 domain has been mapped to amino acid residues 1241–1385 of SRC-1a (Kalkhoven *et al.,* 1998). In contrast, little activation was observed with amino acid 948–1441 or 1241–1441 fragment of SRC-1a (Oñate *et al.,* 1998; Kalkhoven *et al.,* 1998). Therefore, the C-terminal 56 amino acids may inhibit transactivation of AD3 (Kalkhoven *et al.,* 1998). This putative C-terminal suppressor domain contains a LXXLL motif that can interact with liganded receptors; it is conceivable that protein–protein interaction of this LXXLL motif with the receptors might regulate AD3 activity. It is currently unclear whether a similar suppressor domain exists in SRC-2 or SRC-3. However, both SRC-2 and SRC-3 lack a C-terminal LXXLL motif, suggesting that the mechanism of transcriptional activation may differ among members of the SRC coactivator family.

B. Interaction with CBP/p300

The ability of SRC coactivator to activate transcription has been linked, at least in part, to interaction with CBP/p300. The interaction between SRC and CBP/p300 was first revealed by identification of SRC-1 as a CBP/p300 binding protein (Yao *et al.,* 1996; Kamei *et al.,* 1996). A mouse cDNA fragment encoding amino acid residues 789–993 of SRC-1 was isolated in a search for p300-binding proteins in a yeast two-hybrid screen (Yao *et al.,* 1996). The association of mSRC-1 with p300 has been confirmed by GST pull-down, coimmunoprecipitation, and subcellular colocalization assays, and the interaction surface on p300 was mapped to the C-terminal 308 amino acids (Yao *et al.,* 1996). In addition, mSRC-1 was also isolated as a CBP interacting protein in a far-Western-based screening (Kamei *et al.,* 1996). In addition to interaction with nuclear receptors through the N-terminal region of CBP/p300, the C-terminal fragment between amino acids 2058–2163 of CBP appears to interact with SRC coactivators. Similarly, SRC-2 and SRC-3 have also been shown to interact with CBP/p300 (Torchia *et al.,* 1997; Voegel *et al.,* 1998; Li and Chen, 1997). The CBP/p300 interacting domain of TIF2 was mapped to amino acids 1010–1131 by GST pull-down assay (Voegel *et al.,* 1998). Similarly, the CBP interacting domain of RAC3 was mapped to amino acids 1017–1179 by far-Western analysis (Li and Chen, 1997) (Fig. 13), or to amino acids 947–1084 of p/CIP in yeast two-hybrid assay (Torchia *et al.,* 1997).

The CBP/p300 interacting domain of the SRC coactivator contains three conserved regions similar to the LXXLL motifs involved in re-

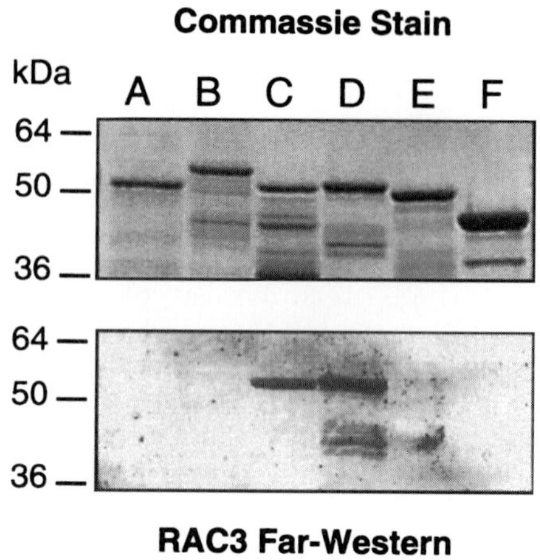

FIG. 13. SRC coactivator interacts with CBP. Coomassie blue staining of the purified GST-CBP fusion proteins (top). The interaction of full-length RAC3 with GST-CBP fragments was probed in a far-Western blot. The A to F fragments of CBP contain amino acid residues 1678–1880, 1801–2000, 1921–2120, 2041–2240, 2161–2360, and 2301–2441, respectively.

ceptor interaction (Fig. 5). The role of these LXXLL motifs in mediating the interaction with CBP/p300 has been analyzed (Voegel *et al.*, 1998; McInerney *et al.*, 1998). In one study, individual deletion of motif iv, motif v, or motif vi of TIF2 has no significant effect on CBP interaction (Voegel *et al.*, 1998). However, mutation of three leucines in the LLXXL core of motif v to alanines, but not alteration of the middle XX residues to alanines, significantly reduces CBP interaction with TIF2 (Voegel *et al.*, 1998), suggesting that motif v is important for CBP interaction. In contrast, mutation of motif iv of NCoA-1 from LXXLL to LAAAA impairs NCoA-1's capability to rescue IgG-inhibited RAR transactivation, while mutation of motif v had no effect (McInerney *et al.*, 1998). Double mutation of motif iv and motif v completely blocks the ability of NCoA-1 to rescue transactivation by RAR, TR, and PPARγ. The motif iv and v double mutant also fails to interact with CBP (McInerney *et al.*, 1998). These data suggest that the conserved LXXLL motifs within the CBP interacting domain of SRC coactivators play an important role in mediating the interaction. It appears that motif iv is more important for SRC-1, while motif v is most critical for TIF2 in CBP binding. Conversely, multiple helices within the SRC interacting domain of CBP are required to various degrees for interaction with NCoA-

1 (McInerney *et al.,* 1998). A single-point mutation K2109A in CBP significantly impairs its interaction with NCoA-1. The predicted structure of the SRC-interacting domain on CBP suggests a hydrophobic binding pocket, analogous to the nuclear receptor-binding pocket, by which the LXXLL motifs of SRC bind to CBP/p300. Because interactions between CREB and the KIX domains of CBP are dependent on phosphorylation, analogous induced-fit events may also regulate the interaction between SRC and CBP/p300.

Comparison of the CBP/p300-interacting domain and the AD2 domain of the SRC coactivator indicates that these two domains overlap with one another (Li and Chen, 1997; McInerney *et al.,* 1996; Voegel *et al.,* 1998). Mutations of TIF2 that affect CBP interaction also inhibit transcriptional activation (Voegel *et al.,* 1998). By analyses of 13 deletion mutants and 2 point mutants generated within residues 1011–1122 of TIF2, all mutants that retain the ability to interact with CBP also activate transcription. In particular, point mutation within motif v of TIF2 that replaces the three leucines with alanines affects both CBP interaction and transcriptional activation. Accordingly, the TIF2 (LLL) mutant showed diminished ER coactivation function (Voegel *et al.,* 1998). These studies suggest that interaction with CBP may underscore the ability of TIF2 to activate transcription.

The requirement of CBP for transcriptional activation by SRC-1 (NCoA-1) and for enhancing transcription by nuclear receptors has also been analyzed by microinjection assay (McInerney *et al.,* 1998). Mutation of the two LXXLL motifs (iv and v) within NCoA1 appears to abolish the function of NCoA-1 in both CBP interaction and coactivation for RAR, TR, and PPARγ. Furthermore, injection of anti-CBP IgG also abolishes transcriptional activation by NCoA-1. Therefore, CBP/p300 interaction is essential for transcriptional activation and coactivation function of SRC-1. Accordingly, microinjection of anti-CBP IgG inhibits RA-dependent transactivation, indicating that CBP is required for RAR-mediated transactivation. Because the N terminus of CBP/p300 also interacts with nuclear receptors, the relative contribution of nuclear receptor interacting domain and the SRC interacting domain on RAR transactivation was tested by antibody microinjection and rescue experiment. It appears that the nuclear receptor interacting domain of CBP is not required to stimulate RAR transactivation. In contrast, the SRC interacting domain is essential for stimulating RAR transactivation (McInerney *et al.,* 1998). These results are in agreement with a bridging hypothesis that SRC coactivators function by recruiting CBP/p300 coactivators to specific promoters.

C. Interaction with P/CAF

In addition to CBP/p300, SRC-1 and ACTR have also been shown to interact with P/CAF, a p300/CBP-associated histone acetyltransferase (Spencer *et al.,* 1997; Chen *et al.,* 1997). P/CAF appears to interact with SRC-1 fragment spanning residues 1027–1139 and 1139–1250, suggesting two independent interaction regions for P/CAF (Spencer *et al.,* 1997). A Gal4-P/ACF fusion protein also interacts with SRC-1 fragments 360–1139, 1138–1441, and 1216–1441 in a mammalian two-hybrid assay. Similar to SRC-1, ACTR was also shown to interact with P/CAF (Chen *et al.,* 1997). The interaction between P/CAF and ACTR was shown by both GST pull-down and yeast two-hybrid assays (Chen *et al.,* 1997). The ability of both SRC-1 and ACTR to interact independently with CBP/p300 and P/CAF provides a molecular scaffold to bridge the HAT protein complex to DNA-bound steroid/nuclear receptors. However, the interaction of SRC coactivators with P/CAF does not correlate with the transcriptional activity of SRC coactivators since the transcriptional activation domain and P/CAF interacting region are separable (Chen *et al.,* 1997). In addition, fusion of P/CAF with Gal4-DBD is unable to activate transcription, suggesting that histone acetylation alone is not sufficient for transcriptional activation by SRC coactivators or P/CAF.

D. Histone Acetylation by SRC Coactivators

Transcriptional coactivators are thought to stimulate transcription by facilitating the assembly of active basal transcriptional machinery. How SRC coactivators gain access to repressed chromatin remains largely unknown. Transcriptionally active chromatin usually contains hyperacetylated histones (Brownell and Allis, 1996; Wade *et al.,* 1997). Accordingly, several transcriptional coactivators including the general coactivators CBP/p300 and its associated protein P/CAF are potent histone acetyltransferases (Yang *et al.,* 1996; Bannister and Kouzarides, 1996; Ogryzko *et al.,* 1996). Interestingly, both SRC-1 and ACTR also exhibit moderate intrinsic histone acetyltransferase activity (Spencer *et al.,* 1997; Chen *et al.,* 1997) and are capable of acetylating free and mononucleosomal histones with substrate preference on histones H3 and H4.

Histone acetylation by SRC-1 was first demonstrated in a filter-binding assay using SRC-1 immunoprecipitates obtained from COS cell extract (IP-HAT) (Spencer *et al.,* 1997) (Fig. 14A). The intrinsic HAT activity of SRC-1 was confirmed by an active gel assay in which the immunoprecipitate was resolved by SDS-PAGE, and the HAT activity of individual polypeptide was determined. Intrinsic HAT activity was also detected with ACTR expressed in a baculovirus system (Chen *et al.,* 1997). The HAT activity of ACTR was confirmed by resolving the purified ACTR on a glycerol gradient and analyzing the HAT activity of each fraction, which revealed cofractionation of HAT activity and ACTR protein. It remains unclear whether SRC-2 (GRIP1/TIF2) also contains HAT activity or interacts with P/CAF (Voegel *et al.,* 1998).

The HAT domains were mapped to the C-terminal fragments at 1107–1441 and 1029–1292 of hSRC-1 and ACTR, respectively (Chen *et al.,* 1997; Spencer *et al.,* 1997). The HAT domain of ACTR lies between two activation domains, indicating that the HAT activity is not directly responsible for transcriptional activation by ACTR. Therefore, histone acetylation by SRC-1, ACTR, and P/CAF appears to be insufficient for transcriptional activation. Transcriptional activation by liganded nuclear receptors may involve a highly coordinated multistep action that could be facilitated by the SRC coactivators. Unlike CBP, which acetylates H2A and H2B in addition to H3 and H4 (Ogryzko *et al.,* 1996; Bannister and Kouzarides, 1996), both SRC-1 and ACTR acetylate only H3 and H4 with a preference for H3 (Chen *et al.,* 1997; Spencer *et al.,* 1997). This acetylation occurs on both free form and mononucleosomal histones, and packing of histones into nucleosome severely reduces the efficacy of acetylation. In addition, the acetylation sites have been identified using synthetic peptides corresponding to the N-terminal tails of H3 and H4 (Spencer *et al.,* 1997). The H3 peptide with preacetylated lysines 9 and 18 remains a good substrate, but preacetylation of lysines 9 and 14 inhibits SRC-1-mediated acetylation. Thus, lysine 14 and perhaps lysine 9 of histone H3 are the preferred sites for acetylation by SRC-1 (Fig. 14B). It is currently unknown whether ACTR acetylates distinct sites on histones, and if acetylation by SRC-1 and ACTR has different functional consequences.

E. Interaction with Cyclin D1

Cyclin D1 forms complexes with CDKs in response to mitogenic stimulation and regulates cell cycle progression through the G_1 phase. Importantly, cyclin D1 is amplified and overexpressed in several human malignancies. Elevated levels of cyclin D1 are observed in up to 50% of human breast cancers (Donnellan and Chetty, 1998; Beijersbergen and Bernards, 1996). Cyclin D1 has been shown to interact and enhance ER-mediated transactivation in a ligand- and CDK-independent manner

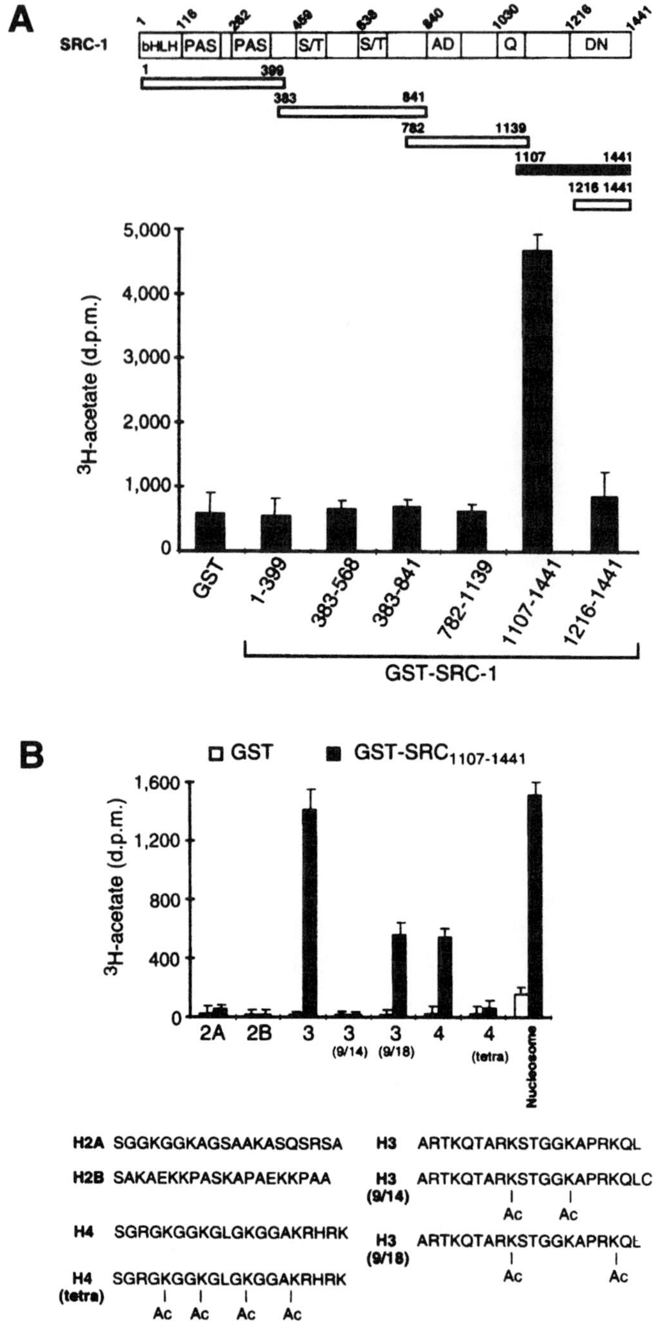
A
SRC-1
bHLH PAS PAS S/T S/T AD Q DN
1 116 262 459 638 840 1030 1216 1441
1 399
383 841
782 1139
1107 1441
1216 1441
3H-acetate (d.p.m.)
5,000
4,000
3,000
2,000
1,000
0
GST
1-399
383-568
383-841
782-1139
1107-1441
1216-1441
GST-SRC-1
B
GST
GST-SRC1107-1441
3H-acetate (d.p.m.)
1,600
1,200
800
400
0
2A 2B 3 3 (9/14) 3 (9/18) 4 4 (tetra) Nucleosome
H2A SGGKGGKAGSAAKASQSRSA
H2B SAKAEKKPASKAPAEKKPAA
H4 SGRGKGGKGLGKGGAKRHRK
H4 (tetra) SGRGKGGKGLGKGGAKRHRK
Ac Ac Ac Ac
H3 ARTKQTARKSTGGKAPRKQL
H3 (9/14) ARTKQTARKSTGGKAPRKQLC
Ac Ac
H3 (9/18) ARTKQTARKSTGGKAPRKQL
Ac Ac

(Neuman *et al.,* 1997; Zwijsen *et al.,* 1997). Recently, cyclin D1 was shown to recruit SRC-1 to ER in the absence of ligand (Zwijsen *et al.,* 1998), accounting at least partly for ligand-independent activation of ER. Cyclin D1 mutant that fails to interact with SRC-1 inhibits cyclin D1-dependent but ligand-independent transactivation by ER. These studies suggest that SRC coactivators may form complexes with cell cycle regulatory proteins to precisely control gene expression at different stages of the cell cycle.

The ability of cyclin D1 to enhance ER transactivation is dependent on a C-terminal region containing a LLXXXL motif, which resembles a motif in the C-terminal AF-2 helix of ER. Disruption of the LLXXXL motif of cyclin D1 impairs cyclin D1-dependent transcriptional activation of ER, although the mutant protein still binds to the unliganded ER. Interestingly, cyclin D1-dependent transactivation does not seem to rely on the AF-2 function of ER, and the SRC-1 dominant negative mutant containing only the C-terminal LXXLL motif markedly repressed the cyclin D1-induced activation of the ER AF-2 mutant. These data suggest a functional interaction between cyclin D1 and SRC-1. As expected, wild-type but not the LLXXXL mutant of cyclin D1 interacts with SRC-1 and AIB1 in a coimmunoprecipitation assay and such interactions appear to be direct. Interestingly, the LXXLL motifs of SRC-1, which mediate the interaction between SRC-1 and nuclear receptors, also interact with cyclin D1. Motif iii of SRC-1 appears to interact with cyclin D1 preferentially (Zwijsen *et al.,* 1998), in contrast to the prefer-

←

FIG. 14. SRC coactivators are histone acetyltransferases. (A) Mapping of the HAT domain of SRC-1. The position of domains for the bHLH, PAS, serine/threonine (S/T)-rich, glutamine(Q)/rich and dominant-negative (DN) regions are as indicated. White and black bars denote regions of SRC-1 without and with HAT activity, respectively, as determined by the filter-binding HAT assay of GST-SRC-1 fusion proteins. The indicated portions of SRC-1 were expressed as GST fusion proteins in *Escherichia coli* (383–568, 383–841, 782–1139, 1107–1441), yeast (1–399, 1216–1441), or insect cells (383–841) and subsequently purified using glutathione-Sepharose beads. The GST control protein was expressed in *E. coli.* About 2 pmol of GST control or indicated GST-SRC-1 fusion proteins was tested for the ability to acetylate free histones in a filter-binding assay using [3H]acetyl-CoA. (B) SRC-1 preferentially acetylates amino-terminal peptide tails of histones H3 and H4. Acetylation of histone N-terminal peptides by GST-SRC (1107–1441) was assessed by measuring 3H-acetate incorporation using the filter binding assay. For each peptide substrate and H1/H5-stripped chicken mononucleosomes, incubations with 2 pmol GST (white bars) or GST-SRC (1107–1441) (black bars) were done in parallel. Sites where *N*-acetyllysine was incorporated during peptide synthesis in order to mimic sites that are acetylated *in vivo* are indicated by (Ac). All peptides were MAP reagents, except diacetyl(9/14)-H3 peptide, which was synthesized with a C-terminal cysteine. [Adapted by permission from Figs. 2 and 3 of Spencer, T. E., *et al.* (1997). Steroid receptor coactivator-1 is a histone acetyltransferase. *Nature* **389,** 194–198.]

ence of motif ii for interaction with ER (Heery *et al.,* 1997). Therefore, various LXXLL motifs of SRC-1 may display specificity for protein–protein interaction with nuclear receptors and cyclin D1, permitting simultaneous interaction of SRC-1 with both cyclin D1 and ER. This observation may explain, at least partly, the multiplicity of the LXXLL motifs in SRC coactivators. The ability of cyclin D1 to bind and activate ER and to interact with SRC-1 suggests an adaptor function for cyclin D1 between unliganded ER and SRC-1 in the absence of ligand. The formation of such a ternary complex also occurs on DNA-bound ER, suggesting that cyclin D1 can promote ligand-independent transactivation of ER by recruiting SRC coactivators to target genes. Physiologically, cyclin D1 mutant that fails to interact with SRC-1 appears to inhibit ER transactivation in breast cancer cells, suggesting an involvement of both cyclin D1 and SRC coactivators on ER-mediated transactivation in breast cancer cells.

The preceding observations suggest that cyclin D1 may provide a single site for interaction with LXXLL motif iii of the SRC coactivator in the absence of ligand. Upon ligand binding of ER, a second binding site is formed on ER for interaction with the LXXLL motif ii of the SRC coactivator. This model may partly explain the synergistic action of estradiol and cyclin D1 on ER activation. However, because unliganded ER also forms complexes with heat shock proteins and liganded ER binds to DNA as homodimer, the precise mechanism of synergism between cyclin D1 and SRC coactivators in ER activation remains unclear. Nevertheless, because cyclin D1 and AIB1 are overexpressed frequently in breast cancer cells, the synergistic action of cyclin D1 and SRC coactivator may have significant roles in ER-dependent cell growth and proliferation of breast cancer cells.

VI. SRC Function and Specificity

The existence of three related SRC coactivators with similar property in transcriptional activation and steroid/nuclear receptor interaction suggests a redundant mechanism for coactivator function. In fact, all three SRC coactivators interact and activate multiple steroid/nuclear receptors. However, several studies have also provided evidence that each SRC coactivator may exhibit a specific mode of function to precisely control transcriptional activation mediated by steroid/nuclear receptors and other classes of transcription factors. This section discusses the possible function and specificity of members of the SRC coactivator family.

A. Expression Patterns

To understand the physiologic function of SRC coactivators, the expression patterns of each SRC coactivator have been analyzed and compared. The SRC-1 message was detected ubiquitously in many tissues (Li and Chen, 1997; Misiti *et al.*, 1998; Zhu *et al.*, 1996; Yao *et al.*, 1996). The expression of SRC-1 is relatively high in skeletal muscle, heart, brain, and pancreas, and low in lung, liver, and kidney (Fig. 15). In a Northern blot assay, two SRC-1 messages of distinct sizes were detected, with the longer form (8 kb) more abundant than the shorter form (7 kb). The identity of these two forms is currently unclear, but they likely represent the SRC-1a and SRC-1e isoforms, respectively. In contrast to the expression of SRC-1, expression of RAC3 (SRC-3) is highly restricted (Fig. 15). The relative abundance of TIF2 message in human tissues is similar to that of RAC3 (Li and Chen, 1997). Both TIF2 and RAC3 are highly expressed in placenta, uterus, mammary gland, pituitary, testis, heart, skeletal muscle, and pancreas, but at lower levels in brain, lung, liver, kidney, and bone marrow (Li and Chen, 1997; Chen *et al.*, 1997; Takeshita *et al.*, 1997; Suen *et al.*, 1998). Interestingly, mouse SRC-2 (GRIP1) and SRC-3 (p/CIP) were detected ubiquitously in many murine tissues, including lung, brain, heart, liver, and testis

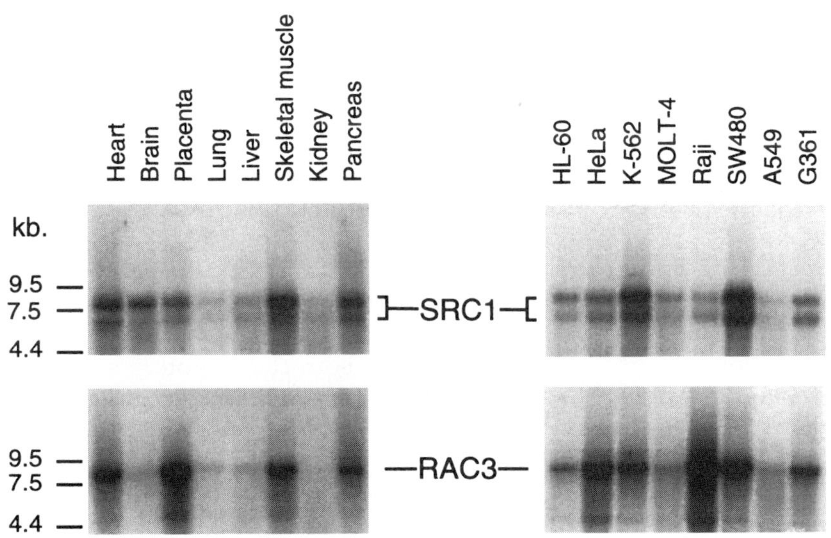

Fig. 15. Expression patterns of SRC-3 coactivators. Human multiple tissue (left) and cancer cell (right) Northern blots (Clontech Inc.) were sequentially hybridized with a ^{32}P-labeled RAC3 and SRC-1 probes.

(Torchia *et al.,* 1997). Therefore, the expression patterns of the SRC coactivators may differ among different species. For instance, TIF2 message is low in human kidney, liver, and lung, but the corresponding mouse tissues express high levels of TIF2. Similarly, SRC-3 appears to be ubiquitously expressed in mouse tissues (Torchia *et al.*, 1997), while expression of SRC-3 is highly restricted in human tissues (Li and Chen *et al.,* 1997; Chen,1997). Consistent with the species-specific distribution, *Xenopus* SRC-3 is highly expressed in adult liver (Kim *et al.,* 1998), where SRC-3 is virtually undetectable in human. These results suggest that the expression of SRC coactivators may differ in different species, reflecting a potential functional difference for specific SRC coactivators in different species.

The expression levels of each SRC coactivator also vary significantly in different cancer cell types. SRC-3 is highly expressed in Burkitt's lymphoma Raji cells, and moderately expressed in epithelioid carcinoma HeLa cells, chronic myelogenous leukemia K-562 cells, colorectal adenocarcinoma SW480 cells, and the melanoma G361 cells (Chen *et al.,* 1997; Li and Chen, 1997; Misiti *et al.,* 1998). The cell-type expression pattern of SRC-2 (TIF2) is similar to that of SRC-3 with the highest expression in the Raji cells. In contrast, SRC-1 is expressed at high levels in K-562 and SW480 cells, with low levels in HL60, HeLa, MOLT-4, Raji, A549, and G361 cells. In addition, SRC-1 message was also detected in many other cell types, including GH3, AtT20, Rat1, NIH3T3, 293, COS7, CHO-K1, and CV-1, with relatively higher levels in the pituitary GH3 cells (Misiti *et al.,* 1998). Both SRC-1a and SRC-1e were also detected in many cell lines analyzed by Rnase protection assay (Kalkhoven *et al.,* 1998). These studies indicate that SRC coactivators are widely expressed in different cell types, suggesting a wide spreading function for SRC coactivators. The differential expression of SRC coactivators suggests that each member of the SRC family might serve as a primary coactivator for a subset of receptors in a given tissue or cell type. Currently, which coactivator is involved in a particular hormonal signaling pathway remains to be determined.

It is likely that the expression level of both the receptors and coactivators and possibly their interactions with other transcriptional regulators will play an important role to control a precise level of gene expression in response to specific hormones. Frequently, the expression of receptor gene is autoregulated by the hormone that binds to and activates the receptor. For instance, expression of RARβ is upregulated by RAR ligand RA (de-Thé *et al.,* 1989). Interestingly, the expression of the coactivator RAC3 appears to be upregulated by RA as well (Li and Chen, 1997). Similarly, T3 treatment also produces an increase in SRC-

1 mRNA level in GH3 cells, as well as in the pituitary gland of adult rats (Misiti *et al.,* 1998). Therefore, it is possible that autoregulation of expression of SRC coactivators may add another level of complexity for cells to control gene expression induced by hormones (Fig. 16).

B. Coactivator Function

Many studies have established the function of SRC coactivators for enhancing ligand-dependent transcriptional activity of steroid/nuclear receptors (Jeyakumar *et al.,* 1997; Henttu *et al.,* 1997; Zhu *et al.,* 1996; McInerney *et al.,* 1996; Smith *et al.,* 1996). Transient transfection has been widely utilized to show the coactivation function of SRC coactivators. For instance, transfection of SRC-1 enhances progesterone-stimulated transactivation by PR, while transfection of SRC-1 has little effect on RU-486 antagonist bound PR (Oñate *et al.,* 1995). It has been shown that overexpression of SRC-1 results in enhancement of ER, GR, TR, PPAR, and RXR transcriptional activities, but has no effect on E2F- or forskolin-stimulated transcription (Oñate *et al.,* 1995; DiRenzo *et al.,* 1997; Zhu *et al.,* 1996). Overexpression of SRC-1 can also reverse the inhibitory effect of E2 on R5020-stimulated transcription, and the C-terminal receptor-interacting domain alone inhibits hormone-stimulated PR and TR transactivation (Oñate *et al.,* 1995). Similarly, SRC-2 (GRIP1/TIF2) and SRC-3 (RAC3/p/CIP/ACTR/AIB1/TRAM-1) also ex-

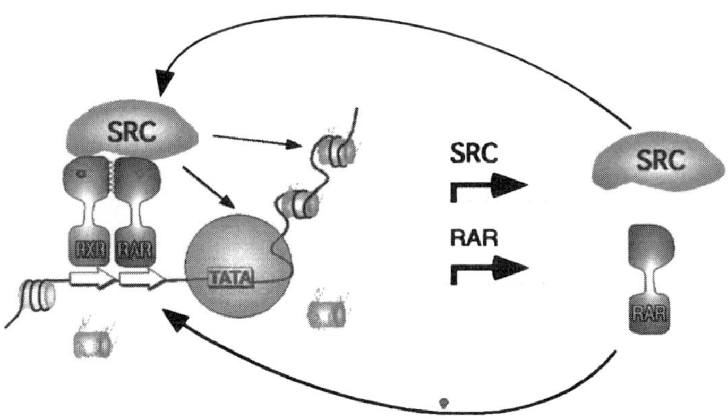

Fig. 16. Model of coactivator autoregulation. After ligand binding, the RXR-RAR heterodimer recruits a coactivator complex that contains members of the SRC family proteins, CBP/p300 and P/CAF. Because both RAR and RAC3 transcripts are elevated by RA treatment, the increased concentration of the two proteins should further amplify the transcriptional responses, leading to a high level of gene induction.

hibit similar coactivation function. Although the relative level of enhancement depends on experimental conditions, these studies suggest that SRC coactivators are limiting cofactors shared by members of the steroid/nuclear receptor family.

The function of SRC coactivators in transcriptional activation by steroid/nuclear receptors and other classes of transcription factors has also been extensively analyzed by microinjection assay. Injection of anti-SRC-1 (NCoA-1) IgG completely inhibits RA stimulated transcription (Fig. 6) (Korzus *et al.,* 1998). Interestingly, such IgG-dependent inhibition could be rescued by coinjection of a NCoA-1 expression vector. Similarly, anti-SRC-1 IgG also inhibits transactivation by troglitazone-stimulated PPARγ (Westin *et al.,* 1998), estradiol-stimulated ER, triac-stimulated TR, or progesterone-stimulated PR (Torchia *et al.,* 1997). In contrast, injection of anti-SRC-1 IgG had no effect on transactivation from Sp1-dependent or CMV promoters (Torchia *et al.,* 1997), or from cAMP-stimulated CREB or interferon γ-stimulated STAT-1-dependent promoters (Korzus *et al.,* 1998), suggesting the specificity of this assay. Paradoxically, injection of anti-NCoA-2 (SRC-2) IgG had no effect on RA-dependent transcription (Torchia *et al.,* 1997), contradicting transient transfection data (Hong *et al.,* 1997; Voegel *et al.,* 1998). Injection of anti-p/CIP (SRC-3) IgG appears to have a profound inhibitory effect on not only all steroid/nuclear receptors tested (RAR, ER, TR, PR), but also on interferon γ and cAMP-dependent transcriptional activation (Torchia *et al.,* 1997; Korzus *et al.,* 1998). These studies suggest a broader role for p/CIP (SRC-3) in different signaling pathways than SRC-1 and SRC-2, consistent with the hypothesis that p/CIP is a component of the CBP/p300 cointegrator complex (Torchia *et al.,* 1997).

By performing IgG microinjection together with a rescuing expression vector for either wild-type or mutant coactivator, the domain requirement and functional redundancy of the three SRC coactivators have been revealed. First, coinjection of wild-type SRC-1 expression vector efficiently restores RA-dependent transcription abrogated by anti-SRC-1 IgG (Fig. 6). It appears that, in addition to SRC-1, SRC-2 (NCoA-2) but not SRC-3 (p/CIP) also restores the anti-SRC-1 IgG-inhibited transcription from a RA-dependent promoter (Torchia *et al.,* 1997), suggesting a functional redundancy between SRC-1 and SRC-2, but not with SRC-3. Interestingly, the inhibition of RA-dependent transcription by p/CIP IgG could only be rescued by coinjection of both p/CIP and CBP expression vectors. Coinjection of individual expression vector for NCoA-1, NCoA-2, or even p/CIP or CBP could not restore RA-dependent transcription abrogated by anti-p/CIP IgG (Torchia *et al.,* 1997). These studies are consistent with the idea that both NCoA-1 and the CBP/p300/p/CIP complex are independently required for gene activation by steroid/nuclear receptors (Fig. 17). In addition, both LXXLL motif ii and motif iii mutants of SRC-1 were unable to restore IgG-inhibited transactivation by RAR, while only motif ii but not motif iii mutation failed to restore ER transactivation. These data indicate that both motif ii and iii are essential for transactivation by RAR, while motif iii is not required for transactivation by ER but motif ii is essential (Torchia *et al.,* 1997). These results also indicate a differential requirement for each LXXLL motif in transcriptional activation by specific steroid/nuclear receptors. Furthermore, the dominant negative effect of either receptor-interacting or transcriptional activation domain of p/CIP alone has also been demonstrated by the microinjection assay. Coinjection of expression vector for p/CIP fragment between amino acids 547 and 1084 inhibited RA-dependent transcription. In contrast, injection of expression vector for p/CIP fragment 947 to 1084 inhibits interferon g-stimulated transcription, which could not be restored by coinjection of CBP (Torchia *et al.,* 1997).

The presence of multiple HAT components of the SRC-CBP/p300-P/CAF coactivator complex raises a question about the requirements for specific HAT activities in transcriptional activation by steroid/nuclear

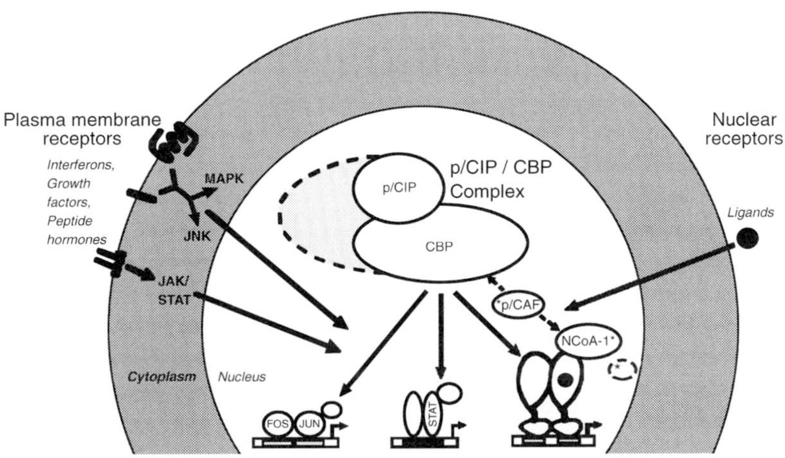

FIG. 17. Model of SRC coactivator function in different signaling pathways. Several signal-transduction pathways that are mediated by specific transcription factors require a functional SRC/CBP/p300 coactivator complex, and potentially p/CAF, with each partner being required, but not sufficient, to mediate transcriptional effects. [Adapted by permission from Fig. 6 of Torchia, J., *et al.* (1997). The transcriptional co-activator p/CIP binds CBP and mediates nuclear receptor function. *Nature* **387,** 677–684.]

receptors. This question was also addressed by the microinjection assay. First, P/CAF was shown to be essential for RAR, TR, and ER-dependent transcriptional activation as microinjection of anti-P/CAF IgG abrogated all transactivation events (Korzus *et al.,* 1998). Similarly, microinjection of anti-CBP IgG also inhibits transcription activities of steroid/nuclear receptors, suggesting that all three classes of coactivators are required for steroid/nuclear receptor function. The requirements for specific acetyltransferase activities of these coactivators were then analyzed by coinjecting rescuing vector for either wild-type or acetylation defective mutants. The results show that only the HAT activity of P/CAF, but not CBP or SRC-1, is required for RAR-mediated transcriptional activation (Korzus *et al.,* 1998). Consistently, the HAT domain of the SRC coactivators is not essential for transcriptional activation by the coactivators (Chen *et al.,* 1997; Voegel *et al.,* 1998). Therefore, although the SRC coactivators are necessary for optimal transcriptional activation by steroid/nuclear receptors, the role of their HAT activity in transcriptional activation remains unclear.

Similar to steroid/nuclear hormones, transforming growth factor β (TGF-β) also regulates cell proliferation and differentiation. Binding of TGF-β to cell surface receptor induces phosphorylation of SMAD2 and SMAD3, which are members of the SMAD family of transcription factos. The phosphorylated form of SMADs forms stable complexes with SMAD4 and these complexes translocate into nucleus where they activate transcription. Recently, TGF-β has been shown to act cooperatively with vitamin D_3, indicating a cross-talk between these two signaling pathways (Yoshizawa *et al.,* 1997; Takeshita *et al.,* 1998). The mechanism of synergism between TGF-β and vitamin D_3 appears to be mediated by SMAD3, but not SMAD2 (Yanagisawa *et al.,* 1999). SMAD3 interacts directly with VDR in a ligand-dependent manner *in vivo,* and this interaction is mediated through the NH_2-terminal Mad homology 1 (MH1) region of SMAD3 and a middle region of the ligand-binding domain of VDR. Interestingly, SMAD3 acts synergistically with SRC-1 to enhance transactivation of VDR (Yanagisawa *et al.,* 1999). It appears that interaction of VDR with SRC-1 is required for the ligand-dependent interaction with SMAD3, since an SRC-1 mutant lacking nuclear receptor interacting motifs inhibits the interaction of VDR with SMAD3. Although, SMAD3 does not appear to interact directly with SRC-1 or TIF2, certain SRC-1-stabilized ligand-dependent conformational changes in VDR may be required for SMAD3 interaction. Alternatively, activation of SMAD3 by TGF-β receptor-mediated phosphorylation may be required for interaction with SRC coactivators. It remains to be established whether the functional interaction between SMAD3 and SRCs plays a role in other TGF-β-mediated signaling pathways.

C. SRC-1 Function in Mice

The *in vivo* biological function of the coactivator SRC-1 has been assessed in mice by gene targeting (J. Xu *et al.,* 1998; Qi *et al.,* 1999). In one study (J. Xu *et al.,* 1998), the endogenous SRC-1 gene was targeted by a vector that deletes all known SRC-1 functional domains except the N-terminal bHLH-PAS region. Although the bHLH-PAS domain is highly conserved among SRC family proteins, it is not essential for transcriptional coactivation by SRC-1. In the knockout mice, both the heterozygous and homozygous mice appear normal and indistinguishable from wild-type mice. Both male and female homozygotes are fertile and develop at a similar rate as the wild-type mice. However, detailed analysis of steroid action in target organs including uterus, prostate, and mammary gland revealed that SRC-1 function is in fact required for maximal response of these organs to steroids *in vivo* (Fig. 18).

First, the uterine response to progesterone appears significantly impaired in SRC-1 null mice. This was measured in ovariectomized mice treated with a high dose of progesterone and a low dose of estrogen, followed by mechanical traumatization (decidual stimulation) of one uterine horn. In wild-type mice, the uterine horn increases in size in response to decidual stimulation. In contrast, the uterine response in SRC-1 null mutant is significantly reduced (2.5-fold) (Fig. 18). Similarly, estrogen-induced uterine growth in SRC-1 null mutants is also significantly reduced. These data suggest that SRC-1 is required for maximal uterine response to steroid hormones *in vivo*. Uterine response to mechanical traumatization is a progesterone receptor (PR)-dependent process; therefore, SRC-1 may be required for efficient transcriptional activation by PR, consistent with its coactivation function in tissue culture cells (Oñate *et al.,* 1995). In addition to progesterone-dependent uterine response, androgen-dependent growth of prostate and testes, and estrogen and progesterone-dependent growth of mammary gland are also significantly inhibited in SRC-1 mutant mice compared to wild-type mice. In castrated male mice with regressed prostates, testosterone-stimulated prostate growth is significantly reduced in SRC-1 mutant mice. Consistently, the size of testes is also smaller in SRC-1 null mutants. Furthermore, the development of mammary ducts and alveoli in virginal and pregnant mice, respectively, are both retarded in SRC-1 null mutants (Fig. 18). Mammary gland development in ovariec-

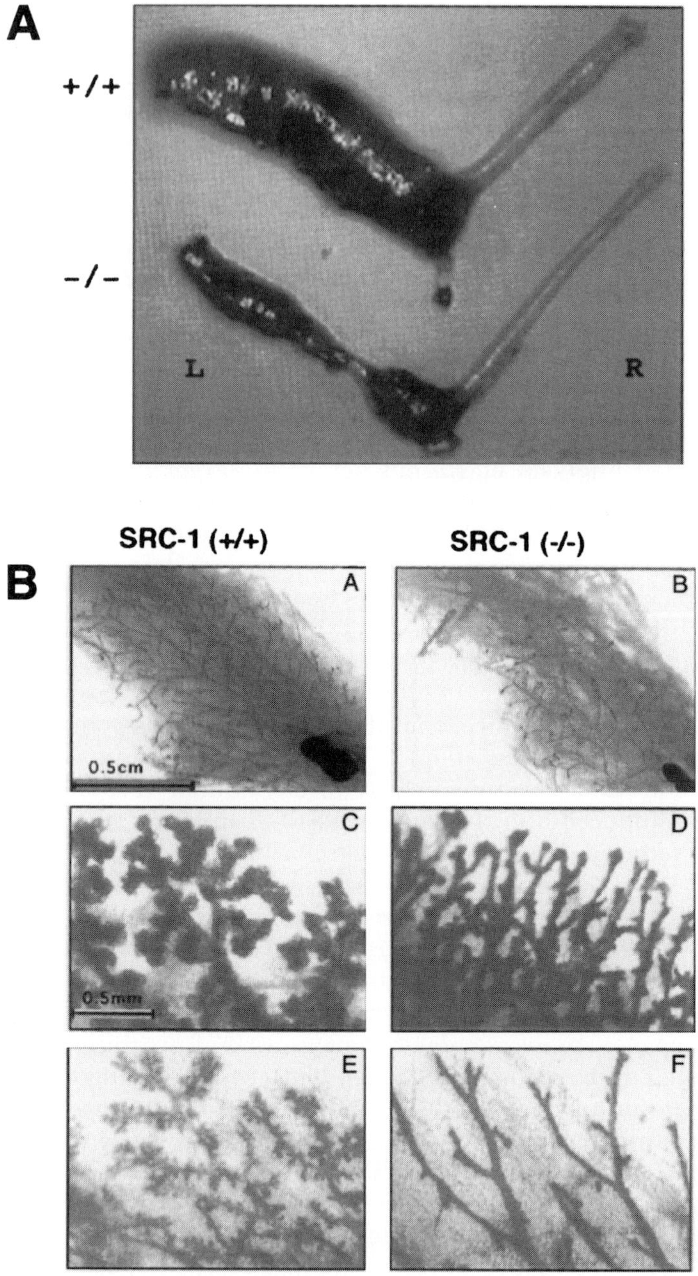
A
+/+
-/-
L
R
B
SRC-1 (+/+)
SRC-1 (-/-)
A
B
0.5cm
C
D
0.5mm
E
F

tomized mice in response to estrogen and progesterone treatments is also significantly affected. Furthermore, estradiol, progesterone, and testosterone concentrations in female null mutants are 1.2 and 1.5 times those in wild-type animals, respectively, consistent with the phenomenon of endocrine feedback regulation. Analysis of the expression of other SRC coactivators in the SRC-1 null mutant mice revealed an elevated level of TIF2, suggesting that other SRC coactivators may redundantly or partially replace the lost function of SRC-1. The partial hormonal response due to impairment of coactivator function might explain certain partial-hormone-resistance syndromes.

The mouse SRC-1 gene was also targeted to replace its central nuclear receptor interacting domain with phosphoglycerate kinase-neomycin gene in another study (Qi *et al.,* 1999). In this case, a correct gene-targeting event would result in a protein that lacks the three critical LXXLL motifs required for interaction with liganded receptors. Similar to the other study (J. Xu *et al.,* 1998), the homozygous SRC-$1^{-/-}$ mice were viable and exhibited no apparent morphologic abnormalities. Both male and female homozygous mice grew normally and were fertile. Extensive analysis of the PPARαB ligand-mediated responses *in vivo* suggests that SRC-1 is not required for PPARα-mediated transcriptional activation. For instance, the SRC-1 null mice response normally to peroxisome proliferators, such as ciprofibrate and Wy-14,643, which induce liver cell proliferation and hepatic peroxisome proliferation. There were also no effects on the expression of PPARα-regulated, fatty acid-metabolizing enzymes in the liver. Because this targeting event may allow expression of a truncated SRC-1 mutant, a

Fig. 18. Uterine and mammary gland development in SRC-$1^{-/-}$ mutant mice. (A) Uterine responses to a decidual stimulus were measured in wild-type (+/+) or SRC-1 null mutant (−/−) females. Eight-week-old females were ovariectomized on day 0, treated with estradiol (0.1 μg per mouse per day) from day 10 to day 12, and treated with progesterone (1 mg per mouse per day) and estradiol (6.7 ng per mouse per day) from day 16 to day 23. Mechanical decidualization in the left uterine horn was done 6 h after hormone injection on day 18. The whole uterus was dissected 6 h after hormone injection on day 23. (B) The fourth pair of mammary glands from 8-week-old virgins with the indicated SRC-1 genotypes (A and B). The ducts and alveolar structures of the fourth pair of mammary glands from mice pregnant for the first time with the indicated genotypes (C and D). The mammary ducts and alveolar structures of the fourth pair of mammary glands from 13-week-old females treated with progesterone and estradiol. Eight-week-old females were ovariectomized one day 0 and then treated with progesterone (1 μg per mouse per day) and estradiol (50 mg per mouse per day) from day 14 to day 34 (E and F). [Reprinted with permission from Xu, J., *et al.* (1998). Partial hormone resistance in mice with disruption of the steroid receptor coactivator-1 (SRC-1) gene. *Science* **279,** 1922–1924. Copyright © 1998 American Association for the Advancement of Science.]

functional SRC-1 mutant might still exist in the homozygous mice. Nonetheless, there is no evidence for such a truncated protein and SRC-1 may indeed be nonessential for PPARα-mediated transcriptional response *in vivo*. Alternatively, loss of SRC-1 function might be fully compensated for by other nuclear receptor coactivators. Reciprocal examination of the steroid hormone responses and PPARα function in different SRC-1 null mutant strains might help to clarify the requirement of SRC-1 in steroid/nuclear receptor function in mice. Apparently, additional studies are necessary to fully understand the role of various coactivators by generating mutant mice with defects in one or more coactivator functions.

VII. SRC Coactivators and Human Diseases

A. MOZ-TIF2 Fusion in Acute Myeloid Leukemia

Recently TIF2 was found in a search for genes involved in inv(8) (p11q13)-associated acute myeloid leukemia (AML) (Carapeti *et al.*, 1998). This subtype of AML contains blast cells of a monocytoid phenotype that have pronounced erythrophagocytic activity. This AML subtype is typically associated with the t(8;16)(p11;p13) translocation, and occasionally with the t(8;22)(p11;q13), t(8;19)(p11;q13), and inv(8) (p11;q13) translocations (Mitelman *et al.*, 1997). The genes involved in the t(8;16) have been identified as the MOZ gene at 8p11 fused to the coactivator CBP gene at 16p13 (Borrow *et al.*, 1996). Although the precise function of the MOZ gene is unknown, it contains a PHD/LAP domain involved in protein–protein interaction, and a histone acetyltransferase homologous domain. Because CBP is also a histone acetyltransferase, the mechanism of leukemogenesis in patients with the t(8;16) may involve aberrant chromatin remodeling due to abnormal histone acetylation. Although TIF2 itself has not been shown to have histone acetylation activity, other TIF2-related SRC coactivators possess histone acetylation function, reinforcing the supposition that abnormal chromatin acetylation may cause leukemia. In the inv(8) (p11q13) translocation, the MOZ-TIF2 fusion retains the N-terminal PHD finger and HAT domains of MOZ, along with the C-terminal CBP-interacting domain and the putative HAT domain of TIF2 (Fig. 19). The fusion does not contain the bHLH-PAS or the steroid/nuclear receptor interacting domains of TIF2. Therefore, the HAT activity of TIF2 or its associated protein CBP might overstimulate expression of genes normally regulated by MOZ (Fig. 20). In addition to CBP and TIF2, other coactivators are also found associated with leukemogenesis or other

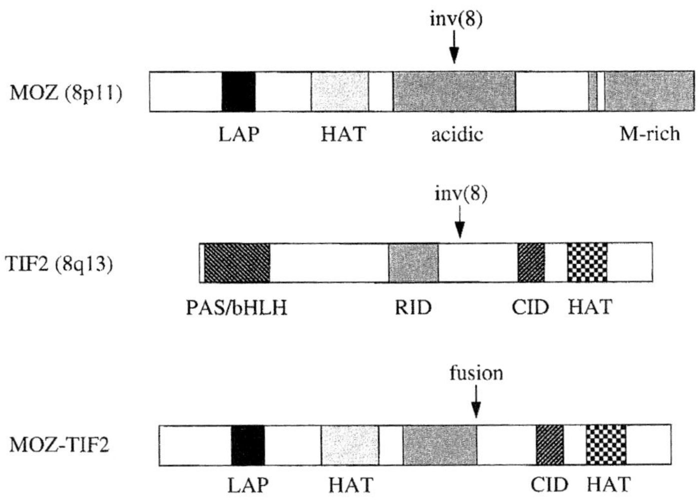

FIG. 19. Schematic representation of MOZ, TIF2, and MOZ-TIF2 fusion proteins. Domains are indicated as follows: LAP, leukemia-associated protein; HAT, histone acetyltransferase; M-rich, methionine-rich; CID and RID, putative CBP and nuclear receptor interacting domains based on homology with SRC-1. The MOZ-TIF2 fusion retains the LAP finger and HAT homology domains of MOZ, along with the CID and HAT domains of TIF2. [Reproduced by permission from Fig. 4 of Carapeti, M., *et al.* (1998). A novel fusion between MOZ and the nuclear receptor coactivator TIF2 in acute myeloid leukemia. *Blood* **91,** 3127–3133.]

cancer types. For instance, p300 is fused to the MLL gene in AML with the t(11;22)(q23;q13), ARA70 is fused to RET in human thyroid papillary carcinoma, and TIF1 fused to B-RAF in the mouse hepatoma-derived oncogene T18. These observations suggest that transcriptional coactivators such as those for nuclear receptors may be widely involved in malignancy.

B. AIB1 Gene Amplification in Cancers

Gene amplification is frequently associated with human cancers for selective overexpression of a subset of genes essential for supporting tumor growth. In breast cancer, several chromosomal regions are commonly amplified, including regions in the long arm of chromosome 20. In a search of target genes amplified from chromosome 20q in breast cancer, AIB1 was cloned by chromosome microdissection and hybrid selection, and mapped to 20q12 (Anzick *et al.,* 1997). Accordingly, AIB1 gene was found highly amplified (>20-fold) in three ER-positive breast carcinoma cell lines (BT-474, MCF-7, and ZR75–1) and in one ovarian

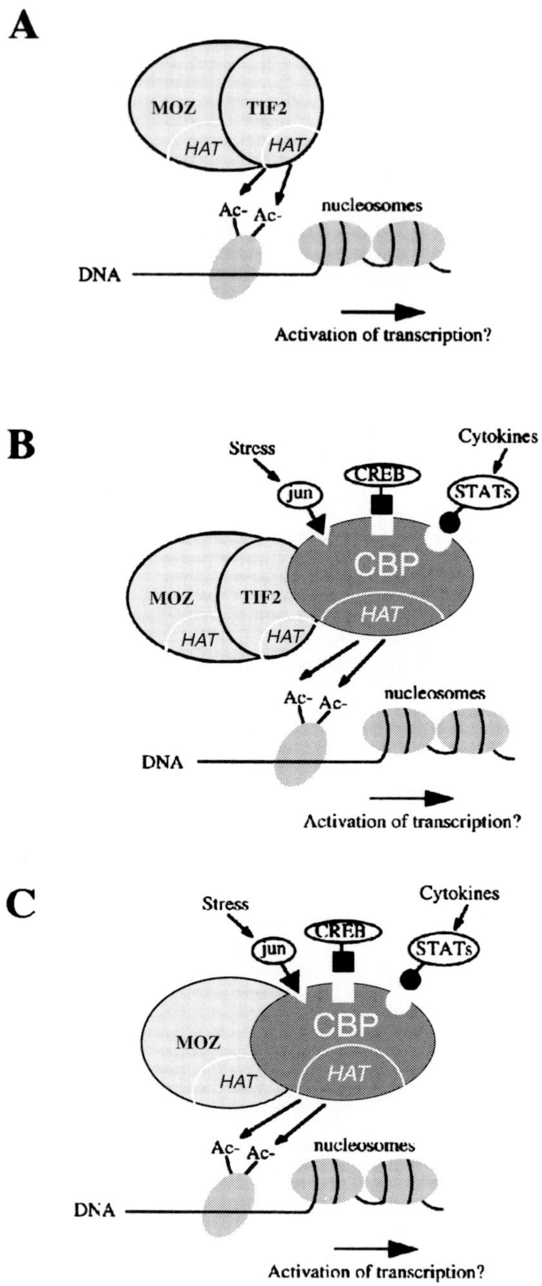
A
MOZ
TIF2
HAT
HAT
Ac-
Ac-
nucleosomes
DNA
Activation of transcription?
B
Stress
Cytokines
jun
CREB
STATs
CBP
MOZ
TIF2
HAT
HAT
HAT
Ac-
Ac-
nucleosomes
DNA
Activation of transcription?
C
Stress
Cytokines
jun
CREB
STATs
CBP
MOZ
HAT
HAT
Ac-
Ac-
nucleosomes
DNA
Activation of transcription?

carcinoma cell line (BG-1) (Fig. 21) (Anzick *et al.,* 1997). In contrast, both SRC-1 and SRC-2 are ubiquitously expressed at low levels in all breast cancer lines analyzed. AIB1 amplification also occurs in primary breast tumors (9.5%), although the amplification levels are not as high as the cell lines (Anzick *et al.,* 1997). As expected, AIB1 gene amplification causes overexpression of AIB1 mRNA and protein. Interestingly, 58% of the mammary tumors that show no AIB1 gene amplification exhibit overexpression of AIB1, as compared with normal mammary epithelium. This suggests that overexpression of AIB1 in breast cancer cells may occur through mechanisms other than gene amplification. In a recent screening of 1157 breast and 122 ovarian tumors by Southern blotting, AIB1 amplification was found in 4.8% of breast cancers and 7.4% of ovarian cancers (Bautista *et al.,* 1998). The degrees of amplification range from 2- to 8-fold in breast tumors and 2- to 10-fold in ovarian tumors. These results indicate that the frequency and level of AIB1 amplification appear higher in ovarian tumors than in breast tumors. In breast tumors, AIB1 amplification appears to correlate positively with either ER or PR expression. In addition, AIB1 amplification is more frequently observed in large tumors (>2 cm) and seems to correlate with MDM2 and FGFR1 amplifications. In contrast, no correlation was found with cyclin D1, Erb-B2, or Myc amplifications. MDM2 is the main repressor of the tumor suppressor p53, thus amplification of MDM2 may result in p53 inactivation. The FGFR1 is a class IV tyrosine kinase receptor that is preferentially activated by FGFs and amplified in 10–15% of breast tumors. The coamplification of AIB1 with MDM2 and FGFR1 suggests possible cooperative pathways of oncogenic activation in breast cancers. Interestingly, although cyclin D1 amplification in breast cancer is clearly associated with ER positivity, cyclin D1 is not coamplified with AIB1. This observation suggests that AIB1 and cyclin D1 amplifications correspond to a distinct subset of ER-positive breast tumors. Recently, cyclin D1 has been shown to interact

←

FIG. 20. Hypothetical models of the mode of action of the MOZ-TIF2 fusion protein. (A) TIF2 may directly modulate the transcriptional activity of genes normally regulated by MOZ through the addition or removal of histone acetyl (Ac) groups by its HAT domain. (B) The TIF2 moiety may serve as a bridge between MOZ and CBP, and it is the HAT or other activities of CBP that leads to leukemogenic alterations in gene expression. Chromatin-associated CBP may be responsive to other cellular signals such as those mediated by Jun, CREB, or STAT proteins. (C) The MOZ-CBP fusion in the t(8;16),7, which is associated with a strikingly similar leukemia cell phenotype to that seen in cases with the inv(8). [Reproduced by permission from Fig. 5 of Carapeti, M., *et al.* (1998). A novel fusion between MOZ and the nuclear receptor coactivator TIF2 in acute myeloid leukemia. *Blood* **91,** 3127–3133.]

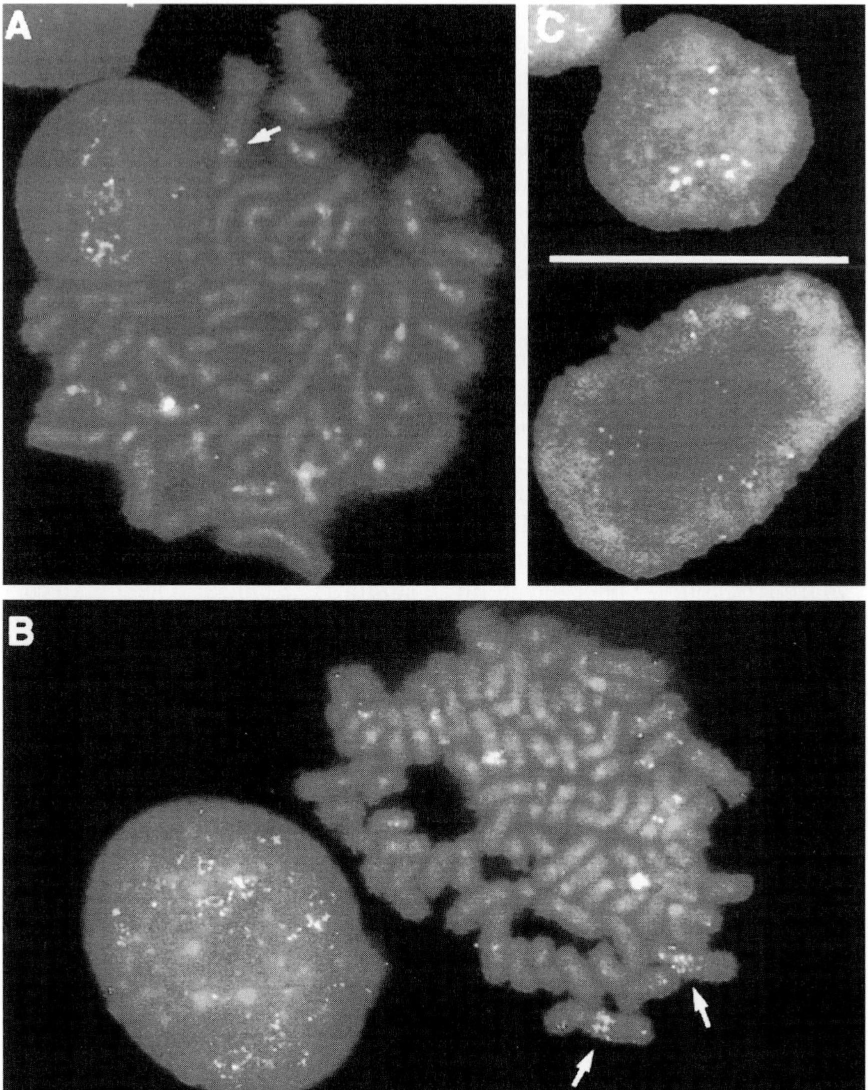

FIG. 21. Amplification of a SRC coactivator in breast cancers. Bicolor FISH analysis demonstrates AIB1 gene amplification in breast cancer cell line ZR75–1 (A), ovarian cancer cell line BG-1 (B), and two uncultured breast cancer samples (C). Intrachromosomal amplification of AIB1 (arrows) is apparent in metaphase chromosomes of ZR75–1 and BG-1, and numerous copies of AIB1 are resolved in the adjacent interphase nuclei. [Reprinted with permission from Anzick, S. L., *et al.* (1997). AIB1, a steroid receptor coactivator amplified in breast and ovarian cancer. *Science* **277,** 965–968. copyright © 1997 American Association for the Advancement of Science.]

with both ER and SRC coactivators (SRC-1 and AIB1) in a ligand-independent manner (Zwijsen *et al.*, 1998). Because the frequency of cyclin D1 amplification correlates with the levels of ER expression, it is conceivable that synergistic action among ER, cyclin D1, and AIB1 may be an indicator of breast tumors.

In addition to breast and ovarian cancers, AIB1 gene amplification and overexpression have also been observed in pancreatic carcinoma (Ghadimi *et al.*, 1999). The incidence of pancreatic cancer is about 0.01 percentage in USA, which contributes to about 20 percentage of cancer deaths, due to poor prognosis of this disease. Cytogenetic studies revealed recurring chromosomal gains on several locations, including chromosome 20q where AIB1 gene is located. Fluorescence *in situ* hybridization (FISH) analysis found that the AIB1 gene is amplified in six out of nine pancreatic carcinoma cell lines, which partially correlate with overexpression of AIB1 mRNA (Ghadimi *et al.*, 1999). These results suggest that AIB1 gene amplification may occur frequently in human tumors and that steroid/nuclear receptors may regulate growth of cells that are not primarily controlled by endocrine stimuli. Alternatively, AIB1 may be involved in signaling pathways other than steroid/nuclear receptors, at least in part, due to its interaction with the general coactivator CBP/p300.

VIII. Conclusions

In conclusion, a novel family of transcriptional coactivators has been identified and shown to play a crucial role in transcriptional activation by steroid/nuclear hormone receptors and possibly other classes of transcriptional regulators. Detailed biochemical and structural analyses have revealed the molecular basis of protein–protein interaction between SRC coactivators and several liganded steroid/nuclear receptors. Additionally, transcription coactivation by SRCs has been linked to histone acetylation, partly by association with general transcriptional coactivators CBP/p300 and P/CAF. It is currently unknown whether these three SRC coactivators share redundant functions or form a protein complex to synergize transcriptional activation. It is important to note that at least two members of the SRC family are directly linked to human malignancies, consistent with a prevailing involvement of steroid/nuclear receptors in human cancers. Future studies are required for understanding the physiologic role of these coactivators in hormone action and the potential development of these genes as drug targets for treating human diseases.

ACKN OWLEDGMENTS

The author thanks Dr. Daniel J. Schroen and Christopher Leo for reading of the manuscript. This work was supported by an American Society of Hematology Scholar Award, an Army Breast Cancer IDEA Award (BC961877), and an R01 grant from the National Institutes of Health (DK52888).

REFERENCES

Anzick, S. L., Kononen, J., Walker, R. L., Azorsa, D. O., Tanner, M. M., Guan, X. Y., Sauter, G., Kallioniemi, O.-P., Trent, J. M., and Meltzer, P. S. (1997). AIB1, a steroid receptor coactivator amplified in breast and ovarian cancer. *Science* **277,** 965–968.

Arany, Z., Seller, W., Livingston, D. M., and Eckner, R. (1994). E1A-associated p300 and CREB-associated CBP belong to a conserved family of coactivators. *Cell (Cambridge, Mass.)* **77,** 799–800.

Bannister, A. J., and Kouzarides, T. (1996). The CBP co-activtor is a histone acetyltransferase. *Nature (London)* **384,** 641–643.

Barettino, D., Vivanco Ruiz, M.d.M., and Stunnenberg, H. G. (1994). Characterization of the ligand-dependent transactivation domain of thyroid hormone receptor. *EMBO J.* **13,** 3039–3049.

Bautista, S., Valles, H., Walker, R. L., Anzick, S., Zeillinger, R., Meltzer, P., and Theillet, C. (1998). In breast cancer, amplification of the steroid receptor coactivator gene AIB1 is correlated with estrogen and progesterone receptor positivity. *Clin. Cancer Res.* **4,** 2925–2929.

Beato, M., Herrlich, P., and Schutz, G. (1995). Steroid hormone receptors: Many actors in search of a plot. *Cell (Cambridge, Mass.)* **83,** 851–857.

Beijersbergen, R. L., and Bernards, R. (1996). Cell cycle regulation by the retinoblastoma family of growth inhibitory proteins. *Biochim. Biophys. Acta* **1287,** 103–120.

Borrow, J., Stanton, V. P., Jr., Andresen, J. M., Becher, R., Behm, F. G., Chaganti, R. S., Civin, C. I., Disteche, C., Dube, I., Frischauf, A. M., Horsman, D., Mitelman, F., Volinia, S., Watmore, A. E., and Housman, D. E. (1996). The translocation t(8;16)(p11;p13) of acute myeloid leukaemia fuses a putative acetyltransferase to the CREB-binding protein. *Nat. Genet.* **14,** 33–41.

Bourguet, W., Ruff, M., Chambon, P., Gronemeyer, H., and Moras, D. (1995). Crystal structure of the ligand-binding domain of the human nuclear receptor RXR-alpha. *Nature (London)* **375,** 477–382.

Brownell, J. E., and Allis, C. D. (1996). Special HATs for special occasions: Linking histone acetylation to chromatin assembly and gene activation. *Curr. Opin. Genet. Dev.* **6,** 176–184.

Brzozowski, A. M., Pike, A. C. W., Dauter, Z., Hubbard, R. E., Bonn, T., Engstrom, O., Ohman, L., Greene, G. L., Gustafsson, J. A., and Carlquist, M. (1997). Molecular basis of agonism and antagonism in the oestrogen receptor. *Nature (London)* **389,** 753–758.

Butler, R., Leigh, P. N., McPhaul, M. J., and Gallo, J. M. (1998). Truncated forms of the androgen receptor are associated with polyglutamine expansion in X-linked spinal and bulbar muscular atrophy. *Hum. Mol. Genet.* **7,** 121–127.

Carapeti, M., Aguiar, R. C., Goldman, J. M., and Cross, N. C. (1998). A novel fusion between MOZ and the nuclear receptor coactivator TIF2 in acute myeloid leukemia. *Blood* **91,** 3127–3133.

Cavaillès, V., Dauvois, S., Danielian, P. S., and Parker, M. G. (1994). Interaction of proteins with transcriptionally active estrogen receptors. *Proc. Natl. Acad. Sci. U.S.A.* **91,** 10009–10013.

Cavaillès, V., Dauvois, S., L'Horset, F., Lopez, G., Hoare, S., Kushner, P. J., and Parker, M. G. (1995). Nuclear factor RIP140 modulates transcriptional activation by the estrogen receptor. *EMBO J.* **14,** 3741–3751.

Chen, H., Lin, R. J., Schiltz, R. L., Chakravarti, D., Nash, A., Nagy, L., Privalsky, M. L., Nakatani, Y., and Evans, R. M. (1997). Nuclear receptor coactivator ACTR is a novel histone acetyltransferase and forms a multimeric activation complex with P/CAF and CBP/p300. *Cell (Cambridge, Mass.)* **90,** 569–580.

Chen, J. D., and Evans, R. M. (1995). A transcriptional co-repressor that interacts with nuclear hormone receptors. *Nature (London)* **377,** 454–457.

Chen, J. D., and Li, H. (1998). Coactivation and corepression in transcriptional regulation by steroid/nuclear hormone receptors. *Crit. Rev. Eukaryotic Gene Expression* **8,** 169–190.

Chiba, H., Muramatsu, M., Nomoto, A., and Kato, H. (1994). Two human homologues of Saccharomyces cerevisiae SWI2/SNF2 and Drosophila brahma are transcriptional coactivators cooperating with the estrogen receptor and the retinoic acid receptor. *Nucleic Acids Res.* **22,** 1815–1820.

Chrivia, J. C., Kwok, R. P. S., Lamb, N., Hagiwara, M., Montminy, M. R., and Goodman, R. H. (1993). Phosphorylated CREB binds specifically to the nuclear protein CBP. *Nature (London)* **365,** 855–859.

Darimont, B. D., Wagner, R. L., Apriletti, J. W., Stallcup, M. R., Kushner, P. J., Baxter, J. D., Fletterick, R. J., and Yamamoto, K. R. (1998). Structure and specificity of nuclear receptor-coactivator interactions. *Genes Dev.* **12,** 3343–3356.

de-Thé, H., Marchio, A., Tiollais, P., and Dejean, A. (1989). Differential expression and ligand regulation of the retinoic acid receptor a and b genes. *EMBO J.* **8,** 429–433.

Ding, X. F., Anderson, C. M., Ma, H., Hong, H., Uht, R. M., Kushner, P. J., and Stallcup, M. R. (1998). Nuclear receptor-binding sites of coactivators glucocorticoid receptor interacting protein 1 (GRIP1) and steroid receptor coactivator 1 (SRC-1): Multiple motifs with different binding specificities. *Mol. Endocrinol.* **12,** 302–313.

DiRenzo, J., Soderstrom, M., Kurokawa, R., Ogliastro, M. H., Ricote, M., Ingrey, S., Horlein, A., Rosenfeld, M. G., and Glass, C. K. (1997). Peroxisome proliferator-activated receptors and retinoic acid receptors differentially control the interactions of retinoid X receptor heterodimers with ligands, coactivators, and corepressors. *Mol. Cell. Biol.* **17,** 2166–2176.

Donnellan, R., and Chetty, R. (1998). Cyclin D1 and human neoplasia. *Mol. Pathol.* **51,** 1–7.

Eggert, M., Mows, C. C., Tripier, D., Arnold, R., Michel, J., Nickel, J., Schmidt, S., Beato, M., and Renkawitz, R. (1995). A fraction enriched in a novel glucocorticoid receptor-interacting protein stimulates receptor-dependent transcription in vitro. *J. Biol. Chem.* **270,** 30755–30759.

Feng, W., Ribeiro, R. C., Wagner, R. L., Nguyen, H., Apriletti, J. W., Fletterick, R. J., Baxter, J. C., Kushner, P. J., and West, B. L. (1998). Hormone-dependent coactivator binding to a hydrophobic cleft on nuclear receptors. *Science* **280,** 1747–1749.

Ghadimi, B. M., Schrock, E., Walker, R. L., Wangsa, D., Jauho, A., Meltzer, P. S., and Ried, T. (1999). Specific chromosomal aberrations and amplification of the AIB1 nuclear receptor coactivator gene in pancreatic carcinomas. *Am. J. Pathol.* **154,** 525–536.

Glass, C. K., Rose, D. W., and Rosenfeld, M. G. (1997). Nuclear receptor coactivators. *Curr. Opin. Cell Biol.* **9,** 222–232.

Halachmi, S., Marden, E., Martin, G., MacKay, H., Abbondanza, C., and Brown, M. (1994). Estrogen receptor-associated proteins: Possible mediators of hormone-induced transcription. *Science* **264,** 1455–1458.

Heery, D. M., Kalkhoven, E., Hoare, S., and Parker, M. G. (1997). A signature motif in transcriptional co-activators mediates binding to nuclear receptors. *Nature (London)* **387,** 733–736.

Henttu, P. M., Kalkhoven, E., and Parker, M. G. (1997). AF-2 activity and recruitment of steroid receptor coactivator 1 to the estrogen receptor depend on a lysine residue conserved in nuclear receptors. *Mol. Cell. Biol.* **17,** 1832–1839.

Hong, H., Kohli, K., Trivedi, A., Johnson, D. L., and Stallcup, M. R. (1996). GRIP1, a novel mouse protein that serves as a transcriptional coactivator in yeast for the hormone binding domains of steroid receptors. *Proc. Natl. Acad. Sci. U.S.A.* **93,** 4948–4952.

Hong, H., Kohli, K., Garabedian, M. J., and Stallcup, M. R. (1997). GRIP1, a transcriptional coactivator for the AF-2 transactivation domain of steroid, thyroid, retinoid, and vitamin D receptors. *Mol. Cell. Biol.* **17,** 2735–2744.

Horlein, A. J., Naar, A. M., Heiinzel, T., Torchia, J., Gloss, B., Kurokawa, R., Ryan, A., Kamei, Y., Soderstrom, M., Glass, C. K., and Rosenfeld, M. G. (1995). Ligand-independent repression by the thyroid hormone receptor mediated by a nuclear receptor co-repressor. *Nature (London)* **377,** 397–404.

Horwitz, K. B., Jackson, T. A., Bain, D. L., Richer, J. K., Takimoto, G. S., and Tung, L. (1996). Nuclear receptor coactivators and corepressors. *Mol. Endocrinol.* **10,** 1167–1177.

Jeyakumar, M., Tanen, M. R., and Bagchi, M. K. (1997). Analysis of the functional role of steroid receptor coactivator-1 in ligand-induced transcription by thyroid hormone receptor. *Mol. Endocrinol.* **11,** 755–767.

Kalkhoven, E., Valentine, J. E., Heery, D. M., and Parker, M. G. (1998). Isoforms of steroid receptor co-activator 1 differ in their ability to potentiate transcription by the oestrogen receptor. *EMBO J.* **17,** 232–243.

Kamei, Y., Xu, L., Heinzel, T., Torchia, J., Kurokawa, R., Gloss, B., Lin, S. C., Heyman, R. A., Rose, D. W., Glass, C. K., and Rosenfeld, M. G. (1996). A CBP integrator complex mediates transcriptional activation and AP-1 inhibition by nuclear receptors. *Cell (Cambridge, Mass.)* **85,** 403–414.

Kastner, P., Mark, M., and Chambon, P. (1995). Nonsteroid nuclear receptors: What are genetic studies telling us about their role in real life? *Cell (Cambridge, Mass.)* **83,** 859–869.

Khavari, P. A., Peterson, C. L., Tamkun, J. W., Mendel, D. B., and Crabtree, G. R. (1993). BRG1 contains a conserved domain of the SWI2/SNF2 family necessary for normal mitotic growth and transcription. *Nature (London)* **366,** 170–174.

Kim, H. J., Lee, S. K., Na, S. Y., Choi, H. S., and Lee, J. W. (1998). Molecular cloning of xSRC-3, a novel transcription coactivator from Xenopus, that is related to AIB1, p/CIP, and TIF2. *Mol. Endocrinol.* **12,** 1038–1047.

Kliewer, S. A., Umesono, K., Noonan, D. J., Heyman, R. A., and Evans, R. M. (1992). Convergence of 9-cis retinoic acid and peroxisome proliferator signalling pathways through heterodimer formation of their receptors. *Nature (London)* **358,** 771–774.

Korzus, E., Torchia, J., Rose, D. W., Xu, L., Kurokawa, R., McInerney, E. M., Mullen, T. M., Glass, C. K., and Rosenfeld, M. G. (1998). Transcription factor-specific requirements for coactivators and their acetyltransferase functions. *Science* **279,** 703–707.

Koshy, B. T., and Zoghbi, H. Y. (1997). The CAG/polyglutamine tract diseases: Gene products and molecular pathogenesis. *Brain Pathol.* **7,** 927–942.

Kurokawa, R., DiRenzo, J., Boehm, M., Sugarman, J., Gloss, B., Rosenfeld, M. G., Heyman, R. A., and Glass, C. K. (1994). Regulation of retinoid signalling by receptor polarity and allosteric control of ligand binding. *Nature (London)* **371,** 528–531.

Kurokawa, R., Soderstrom, M., Horlein, A., Halachmi, S., Brown, M., Rosenfeld, M. G., and Glass, C. K. (1995). Polarity-specific activities of retinoic acid receptors determined by a co-repressor. *Nature (London)* **377,** 451–454.

Kurokawa, R., Kalafus, D., Ogliastro, M. H., Kioussi, C., Xu, L., Torchia, J., Rosenfeld, M. G., and Glass, C. K. (1998). Differential use of CREB binding protein-coactivator complexes. *Science* **279,** 700–703.

Kwon, H., Imbalzano, A. N., Khavari, P. A., Kingston, R. E., and Green, M. R. (1994). Nucleosome disruption and enhancement of activator binding by a human SW1/SNF complex. *Nature (London)* **370,** 477–481.

Le Douarin, B., Zechel, C., Garnier, J. M., Lutz, Y., Tora, L., Pierrat, B., Heery, D., Gronemeyer, H., Chambon, P., and Losson, R. (1995). The N-terminal part of TIF-1, a putative mediator of the ligand-dependent activation function (AF-2) of nuclear receptors, is fused to B-raf in the oncogenic protein T18. *EMBO J.* **14,** 2020–2033.

Le Douarin, B., Nielsen, A. L., Garnier, J. M., Ichinose, H., Jeanmougin, F., Losson, R., and Chambon, P. (1996). A possible involvement of TIF1 alpha and TIF1 beta in the epigenetic control of transcription by nuclear receptors. *EMBO J.* **15,** 6701–6715.

Leers, J., Treuter, E., and Gustafsson, J. A. (1998). Mechanistic principles in NR box-dependent interaction between nuclear hormone receptors and the coactivator TIF2. *Mol. Cell. Biol.* **18,** 6001–6013.

Li, H., and Chen, J. D. (1997). Nuclear receptor coactivator RAC3 amplifies transcription through CBP recruitment and autoregulation. *J. Biol. Chem.* **273,** 5948–5954.

Li, H., Gomes, P. J., and Chen, J. D. (1997). RAC3, a steroid/nuclear receptor-associated coactivator that is related to SRC1 and TIF2. *Proc. Natl. Acad. Sci. U.S.A.* **94,** 8479–8484.

Lindebro, M. C., Poellinger, L., and Whitelaw, M. L. (1995). Protein-protein interaction via PAS domains: Role of the PAS domain in positive and negative regulation of the bHLH/PAS dioxin receptor-Arnt transcription factor complex. *EMBO J.* **14,** 3528–3539.

Mangelsdorf, D. J., and Evans, R. M. (1995). The RXR heterodimers and orphan receptors. *Cell (Cambridge, Mass.)* **83,** 841–850.

Mangelsdorf, D. J., Thummel, C., Beato, M., Herrlich, P., Schütz, G., Umesono, K., Blumberg, B., Kastner, P., Mark, M., Chambon, P., and Evans, R. M. (1995). The nuclear receptor superfamily: The second decade. *Cell (Cambridge, Mass.)* **83,** 835–839.

McInerney, E. M., Tsai, M. J., O'Malley, B. W., and Katzenellenbogen, B. S. (1996). Analysis of estrogen receptor transcriptional enhancement by a nuclear hormone receptor coactivator. *Proc. Natl. Acad. Sci. U.S.A.* **93,** 10069–10073.

McInerney, E. M., Rose, D. W., Flynn, S. E., Westin, S., Mullen, T. M., Krones, A., Inostroza, J., Torchia, J., Nolte, R. T., Assa-Munt, N., Milburn, M. V., Glass, C. K., and Rosenfeld, M. G. (1998). Determinants of coactivator LXXLL motif specificity in nuclear receptor transcriptional activation. *Genes Dev.* **12,** 3357–3368.

Meyer, M., Gronemeyer, H., Turcotte, B., Bocquel, M., Tasset, D., and Chambon, P. (1989). Steroid hormone receptors compete for factors that mediate their enhancer function. *Cell (Cambridge, Mass.)* **57,** 433–442.

Misiti, S., Schomburg, L., Yen, P. M., and Chin, W. W. (1998). Expression and hormonal regulation of coactivator and corepressor genes. *Endocrinology (Baltimore)* **139,** 2493–2500.

Mitelman, F., Mertens, F., and Johansson, B. (1997). A breakpoint map of recurrent chromosomal rearrangements in human neoplasia. *Nat. Genet.* **15** (Spec. No.), 417–474.

Muchardt, C., and Yaniv, M. (1993). A human homologue of *Saccharomyces cerevisiae* SNF2/SWI2 and Drosophila brm genes potentiates transcriptional activation by the glucocorticoid receptor. *EMBO J.* **12,** 4279–4290.

Murre, C., McCaw, P. S., and Baltimore, D. (1989a). A new DNA binding and dimerization motif in immunoglobulin enhancer binding, daughterless, MyoD, and myc proteins. *Cell (Cambridge, Mass.)* **56,** 777–783.

Murre, C., McCaw, P. S., Vaessin, H., Caudy, M., Jan, L. Y., Jan, Y. N., Cabrera, C. V., Buskin, J. N., Hauschka, S. D., Lassar, A. B., Weintraub, H., and Baltimore, D. (1989b). Interactions between heterologous helix-loop-helix proteins generate complexes that bind specifically to a common DNA sequence. *Cell (Cambridge, Mass.)* **58,** 537–544.

Neuman, E., Ladha, M. H., Lin, N., Upton, T. M., Miller, S. J., DiRenzo, J., Pestell, R. G., Hinds, P. W., Dowdy, S. F., Brown, M., and Ewen, M. E. (1997). Cyclin D1 stimulation of estrogen receptor transcriptional activity independent of cdk4. *Mol. Cell. Biol.* **17,** 5338–5347.

Nolte, R. T., Wisely, G. B., Westin, S., Cobb, J. E., Lambert, M. H., Kurokawa, R., Rosenfeld, M. G., Willson, T. M., Glass, C. K., and Milburn, M. V. (1998). Ligand binding and co-activator assembly of the peroxisome proliferator-activated receptor-gamma. *Nature (London)* **395,** 137–143.

Ogryzko, V. V., Schiltz, R. L., Russanova, V., Howard, B. H., and Nakatani, Y. (1996). The transcriptional coactivators p300 and CBP are histone acetyltransferases. *Cell (Cambridge, Mass.)* **87,** 953–959.

Oñate, S. A., Tsai, S. Y., Tsai, M. J., and O'Malley, B. W. (1995). Sequence and characterization of a coactivator for the steroid hormone receptor superfamily. *Science* **270,** 1354–1357.

Oñate, S. A., Boonyaratanakornkit, V., Spencer, T. E., Tsai, S. Y., Tsai, M. J., Edwards, D. P., and O'Malley, B. W. (1998). The steroid receptor coactivator-1 contains multiple receptor interacting and activation domains that cooperatively enhance the activation function 1 (AF1) and AF2 domains of steroid receptors. *J. Biol. Chem.* **273,** 12101–12108.

Qi, C., Zhu, Y., Pan, J., Yeldandi, A. V., Rao, M. S., Maeda, N., Subbarao, V., Pulikuri, S., Hashimoto, T., and Reddy, J. K. (1999). Mouse steroid receptor coactivator-1 is not essential for peroxisome proliferator-activated receptor alpha-regulated gene expression. *Proc. Natl. Acad. Sci. U.S.A.* **96,** 1585–1590.

Reddy, P. S., and Housman, D. E. (1997). The complex pathology of trinucleotide repeats. *Curr. Opin. Cell Biol.* **9,** 364–372.

Renaud, J. P., Rochel, N., Ruff, M., Vivat, V., Chambon, P., Gronemeyer, H., and Moras, D. (1995). Crystal structure of the RAR-gamma ligand-binding domain bound to all-trans retinoic acid. *Nature (London)* **378,** 681–689.

Seol, W., Choi, H. S., and Moore, D. D. (1995). Isolation of proteins that interact specifically with the retinoid X receptor: Two novel orphan receptors. *Mol. Endocrinol.* **9,** 72–85.

Shiau, A. K., Barstad, D., Loria, P. M., Cheng, L., Kushner, P. J., Agard, D. A., and Greene, G. L. (1998). The structural basis of estrogen receptor/coactivator recognition and the antagonism of this interaction by tamoxifen. *Cell (Cambridge, Mass.)* **95,** 927–937.

Shirazi, S. K., Bober, M. A., and Coetzee, G. A. (1998). Polymorphic exonic CAG microsatellites in the gene amplified in breast cancer (AIB1 gene). *Clin. Genet.* **54,** 102–103.

Smith, C. L., Onate, S. A., Tsai, M. J., and O'Malley, B. W. (1996). CREB binding protein acts synergistically with steroid receptor coactivator-1 to enhance steroid receptor-dependent transcription. *Proc. Natl. Acad. Sci. U.S.A.* **93,** 8884–8888.

Spencer, T. E., Jenster, G., Burcin, M. M., Allis, C. D., Zhou, J., Mizzen, C. A., McKenna,

N. J., Oñate, S. A., Tsai, S. Y., Tsai, M. J., and O'Malley, B. W. (1997). Steroid receptor coactivator-1 is a histone acetyltransferase. *Nature (London)* **389,** 194–198.

Suen, C. S., Berrodin, T. J., Mastroeni, R., Cheskis, B. J., Lyttle, C. R., and Frail, D. E. (1998). A transcriptional coactivator, steroid receptor coactivator-3, selectively augments steroid receptor transcriptional activity. *J. Biol. Chem.* **273,** 27645–27653.

Swanson, H. I., Chan, W. K., and Bradfield, C. A. (1995). DNA binding specificities and pairing rules of the Ah receptor, ARNT, and SIM proteins. *J. Biol. Chem.* **270,** 26292–26302.

Takeshita, A., Yen, P. M., Misiti, S., Cardona, G. R., Liu, Y., and Chin, W. W. (1996). Molecular cloning and properties of a full-length putative thyroid hormone receptor coactivator. *Endocrinology (Baltimore)* **137,** 3594–3597.

Takeshita, A., Cardona, G. R., Koibuchi, N., Suen, C.-S., and Chin, W. W. (1997). TRAM-1, a novel 160-kDa thyroid hormone receptor activator molecule, exhibits distinct properties from steroid receptor coactivator-1. *J. Biol. Chem.* **272,** 27629–27634.

Takeshita, A., Imai, K., Kato, S., Kitano, S., and Hanazawa, S. (1998). 1alpha,25-dehydroxyvitamin D3 synergism toward transforming growth factor-beta1-induced AP-1 transcriptional activity in mouse osteoblastic cells via its nuclear receptor. *J. Biol. Chem.* **273,** 14738–14744.

Tanenbaum, D. M., Wang, Y., Williams, S. P., and Sigler, P. B. (1998). Crystallographic comparison of the estrogen and progesterone receptor's ligand binding domains. *Proc. Natl. Acad. Sci. U.S.A.* **95,** 5998–6003.

Thummel, C. S. (1995). From embryogenesis to metamorphosis: The regulation and function of Drosophila nuclear receptor superfamily members. *Cell (Cambridge, Mass.)* **83,** 871–877.

Torchia, J., Rose, D. W., Inostroza, J., Kamei, Y., Westin, S., Glass, C. K., and Rosenfeld, M. G. (1997). The transcriptional co-activator p/CIP binds CBP and mediates nuclear-receptor function. *Nature (London)* **387,** 677–684.

Vogel, J. J., Heine, M. J. S., Zechel, C., Chambon, P., and Gronemeyer, H. (1996). TIF2, a 160 kDa transcriptional mediator for the ligand-dependent activation function AF-2 of nuclear receptors. *EMBO J.* **15,** 3667–3675.

Voegel, J. J., Heine, M. J., Tini, M., Vivat, V., Chambon, P., and Gronemeyer, H. (1998). The coactivator TIF2 contains three nuclear receptor-binding motifs and mediates transactivation through CBP binding-dependent and -independent pathways. *EMBO J.* **17,** 507–519.

Wade, P. A., Pruss, D., and Wolfe, A. P. (1997). Histone acetylation: Chromatin in action. *Trends Biochem. Sci.* **22,** 128–132.

Wagner, R. L., Apriletti, J. W., West, B. L., Baxter, J. D., and Fletterick, R. J. (1995). A structural role for hormone in the thyroid hormone receptor ligand-binding domain. *Nature (London)* **378,** 690–697.

Westin, S., Kurokawa, R., Nolte, R. T., Wisely, G. B., McInerney, E. M., Rose, D. W., Milburn, M. V., Rosenfeld, M. G., and Glass, C. K. (1998). Interactions controlling the assembly of nuclear-receptor heterodimers and co-activators. *Nature (London)* **395,** 199–202.

Xu, J., Qiu, Y., DeMayo, F. J., Tsai, S. Y., Tsai, M. J., and O'Malley, B. W. (1998). Partial hormone resistance in mice with disruption of the steroid receptor coactivator-1 (SRC-1) gene. *Science* **279,** 1922–1925.

Xu, L., Lavinsky, R. M., Dasen, J. S., Flynn, S. E., McInerney, E. M., Mullen, T. M., Heinzel, T., Szeto, D., Korzus, E., Kurokawa, R., Aggarwal, A. K., Rose, D. W., Glass, C. K., and Rosenfeld, M. G. (1988). Signal-specific co-activator domain requirements for Pit-1 activation. *Nature (London)* **395,** 301–306.

Yanagisawa, J., Yanagi, Y., Masuhiro, Y., Suzawa, M., Watanabe, M., Kashiwagi, K., Toriyabe, T., Kawabata, M., Miyazono, K., and Kato, S. (1999). Convergence of transforming growth factor-beta and vitamin D signaling pathways on SMAD transcriptional coactivators. *Science* **283,** 1317–1321.

Yang, X. J., Ogryzko, V. V., Nishikawa, J., Howard, B. H., and Nakatani, Y. (1996). A p300/CBP-associated factor that competes with the adenoviral oncoprotein E1A. *Nature (London)* **382,** 319–324.

Yao, T. P., Ku, G., Zhou, N., Scully, R., and Livingston, D. M. (1996). The nuclear hormone receptor coactivator SRC-1 is a specific target of p300. *Proc. Natl. Acad. Sci. U.S.A.* **93,** 10626–10631.

Yeh, S., and Chang, C. (1996). Cloning and characterization of a specific coactivator, ARA70, for the androgen receptor in human prostate cells. *Proc. Natl. Acad. Sci. U.S.A.* **93,** 5517–5521.

Yoshizawa, T., Handa, Y., Uematsu, Y., Takeda, S., Sekine, K., Yoshihara, Y., Kawakami, T., Arioka, K., Sato, H., Uchiyama, Y., Masushige, S., Fukamizu, A., Matsumoto, T., and Kato, S. (1997). Mice lacking the vitamin D receptor exhibit impaired bone formation, uterine hypoplasia and growth retardation after weaning. *Nat. Genet.* **16,** 391–396.

Zelzer, E., Wappner, P., and Shilo, B.-Z. (1997). The PAS domain confers target gene specificity of Drosophila bHLH/PAS proteins. *Genes Dev.* **11,** 2065–2079.

Zhu, Y., Qi, C., Calandra, C., Rao, M. S., and Reddy, J. K. (1996). Cloning and identification of mouse steroid receptor coactivator-1 (mSRC-1), as a coactivator of peroxisome proliferator-activated receptor gamma. *Gene Expression* **6,** 185–195.

Zwijsen, R. M., Wientjens, E., Klompmaker, R., van der Sman, J., Bernards, R., and Michalides, R. J. (1997). CDK-independent activation of estrogen receptor by cyclin D1. *Cell (Cambridge, Mass.)* **88,** 405–415.

Zwijsen, R. M. L., Buckle, R. S., Hijmans, E. M., Loomans, C. J. M., and Bernards, R. (1998). Ligand-independent recruitment of steroid receptor coactivators to estrogen receptor by cyclin D1. *Genes Dev.* **12,** 3488–3498.

VITAMINS AND HORMONES, VOL. 58

Thyroid Hormone Receptor, v-ErbA, and Chromatin

ALAN P. WOLFFE, TREVOR N. COLLINGWOOD, QIAO LI, JANET YEE, FYODOR URNOV, AND YUN-BO SHI

Laboratory of Molecular Embryology, National Institute of Child Health and Human Development, Bethesda, Maryland 20892–5431

The thyroid hormone receptor and the highly related viral oncoprotein v-erbA are found exclusively in the nucleus as stable constituents of chromatin. Unlike most transcriptional regulators, the thyroid hormone receptor binds with comparable affinity to naked and nucleosomal DNA. *In vitro* reconstitution experiments and *in vivo* genomic footprinting have delineated the chromatin structural features that facilitate association with the receptor. Chromatin bound thyroid hormone receptor and v-erbA generate Dnase I hypersensitive sites independent of ligand. The unliganded thyroid hormone receptor and v-erbA associate with a corepressor complex containing NCoR, SIN3, and histone deacetylase. The

0083-6729/00 $30.00

enzymatic activity of the deacetylase and a chromatin environment are essential for the dominant repression of transcription by both the unliganded thyroid hormone receptor and v-erbA. In the presence of ligand, the thyroid hormone receptor undergoes a conformational change that weakens interactions with the corepressor complex while facilitating the recruitment of transcriptional coactivators such as p300 and PCAF possessing histone acetyltransferase activity. The ligand-bound thyroid hormone receptor directs chromatin disruption events in addition to histone acetylation. Thus, the thyroid hormone receptor and v-erbA make very effective use of their stable association with chromatin and their capacity to alter the chromatin environment as a major component of the transcription regulation process. This system provides an exceptionally useful paradigm for investigating the structural and functional consequences of targeted chromatin modification.

I. Introduction

The thyroid hormone receptor (TR) has the capacity to regulate gene activity through alternatively silencing or activating transcription dependent on the absence or presence of thyroid hormone (T_3) (Evans, 1988; Glass *et al.*, 1989; Graupner *et al.*, 1989; Brent *et al.*, 1993). TR exerts important functions in development (Wikstrom *et al.*, 1998; Fraichard *et al.*, 1997; Forrest *et al.*, 1996a,b) and mutations in TR can lead to the clinical disorder of thyroid hormone resistance (Chatterjee, 1997; Tenbaum and Baniahmad, 1997). An understanding of the molecular mechanisms through which the TR controls transcription is likely to provide insight into regulatory events in development and suggest new avenues for therapeutic intervention.

Interest in TR regulated developmental pathways is enhanced by the existence of a highly related viral oncoprotein, v-erbA. The avian erythroblastosis virus (AEV) induces fatal erythroleukemia dependent on the activity of two oncogenes *v-erbA* and *v-erbB*. Both v-erbA and v-erbB subvert normal cell regulatory pathways to promote proliferation rather than differentiation. v-ErbB is a mutant form of the epidermal growth factor receptor that promotes cell proliferation, and v-erbA is a mutant form of TR that inhibits differentiation (reviewed by Beug *et al.*, 1994; Gandrillon *et al.*, 1995). v-ErbA inhibits the differentiation of erythroblast progenitors by targeting the repression of erythroid-specific genes (Zenke *et al.*, 1990; Disela *et al.*, 1991). In contrast to TR, v-erbA does not bind T_3 and cannot activate transcription (Munoz *et al.*, 1988;

Zenke *et al.,* 1990). The capacity for diverse variant forms of nuclear receptors to inhibit differentiation appears to have a contributory role in other cancers (Lin *et al.,* 1998; Grignani *et al.,* 1998).

In this review we summarize experiments that link the regulation of gene activity by TR and v-erbA to the enzymatic modification of chromatin. We begin with discussion of classical experiments from the laboratory of Dr. H. H. Samuels that established that the thyroid hormone receptor was an integral component of chromatin. We describe our own work using *in vitro* and *in vivo* chromatin assembly systems that confirm and extend these early observations. The explosion of information concerning the definition of transcriptional coactivators and corepressors as mediators of transcriptional control by TR is simplified by the recognition that these diverse enzymatic complexes respectively acetylate or deacetylate chromatin. A working model for gene regulation by TR is summarized in Fig. 1. The significance of histone modification for

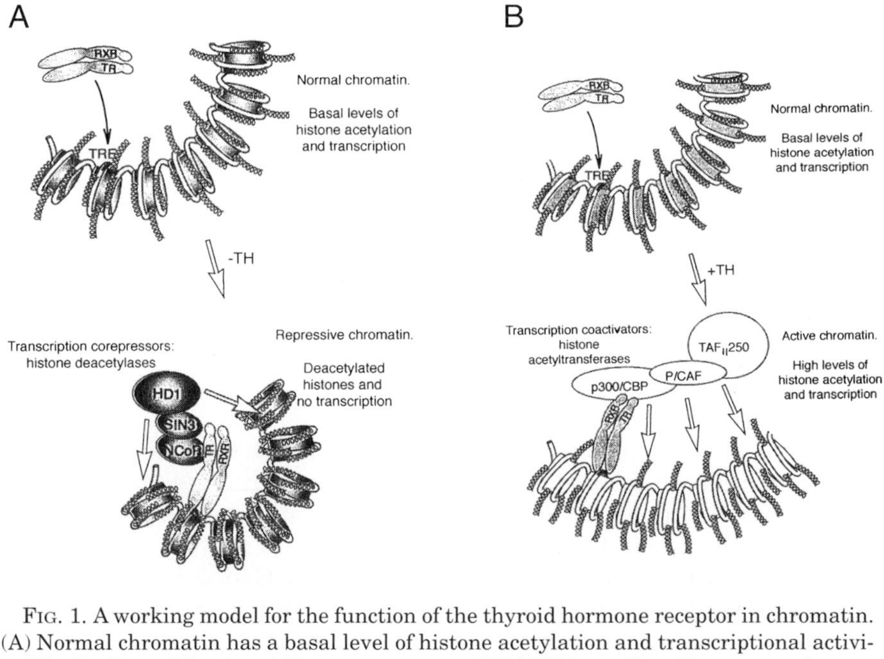

FIG. 1. A working model for the function of the thyroid hormone receptor in chromatin. (A) Normal chromatin has a basal level of histone acetylation and transcriptional activity. The binding of the thyroid hormone receptor (TR/RXR) to a thyroid response element (TRE) on a positioned nucleosome in the absence of thyroid hormone (TH) leads to the recruitment of a corepressor complex (NCoR, Sin3, HD1) to direct histone deacetylation and transcriptional repression. (B) The binding of the thyroid hormone receptor to a TRE in the presence of ligand leads to the recruitment of the coactivator complex (p300/CBP, p/CAF, TAF$_{II}$250) that directs histone acetylation and facilitates transcription.

structure and function of chromatin in the vicinity of the bound TR is discussed later in the text. We speculate on how these pathways of gene control utilizing alterations to the chromatin infrastructure as a driving force might contribute to the execution of developmental programs by TR, how mutations in TR might contribute to human disease, and how v-erbA contributes to the transformed phenotype.

II. Thyroid Hormone Receptor as a Stable Component of Chromatin

If a transcriptional regulator is to regulate gene expression through the targeted modification of chromatin, it has to be able to bind to DNA in a chromatin context. There are *in vivo* and *in vitro* approaches to determining whether a regulator can cope with a chromatin environment. Endogenous chromosomes provide the most natural *in vivo* substrate to test for the binding of a particular transcription factor, however, chromosomal sites are difficult to manipulate. Alterations to endogenous DNA sequences require transgenic approaches and homologous recombination to achieve a true chromosomal environment with the appropriate developmental history. A solution to this problem is to introduce plasmid DNA into cells under conditions such that it is assembled into chromatin (Reeves *et al.,* 1985); however, a common problem with this approach is that large amounts of DNA either do not enter the cells or are degraded in the cytoplasm. In contrast, plasmid DNA microinjected into cell nuclei is efficiently assembled into minichromosomes, thereby providing a surrogate and easily manipulable chromatin substrate (Buschhausen *et al.,* 1987; Kass *et al.,* 1997). TR itself is relatively easy to detect within chromatin using radiolabeled T_3, using antibodies or by genomic footprinting.

In vitro studies of transcription factor association with recognition elements usually make use of naked DNA. In general, reconstitution of these recognition elements into nucleosomes using purified histones *in vitro* reduces the binding affinity of these factors by 100- to 1000-fold (Hayes and Wolffe, 1992; Blomquist *et al.,* 1996; Godde *et al.,* 1995; reviewed by Wolffe, 1998). However certain transcription factors including nuclear receptors can bind efficiently to nucleosomal DNA despite the wrapping of their recognition elements on the surface of the histones (Lefebvre *et al.,* 1998; Perlmann and Wrange, 1988; Pina *et al.,* 1990a,b; Wong *et al.,* 1995, 1997a). This association is dependent on the precise orientation and position of the DNA sequences that comprise the recognition element for the receptor within the nucleosome (Li and

Wrange, 1993, 1995; Wong *et al.,* 1997b). Thus, the architecture of chromatin itself provides an important determinant of hormonal responsiveness and gene control.

A. *In Vivo* Analysis of the Thyroid Hormone Receptor in the Chromatin of Mammalian Cell Nuclei

By 1982, Samuels and colleagues had established that the thyroid hormone receptor could form a stable complex with the nucleosome (Samuels *et al.,* 1982). Their strategy (Fig. 2) used controlled micrococcal nuclease digestion of chromatin isolated from GH1 cells, a rat somatotrophic cell line and the presence of TR in the chromatin digests was assayed using radiolabeled T_3 (Samuels *et al.,* 1980; Fig. 1A). Micrococcal nuclease preferentially digests the linker DNA between nucleosome cores and prolonged digestion leads to the accumulation of a stable kinetic intermediate containing only the core histones and 146 bp of DNA. This nucleosome core sediments in low-salt isokinetic sucrose gradients at 11.5S, while the TR itself sediments at a much smaller size of 3.5–3.8S (Samuels *et al.,* 1980). The chromatin digests released from GH1 cell nuclei revealed that 80% of the TR sedimented at 6.5S and 10% at 12.5S (Fig. 2C). These values exclude the TR associated with the 90% of high molecular weight chromatin that pelleted in the sucrose gradients. Samuels interpreted this result to suggest that the 6.5S form of TR was bound to residual linker DNA and that the 12.5S form of TR was bound to nucleosome cores (Fig. 2B). Assuming one TR per nucleosome, the subset of nucleosomes associated with receptor is 0.006% of total or approximately 2000 per cell nucleus. Because there are approximately 3×10^7 nucleosomes in each mammalian cell nucleus, this implies that considerable specificity exists in the recognition of nucleosomal DNA by TR.

In subsequent experiments these investigators found that the excision of the 12.5S form of TR-nucleosome core complex from chromatin by micrococcal nuclease occurred more rapidly than that of the 6.5S form of receptor bound to DNA or of bulk nucleosome cores. This result was interpreted to suggest that TR bound to nucleosomes in the presence of thyroid hormone creates a chromatin environment that is preferentially sensitive to nuclease digestion. It is now recognized that ligand-bound TR has the capacity to disrupt nucleosomal arrays, rendering them more accessible to micrococcal nuclease (Wong *et al.,* 1995, 1997a). Extensive Dnase I digestion converted the 6.5S form of TR to 3.8S, consistent with the presence of DNA complexed with TR (Perlman *et al.,* 1982). These authors further suggested that the 6.5S form of TR

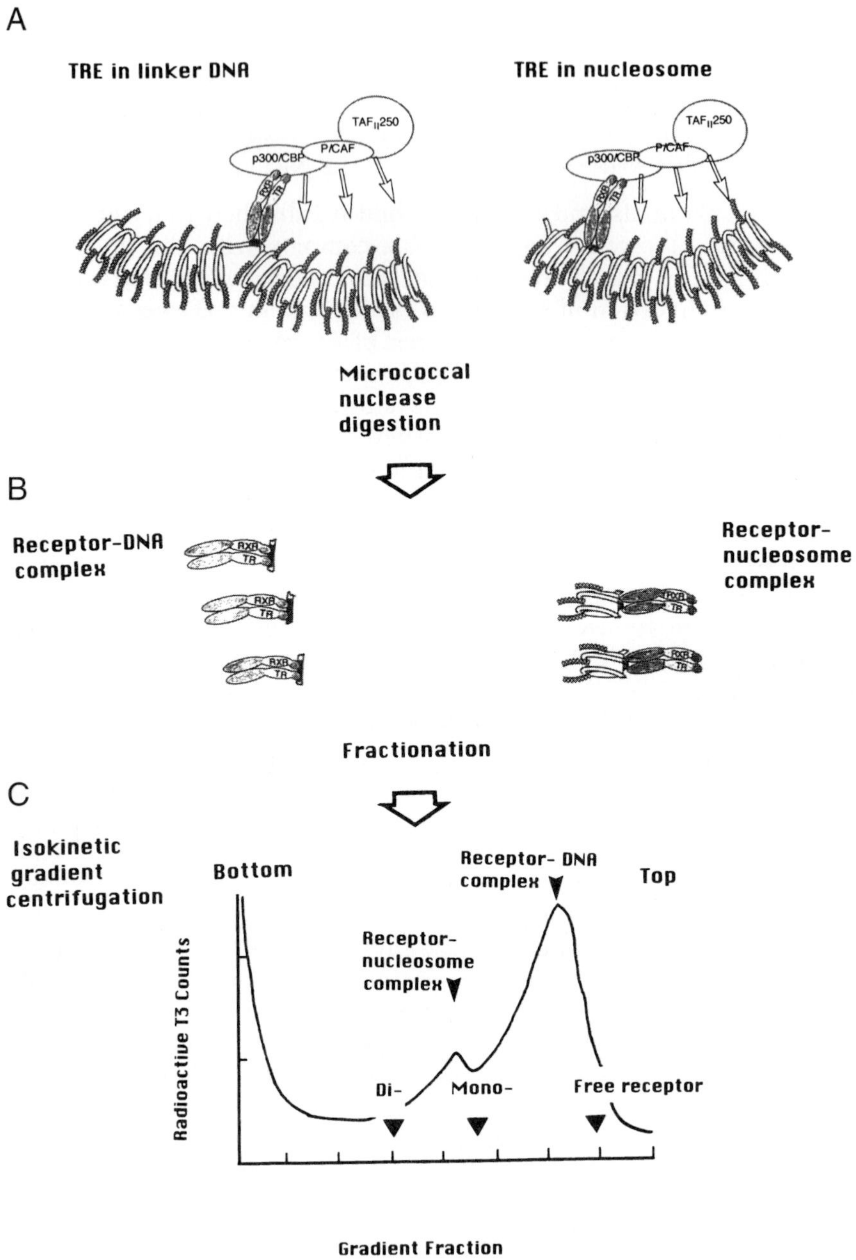
A
TRE in linker DNA
TRE in nucleosome
TAF$_{II}$250
P/CAF
p300/CBP
RXR
TR
Micrococcal nuclease digestion
B
Receptor-DNA complex
Receptor-nucleosome complex
Fractionation
C
Isokinetic gradient centrifugation
Bottom
Top
Receptor- DNA complex
Receptor-nucleosome complex
Radioactive T3 Counts
Di-
Mono-
Free receptor
Gradient Fraction

existed in chromatin as a multimer (or heterodimer) in association with 35–40 bp of DNA. These observations, having since been confirmed and extended in detail, provide independent and compelling evidence for considering the TR as an architectural component of chromatin.

The stability of TR within the chromatin of GH1 cells was assayed by labeling of TR with dense amino acids (Raaka and Samuels, 1981). Newly synthesized proteins are of higher density than preexisting proteins and can be separated using gradient centrifugation techniques. The use of radioactive T_3 allows the position of normal and dense receptors to be independently assayed in sucrose gradients. By quantitating the abundance of normal receptor at different times following the addition of dense amino acids to the GH1 cells, the receptor half-life can be assayed. The accumulation of dense receptor allows the rate of receptor synthesis to be determined. TR has a half-life of 4.5 h in GH1 cells cultured in the absence of T_3, and of 3 h in the presence of hormone. Importantly, the vast majority (95%) of the receptor decays with an identical half-life, indicating that no appreciable nuclear associated pools exist that have distinct stabilities (Fig. 3). Moreover the presence or absence of T_3 does not influence the rate of TR decay. This suggests that the TR is exchanged from chromatin with equivalent efficiency whether it is engaged in repressing or activating transcription. In addition, more than 1700 molecules of TR were synthesized per hour in each cell to replace those being degraded. These kinetics imply that more than 95% of the receptor will be replaced within chromatin in 24 h. Thus the receptor needs to be continually replaced in the chromo-

---

FIG. 2. The thyroid hormone receptor functions as an integral component of the chromatin infrastructure. (A) Depending on nucleosome positioning, a TRE could be within the linker DNA between two adjacent nucleosome cores (TRE in linker DNA) or it could be positioned within a nucleosome core itself (TRE in nucleosome). In the presence of thyroid hormone, TR/RXR will recruit transcriptional coactivators that will modify nucleosomal arrays to make them more accessible to nuclease digestion. (B) Controlled micrococcal nuclease digestion of chromatin leads to the accumulation of mononucleosomes and dinucleosomes, together with the TR as a heterodimer with RXR bound either to fragments of linker DNA (receptor–DNA complex) or to nucleosome cores (receptor–nucleosome complex). (C) Isokinetic sucrose gradient centrifugation allows the fractionation of chromatin fragments from the micrococcal nuclease digestion. Free receptor has a sedimentation coefficient of 3.8S, mononucleosomes of 11.5S, and dinucleosomes of 16S. Their peaks of distribution in the fractionated sucrose gradient are indicated by the closed triangles. The receptor–DNA complex has a sedimentation coefficient of 6.5S and the receptor–nucleosome complex of 12.5S. Their distributions within a representative fractionated sucrose gradient are shown (see Samuels *et al.*, 1982, for examples of experimental data). The peaks of distribution for the receptor–DNA complex and the receptor–nucleosome complex are indicated by arrowheads.

some independent of cell division cycles (Fig. 3). As we will discuss, the organization of the thyroid hormone receptor recognition elements in chromatin facilitates this continual access (Wong *et al.,* 1995). However these observations also imply that the receptor is also able to exchange from coactivator and corepressor complexes without substantial impediment.

Note that the thyroid hormone receptor is also likely to occupy nuclear compartments in addition to those immediately engaged in transcriptional activation and repression. For example, morphologic and extraction criteria indicate that a substantial fraction of the glucocorticoid receptor associates with the nuclear matrix even after chromatin

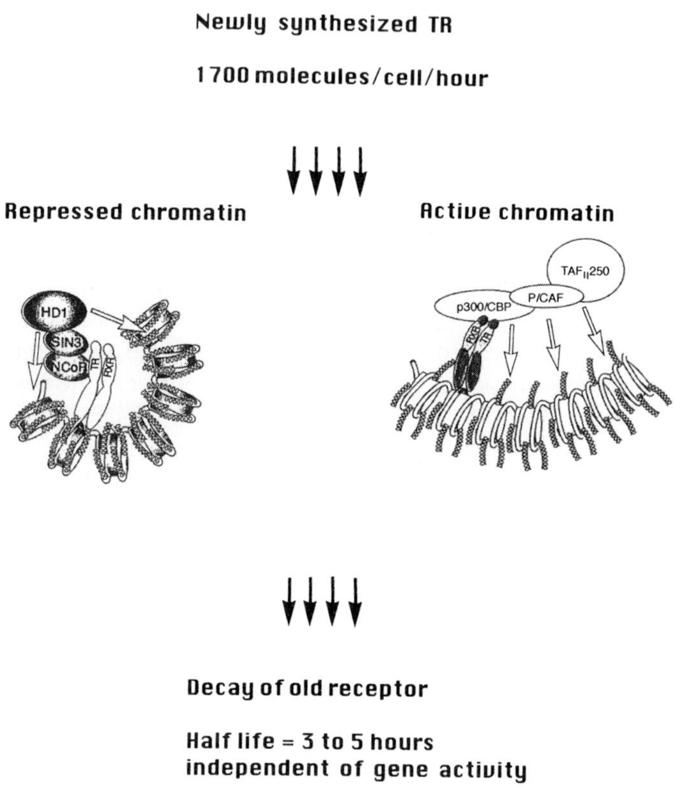

FIG. 3. The thyroid hormone receptor is subject to continual turnover within a single cell cycle. TR is synthesized at a rate of 1700 molecules per cell per hour. The TR is then engaged in repression or activation of transcription as indicated. It is also probable that other pools of TR exist in the nucleus as sites of storage or degradation. The TR decays with a half-life of 3–5 h independent of gene activity (Raaka and Samuels, 1981).

has been extracted (Tang and De Franco, 1996; Yang *et al.,* 1996). The nuclear matrix might of course represent the large coactivator and corepressor complexes. Cytologically the glucocorticoid receptor is visualized as present in discrete foci within the nucleus (van Steensel *et al.,* 1995). The function of either the foci or the nuclear matrix association is unresolved. The foci might represent sites of glucocorticoid receptor-mediated transcription, association with coactivators or corepressors. Alternatively, they might represent storage sites or sites of degradation. These are important future issues to examine with the thyroid hormone receptor.

B. *In Vivo* Analysis of the Thyroid Hormone Receptor in a Minichromosome

To understand the significance of the capacity of TR to associate with chromatin it is important to use an experimentally manipulable system. The microinjection of DNA into *Xenopus* oocyte nuclei leads to the assembly of chromatin that can be alternately transcriptionally repressed or active (Almouzni and Wolffe, 1993; Landsberger *et al.,* 1995; Landsberger and Wolffe, 1995; Wong *et al.,* 1995). These active or repressed minichromosomes provide a powerful tool to dissect transcriptional regulation in an *in vivo* context.

In vivo chromatin assembly normally occurs during S phase; this is when DNA and the histones are synthesized and when the molecular chaperones that direct the deposition of histones onto nascent DNA are maximally effective (Kaufmann and Botchan, 1994; Wolffe, 1998). Microinjection of single-stranded DNA into a *Xenopus* oocyte nucleus leads to rapid second strand synthesis, an event to which chromatin assembly is coupled (Almouzni and Wolffe, 1993). This rapid and efficient chromatin assembly pathway leads to the repression of basal transcription in the fully assembled minichromosome. Repression occurs because the molecular mechanisms driving histone deposition outcompete those that direct the association of the basal transcriptional machinery with the promoter. The second pathway of chromatin assembly occurs when double-stranded DNA is microinjected into oocyte nuclei. This assembly pathway is not coupled to replication and resembles the consequence of transiently transfecting or microinjecting plasmid DNA molecules into mammalian tissue culture cells. Under these conditions chromatin assembly will still occur on the fraction of DNA that is taken up into the nucleus, but does so with much slower kinetics in a process occupying at least 4 h in the oocyte (which has large stores of histones) and as long as 24 h in a mammalian tissue culture cell (Reeves *et al.,* 1985; Almouzni and Wolffe, 1993; Wong *et al.,* 1998). This provides plenty of time for the basal transcriptional machinery to gain access to promoter elements and to assemble an active minichromosome.

The *Xenopus laevis* TRβA promoter is transcriptionally repressed when assembled into chromatin in a replicative pathway following injection of single-stranded templates into oocyte nuclei, whereas it is active when double-stranded templates are injected (Fig. 4). In both instances the final density of nucleosomes is equivalent, only the pathway of chromatin assembly differs (Wong *et al.,* 1995). This illustrates the potential importance of chromatin context and assembly pathway on transcriptional control. The *Xenopus* oocyte lacks adequate TR to regulate transcription in a T_3 responsive manner. However, microinjection of TR mRNA into the oocyte cytoplasm leads to TR synthesis and the acquisition of T_3 responsive transcription. Hormonal responsiveness is greatly enhanced in the presence of the retinoic X receptor RXR, which indicates that the primary functional form of TR in the oocyte is as a heterodimer with RXR (Wong and Shi, 1995). In the absence of T_3, TR-RXR will repress transcription from the active minichromosome and will further reduce transcription from the already repressed minichromosome assembled during replication (Wong *et al.,* 1995). Consistent with these functional studies, *in vivo* Dnase I footprinting indicates that TR-RXR will bind to a thyroid response element (TRE) in the TRβA gene independent of the prior assembly of this element into chromatin and independent of the presence of T_3 (Wong *et al.,* 1995; see Fig. 5).

Several unanticipated consequences result from the TR-RXR binding to chromatin within the minichromosomal context. In the absence of T_3 the underlying nucleosomal array with which TR-RXR is associated remains very regular upon binding of the receptors with no increase in accessibility to micrococcal nuclease (Wong *et al.,* 1997a). This regularity is surprising because the TR-RXR generates a major site of preferential cleavage by Dnase I (Wong *et al.,* 1997a,b; Fig. 6, lanes 8–10). We interpret these results to indicate that the unliganded TR-RXR binds to the minichromosome and functions to repress transcription within a regular nucleosomal infrastructure. With respect to the Dnase I hypersensitive site, it is clear that neither disruption of a nucleosomal array nor transcriptional activity are necessary to generate such a site. We speculate that TR-RXR may recruit Dnase I through protein–protein interactions much as transcriptional coactivators and corepressors might be recruited. Dnase I hypersensitive sites are therefore useful markers for chromatin-bound TR and also potentially for v-erbA bound in chromatin. Indeed, several transcriptional silencers that contain v-erbA sites are hypersensitive to Dnase I *in vivo* (Baniahmad *et al.,* 1990) consistent with stable association of the v-erbA protein.

On the addition of T_3, transcriptional repression is relieved on a minichromosome generated under replication-coupled chromatin as-

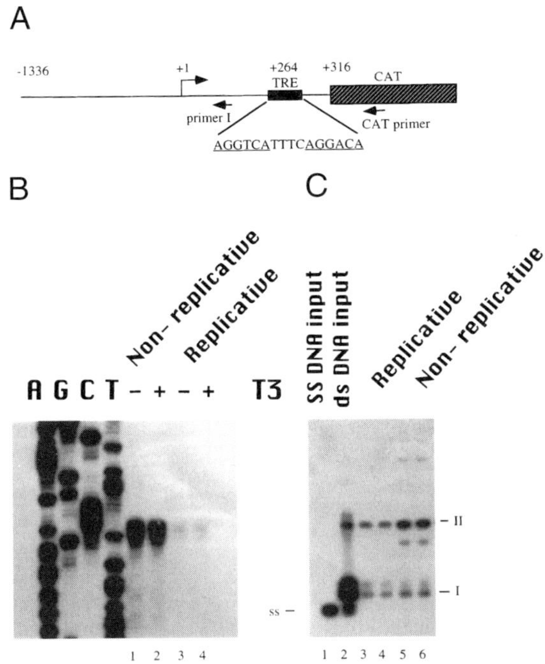

FIG. 4. Replication coupled assembly of a repressed minichromosome and the assembly of a transcriptionally active minichromosome on nonreplicating double-stranded DNA. (A) Structure of the TRβA promoter construct. The transcriptional start site in oocytes is designated as +1 (see part B). The TRE identified previously (Ranjan *et al.,* 1994) is located at +264 (underlined). Also indicated are the relative positions of primer I (+107) and CAT primer (+397), which were used for primer extension analysis. (B) Replication-coupled chromatin assembly effectively represses the transcription from the TRβA promoter. Forty oocytes were injected into the nuclei with double-stranded (ds) (2.3 ng/oocyte) or single-stranded (ss) (1.15 ng/oocyte) pTRβA templates. The injected oocytes were then divided into two groups with equal number of oocytes and incubated at 18°C overnight with (+) or without (−) 50 n*M* of T_3. The transcription from the TRβA promoter was analyzed by primer extension using primer I. DNA sequencing markers using the TRβA promoter and end-labeled primer I indicate +1 as the start site. (C) Recovery of DNA after either ssDNA or dsDNA are injected into oocytes. DNAs recovered from the groups of oocytes in part B were resolved on a 1% agarose gel, blotted, and hybridized with random primer-labeled pTRβA. Lane 1, uninjected ssDNA; lane 2, uninjected dsDNA. Double-stranded form I (closed circle supercoiled), form II (nicked circular), and ssDNA (ss) are indicated.

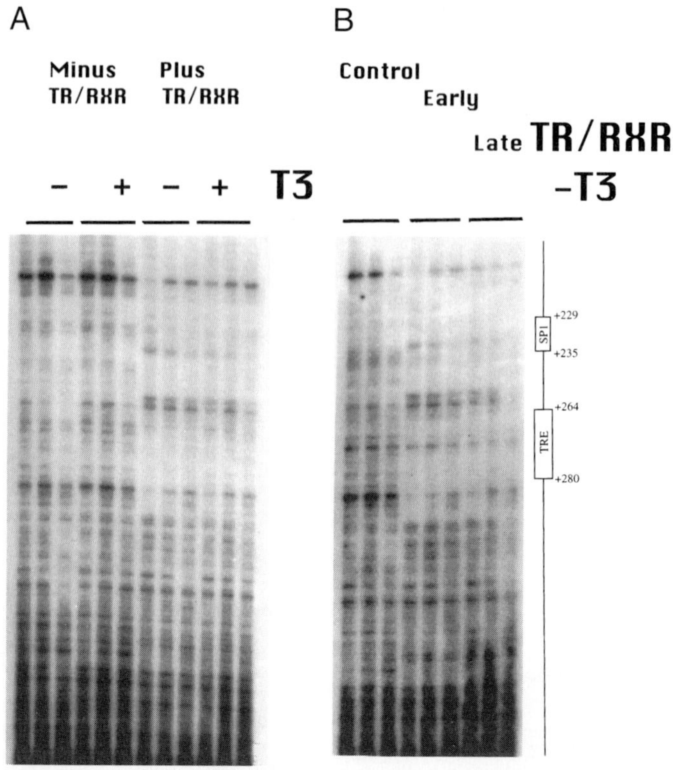

FIG. 5. TRβ/RXRα heterodimer can bind to the TRE in a T_3-independent manner and even when the TRE is preassembled into chromatin *in vivo.* (A) TRβ/RXRα heterodimer binds to the TRE independently of T_3. Groups of 30 oocytes were injected with (+, lanes 7–12) or without (−, lanes 1–6) TRβ/RXRα mRNAs (2.7 ng/oocyte) 6 h before single-stranded pTRβA was injected and treated with (+, lanes 4–6, 10–12) or without (−, lanes 1–3, 7–9) T_3 as indicated. After overnight incubation, the oocytes were collected and processed for PCR-mediated Dnase I footprinting analysis with increasing amounts of Dnase I (400, 600, and 800 U/mL, respectively). The diagram at right of part B illustrates the position of the TRE in TRβA promoter. Note that the protection of the region further upstream (top of the gel) is also dependent on the presence of the TRβ/RXRα. This region contains binding sites for SP1. (B) TRβ/RXRα heterodimer binds to the TRE preassembled into chromatin *in vivo.* Groups of 30 oocytes were injected with mRNAs (lanes 4–6) so that the receptor was available early in the chromatin assembly process or single-stranded pTRβA first (lanes 7–9) followed by the injection of single-stranded pTRβA or mRNAs, respectively, 6 h later. Thus the receptor was only available late in the chromatin assembly process. Control indicates no mRNA injection (lanes 1–3). Dnase I footprinting was processed as in part A. Note that the region upstream of the TRE was also protected in a TRβ/RXRα heterodimer-dependent manner under both injection conditions.

sembly conditions (injection of single-stranded template). This relief of repression leads to an apparent activation of transcription because the final transcriptional levels are more than 200 times those measured under repressed conditions (Wong *et al.,* 1995). However, for the TRβA gene little true activation occurs because the level of transcriptional activity in the presence of TR/RXR and T_3 only approximates to that seen on chromatin templates assembled in a nonreplicative pathway (injection of double-stranded template), the extended dynamic range of regulation being due to a lower level of repressed transcription. When chromatin assembly occurs through a nonreplicative pathway the basal transcriptional machinery functions as if associated with naked DNA in the absence of TR-RXR. Thus the TR-RXR might regulate gene activity entirely through the establishment of a repressed transcriptional state.

From a teleological perspective, it is useful to consider the physiologic

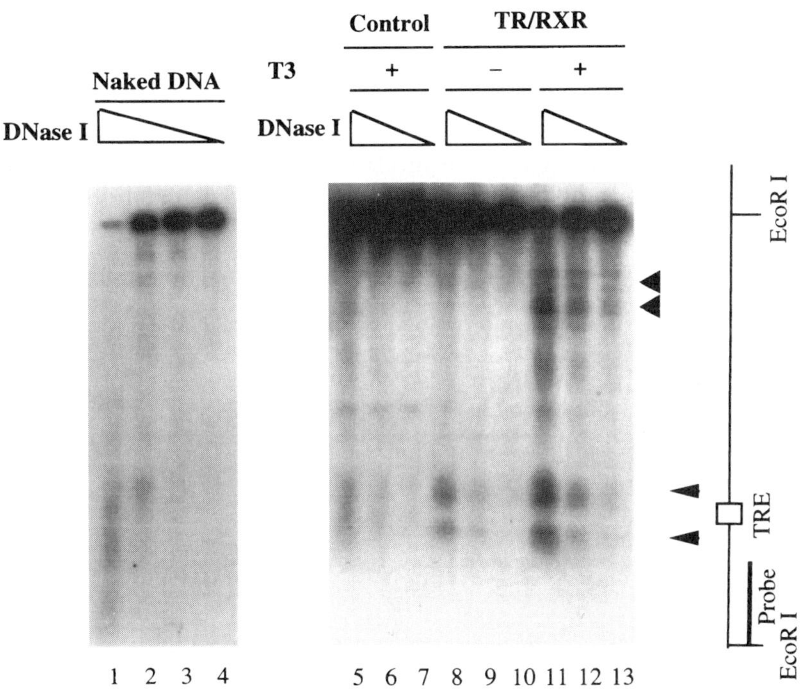

FIG. 6. TRβ/RXRα generates a Dnase I hypersensitive site *in vivo* in the presence or absence of hormone. Dnase I hypersensitivity over the promoter only occurs in the presence of hormone. Horizontal arrows indicate the Dnase I hypersensitivity. The triangles indicate Dnase I cleavage over the promoter in the presence of hormone-bound receptor.

rationale behind unliganded receptor occupancy of target genes and concomitant repression. Unliganded TR has been implicated in maintenance of the proliferative state during vertebrate ontogeny (Bauer *et al.*, 1998), while an important consequence of induction by T_3 is the initiation of cell cycle arrest and differentiation (Bauer *et al.*, 1998; Brown *et al.*, 1995; Tata, 1996). The transient immaturity of chromatin following passage of the DNA replication fork (Seale, 1975, 1978) offers the opportunity for spurious activation of genes by the basal transcription machinery. Thus, the persistent ligand-independent interaction of TR with target genes within chromatin may represent a mechanism by which inappropriate gene activation during postreplicative chromatin maturation is suppressed.

The TR-RXR remains bound to the TRE in chromatin following transcriptional activation in the presence of hormone but the underlying nucleosomal infrastructure becomes much more accessible to micrococcal nuclease (Wong *et al.*, 1997a,b). The TRE itself is hypersensitive to Dnase I digestion and additional sites of Dnase I cleavage are generated in the vicinity of the promoter on addition of hormone (Fig. 6, lanes 11–13). The increased accessibility of chromatin to nucleases in the presence of ligand bound TR is consistent with the earlier observations of Samuels *et al.* (1982). The important point is that the TRE remains occupied with TR-RXR when chromatin structure is disrupted and the TRβA gene becomes active (Wong *et al.*, 1995). The continual presence of TR-RXR offers the opportunity to down-regulate transcription when hormone concentrations become limiting.

These experiments in *Xenopus* oocytes indicate that TR-RXR is stably bound in chromatin both when transcription is active and under repressive conditions. This stable association allows TR-RXR to provide a focal point for the recruitment of transcriptional coactivators and corepressors (Fig. 1). *In vitro* reconstruction of these regulatory nucleoprotein structures allow the determinants of TR-RXR association with chromatin to be defined.

C. *In Vitro* Analysis of TR-RXR Association with Nucleosomal DNA

Nucleosome positioning occurs when a particular DNA sequence is organized in a nonrandom manner when wrapped around the core histones. Within a translationally positioned nucleosome, histone–DNA contacts begin and end at a particular sequence. In the rotational positioning of DNA in a nucleosome, a particular major or minor groove of the double helix is facing out toward solution, while another is facing the histones (Wolffe, 1998). Several acutely inducible promoters are assembled into highly organized chromatin structures that include positioned nucleosomes. These include the *Saccharomyces cerevisiae PHO5* promoter (Almer and Horz, 1986; Almer *et al.,* 1986), the mouse mammary tumor virus long terminal repeat (MMTV LTR; Richard-Foy and Hager, 1987), and the human immunodeficiency virus long terminal repeat (HIV-1 LTR; Verdin *et al.,* 1993). Intrinsic features of DNA sequence recognized by the histones (Wolffe and Drew, 1989; Hayes *et al.,* 1990; Godde and Wolffe, 1996) and histone interactions with sequence-specific DNA binding proteins (Roth *et al.,* 1990; Edmondson *et al.,* 1996) can all contribute to the precise organization of chromatin. The translational positioning of nucleosomes (Truss *et al.,* 1995; Chavez and Beato, 1997) and the rotational positioning of DNA on the surface of the histones (Perlmann and Wrange, 1988; Pina *et al.,* 1990a,b; Archer *et al.,* 1991) have been suggested as important contributory factors in the regulation of MMTV LTR transcription by the glucocorticoid receptor. *In vitro* experiments have shown that both the rotational positioning and translational positioning of the glucocorticoid receptor recognition element within a nucleosome can influence the binding of the receptor to chromatin (Li and Wrange, 1993, 1995).

We have examined whether the thyroid receptor recognition element (TRE) present in the *Xenopus* TRβA promoter would be positioned within nucleosomes assembled using this DNA. We find that the DNA sequences within the first 500 bp of the TRβA gene have the capacity to direct the translational positioning of histone octamers (Wong *et al.,* 1997b; Guschin *et al.,* 1998). The type of thyroid receptor recognition element present in the TRβA promoter is a direct repeat separated by 4 bp (Glass, 1994). The rotational position adopted by the TRE on the surface of the nucleosome assembled on the wild-type TRβA gene appears optimal for continued access of the TR/RXR within chromatin (Fig. 7). The key contacts made by the receptor with the TRE through the major groove are predicted to remain accessible from solution. The TR-RXR heterodimer is able to bind to the TRE present in the TRβA gene in a nucleosome without significant impediment (Fig. 8). The importance of rotational positioning for the continued capacity of TR-RXR to bind to nucleosomal DNA is shown by a small 3-bp alteration in placement of the TRE, leading to an approximate 102°—change in orientation with respect to the histone surface. This rotation leads to a reduction in the capacity of TR/RXR to bind to nucleosomal DNA *in vitro,* to a reduction in the capacity of TR/RXR to bind to polynucleosomal arrays *in vivo,* and to a reduction in the capacity of TR/RXR to activate transcription (Wong *et al.,* 1997b). In contrast, the manipulation of translational position of

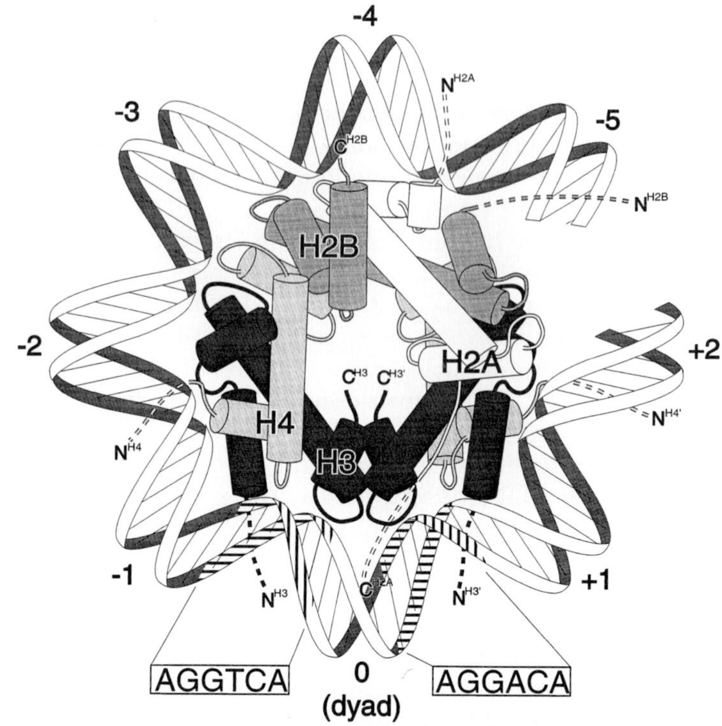

FIG. 7. A model for the nucleosome assembled on the TRβA gene that contains the TRE. The TRE is indicated by the hatched DNA and the boxed sequences. The dyad position in the nucleosome is indicated as are integral turns of DNA away from the dyad axis (negative to 5′, positive to 3′). The histones are shown in the cartoon (see Arents *et al.*, 1991; Arents and Moudrianakis, 1993; Pruss *et al.*, 1995, for details of histone positions).

nucleosomes over the TRβA promoter is without major consequences for TR/RXR association. Thus considerable flexibility might exist in terms of exactly where in the nucleosome the TRE is positioned for transcriptional regulation as long as the TRE is orientated on the histone surface with the appropriate rotational organization. An important goal for future research is to extend our observations to other TR response elements and to investigate the extent to which their rotational positioning on the surface of the octamer contributes to establishing the receptor as a stable component of chromatin.

D. HISTONE H1 AND TR-RXR BINDING TO CHROMATIN

In a natural chromatin context the TR-RXR not only has to contend with the wrapping of DNA around the core histones, but also with the structural consequences of incorporating linker histones into nucleosomes. Oocytes have a specialized form of linker histone known as histone B4 that organizes nucleosomal DNA in a comparable manner to histone H1 (Dimitrov *et al.,* 1994). Histone B4 binds to nucleosomal DNA with significantly lower affinity than histone H1 and might be expected to assemble a more permissive chromatin environment

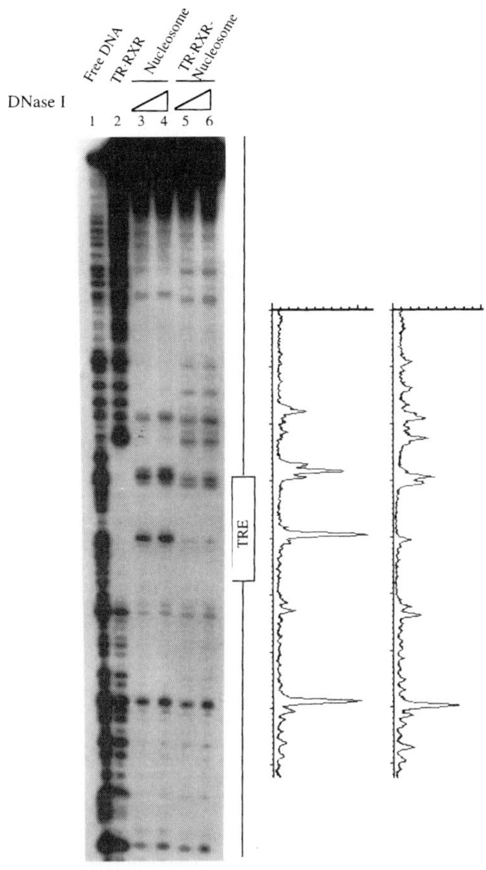

FIG. 8. Dnase I footprinting reveals that the TR/RXR heterodimer binds to a rotationally positioned nucleosome *in vitro.* TR/RXR was bound to a nucleosome treated with Dnase I and the complexes resolved on a nondenaturing polyacrylamide gel. The complexes correspond to nucleosome (lanes 3, 4) and TR/RXR-nucleosome (lanes 5, 6) were then analyzed with a 6% sequencing gel. As controls, free probes incubated either with 2 μL of control oocyte extract (lane 1) or with 2 μL of oocyte extract with TR/RXR (lane 2) in 25 μL of binding reaction were digested with Dnase I and analyzed on the same gel. PhosphorImager scanning of lanes 4 and 6 demonstrates strong protection of the TRE region by TR/RXR in the nucleosome.

(Nightingale *et al.,* 1996; Ura *et al.,* 1996). Most of our functional studies use minichromosomes assembled in oocytes. We have examined the *in vitro* and *in vivo* consequences for TR-RXR association of reconstituting nucleosomes containing the TRE with histone H1.

Incorporation of histone H1 into the TRβA nucleosome that contains the TRE leads to an asymmetric protection of DNA sequences flanking the nucleosome core from micrococcal nuclease digestion (Wong *et al.,* 1997b; Guschin *et al.,* 1998). A comparable asymmetric protection of DNA sequence occurs when histone H1 is incorporated into the adjacent nucleosome. These results are similar to those obtained with other defined DNA sequences (Hayes and Wolffe, 1993; Ura *et al.,* 1995) leading to the suggestion that an asymmetric association of linker histones within the nucleosome (Pruss *et al.,* 1996; Hayes, 1996) may be a common feature of nucleosomal architecture (Pruss *et al.,* 1995). Incorporation of histone H1 leads to no significant change in either hydroxyl radical or Dnase I cleavage of nucleosomal DNA (Wong *et al.,* 1997b; Guschin *et al.,* 1998). This is also consistent with a model for the nucleosome in which linker histones occupy a site internal to the turns of DNA (Pruss *et al.,* 1995). The TRE remains accessible to TR-RXR in nucleosomes containing H1 and there are no functional consequences for TRβA promoter activity. *In vivo* ablation experiments (Bouvet *et al.,* 1994; Steinbach *et al.,* 1997) and genetic knockouts of histone H1 (Shen and Gorovsky, 1996) indicate that histone H1 has a highly selective role in directing transcriptional repression. Thus the failure of histone H1 to influence TRβA gene expression has substantial precedent.

III. Transcriptional Coactivators and Corepressors for TR-RXR and v-ErbA That Modify Chromatin and Regulate Transcription

Biochemical approaches using the ligand binding domains of nuclear receptors to screen for interacting factors have identified a number of interesting proteins (reviewed by Torchia *et al.,* 1998). The thyroid hormone receptor has the capacity to interact with diverse coactivators and corepressors in a manner that is dependent on the presence or absence of T_3. In the absence of ligand, TR has been shown to interact with the corepressors NCoR (Horlein *et al.,* 1995) and SMRT (Chen and Evans, 1995), while in the presence of ligand the receptor recruits a number of transcriptional coactivators such as ERAP160 (Halachmi *et al.,* 1994), RIP140 (Cavaillès *et al.,* 1995), SRC-1 (Kamei *et al.,* 1996; Oñate *et al.,* 1995), TIF2 (Voegel *et al.,* 1996), ACTR (Chen *et al.,* 1997),

PCAF (Chen *et al.,* 1997), and p300/CBP (Chakravarti *et al.,* 1996). As discussed later, the corepressors further recruit histone deacetylase, while the coactivators themselves possess or recruit histone acetyltransferase activity. Interestingly, v-erbA, which lacks the ability to bind hormone and recruit coactivators, is presumed to only recruit corepressors and may function entirely by silencing gene expression. Thus the capacity to modify chromatin structure emerges as a unifying theme for gene regulation by TR (Wolffe, 1997; see Fig. 1).

A. TR and v-ErbA: Functional Domains

As with all members of the nuclear receptor superfamily, TR has a modular structure that consists of a central DNA-binding domain flanked by a C-terminal region comprising the ligand binding domain (LBD). The LBD is responsible for the ligand-dependent transcriptional activation function (AF-2), which accounts for the major component of TR-mediated activation. The N-terminal region of the receptor is implicated both in conferring a constitutive ligand-independent transcriptional activation function (AF-1) and modulating the AF-2 activity of the C-terminal LBD (Tomura *et al.,* 1995; Hollenberg *et al.,* 1995; Hadzic *et al.,* 1995). In addition to binding hormone, the LBD is involved in diverse interdependent functions including receptor dimerization, nuclear localization, and the binding of coactivators and corepressors (reviewed by Chatterjee *et al.,* 1997). Recent structural studies have shown the TR LBD to be comprised of 12 α-helices arranged in a triple layered fold surrounding a hydrophobic pocket into which the ligand is drawn (Wagner *et al.,* 1995). The extreme carboxy-terminal α-helix (H12) has been shown to be critical for AF-2 activity and the recruitment of coactivators (Barrettino *et al.,* 1994; Tone *et al.,* 1994; Collingwood *et al.,* 1997). Based on studies of the retinoid receptors, H12 is believed to undergo a marked realignment of its position within the LBD (Renaud *et al.,* 1995; Bourguet *et al.,* 1995) such that, in association with helices H3 and H5, a hydrophobic cleft is formed into which the LXXLL recognition motif of coactivators can bind (Heery *et al.,* 1997; Collingwood *et al.,* 1998; Darimont *et al.,* 1998; Feng *et al.,* 1998). Transcriptional silencing has also been mapped to the LBD (Baniahmad *et al.,* 1992) and this function is mediated by at least two corepressor proteins, NCoR and SMRT, which bind to the receptor in the absence of hormone but are released upon the addition of T3 (Horlein *et al.,* 1995; Chen and Evans, 1995). Initial analyses revealed that the "CoR box" region immediately C-terminal to the DNA binding domain is crucial for corepressor interaction and that mutations in this region could abrogate both repression and corepressor recruitment. Further studies have shown that H11 and H12 at the opposite end of the LBD also play a major role in repression. Lin *et al.* (1997) showed that mutations which disrupt the integrity of this domain can inhibit ligand-dependent corepressor release, despite preservation of ligand binding. It has been suggested that in certain instances the association of coactivators and corepressors with TR may be mutually exclusive (Blanco *et al.,* 1998). It is also conceivable that coactivators and corepressors might associate with each other in the context of the chromatin-bound receptor.

v-ErbA lacks the capacity to bind ligand and is devoid of the extreme carboxy-terminal 9-amino-acid residues that comprise the AF-2 core domain of TR and are critical for coactivator recruitment (Zenke *et al.,* 1990; Saatcioglu *et al.,* 1993). However, v-erbA retains the capacity to bind to corepressors (Chen and Evans, 1995). Association with corepressors through contacts made with the CoR box appear essential for oncogenic activity (Damm *et al.,* 1987; Damm and Evans, 1993). As compared to TR, the v-erbA protein recognizes TREs with a distinct, but highly related, selectivity and also appears to differ from TR in its requirement for heterodimerization with RXR when binding to particular target response elements (Chen *et al.,* 1993; Wahlstrom *et al.,* 1996; Judelson and Privalsky, 1996). Differences in the DNA binding domain between v-erbA and TR also appear important in allowing v-erbA to transform erythroid cells (Bonde *et al.,* 1991).

The detailed structural and functional information available for TR and v-erbA emphasizes the linkages between different domains and functions of the proteins. Everything from the recognition of a particular TRE, through heterodimerization, association with ligand, release of corepressors and recruitment of coactivators appears to have mutual interdependence. Biologically, the capacity of TR and v-erbA to selectively recognize particular TREs might be important for determining the spectrum of genes regulated in a particular cell lineage. For example, it has been proposed that v-erbA might need to alter the physiology of AEV-transformed erythroblasts to make them more tolerant to wide variations in pH or bicarbonate anion by repressing the erythrocyte anion transporter and carbonic anhydrase II genes (Zenke *et al.,* 1990).

B. Transcriptional Corepressors Recruit Histone Deacetylase

In addition to their recruitment by TR and v-erbA, NCoR and SMRT also interact with the mammalian proteins mSin3A and mSin3B that have striking homology to *S. cerevisiae* Sin3p (Heinzel *et al.*, 1997; Alland *et al.*, 1997; Nagy *et al.*, 1997; Laherty *et al.*, 1997). Mammalian Sin3 contains four paired amphipathic helices (PAH1–4). NCoR associates with the most amino terminal of these domains, PAH1. *S. cerevisiae* Sin3p interacts with the histone deacetylase RPD3 (Kadosh and Struhl, 1997) and likewise the mammalian Sin3 proteins associate with the RPD3 homolog using contacts between the PAH3 and 4 domains (Alland *et al.*, 1997). These interconnections suggested that TR-RXR might target transcriptional repression by recruiting histone deacetylase to modify the local chromatin environment (Wolffe, 1997). However it remained possible that components of the recruited corepressor complex such as NCoR, SMRT, or Sin3 might exert repressive activities on components of the basal machinery or organize repressive chromatin structures directly without the requirement for histone deacetylase (for example, Fondell *et al.*, 1993; Le Douarin *et al.*, 1995). We were able to discriminate between these possibilities using the microinjection of double-stranded DNA templates into *Xenopus* oocyte nuclei. The slow kinetics of chromatin assembly on double-stranded DNA were especially advantageous for following the role of nucleosomes in transcriptional silencing. We found that histone deacetylase alone could repress transcription of the *Xenopus* TRβA gene dependent on the assembly of a nucleosomal template (Wong *et al.*, 1998) and that this repression could be overcome either by hormone-bound TR-RXR or by inhibition of the histone deacetylase using TSA. Interestingly there are distinct requirements for chromatin assembly in the establishment of transcriptional repression by the histone deacetylase alone compared with those needed in the presence of unliganded TR-RXR. The receptor can direct the repression of transcription under conditions where low levels of chromatin assembly have occurred on the template, while the deacetylase alone requires much higher levels of chromatin assembly to repress transcription. However, it is important to note that even when unliganded TR-RXR represses transcription during chromatin assembly, inhibition of histone deacetylase using Trichostatin A blocks this repression. Possible explanations for the repression of transcription by unliganded TR-RXR occurring under conditions of incomplete chromatin assembly, yet still requiring histone deacetylase include (1) the active recruitment of histones to the transcriptional corepressor complex so that they are used as part of the regulatory nucleoprotein architecture to repress transcription after deacetylation (Roth *et al.*, 1990; Edmondson *et al.*, 1996; Wolffe and Pruss, 1996); (2) deacetylase may act to modify basal transcription factors and not histones (Imhof *et al.*, 1997), generating acetylated components of the basal transcriptional machinery with enhanced functional properties; and (3) the initial process of establishing transcriptional repression might involve interactions between TR-RXR and/or the corepressor complex with the basal transcriptional machinery that might interfere with transcription initiation (Baniahmad *et al.,* 1993; Fondell *et al.,* 1993). This state of repression would be subsequently stabilized by other events requiring deacetylase activity, potentially including nucleosome assembly.

Exactly how the unliganded TR-RXR uses histone deacetylase to repress transcription remains unclear, nevertheless the major enzymatic consequence is the targeting of deacetylase activity to the vicinity of the promoter. In the presence of ligand this situation is reversed, and histone acetyltransferases are recruited.

C. Transcription Coactivators Possess or Recruit Histone Acetyltransferase Activity

A unifying feature of several of the diverse coactivator proteins described earlier is their capacity to function as histone acetyltransferases (Yang *et al.,* 1996; Ogrzyko *et al.,* 1996; Chen *et al.,* 1997). Thus a common theme for transcriptional activation by ligand-bound nuclear receptors is the recruitment of histone acetyltransferase activity. This recruitment might be expected to counteract the repressive influence of histone deacetylase. Consistent with this hypothesis, the addition of histone deacetylase inhibitors can relieve the requirement for transcriptional coactivators in chromatin assembled on the TRβA promoter in *Xenopus* oocyte nuclei (Wong *et al.,* 1998; Li *et al.,* 1998; Li *et al.,* 1999). Histone acetylation states are dynamic, with the acetylated lysines of hyperacetylated histones turning over rapidly with half-lives of minutes within transcriptionally active chromatin, but much less rapidly for the hypoacetylated histones of transcriptionally silent regions (Covault and Chalkley, 1980; Zhang and Nelson, 1988). This observation implies that transcription coactivators must continually exert their acetyltransferase functions in the vicinity of a promoter. The dynamics of histone acetylation provides an attractive mechanistic foundation for the reversible activation and repression of transcription.

D. Chromatin Disruption by the Ligand-Bound TR-RXR

We have discussed how the TR-RXR can bind to nucleosomal DNA to target both transcriptional repression and activation. The TR-RXR silences transcription effectively in the context of a positioned nucleosomal array (Wong *et al.,* 1995, 1997a). On the addition of ligand this regular nucleosomal array is disrupted. The requirements for this disruption offer additional insight into the potential roles for histone acetylation and chromatin modification in gene control.

The TR-RXR alone and its recognition element are sufficient to initiate a process of chromatin disruption over an extended segment of DNA sequence as assayed by micrococcal nuclease cleavage (Wong *et al.*, 1997a). Each TRE can target the loss of topological constraint found in three to four nucleosomes, provided the TREs are separated by at least 400 bp of DNA. Transcription is not required for chromatin disruption and additional proximal promoter elements are necessary to facilitate hormone-dependent activation of transcription by the TR-RXR. The T_3-dependent chromatin disruption depends on the AF-2 activity of the LBD (Figs. 9 and 10). Thus we can dissect the process of gene activation by TR-RXR into at least three steps: (1) binding of the receptor to chromatin and the targeted assembly of a repressive chromatin structure; (2) the disruption of chromatin on the addition of ligand; and (3) subsequent transcriptional activation (Wong *et al.*, 1997a). The separation of chromatin disruption from transcription would be consistent with the opening up of a promoter from a repressive chromatin environment in order to facilitate the subsequent activity of the basal transcription machinery (Wolffe and Pruss, 1996). An important, yet unresolved question concerns the role of histone acetyltransferases in this disruption process. At this time, the role of transcriptional coactivators such as PCAF, p300, and SRC-1 in chromatin disruption has not been assayed; however, the structural and functional consequences of histone acetylation have been intensively investigated.

Chromatin disruption targeted by ligand-bound TR-RXR leads to major topological changes in minichromosomes consistent with a loss of wrapping of DNA around the histones (Germond *et al.*, 1975; Bauer *et al.*, 1994; Wong *et al.*, 1997a). Although histone acetyltransferases are recruited to the ligand-bound TR-RXR, it is difficult to account for the large topological change observed through acetylation alone (Norton *et al.*, 1989; Bauer *et al.*, 1994). In fact transcription can be activated without significant topological change from a TRβA promoter complexed in chromatin with unliganded TR-RXR simply by the addition of the deacetylase inhibitor TSA (Wong *et al.*, 1998). Thus, histone hyperacetylation in isolation seems unlikely to account for chromatin disruption. More importantly, the chromatin disruption phenomenon in a minichromosomal context as assayed by topological change and increases in micrococcal nuclease digestion is not necessary for transcriptional activation.

Chromatin disruption as assayed by topological change and increas-

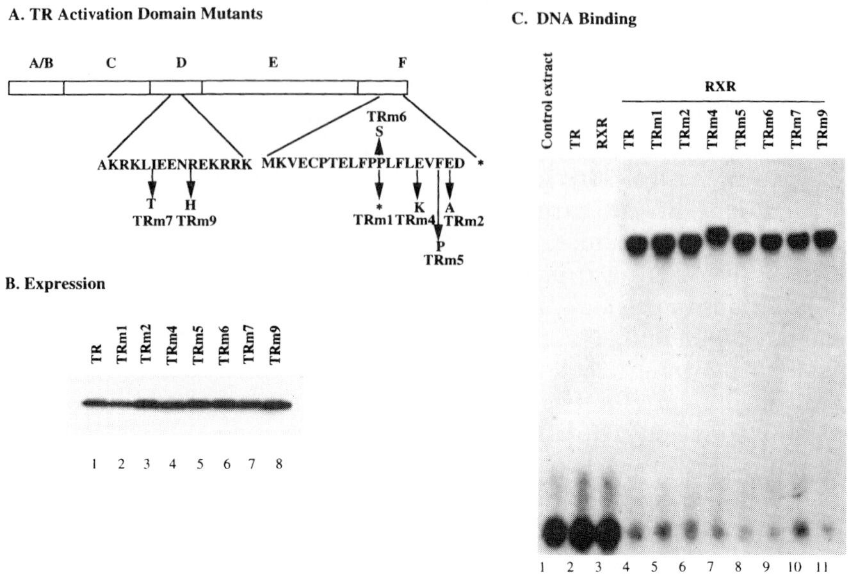

FIG. 9. Characterization of TR mutants. (A) The diagram shows the structural and functional domains of TR. A/B, N-terminal domain; C, DNA-binding domain; D, hinge region; E and F, hormone-binding and T_3-dependent transactivation domain. The amino acid change in each TR mutant is indicated. TRm1 has the change of proline to a stop codon. (B) Western blotting analysis of TR mutants expressed in oocytes. mRNAs encoding mutant TRs were synthesized *in vitro* and injected into groups of oocytes. After overnight incubation, the groups of oocytes were collected and homogenized, and half an oocyte equivalent was resolved with a 10% SDS-PAGE, blotted nitrocellulose, and vizualized with antibody against recombinant TR. (C) All TR mutants can form heterodimers with RXR and bind TRE as effectively as wild-type TR. For the gel retardation assay, 1 μL of TR or TR mutant oocyte extract (lanes 4–11) was mixed with 1 μL of RXR oocyte extract and incubated with an end-labeled double-stranded oligonucleotide containing the TRE from the TRβA promoter in 10 μL of binding buffer. As controls, uninjected oocyte extract, TR or RXR oocyte extract alone also were used (lanes 1–3). Note that TR alone does not bind to the probe.

es in micrococcal nuclease cleavage has generally been interpreted as essential for transcriptional activation (Hirschhorn *et al.,* 1992; Imbalzano *et al.,* 1994; Tsukiyama *et al.,* 1994). However, transcriptional activation can occur with minimal changes in DNA topology (Pederson and Morse, 1990; Drabik *et al.,* 1997). This would be anticipated if histone acetylation were the only alteration to chromatin structure necessary for transcriptional activation. It is possible that the pathways directing chromatin disruption in the oocyte nucleus in response to ligand-bound TR-RXR do not involve histone acetylation and that al-

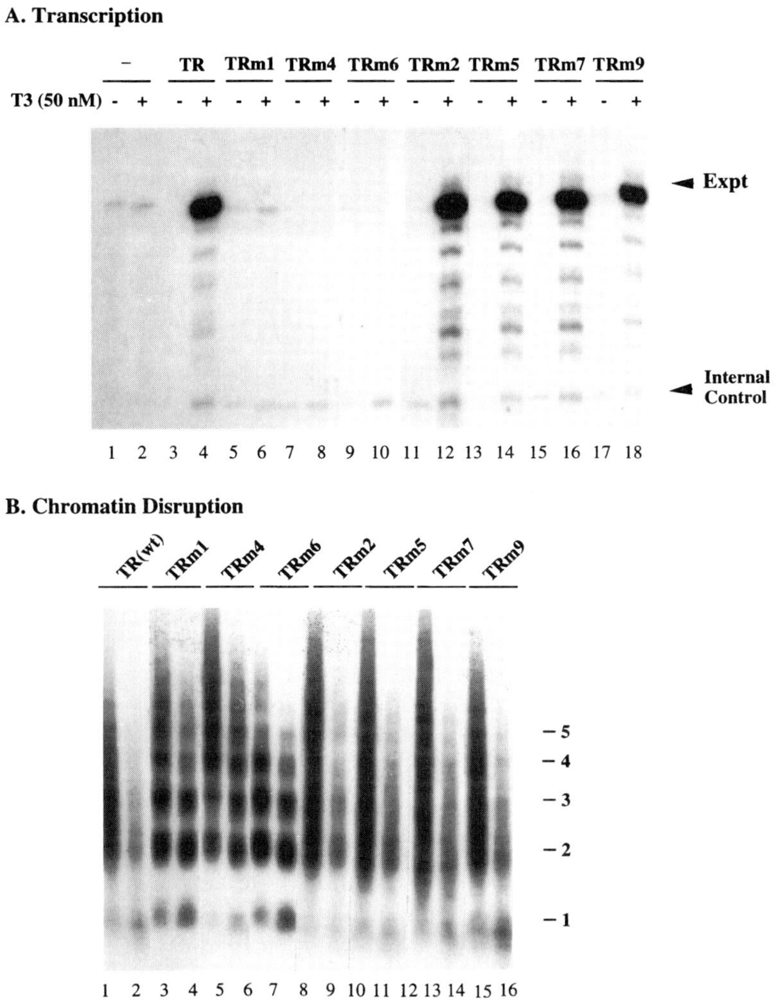

FIG. 10. The transcriptional activation domain is required for both hormone-dependent transcriptional activation and chromatin disruption. (A) The transcriptional activation of the TRβA promoter by TR mutants. The groups of oocytes were injected with ssDNA of pTRβA (100 ng/μL, 23 nL/oocyte) and with mRNAs encoding RXR and TR or TR mutants (100 ng/μL each, 23 nL/oocyte) and treated with T_3 (50 nM) overnight. Half of the oocytes in each group were used for transcriptional analysis by primer extension and half for chromatin disruption analysis by Mnase assay. Note that TRm1, TRm4, and TRm6 failed to activate transcription from the TRβA promoter (compare lanes 6, 8 and 10 with lane 4). The internal control represented the extension product of the endogenous oocyte H4 mRNA. (B) The mutants impaired in transcriptional activation also failed to disrupt canonical chromatin structure as assayed by Mnase. Partial Mnase digestions of minichromosomes were carried out with 10 U of Mnase in the even lanes and 5 U of Mnase in the odd lanes at room temperature for 20 min. The DNAs were purified, resolved with a 1.5% agarose gel, blotted to nitrocellulose membrane and probed with random primer-labeled DNA fragment including the TRE (from +218 to +314). Also indicated on the right are the positions of mono-, di-, tri-, tetra-, and penta-nucleosomal DNAs.

ternative means of disrupting histone–DNA contacts are employed leading to similar transcriptional consequences. Candidate mechanisms include recruitment of molecular machines such as SWI/SNF (Coté *et al.,* 1994; Imbalzano *et al.,* 1994; Wang *et al.,* 1996a,b; Fryer and Archer, 1998) or RNA polymerase itself (Wilson *et al.,* 1996; Gaudreau *et al.,* 1997). Recent work supports this idea (Li *et al.,* 1999).

Chromatin disruption directed by TR-RXR might serve other purposes aside from the activation of transcription on minichromosomes. It might be necessary to destabilize higher order chromatin structures (Hansen *et al.,* 1991; Schwarz and Hansen, 1994). Such substantial alterations in chromatin structure, including loss of DNA wrapping in the nucleosome, might serve as an epigenetic mark for the propagation of gene activity states through replication and chromosomal duplication (Weintraub, 1985), or for the recruitment of the promoter to a nuclear site competent for transcription in a true chromosomal environment (Cook, 1994).

E. How Might Histone Acetylation Alter Chromatin and Facilitate Transcription?

Experimental results strongly support an active role for deacetylase and acetyltransferase activities in transcriptional regulation by TR-RXR. Here we summarize existing knowledge concerning the consequences of histone acetylation for chromatin structure and function. About 25% of the mass of the core histones is contained within the "tail" domains (Arents *et al.,* 1991; Arents and Moudrianakis, 1993, 1995). These domains, located at the N termini of all four core histone proteins and the C terminus of histone H2A, were initially defined by their sensitivity to proteases. Proteolytic removal of the tail domains does not drastically alter the conformation or hydrodynamic properties of individual nucleosomes, and the tails do not play a role in nucleosome positioning or the correct assembly of nucleosomes *in vitro* (Dong *et al.,* 1989; Hayes *et al.,* 1991). These N termini, if fully extended, can project well beyond the superhelical turns of DNA in the nucleosome (Wolffe and Hayes, 1999). Consistent with their length, centrifugation studies with nucleosomal arrays lacking linker histones indicate that the histone tails mediate internucleosomal contacts as extended chains of nucleosomes are compacted to form the 30-nm chromatin fiber (Garcia-Ramirez *et al.,* 1992, 1995). Further, the tails are critical for the self-assembly of condensed fibers into higher order structures (Tse and Hansen, 1997; Tse *et al.,* 1998). Interestingly, histone tail interactions change as the chromatin fiber undergoes folding or compaction, suggesting that specific tail interactions are correlated with specific conformations of the fiber (Fletcher and Hansen, 1996). Thus, certain posttranslational modifications may evoke specific functional and/or conformational states of the chromatin fiber by inducing a defined alteration in the array of histone tail interactions.

Although the exact mechanism by which acetylation affects the biophysical properties of chromatin remains somewhat undefined, it is clear that acetylation of the core histone N termini affects the transcriptional properties of chromatin at several levels of chromatin structure. Acetylation can facilitate the binding of transcription factors to their recognition elements within isolated nucleosomes (Lee *et al.,* 1993; Lefebvre *et al.,* 1998; Vetesse-Dadey *et al.,* 1996; Wolffe *et al.,* 1993). Proteolytic removal of the N termini of the core histones leads to comparable increases in transcription factor access to nucleosomal DNA and transcription of chromatin templates as histone acetylation (Nightingale *et al.,* 1998; Ura *et al.,* 1997; Vetesse-Dadey *et al.,* 1994), consistent with acetylation reducing the stability of interaction of the histone tails with nucleosomal DNA (Hong *et al.,* 1993). It should nevertheless be noted that the N termini of the core histones always make contact with DNA even when they are acetylated (Mutskov *et al.,* 1998). Acetylated histones wrap DNA less tightly in mononucleosomes, which may result in a decrease in the amount of DNA superhelical writhe constrained by the nucleosome (Norton *et al.,* 1989; Krajewski and Becker, 1998). These changes might be due to the fact that the acetylated N-terminal histone tails bind DNA with reduced affinity (Hong *et al.,* 1993; Puig *et al.,* 1998) and are more mobile with respect to the DNA surface than unmodified tails (Cary *et al.,* 1982). Another interesting possibility is that acetylation disrupts the secondary structures that are known to exist within the H3 and H4 N termini when they are bound to nucleosomal DNA (Baneres *et al.,* 1997; Hansen *et al.,* 1999). This might further destabilize interactions with DNA and the nucleosome itself.

Beyond effects on individual nucleosomes, acetylation facilitates factor access and transcription from nucleosomal arrays by decreasing the stability of the completely compacted 30-nm fiber (reviewed by Tse *et al.,* 1998). It is also likely that acetylation leads to the destabilization of long-range structures through which the chromatin fiber is folded into the chromosome itself (Annunziato *et al.,* 1988). Interactions between adjacent nucleosomal arrays are reduced when they are reconstituted with acetylated histones and chromatin solubility is increased (Tse *et al.,* 1998; Perry and Chalkley, 1982). Within the 75 mg/ml nucleoprotein environment of the eukaryotic nucleus, all of these structural transitions might contribute to facilitating transcription.

Interestingly, the level of histone modification required to facilitate the transcription process is relatively low, and a total of 12 acetylated lysines per histone octamer (out of 28 potential acetylated lysines) will promote *in vitro* transcription more than 15-fold. This level of modification reduces chromatin compaction to the same extent as proteolytic removal of the N termini, again suggesting that the primary consequence of hyperacetylation is to reduce the interaction of the tails with the other components of chromatin including nucleosomal DNA, linker DNA, and the histones of adjacent nucleosomes. However, the level of charge neutralization necessary to facilitate the destabilization of chromatin higher order structure is so low that other structural features must amplify the consequences of acetylation. As discussed, these might include alterations to secondary structure in the tail domains and/or changes in the association of the tails with other nonhistone proteins. Acetylation of the histones probably serves to illuminate particular nucleosomes and/or segments of chromatin for interaction with other chromatin remodeling factors or components of the transcriptional machinery. The potential combination of direct chromatin structural transitions and modulation of protein–protein interactions following acetylation or deacetylation of the histone tails provides a powerful means of regulating transcription.

IV. Insight into the Functional Roles of TR and v-ErbA in Development, Transformation, and Human Disease

A. Gene Knockouts of TRα and TRβ

In *Xenopus,* constitutive expression of TRβ and acute developmental regulation of TRβ genes has been shown to occur during development (Shi and Brown, 1992; Kanamori and Brown, 1992; Machuca *et al.,* 1995; Kawahara *et al.,* 1991; Yaoita *et al.,* 1990; Fig. 11), while in mammalian systems receptor isoform distribution also exhibits a temporally regulated tissue specificity that is itself influenced by thyroid hormone (Forrest *et al.,* 1990; Strait *et al.,* 1990; Bradley *et al.,* 1994). Recent studies involving the disruption of both TRα and TRβ genes have provided insight into the specific mammalian developmental roles of each receptor subtype. Targeted disruption of TRβ in the mouse resulted in a surprisingly subtle phenotype. $TR\beta^{-/-}$ mice were viable and fertile with normal growth rates and no detectable organ abnormalities except a variably enlarged thyroid gland (Forrest *et al.*, 1996a,b). However, failure to down-regulate the pituitary hormone TSH in response

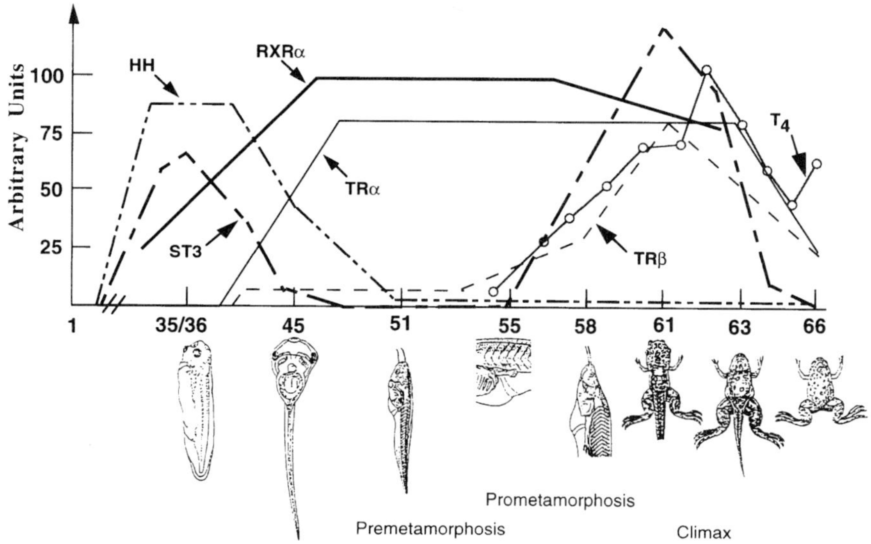

FIG. 11. Correlation of the hedgehog (HH), stromelysin3 (ST3), levels of endogenous TH and the mRNAs of TRα, TRβ, and RXRα genes with stages in *X. laevis* development. The plasma concentrations (arbitrary units) of thyroid hormone T_3 (crosses) and T_4 (open circles) are from Leloup and Buscaglia (1977). The mRNA levels for HH, ST_3, TRα, TRβ, and RXRα (bold line) are based on published data (Yaoita *et al.,* 1990; Kawahara *et al.,* 1991; Eliceiri and Brown, 1994; Patterton *et al.,* 1995; Tata, 1993) and for clarity, plotted on different scales.

to the elevated levels of T3 was is keeping with the negative regulatory role of TRβ in the hypothalamic–pituitary–thyroid axis. Interestingly, analysis of the CNS showed no developmental deffects, even in structures known to be sensitive to T3, such as the cerebellum and hippocampus. In contrast to the TRβ gene knockout, disruption of the TRα gene had significant developmental consequences. At 2 weeks postnatal the TRα$^{-/-}$ mice exhibited a hypothyroid state with reduced T3, T4, and TSH levels and thyroid gland hypoplasia, suggesting a role for TRα isoforms in the up-regulation of TSH. The mice also displayed growth arrest and delayed maturation of the small intestine and bones and subsequently died at 5 weeks of age (Fraichard *et al.,* 1997). In a similar study TRα$^{-/-}$ mice also showed a 20% reduction in heart rate and 0.5°C reduction in body temperature (Wikstrom *et al.*, 1998). However, injection of T3 at 3 weeks of age partially restored growth and organ maturation was normal, although both males and females appeared infertile (Fraichard *et al.,* 1997). These studies indicate a significant developmental role for TRα but that the role of TRβ is more tissue and temporally specific.

Requirements for TR are in many ways similar to those seen in the transition from egg to hatched chick and from tadpole to frog (Forrest *et al.,* 1990; Shi, 1994). These relatively minor determinative consequences, but more severe differentiative or adaptive requirements, might be explained by TR exerting gene control predominatly by binding in chromatin and repressing gene activity. Thus, ablation of TR will have the same consequence as adding hormone in that repression will be relieved and transcription will return to high basal levels. If TR function is not to activate transcription much above basal levels in chromatin then removing the receptor will provide a phenotype similar to that of an animal constitutively exposed to thyroid hormone. TRα and TRβ might be expected to be required to activate a limited spectrum of genes in the presence of thyroid hormone. Similar conclusions follow from the analysis of clinical syndromes in which afflicted individuals express dominant negative forms of the thyroid hormone receptor.

B. Resistance to Thyroid Hormone

A diverse array of naturally occurring mutations in the human TRβ gene has given rise to the syndrome of resistance to thyroid hormone (RTH), a disorder that is usually inherited as an autosomal dominant and is characterized by elevated levels of circulating thyroid hormone with nonsuppressed TSH secretion and variable peripheral response to hormone (reviewed by Chatterjee, 1997). Affected individuals are heterozygous for mutations that result in aberrant TRs that inhibit the activity of wild-type receptor in a dominant negative manner. The associated phenotype varies from relatively mild "generalized" resistance (GRTH), in which individuals appear asymptomatic, to more severe "pituitary-specific" resistance (PRTH), characterized by growth and neural abnormalities, as well as cardiac dysfunction. Interestingly, in accordance with the mouse model for the TRβ gene knockout, an individual who was homozygous for TRβ deletion also exhibited impaired hearing loss (Refetoff *et al.,* 1967; Takeda *et al.,* 1992).

To date, in excess of 50 different mutations have been reported—all of which lie within the receptor LBD. Almost all mutations result in variable reduction in ligand-binding affinity, while DNA binding and heterodimerization with RXR are preserved, although a few select mutants exhibit impaired homodimer formation that is TRE configuration dependent (Collingwood *et al.,* 1994, 1997, 1998). However, some mutants are unable to fully transactivate despite saturating levels of T3 and further studies directly implicate aberrant interaction with chromatin remodeling factors. Impaired release of the corepressor SMRT in response to T3 has been demonstrated for a number of natural mutants (Yoh *et al.,* 1997; Lin *et al.,* 1997; Clifton-Bligh *et al.,* 1998)—the consequence of which being continued recruitment of histone deactylase activity to the promoter and retention of the nontranscriptionally permissive repressed chromatin state. Similarly, other studies have shown that natural mutations of residues integral to the coactivator recruitment interface of TR specifically impair the ligand-dependent recruitment of coactivators, such as SRC-1, ACTR, and RIP140, which mediate acetyltransferase activity and the generation of more transcriptionally permissive acetylated chromatin (Collingwood *et al.,* 1997, 1998). It is possible that the recruitment of coactivators serves simply to more effectively relieve the transcriptionally repressed state dependent on association with corepressors and that release of the corepressor complex on addition of hormone might be sufficient for most of the gene regulatory phenomena. The significance of targeted gene repression mediated by TR is further emphasized by studies on the functional properties of the oncoprotein v-erbA.

C. v-ErbA: An Oncoprotein That Inhibits Cell Differentiation and Promotes Proliferation

v-ErbA lacks the capacity to activate transcription and to interact with thyroid hormone as a consequence of small deletions in the ligand-binding domain (Sap *et al.,* 1986; Weinberger *et al.,* 1986). Neoplastic transformation by v-erbA appears to be due in part to the inhibition of genes regulated by TR and RAR (Zenke *et al.,* 1990; Scharif and Privalksky, 1991). v-ErbA interacts with corepressors constitutively (Chen and Evans, 1995) and requires histone deacetylase activity to silence transcription (F. Urnov, J. Yee, T. N. Collingwood, and A. P. Wolffe, personal communication). Transformation defective variants of v-erbA that fail to repress transcription (Damm *et al.,* 1987; Damm and Evans, 1993) also fail to interact with corepressors (Chen and Evans, 1995). The capacity of v-erbA to repress specific transcription, presumably via the recruitment of histone deacetylase, inhibits cell differentiation and allows cell proliferation to proceed. This mode of action of nuclear receptors appears to be coopted by diverse oncoproteins containing nuclear receptor corepressor binding domains (Lin *et al.,* 1998; Grignani *et al.,* 1998). v-ErbA also has the capacity to influence cross-talk between various transcription activation pathways. For example, v-erbA can enhance transcriptional activation mediated by AP1. The capacitation of the activity of the protooncogenes c-jun and c-fos may also contribute to cell transformation (Tenbaum and Baniahmad, 1997).

Recent results (Bauer *et al.,* 1998) demonstrate that in media depleted of T_3 and retinoids, overexpressed TRα functions exactly like v-ErbA. These experiments establish that unliganded TRα could impede differentiation, associate with histone deacetylase and repress carbonic anhydrase and *c-myb.* These phenomena were reversed on the addition of hormone. These authors further suggest that TRα can function as a ligand-operated molecular switch, regulating the balance between erythroblast self-renewal and differentiation. It is attractive to speculate that TRα in *Xenopus* might have a similar function during tadpole development prior to the differentiation of a thyroid gland and the synthesis of thyroid hormone (Wolffe and Shi, 1999). The unliganded *Xenopus* TRα accumulates early in larval development (Fig. 11) when it may serve to repress the expression of genes such as those encoding signaling molecules (hedgehog) or matrix metalloproteinases (stomolysin 3) that are not required until metamorphosis. Thus, TR may facilitate the proliferation of adult stem cell lineages in the absence of differentiation. Once thyroid hormone is synthesized the same TRα bound within chromatin will serve to activate gene expression, including the TRβ gene. The surge in TRβ gene expression may be important in facilitating metamorphic climax by activating genes. In this scenario TRβ is the immediate early gene whose transcription factor product initiates the cascade of gene activity associated with metamorphosis (Shi *et al.,* 1996).

V. Conclusions

The thyroid hormone receptor and the related oncoprotein v-erbA provide wonderful tools to investigate the functional capabilities of chromatin architecture and modification. The TR has the capacity to have regulated interactions with a rich variety of transcriptional coactivators and corepressors. A unifying feature of these interactions is the capacity of coactivators to acetylate histones and of corepressors to deacetylate histones. TR also facilitates other structural modifications to chromatin with unknown functional consequences at the present time. For the vast majority of genes regulated by TR, acetylation of proteins in the vicinity of the TRE appears to activate transcription while deacetylation represses. Exactly how this comes about is presently unknown. However, it is clear that chromatin structure is destabilized as a consequence of acetylating the N termini of the core histones and that this modification facilitates the accessibility of DNA in chromatin to the transcriptional machinery. It is important to recognize that chromatin structure can also activate transcription dependent on architectural context (Schild *et al.,* 1993; Stunkel *et al.,* 1997). Thus chromatin structural transitions also have the capacity to repress transcription. Transcriptional corepressors are known to activate transcription on the negatively regulated promoters, which drive expression of the pituitary thyroid stimulating hormone α- and β-subunits (TSHα, TSHβ) and of the hypothalamic thyrotropin releasing hormone (Tagami *et al.,* 1997). An understanding of nucleoprotein architecture is undoubtedly essential to interpret these results; however, the key point is that the activation or repression of a particular promoter will depend on the recruitment of a molecular machinery that modifies or disrupts the chromatin infrastructure. Large gaps in our knowledge as to how this occurs exist at both a structural and functional level. Filling these gaps will contribute greatly to our understanding of the molecular mechanisms during cell transformation and also in generating targeted therapeutic strategies for intervention in human clinical disorders.

REFERENCES

Alland, L., Muhle, R., Hou, H., Jr., Potes, J., Chin, L., Schreiber-Agus, N., and De Pinho, R.A. (1997). Role of NCoR and histone deacetylase in Sin3-mediated transcriptional and oncogenic repression. *Nature (London)* **387**, 49–55.

Almer, A., and Horz, W. (1986). Nuclease hypersensitive regions with adjacent positioned nucleosomes mark the gene boundaries of the PHO5/PHO3 locus in yeast. *EMBO J.* **5**, 2681.

Almer, A., Rudolph, H., Hinnen, A., and Horz, W. (1986). Removal of positioned nucleosomes from the yeast PHO5 promoter upon PHO5 induction releases additional activating DNA elements. *EMBO J.* **5**, 2689–2696.

Almouzni, G., and Wolffe, A. P. (1993). Replication coupled chromatin assembly is required for the repression of basal transcription *in vivo. Genes Dev.* **7**, 2033–2047.

Annunziato, A. A., Frado, L.-L. Y., Seale, R. L., and Woodcock, C. L. F. (1988). Treatment with sodium butyrate inhibits the complete condensation of interphase chromatin. *Chromosoma* **96**, 132–138.

Archer, T. K., Cordingley, M. G., Wolford, R. G., and Hager, G. L. (1991). Transcription factor access is mediated by accurately positioned nucleosomes on the mouse mammary tumor virus promoter. *Mol. Cell. Biol.* **11**, 688–698.

Arents, G., and Moudrianakis, E. N. (1993). Topography of the histone octamer surface: Repeating structural motifs utilized in the docking of nucleosomal DNA. *Proc. Natl. Acad. Sci. U.S.A.* **90**, 10489–10493.

Arents, G., and Moudrianakis, E. N. (1995). The histone fold: A ubiquitous architectural motif utilized in DNA compaction and protein dimerization. *Proc. Natl. Acad. Sci. U.S.A.* **92**, 11170–11174.

Arents, G., Burlingame, R. W., Wang, B. W., Love, W. E., and Moudrianakis, E. N. (1991). The nucleosomal core histone octamer at 3.1Å resolution: A tripartite protein assembly and a left-handed superhelix. *Proc. Natl. Acad. Sci. U.S.A.* **88**, 10148–10152.

Baneres, J. L., Martin, A., and Parello, J. (1997). The N tails of histones H3 and H4 adopt a highly structured conformation in the nucleosome. *J. Mol. Biol.* **273**, 503–508.

Baniahmad, A., Steiner, C., Kohne, A. C., and Renkawitz, R. (1990). Modular structure of a chicken lysozyme silencer: Involvement of an unusual thyroid hormone receptor binding site. *Cell* **61**, 505–514.

Baniahmad, A., Kohne, A. C., and Renkawitz, R. (1992). A transferable silencing domain is present in the thyroid hormone receptor, in the v-erbA oncogene product and in the retinoic acid receptor. *EMBO J.* **11**, 1015–1023.

Baniahmad, A., Ha, I., Reinberg, D., Tsai, S. Y., Tsai, M. J., and O'Malley, B. W. (1993). Interaction of a receptor β with transcription factor TFIIB may mediate target gene activation and derepression by thyroid hormone. *Proc. Natl. Acad. Sci. U.S.A.* **90**, 8832–8836.

Barrettino, D., Ruiz, M. V., and Stunnenberg, H. G. (1994). Characterization of the ligand dependent transactivation domain of thyroid hormone receptor. *EMBO J.* **13**, 3039–3049.

Bauer, A., Mikulits, W., Lagger, G., Stengl, G., Brosch, G., and Beug, H. (1998). The thyroid hormone receptor functions as a ligand-operated developmental switch between proliferation and differentiation of erythroid progenitors. *EMBO J.* **17**, 4291–4303.

Bauer, W. R., Hayes, J. J., White, J. H., and Wolffe, A. P. (1994). Nucleosome structural changes due to acetylation. *J. Mol. Biol.* **236**, 685–690.

Beug, H., Mullner, E. W., and Hyman, M. J. (1994). Insights into erythroid differentiation obtained from studies on avian erythroblastosis virus. *Curr. Opin. Cell Biol.* **6**, 816–824.

Blanco, J. C., Minucci, S., Lu, J., Zang, X. J., Walker, K. K., Chen, H., Evans, R. M., Nakatani, Y., and Ozato, K. (1998). The histone acetylase PCAF is a nuclear receptor coactivator. *Genes Rev.* **12,** 1638–1651.

Blomquist, P., Li, Q., and Wrange, O. (1996). The affinity of nuclear factor 1 for its DNA site is drastically reduced by nucleosome organization irrespective of its rotational and translational position. *J. Biol. Chem.* **271**, 154–159.

Bonde, B. G., Sharif, M., and Privalsky, M. L. (1991). Ontogeny of the v-erbA oncoprotein from the thyroid hormone receptor: An alteration in the DNA binding domain plays a role crucial for v-erbA function. *J. Virol.* **65**, 2037–2046.

Bourguet, W., Ruff, M., Chambon, P., Gronemeyer, H., and Moras, D. (1995). Crystal structure of the ligand-binding domain of the human nuclear receptor RXR-alpha. *Nature (London)* **375**, 377–382.

Bouvet, P., Dimitrov, S., and Wolffe, A.P. (1994). Specific regulation of chromosomal 5S rRNA gene transcription *in vivo* by histone H1. *Genes Dev.* **8**, 1147–1159.

Bradley, D. J., Towle, H. C., and Young, W. S. (1994). Alpha and beta thyroid hormone receptor (TR) gene expression during auditory neurogenesis: Evidence for TR isoform-specific transcriptional regulation in vivo. *Proc. Natl. Acad. Sci. U.S.A.* **91**, 439–443.

Brent, G. A, Dunn, M. K., Harney, J. W., Gulick, T., Larsen, P. R., and Moore, D. D. (1993). Thyroid hormone aporeceptor represses T_3-inducible promoters and blocks activity of the retinoid acid receptor. *New Biol.* **1**, 329–336.

Brown, D. D., Wang, Z., Kanamori, A., Eliceiri, B., Furlow, J. D., and Schwartzman, R. (1995). Amphibian metamorphosis: A complex program of gene expression changes controlled by the thyroid hormone. *Recent Prog. Horm. Res.* **50**, 309–315.

Buschhausen, G., Wittig, B., Graessmann, M., and Graessman, A. (1987). Chromatin structure is required to block transcription of the methylated herpes simplx virus thymidine kinase gene. *Proc. Natl. Acad. Sci. U.S.A.* **84**, 1177–1181.

Cary, P. D., Crane-Robinson, C., Bradbury, E. M., and Dixon, G. H. (1982). Effect of acetylation on the binding of N-terminal peptides of histone H4 to DNA. *Eur. J. Biochem.* **127**, 137–143.

Cavaillès, V., Dauvois, S., L'Horset, F., Lopez, G., Hoare, S., Kushner, P. J., and Parker, M. G. (1995). Nuclear factor RIP140 modulates transcriptional activation by the estrogen receptor. *EMBO J.* **14**, 3741–3751.

Chakravarti, D., LaMorte, V. J., Nelson, M. C., Nakajima, T., Juguilon, H., Montminy, M., and Evans, R. M. (1996). Mediation of nuclear receptor signalling by CBP/p300. *Nature (London)* **383**, 99–103.

Chatterjee, V. K. K. (1997). Molecular genetics and pathophysiology of thyroid hormone resistance. *Curr. Opin. Endocrinol. Diabetes* **4**, 371–376.

Chatterjee, V. K. K., Clifton-Bligh, R. J., and Matthews, C. (1997). The steroid hormone superfamily of receptors. *Mol. Endocrinol.* **4**, 73–101.

Chavez, S., and Beato, M. (1997). Nucleosome-mediated synergism between transcription factors on the mouse mammary tumor virus promoter. *Proc. Natl. Acad. Sci. U.S.A.* **94**, 2885–2890.

Chen, H., Smit-McBride, Z., Lewis, S., Sharif, M., and Privalsky, M. L. (1993). Nuclear hormone receptors involved in neoplasia: Erb A exhibits a novel DNA sequence specificity determined by amino acids outside of the zinc-finger domain. *Mol. Cell. Biol.* **13**, 2366–2376.

Chen, H., Lin, R. J., Schiltz, R. L., Chakravarti, D., Nash, A., Nagy, L., Privalsky, M. L., Nakatani, Y., and Evans, R. M. (1997). Nuclear receptor coactivator ACTR is a novel histone acetyltransferase and forms a multimeric activation complex with P/CAF and CBP/p300. *Cell (Cambridge, Mass.)* **90**, 569–580.

Chen, J. D., and Evans, R. M. (1995). A transcriptional co-repressor that interacts with nuclear hormone receptor. *Nature (London)* **377**, 454–457.

Clifton-Bligh, R. J., de Zegher, F., Wagner, R. L., Collingwood, T. N., Francois, I., Van Helvoirt, M., Fletterick, R. J., and Chatterjee, C. K. (1998). A novel TR beta mutation (R383H) in resistance to thyroid hormone syndrome predominantly impairs corepressor release and negative transcriptional regulation. *Mol. Endocrinol.* **12**, 609–621.

Collingwood, T. N., Adams, M., Tone, Y., and Chatterjee, V. K. K. (1994). Spectrum of transcriptional, dimerization, and domainant negative properties of twenty different mutant thyroid hormone β-receptors in thyroid hormone resistance syndrome. *Mol. Endocrinol.* **8**, 1262–1277.

Collingwood, T. N., Rajanayagam, O., Adams, M., Wagner, R., Cavaillès, V., Kalkhoven, E., Matthews, C., Nystrom, E., Stenlof, K., Lindstedt, G., Tisell, L., Fletterick, R. J., Parker, M. G., and Chatterjee, V. K. (1997). A natural transactivation mutation in the thyroid hormone beta receptor: Impaired interaction with putative transcriptional mediators. *Proc. Natl. Acad. Sci. U.S.A.* **94**, 248–253.

Collingwood, T. N., Wagner, R., Matthews, C. H., Clifton-Bligh, R. J., Gurnell, M., Rajanayagam, O., Agostini, M., Fletterick, R. J., Beck-Peccoz, P., Reinhardt, W., Binder, G., Ranke, M. B., Hermus, A., Hesch, R. D., Lazarus, J., Newrick, P., Parfitt, V., Raggatt, P., de Zegher, F., and Chatterjee, V. K. (1998). A role for helix 3 of the TRβ ligand-binding domain in coactivator recruitment identified by characterization of a third cluster of mutations in resistance to thyroid hormone. *EMBO J.* **17**, 4760–4770.

Cook, P. R. (1994). RNA polymerase: Structural determinant of the chromatin loop and the chromosome. *BioEssays* **16**, 425–430.

Coté, J., Quinn, J., Workman, J. L., and Peterson, C. L. (1994). Stimulation of GAL4 derivative binding to nucleosomal DNA by the yeast SWI/SNF complex. *Science* **265**, 53–60.

Covault, J., and Chalkley, R. (1980). The identification of distinct populations of acetylated histone. *J. Biol. Chem.* **255**, 9110–9116.

Damm, K., and Evans, R. M. (1993). Identification of a domain required for oncogenic activity and transcriptional suppression by v-erbA and thyroid-hormone receptor α. *Proc. Natl. Acad. Sci. U.S.A.* **90**, 10668–10672.

Damm, K., Beug, H., Graf, T., and Vennstrom, B. (1987). A single point mutation in erbA restores the erythroid transforming potential of a mutant avian erythroblastosis virus (AEV) defective in both erbA and erbB oncogenes. *EMBO J.* **6**, 375–382.

Darimont, B. D., Wagner, R. L., Apriletti, J. W., Stallcup, M. R., Kushner, P. J., Baxter, J. D., Fletterick, R. J., and Yamamoto, K. R. (1998). Structure and specificity of nuclear receptor-coactivator interactions. *Genes Dev.* **12**, 3343–3356.

Dimitrov, S., Dasso, M. C., and Wolffe, A. P. (1994). Remodeling sperm chromatin in *Xenopus laevis* egg extracts: The role of core histone phosphorylation and linker histone B4 in chromatin assembly. *J. Cell Biol.* **126**, 591–601.

Disela, C., Glineur, C., Bugge, T., Sap, J., Stengl, G., Dogson, J., Stunnenberg, H., Beug, H., and Zenke, M. (1991). v-erbA overexpression is required to extinguish c-erebA function in erythroid cell differentiation and regulation of the erbA target gene CAII. *Genes Dev.* **5**, 2033–2047.

Dong, F., Hansen, J. C., and van Holde, K. E. (1989). DNA and protein determinants of nucleosome positioning on sea urchin 5S rRNA gene sequences in vitro. *Proc. Natl. Acad. Sci. U.S.A.* **87**, 5724–5728.

Drabik, C. E., Nicita, C. A., and Lutter, L. C. (1997). Measurement of linker number change in transcribing chromatin. *J. Mol. Biol.* **267**, 794–806.

Edmondson, D. G., Smith, M. M., and Roth, S. Y. (1996). Repression domain of the yeast global repressor Tup1 interacts directly with histones H3 and H4. *Genes Dev.* **10**, 1247–1259.

Eliceiri, B. P., and Brown, D. D. (1994). Quantitation of endogenous thyroid hormone receptors alpha and beta during embryogenesis and metamorphosis in *Xenopus laevis. J. Biol. Chem.* **269**, 24459–24465.

Evans, R. M. (1988). The steroid and thyroid hormone receptor superfamily. *Science* **240**, 889–895.

Feng, W., Ribeiro, R. C. J., Wagner, R. L., Nguyen, H., Apriletti, J. W., Fletterick, R. J., Baxter, J. D., Kushner, P. J., and West, B. L. (1998). Hormone-dependent coactivator binding to a hydrophobic cleft on nuclear receptors. *Science* **280**, 1747–1749.

Fletcher, T. M., and Hansen, J. C. (1996). The nucleosomal array: Structure/function relationships. *Crit. Rev. Eukaryotic Gene Expression* **6**, 149–188.

Fondell, J. D., Roy, A. L., and Roeder, R. G. (1993). Unliganded thyroid hormone receptor inhibits formation of a functional preinitiation complex: Implications for active repression. *Genes Dev.* **7**, 1400–1410.

Forrest, D., Sjöberg, M., and Vennstrom, B. (1990). Contrasting developmental and tissue-specific expression of α and β thyroid hormone receptor genes. *EMBO J.* **9**, 1519–1528.

Forrest, D., Hanebuth, E., Smeyne, R. J., Everds, N., Stewart, C. L., Wehner, J. M., and Curran, T. (1996a). Recessive resistance to thyroid hormone in mice lacking thyroid hormone receptor β: Evidence for tissue-specific modulation of receptor function. *EMBO J.* **15**, 3006–3015.

Forrest, D., Erway, L. C., Ng, L., Altschuler, R., and Curran, T. (1996b). Thyroid hormone receptor β is essential for development of auditory function. *Nat. Gen.* **13**, 354–357.

Fraichard, A., Chassande, O., Plateroti, M., Roux, J. P., Trouillas, J., Dehay, C., Legrand, C., Gauthier, K., Kedinger, M., Malaval, L., Rousset, B., and Samarut, J. (1997). The T3Rα gene encoding a thyroid hormone receptor is essential for post-natal development and thyroid hormone production. *EMBO J.* **16**, 4412–4420.

Fryer, C. J., and Archer, T. K. (1998). Chromatin remodeling by the glucocorticoid receptor requires the BRG1 complex. *Nature (London)* **393**, 88–91.

Gandrillon, O., Rascle, A., and Samarut, J. (1995). The v-erbA oncogene: A superb tool for dissecting the involvement of nuclear hormone receptors in differentiation and neoplasia. *Int. J. Oncol.* **6**, 215–234.

Garcia-Ramirez, M., Dong, F., and Ausio, J. (1992). Role of the histone 'tails' in the folding of oligonucleosomes depleted of histone H1. *J. Biol. Chem.* **267**, 19587–19595.

Garcia-Ramirez, M., Rocchini, C., and Ausio, J. (1995). Modulation of chromatin folding by histone acetylation. *J. Biol. Chem.* **270**, 17923–17928.

Gaudreau, L., Schmid, A., Blaschke, D., Ptashne, M., and Horz, W. (1997). RNA polymerase II holoenzyme recruitment is sufficient to remodel chromatin at the yeast *PHO5* promoter. *Cell (Cambridge, Mass.)* **89**, 55–62.

Germond, J. E., Hirt, B., Oudet, P., Gross-Bellard, M., and Chambon, P. (1975). Folding of the double helix in chromatin like structures from simian virus 40. *Proc. Natl. Acad. Sci. U.S.A.* **72**, 1843–1847.

Glass, C. K. (1994). Differential recognition of target genes by nuclear receptor monomers, dimers, and heterodimers. *Endocr. Rev.* **15**, 391–407.

Glass, C. K., Lipkin, S. M., Devary, O. V., and Rosenfeld, M. G. (1989). Positive and negative regulation of gene transcription by a retinoic acid-thyroid hormone receptor heterodimer. *Cell (Cambridge, Mass.)* **59**, 697–708.

Godde, J. S., and Wolffe, A. P. (1996). Nucleosome assembly of CTG triplet repeats. *J. Biol. Chem.* **271**, 15222–15229.

Godde, J. S., Nakatani, Y., and Wolffe, A. P. (1995). The aminio-terminal tails of the core histones and the translational position of the TATA box determine TBP/TFIIA association with nucleosomal DNA. *Nucleic Acids Res.* **23**, 4557–4564.

Graupner, G., Wills, K. N., Tzukerman, M., Zhang, X. K., and Pfahl, M. (1989). Dual regulatory role for thyroid-hormone receptors allows control of retinoic-acid receptor activity. *Nature (London)* **340**, 653–656.

Grignani, F., De Matteis, S., Nervi, C., Tomassoni, L., Gelmetti, V., Cioce, M., Fanelli, M., Ruthardt, M., Ferrara, F. F., Zamir, I., Seiser, C., Grignani, F., Lazar, M. A., Minucci, S., and Pelicci, P. G. (1998). Fusion proteins of the retinoic acid receptor-alpha recruit histone deacetylase in promyelocytic leukaemia. *Nature (London)* **391**, 815–818.

Guschin, D., Chandler, S., and Wolffe, A. P. (1998). Asymmetric linker histone association directs the asymmetric rearrangement of core histone interactions in a positioned nucleosome containing a thyroid hromone response element. *Biochemistry* **37**, 8629–8636.

Hadzic, E., Desai-Yajnik, V., Helmer, E., Guo, S., Wu, S., Koudinova, N., Casanova, J., Raaka, B. M., and Samuels, H. H. (1995). A 10-amino-acid sequence in the N-terminal A/B domain of thyroid hormone receptor α is essential for transcriptional activation and interaction with the general transcription factor TFIIB. *Mol. Cell. Biol.* **15**, 4507–4517.

Halachmi, S., Marden, E., Martin, G., MacKay, H., Abbondanza, C., and Brown, M. (1994). Estrogen receptor-associated proteins: Possible mediators of hormone-induced transcription. *Science* **264**, 1455–1458.

Hansen, J. C., van Holde, K. E., and Lohr, D. (1991). The mechanism of nucleosome assembly onto oligomers of the sea urchin 5S DNA positioning sequence. *J. Biol. Chem.* **266**, 4276–4282.

Hansen, J. C., Tse, C., and Wolffe, A. P. (1999). Structure and function of the core histone N-termini: More than meets the eye. *Biochemistry* **37,** 17637–17641.

Hayes J. J (1996). Site-directed cleavage of DNA by a linker histone-Fe(II) EDTA coinugate: Localization of a globular domain binding site within a nucleosome. *Biochemistry* **35**, 11931–11937.

Hayes, J. J., and Wolffe, A. P. (1992). Histones H2A/H2B inhibit the interactions of transcription factor IIIA with the *Xenopus borealis* somatic 5S RNA gene in a nucleosome. *Proc. Natl. Acad. Sci. U.S.A.* **89**, 1229–1233.

Hayes, J. J., and Wolffe, A. P. (1993). Preferential and asymmetric interaction of linker histones with 5S DNA in the nucleosome. *Proc. Natl. Acad. Sci. U.S.A.* **90**, 6415–6419.

Hayes, J. J., Tullius, T. D., and Wolffe, A. P. (1990). The structure of DNA in a nucleosome. *Proc. Natl. Acad. Sci. U.S.A.* **87**, 7405–7409.

Hayes, J. J., Clark, D. J., and Wolffe, A. P. (1991). Histone contributions to the structure of DNA in the nucleosome. *Proc. Natl. Acad. Sci. U.S.A.* **88**, 6829–6833.

Heery, D. M., Kalkhoven, E., Hoare, S., and Parker, M. G. (1997). A signature motif in transcriptional co-activators mediates binding to nuclear receptors. *Nature (London)* **387**, 733–736.

Heinzel, T., Laviusky, R. M., Mullen, T. M., Soderstrom, M., Laherty, C. D., Torchia, J. T., Yang, W.-M., Brard, C., Ngo, S. G., Davie, J. R., Seto, E., Eisenman, R. M., Rose, D. W., Glass, C. K., and Rosenfeld, M. G. (1997). N-CoR, mSIN3, and histone deacetylase in a complex required for repression by nuclear receptors and Mad. *Nature (London)* **387**, 43–48.

Hirschhorn, J. N., Brown, S. A., Clark, C. D., and Winston, F. (1992). Evidence that SNF2/SWI2 and SNF activate transcription in yeast by altering chromatin structure. *Genes Dev.* **6**, 2288–2298.

Hollenberg, A. N., Monden, T., and Wondisford, F. E. (1995). Ligand-independent and -dependent functions of thyroid hormone receptor isoforms depend upon their distinct amino termini. *J. Biol. Chem.* **270**, 14274–14280.

Hong, L., Schroth, G. P., Matthews, H. R., Yau, P., and Bradbury, E. M. (1993). Studies of the DNA binding properties of the histone H4 amino terminus. *J. Biol. Chem.* **268**, 305–314.

Horlein, A. J., Naar, A. M., Heinzel, T., Torchia, J., Gloss, B., Kurokawa, R., Ryan, A., Kamei, Y., Soderstrom, M., Glass, C. K., and Rosenfeld, M. G. (1995). Ligand-independent repression by the thyroid hormone receptor mediated by a nuclear receptor co-repressor. *Nature (London)* **377**, 397–404.

Imbalzano, A. M., Kwon, H., Green, M. R., and Kingston, R. E. (1994). Facilitated binding of TATA-binding protein to nucleosomal DNA. *Nature (London)* **370**, 481–485.

Imhof, A., Yang, X. J., Ogryzko, V. V., Nakatani, Y., Wolffe, A. P., and Ge, H. (1997). Acetylation of general transcription factors by histone acetyltransferases: Identification of a major site of acetylation in TFIIEβ. *Curr. Biol.* **7**, 689–692.

Judelson, C., and Privalsky, M. L. (1996). DNA recognition by normal and oncogenic thyroid hormone receptors. *J. Biol. Chem.* **18**, 10800–10805.

Kadosh, D., and Struhl, K. (1997). Repression by Ume6 involves recruitment of a complex containing Sin3 corepressor and Rpd3 histone deacetylase to target promoters. *Cell (Cambridge, Mass.)* **89**, 365–571.

Kamei, Y., Xu, L., Heinzel, T., Torchia, J., Kurokama, R., Gloss, B., Lin, S.-C., Heyman, R. A., Rose, D. W., Glass, C. K., and Rosenfeld, M. G. (1996). A CBP integrator complex mediates transcriptional activation and AP-1 inhibition by nuclear receptors. *Cell (Cambridge, Mass.)* **85**, 403–414.

Kanamori, A., and Brown, D. D. (1992). The regulation of thyroid hormone receptor β genes by thyroid hormone in *Xenopus laevis*. *J. Biol. Chem.* **267**, 739–745.

Kass, S. U., Landsberger, N., and Wolffe, A. P. (1997). DNA methylation directs a time-dependent repression of transcription initiation. *Curr. Biol.* **7**, 157–165.

Kaufmann, P. D., and Botchan, M. R. (1994). Assembly of nucleosomes: Do multiple assembly factors mean multiple mechanisms? *Curr. Opin. Genet. Dev.* **4**, 229–235.

Kawahara, A., Baker, B. S., and Tata, J. R. (1991). Developmental and regional expression of thyroid hormone receptor genes during Xenopus metamorphosis. *Development (Cambridge, UK)* **112**, 933–943.

Krajewski, W. A., and Becker, P. B. (1998). Reconstitution of hyperacetylated, DNase I-sensitive chromatin chracterized by high conformational flexibility of nucleosomal DNA. *Proc. Natl. Acad. Sci. U.S.A.* **95**, 1540–1545.

Laherty, C. D., Yang, W. M., Sun, J. M., Davie, J. R., Seto, E., and Eisenman, R. M. (1997). Histone deacetylase associated with the mSin3 corepressor mediate Mad transcriptional repression. *Cell (Cambridge, Mass.)* **89**, 349–356.

Landsberger, N., and Wolffe, A. P. (1995). The role of chromatin and *Xenopus* heat shock transcription factor (XHSF1) in the regulation of the *Xenopus* hsp70 promoter *in vivo*. *Mol. Cell. Biol.* **15**, 6013–6024.

Landsberger, N., Ranjan, M., Almouzni, G., Stump, D., and Wolffe, A. P. (1995). The heat shock response in *Xenopus* oocytes, embryos and somatic cells: A regulatory role for chromatin. *Dev. Biol.* **170**, 62–74.

Le Douarin, B., Zechel, C., Garnier, J.-M., Lutz, Y., Tora, L., Pierrat, B., and Heery, D. (1995). The N-terminal part of TIF1, a putative mediator of the ligand-dependent activation function (AF-2) of nuclear receptors, is fused to B-raf in the oncogenic protein T18. *EMBO J.* **14**, 2020–2033.

Lee, D. Y., Hayes, J. J., Pruss, D., and Wolffe, A. P. (1993). A positive role for histone acetylation in transcription factor binding to nucleosomal DNA. *Cell (Cambridge, Mass.)* **72**, 73–84.

Lefebvre, D., Mouchon, A., Lefebvre, B., and Formstecher, P. (1998). Binding of retinoic acid receptor heterodimers to DNA-role for histones NH_2 termini. *J. Biol. Chem.* **273**, 12288–12295.

Leloup, J., and Buscaglia, M. (1977). La thiiodothyranine, hormone de la métamorphose des Amphibiens. *C. R. Hebd. Seances Acad. Sci. Ser. D* **284,** 2261–2263.

Li, Q., Imhof, A., Collingwood, T. N., Urnov, F. D., and Wolffe, A. P. (1999). p300 Stimulates transcription instigated by ligand-bound thyroid hormone receptor at a step subsequent Fc chromatin disruption. *EMBO J.* (in press).

Li, Q., and Wrange, O. (1993). Translational positioning of a nucleosomal glucocorticoid response element modulates glucocorticoid receptor affinity. *Genes Dev.* **7**, 2471–2482.

Li, Q., and Wrange, O. (1995). The accessibility of a glucocorticoid response element dependent on its rotational positioning. *Mol. Cell. Biol.* **15**, 4375–4384.

Li, Q., Herrler, M., Landsberger, N., Kaludov, N., Ogrzyko, V. V., Nakatani, Y., and Wolffe, A. P. (1998). *Xenopus* NF-Y pre-sets chromatin and potentiates acetylation responsive transcription from the *Xenopus* hsp70 promoter *in vivo*. *EMBO J.* (in press).

Lin, B. C., Hong, S.-H., Krig, S., Yoh, S. M., and Privalsky, M. L. (1997). A conformational switch in nuclear hormone receptors is involved in coupling hormone binding to corepressor release. *Mol. Cell Biol.* **17**, 6131–6138.

Lin, R. J., Nagy, K., Inoue, S., Shao, W., Miller, W. H., Jr., and Evans, R. M. (1998). Role of the histone deacetylase complex in acute promyelocytic leukaemia. *Nature (London)* **391**, 811–814.

Machuca, I., Esslemont, G., Fairclough, L., and Tata, J. R. (1995). Analysis of structure and expression of the *Xenopus* thyroid hormone receptor β gene to explain its autoregulation. *Mol. Endocrinol.* **9**, 96–107.

Munoz, A., Zenke, M., Gehring, U., Sap, J., Beug, H., and Vennstrom, B. (1988). Characterization of the hormone-binding domain of the chicken c-erbA/thyroid hormone receptor protein. *EMBO J.* **7**, 155–159.

Mutskov, V., Gerber, D., Angelov, D., Ausio, J., Workman, J., and Dimitrov, S. (1998). Persistant interactions of core histone tails with nucleosomal DNA following acetylation and transcription factor binding. *Mol. Cell. Biol.* **18**, 6293–6304.

Nagy, L., Kao, H. Y., Chakravarti, D., Lin, R. J., Hassig, C. A., Ayer, D. E., Schreiber, S. L., and Evans, R. M. (1997). Nuclear receptor repression mediated by a complex containing SMRT, mSin3A, and histone deacetylase. *Cell (Cambridge, Mass.)* **89**, 373–380.

Nightingale, K. P., Dimitrov, S., Reeves, R., and Wolffe, A. P. (1996). Evidence for a shared structural role for HMG1 and linker histones B4 and H1 in organizing chromatin. *EMBO J.* **15**, 548–561.

Nightingale, K. P., Wellinger, R. E., Sogo, J. M., and Becker, P. B. (1998). Histone acetylation facilitates RNA polymerase II transcription of the *Drosophila hsp26* gene in chromatin. *EMBO J.* **17**, 2865–2876.

Norton, V. G., Imai, B. S., Yau, P., and Bradbury, E. M. (1989). Histone acetylation reduces nucleosome core particle linking number change. *Cell (Cambridge, Mass.)* **57**, 449–457.

Ogryzko, V. V., Schiltz, R. L., Russanova, V., Howard, B. H., and Nakatani, Y. (1996). The transcriptional coactivators p300 and CBP are histone acetyltransferases. *Cell (Cambridge, Mass.)* **87**, 953–959.

Oñate, S. A., Tsai, S. Y., Tsai, M.-J., and O'Malley, B. W. (1995). Sequence and characterization of a coactivator for the steroid hormone receptor superfamily. *Science* **270**, 1354–1357.

Patterton, D., Hayes, W. P., and Shi, Y.-B. (1995). Transcriptional activation of the matrix metalloproteinase gene stromelysin-3 coincides with thyroid hormone induced cell death during frog metamorphosis. *Dev. Biol.* **167**, 252–262.

Pederson, D. S., and Morse, R. H. (1990). Effect of transcription of yeast chromatin on DNA topology *in vivo*. *EMBO J.* **9**, 1873–1881.

Perlman, A. J., Stanley, F., and Samuels, H. H. (1982). Thyroid hormone nuclear receptor: Evidence for multimeric organization in chromatin. *J. Biol. Chem.* **257**, 930–938.

Perlmann, T., and Wrange, O. (1988). Specific glucocorticoid receptor binding to DNA reconstituted in a nucleosome. *EMBO J.* **7**, 3073–3083.

Perry, M., and Chalkley, R. (1982). Histone acetylation increases the solubility of chromatin and occurs sequentially over most of the chromatin. *J. Biol. Chem.* **257**, 7336–7347.

Pina, B., Bruggemeier, U., and Beato, M. (1990a). Nucleosome positioning modulates accessibility of regulatory proteins to the mouse mammary tumor virus promoter. *Cell (Cambridge, Mass.)* **60**, 719–731.

Pina, B., Barettino, D., Truss, M., and Beato, M. (1990b). Structural features of a regulatory nucleosome. *J. Mol. Biol.* **216**, 975–990.

Pruss, D., Hayes, J. J., and Wolffe, A. P. (1995). Nucleosomal anatomy—where are the histones? *BioEssays* **17**, 161–170.

Pruss, D., Bartholomew, B., Persinger, J., Hayes, J., Arents, G., Moudrianakis, E. N., and Wolffe, A. P. (1996). An asymmetric model for the nucleosome: A binding site for linker histones inside the DNA gyres. *Science* **274**, 614–617.

Puig, O. M., Belles, E., Lopez-Rodas, G., Sendra, R., and Tordera, V. (1998). Interaction between N-terminal domain of H4 and DNA is regulated by the acetylation degree. *Biochim. Biophys. Acta* **1397**, 79–80.

Raaka, B. M., and Samuels, H. H. (1981). Regulation of thyroid hormone nuclear receptor levels in GH1 cells by 3, 5, 3'-triiodo-L-thyronine. Use of dense amino acid labeling to determine the influence of hromone on the receptor half-life and the rate of appearance of newly synthesized receptor. *J. Biol. Chem.* **256**, 6883–6889.

Ranjan, M., Wong, J., and Shi, Y.-B. (1994). Transcriptional repression of *Xenopus* TRβ gene is mediated by a thyroid hormone response element located near the start site. *J. Biol. Chem.* **269**, 24699–24705.

Reeves, R., Gorman, C. M., and Howard, B. (1985). Minichromosome assembly of non-integrated plasmid DNA transfected into mammalian cells. *Nucleic Acids Res.* **13**, 3599–3615.

Refetoff, S., DeWind, L. T., and de Groot, L. J. (1967). Familial syndrome combining deaf-mutism, stippled epiphyses, goiter and abnormally high PBI: Possible target organ refractoriness to thyroid hormone. *J. Clin. Endocrinol. Metab.* **27**, 279–294.

Renaud, J.-P., Rochel, N., Ruff, M., Vivat, V., Chambon, P., Gronemeyer, H., and Moras, D. (1995). Crystal structure of the RAR-γ ligand-binding domain bound to all-trans retinoic acid. *Nature (London)* **378**, 681–689.

Richard-Foy, H., and Hager, G. L. (1987). Sequence-specific positioning of nucleosomes over the steroid-inducible MMTV promoter. *EMBO J.* **6**, 2321–2328.

Roth, S. Y., Dean, A., and Simpson, R. T. (1990). Yeast α2 repressor positions nucleosomes in TRP1/ARS1 chromatin. *Mol. Cell. Biol.* **10**, 2247–2260.

Saatcioglu, F., Bartunek, P., Deng, T., Zenke, M., and Karin, M. (1993). A conserved C-terminal sequence that is deleted in v-ErbA is essential for the biological activities of c-ErbA. *Mol. Cell. Biol.* **13**, 3675–3685.

Samuels, H. H., Stanley, F., Casanova, J., and Shao, T. C. (1980). Thyroid hormone nuclear receptor levels are influenced by the acetylation of chromatin-associated proteins. *J. Biol. Chem.* **255**, 2499–2508.

Samuels, H. H., Perlman, A. J., Raaka, B. M., and Stanley, F. (1982). Organization of the thyroid hormone receptor in chromatin. *Recent Prog. Horm. Res.* **38**, 557–599.

Sap, J., Munoz, A., Damm, K., Goldberg, Y., Ghysdael, J., Leutz, A., Beug, H., and Vennstrom, B. (1986). The c-erb-A protein is a high-affinity receptor for thyroid hormone. *Nature (London)* **324**, 635–640.

Scharif, M., and Privalsky, M.L. (1991). V-erbA oncogene function in neoplasia correlates with its ability to repress retinoic acid receptor action. *Cell (Cambridge, Mass.)* **66**, 885–893.

Schild, C., Claret, F. X., Wahli, W., and Wolffe, A. P. (1993). A nucleosome-dependent static loop potentiates estrogen-regulated transcription from the Xenopus vitellogenin B1 promoter in vitro. *EMBO J.* **12**, 423–433.

Schwarz, P. M., and Hansen, J. C. (1994). Formation and stability of higher order chromatin structures. Contributions of the histone octamer. *J. Biol. Chem.* **269**, 16284–16298.

Seale, R. L. (1975). Assembly of DNA and protein during replication in HeLa cells. *Nature (London)* **255**, 247–249.

Seale, R. L. (1978). Nucleosomes associated with newly replicated DNA have an altered conformation. *Proc. Natl. Acad. Sci. U.S.A.* **75**, 2717–2721.

Shen, X., and Gorovsky, M. A. (1996). Linker histone H1 regulates specific gene expression but not global transcription *in vivo*. *Cell (Cambridge, Mass.)* **86**, 475–483.

Shi, Y.-B. (1994). Thyroid hormone-regulated early and late genes during amphibian metamorphosis. *In* "metamorphosis: Post Embryonic Reprogramming of Gene Expression in Amphibian and Insect Cells" (L. I. Gilbert, B. G. Atkinson, and J. R. Tata, eds.). Academic Press, San Diego, CA.

Shi, Y.-B., and Brown, D. D. (1992). Genomic organization and alternative promoter usage of the thyroid hormone receptor βgenes in *Xenopus laevis. J. Biol. Chem.* **267**, 733–738.

Shi, Y.-B., Wong, J., Puzianowska-Kuznicka, M., and Stolow, M. A. (1996). Tadpole competence and tissue-specific temporal regulation of amphibian metamorphosis: Roles of thyroid hormone and its receptors. *BioEssays* **18**, 391–399.

Steinbach, O. C., Wolffe, A. P., and Rupp, R. (1997). Somatic linker histones causes loss of mesodermal competence in *Xenopus. Nature (London)* **389**, 395–399.

Strait, K. A., Schwartz, H. L., Perez-Castillo, A., and Oppenheimer, J. H. (1990). Relationship of c-erbA mRNA content to tissue triiodothyronine nuclear binding capacity and function in developing and adult rats. *J. Biol. Chem.* **265**, 10514–10521.

Stunkel, W., Kober, I., and Seifart, K. H. (1997). A nucleosome positioned in the distal promoter region activates transcription of the human U6 gene. *Mol. Cell. Biol.* **17**, 4397–4405.

Tagami, T., Madison, L. D., Nagaya, T., and Jameson, J. L. (1997). Nuclear receptor corepressors activate rather than suppress basal transcription of genes that are negatively regulated by thyroid hormone. *Mol. Cell. Biol.* **17**, 2642–2648.

Takeda, K., Sakurai, A., de Groot, L. J., and Refetoff, S. (1992). Recessive inheritance of thyroid hormone resistance caused by a complete deletion of the protein-coding region of the thyroid hormone receptor-beta gene. *J. Clin. Endocrinol. Metab.* **74**, 49–55.

Tang, Y., and DeFranco, D. B. (1996). ATP dependent release of glucocorticoid receptors from the nuclear matrix. *Mol. Cell Biol.* **16**, 1989–2001.

Tata, J. R. (1993). Gene expression during metamorphosis: An ideal model for post embryonic development. *BioEssays* **15**, 239–248.

Tata, J. R. (1996). Metamorphosis: An exquisite model for hormonal regulation of postembryonic development. *Biochem. Soc. Symp.* **62**, 123–136.

Tenbaum, S., and Baniahmad, A (1997) Nuclear receptors: Structure, function and involvement in disease. *Int. J. Biochem. Cell. Biol.* **29**, 1325–1342.

Tomura, H., Lazar, J., Phyillaier, M., and Nikodem, V. M. (1995). The N-terminal region (A/B) of rat thyroid hormone receptors α1, β1, but not β2 contains a strong thyroid hormone-dependent transactivation function. *Proc. Natl. Acad. Sci. U.S.A.* **92**, 5600–5604.

Tone, Y., Collingwood, T. N., Adams, M., and Chatterjee, V. K. (1994). Functional analysis of a transactivation domain in the thyroid hormone β receptor. *J. Biol. Chem.* **269**, 31157–31161.Torchia, J., Glass, C., and Rosenfeld, M. (1998). Co-activators and corepressors in the integration of transcriptional responses. *Curr. Opin. Cell Biol.* **10**, 373–383.

Truss, M., Bartsch, J., Schelbert, A., Haché, R. J. G., and Beato, M. (1995). Hormone induces binding of receptors and transcription factors to a rearranged nucleosome on the MMTV promoter *in vivo*. *EMBO J.* **14**, 1737–1751.

Tse, C., and Hansen, J. C. (1997). Hybrid trypsinized nucleosomal arrays: Identification of multiple functional roles of the H2A/H2B and H3/H4 N-termini in chromatin fiber compaction. *Biochemistry* **36**, 11381–11388.

Tse, C., Sera, T., Wolffe, A. P., and Hansen, J. C. (1998). Disruption of higher order folding by core histone acetylation dramatically enhances transcription of nucleosomal arrays by RNA polymerase III. *Mol. Cell. Biol.* **18**, 4629–4638.

Tsukiyama, T., Becker, P. B., and Wu, C. (1994). ATP-dependent nucleosome disruption at a heat-shock promoter mediated by binding of GAGA transcription factor. *Nature (London)* **367**, 525–532.

Ura, K., Hayes, J. J., and Wolffe, A. P. (1995). A positive role for nucleosome mobility in the transcriptional activity of chromatin templates: Restriction by linker histones. *EMBO J.* **14**, 3752–3765.

Ura, K., Nightingale, K., and Wolffe, A. P. (1996). Differential association of HMG1 and linker histones B4 and H1 with dinucleosomal DNA: Structural transitions and transcriptional repression. *EMBO J.* **15**, 4959–4969.

Ura, K., Kurumizaka, H., Dimitrov, S., Almouzni, G., and Wolffe, A. P. (1997). Histone acetylation: Influence on transcription by RNA polymerase, nucleosome mobility and positioning, and linker histone dependent transcriptional repression. *EMBO J.* **16**, 2096–2107.

van Steensel, B., Brink, M., van der Meulen, K., van Binnendijl, E. P., Wansink, D. G., de Jong, L., de Kloet, E. R., and van Driel, R. (1995). Localization of the glucocorticoid receptor in discrete clusters in the cell nucleus. *J. Cell Sci.* **108**, 3003–3011.

Verdin, E., Paras, P., and van Lint, C. (1993). Chromatin disruption in the promoter of human immunodeficiency virus type 1 during transcriptional activation. *EMBO J.* **12**, 3249–3259.

Vetesse-Dadey, M., Walter, P., Chen, H., Juan, L.-J., and Workman, J. L. (1994). Role of the histone amino termini in facilitated binding of a transcription factor, GAL4-AH to nucleosome cores. *Mol. Cell. Biol.* **14**, 970–981.

Vetesse-Dadey, M., Grant, P. A., Hebbes, T. R., Crane-Robinson, C., Allis, C. D., and Workman, J. L. (1996). Acetylation of histone H4 plays a primary role in enhancing transcription factor binding to nucleosomal DNA *in vitro*. *EMBO J.* **15**, 2508–2518.

Voegel, J. J., Heine, M. J. S., Zechel, C., Chambon, P., and Gronemeyer, H. (1996). TIF2, a 160 KDa transcription mediator for the ligand-dependent activation function AF-2 of nuclear receptors. *EMBO J.* **15**, 3667–3675.

Wagner, R. L., Apriletti, J. W., McGrath, M. E., West, B. L., Baxter, J. D., and Fletterick, R. J. (1995). A structural role for hormone in the thyroid hormone receptor. *Nature (London)* **378**, 690–697.

Wahlstrom, G. M., Harbers, M., and Vennstrom, B. (1996). The oncoprotein P75gag-v-erbA represses thyroid hormone induced transcriptional only via response elements containing palindromic half-sites. *Oncogene* **13**, 843–852.

Wang, W., Côté, J., Xue, Y., Zhou, S., Khavara, P. A., Biggar, S. R., Muchardt, C., Kalpana, G. V., Goff, S. P., Yaniv, M., Workman, J. L., and Crabtree, G. R. (1996a). Purification and biochemical heterogeneity of the mammalian SWI/SNF complex. *EMBO J.* **15**, 5370–5382.

Wang, W., Xue, Y., Zhou, S., Kuo, A., Cairns, B. R., and Crabtree, G. R. (1996b). Diversity and specialization of mammalian SWI/SNF complexes. *Genes Dev.* **10**, 2117–2130.

Weinberger, C., Thompson, C. C., Ong, E. S., Lebo, R., Gruol, D. J., and Evans, R. M. (1986). The c-erbA gene encodes a thyroid hormone receptor. *Nature (London)* **324**, 641–646.

Weintraub, H. (1985). Assembly and propagation of repressed and derepressed chromosomal states. *Cell (Cambridge, Mass.)* **42**, 705–711.

Wikstrom, L., Johansson, Ca., Salto, C., Barlow, C., Barros, A. C., Baas, F., Forrest, D., Thoren P., and Vennstrom, B. (1998). Abnormal heart rate and body temperature in mice lacking thyroid hormone receptor $\alpha 1$. *EMBO J.* **17**, 455–461.

Wilson, C. J., Chao, D. M., Imbalzano, A. N., Schnitzer, G. R., Kingston, R., and Young, R. A. (1996). RNA polymerase II holoenzyme contains SWI/SNF regulators involved in chromatin remodeling. *Cell (Cambridge, Mass.)* **84**, 235–244.

Wolffe, A. P. (1997). Sinful repression. *Nature (London)* **387,** 16–17.

Wolffe, A. P. (1998). "Chromatin Structure and Function," 3rd ed. Academic Press, London.

Wolffe, A. P., and Drew, H. R. (1989). The initiation of transcription on nucleosomal templates. *Proc. Natl. Acad. Sci. U.S.A.* **86**, 9817–9821.

Wolffe, A. P., and Hayes, J. J. (1999). Chromatin disruption and modification. *Nucleic Acids Res.* **27**, 711–720.

Wolffe, A. P., and Pruss, D. (1996). Targeting chromatin disruption: Transcription regulators that acetylate histones. *Cell (Cambridge, Mass.)* **84**, 817–819.

Wolffe, A. P., and Shi, Y.-B. (1999). A hypothesis for the transcriptional control of amphibian metamorphosis by the thyroid hormone receptor. *Am. Zool.* **38,** 253–264.

Wolffe, A. P., Almouzni, G., Ura, K., Pruss, D., and Hayes, J. J. (1993). Transcription factor access to DNA in the nucleosome. *Cold Spring Harbor Symp. Quant. Biol.* **58,** 225–236.

Wong, J., and Shi, Y-B. (1995). Coordinated regulation of and transcriptional activation by *Xenopus* thyroid hormone and retinoid X receptors. *J. Biol. Chem.* **270,** 18479–18483.

Wong, J., Shi, Y.-B., and Wolffe, A. P. (1995). A role for nucleosome assembly in both silencing and activation of the *Xenopus* TRβA gene by the thyroid hormone receptor. *Genes Dev.* **9**, 2696–2711.

Wong, J., Shi, Y.-B., and Wolffe, A. P. (1997a). Determinants of chromatin disruption and transcriptional regulation instigated by the thyroid hormone receptor: Hormone regulated chromatin disruption is not sufficient for transcriptional activation. *EMBO J.* **16**, 3158–3171.

Wong, J., Li, Q., Levi, B.-Z., Shi, Y.-B., and Wolffe, A. P. (1997b). Structural and functional features of a specific nucleosome containing a recognition element for the thyroid hormone receptor. *EMBO J.* **16**, 7130–7145.

Wong, J., Patterton, D., Imhof, A., Shi, Y.-B., and Wolffe, A. P. (1998). Distinct requirements for chromatin assembly in transcriptional repression by thyroid hormone receptor and histone deacetylase. *EMBO J.* **17**, 520–534.

Yang, X.-J., Ogryzko, V. V., Nishikawa, J.-I., Howard, B., and Nakatani, Y. (1996). A p300/CBP-associated factor that competes with the adenoviral E1A oncoprotein. *Nature (London)* **382**, 319–324.

Yaoita, Y., Shi, Y.-B., and Brown, D. D. (1990). *Xenopus laevis* α and β thyroid hormone receptors. *Proc. Natl. Aacd. Sci. U.S.A.* **87**, 7090–7094.

Yoh, S. M., Chatterjee, V. K. K., and Privalsky, M. L. (1997). Thyroid hormoner esistance syndrome manifests as an aberrant interaction between mutant T_3 receptors and transcriptional corepressors. *Mol. Endocrinol.* **11**, 470–480.

Zenke, M., Munoz, A., Sap, J., Vennstrom, B., and Beug, H. (1990). v-erbA oncogene activation entails the loss of hormone-dependent regulator activity of c-erbA. *Cell (Cambridge, Mass.)* **61**, 1035–1049.

Zhang, D.-E., and Nelson, D. A. (1988). Histone acetylation in chicken erythrocytes. Rates of acetylation and evidence that histones in both active and potentially active chromatin are rapidly modified. *Biochem. J.* **250**, 233–240.

Index

A

M

N

R

S

T

U

V

W

X

ISBN 0-12-709858-5

90051

9 780127 098586

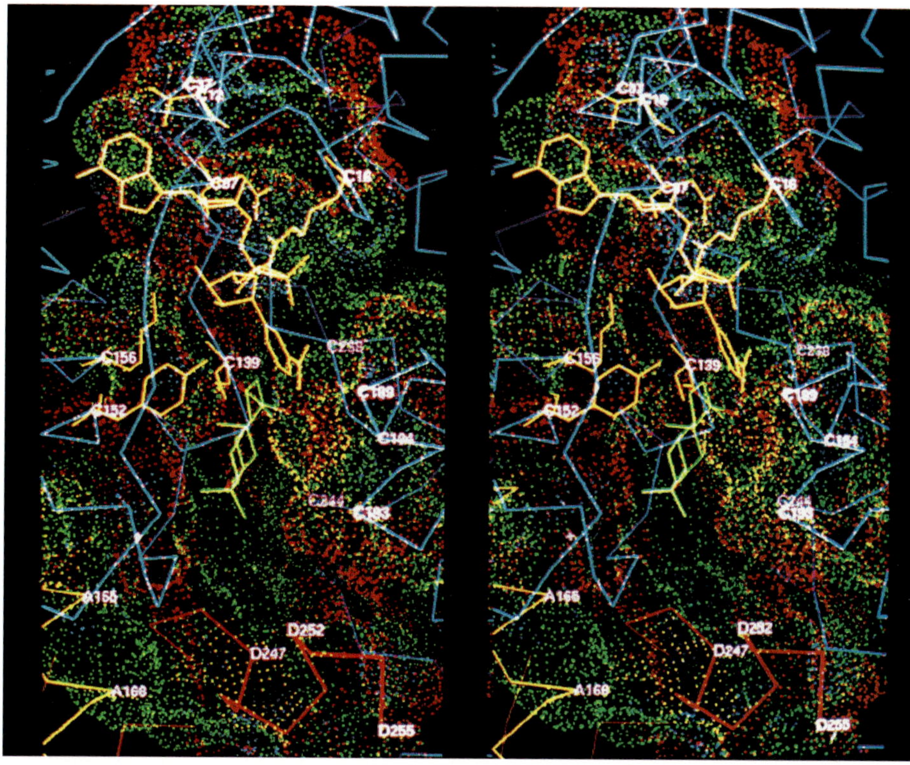

CHAPTER 5, FIG. 4. A cortisone molecule modeled in the putative catalytic cleft of 3α,20β-HSD. The Cα chains of the enzyme, NADH cofactor, and residues having possible roles in catalysis are shown. (From Ghosh *et al.,* 1994a.)

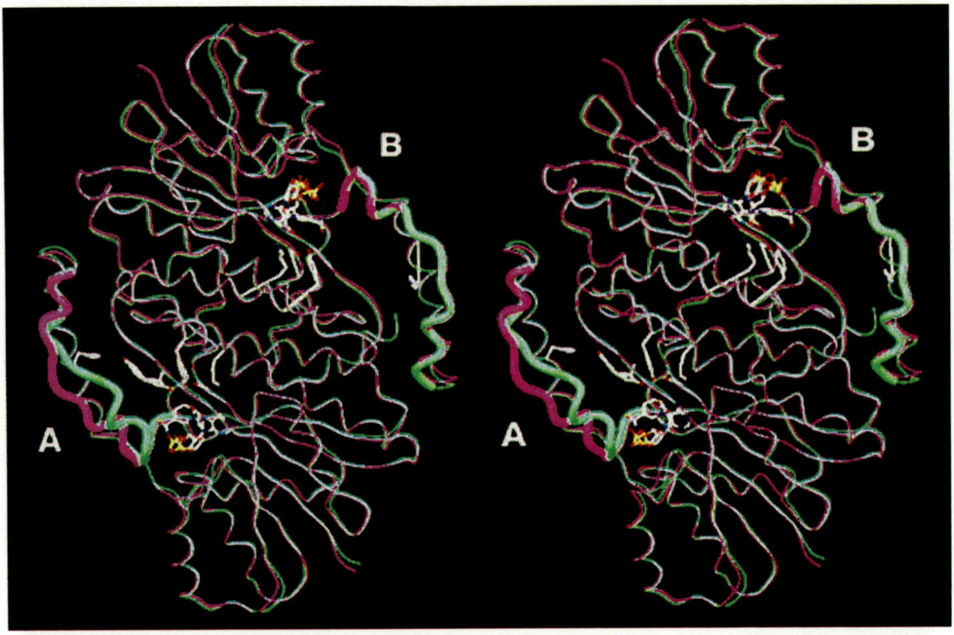

CHAPTER 5, FIG. 15. Least-squares superposition of the Cα backbones of the dimers of the 17β-HSD1–equilin complex (blue), the apo–17β-HSD1 complex (magenta), and E2 complex (yellow) illustrating the differences of the substrate-entry loop structures of the A- and B-subunits. The substrate-entry loops (residues 186–201) are drawn in wider cross-section. The loop in the A-subunit of 17β-HSD1–equilin complex is in the closed conformation, whereas that in the β-subunit has been modeled after the apo-form. NADP+ and equilin of the 17β-HSD1–equilin complex are shown. (From Sawicki *et al.*, 1999a.)

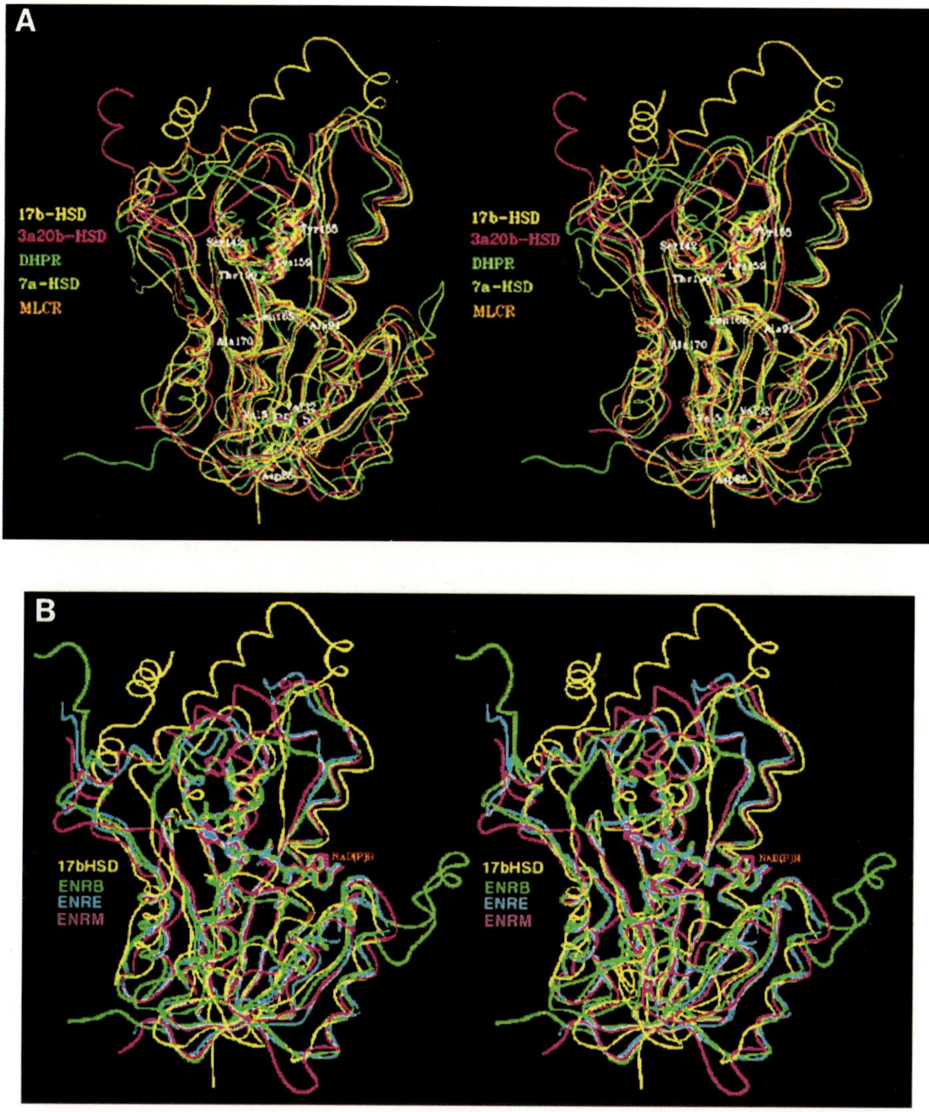

CHAPTER 5, FIG. 16. Superposition stereo comparison of (A) three steroid dehydrogenases (3α,20β-HSD, 17β-HSD, 7α-HSD), and DHPR and MLCR (from Duax *et al.,*1996); (B) 17β-HSD and three enoyl acyl cancer protein reductase enzymes, ENRE, ENRB, and ENRM. Next page: (C) 17β-HSD, BDDH, 7 HNR, and SEP; and (D) 17β-HSD, ADH, and UDPE.

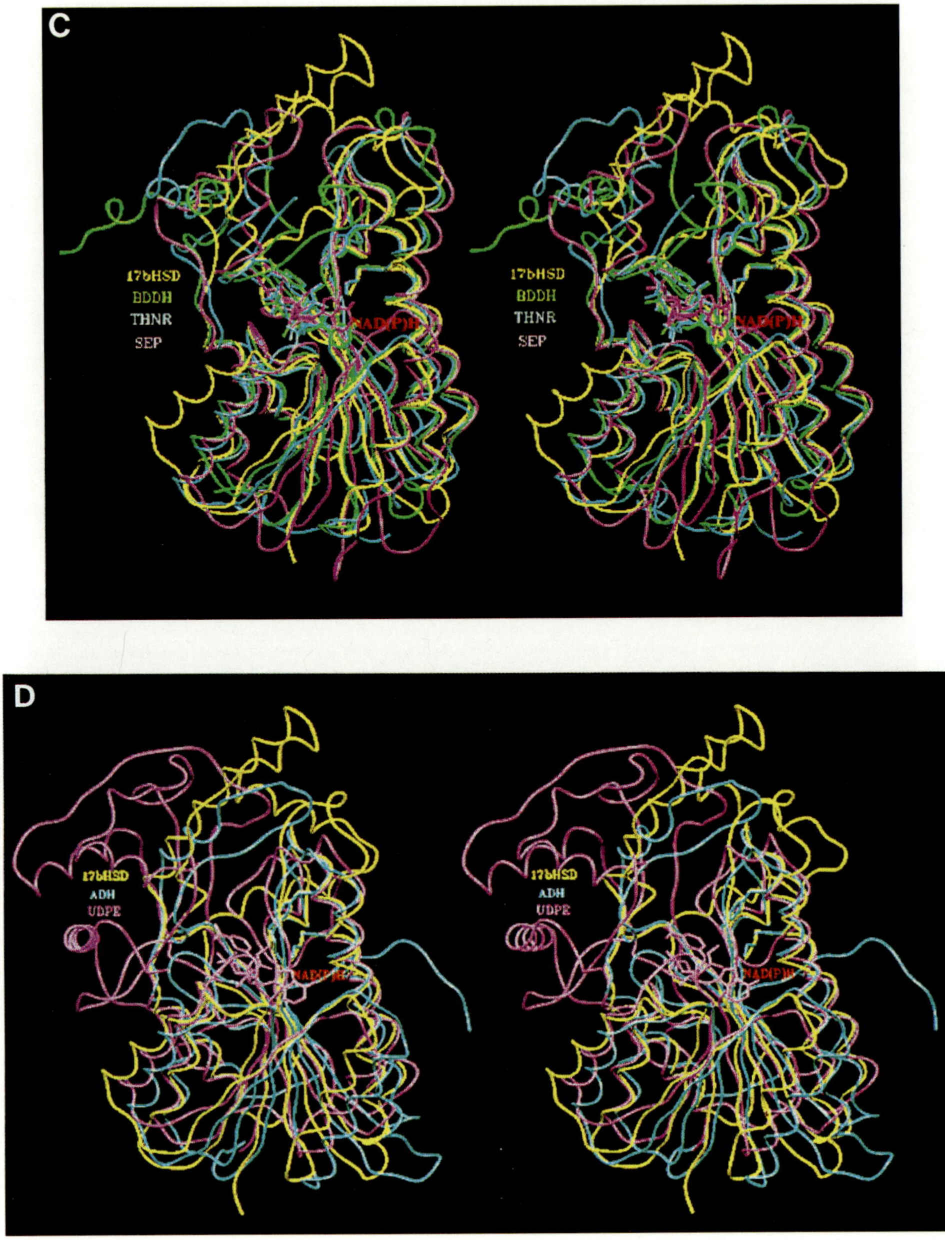

CHAPTER 5, FIG. 16C AND 16D

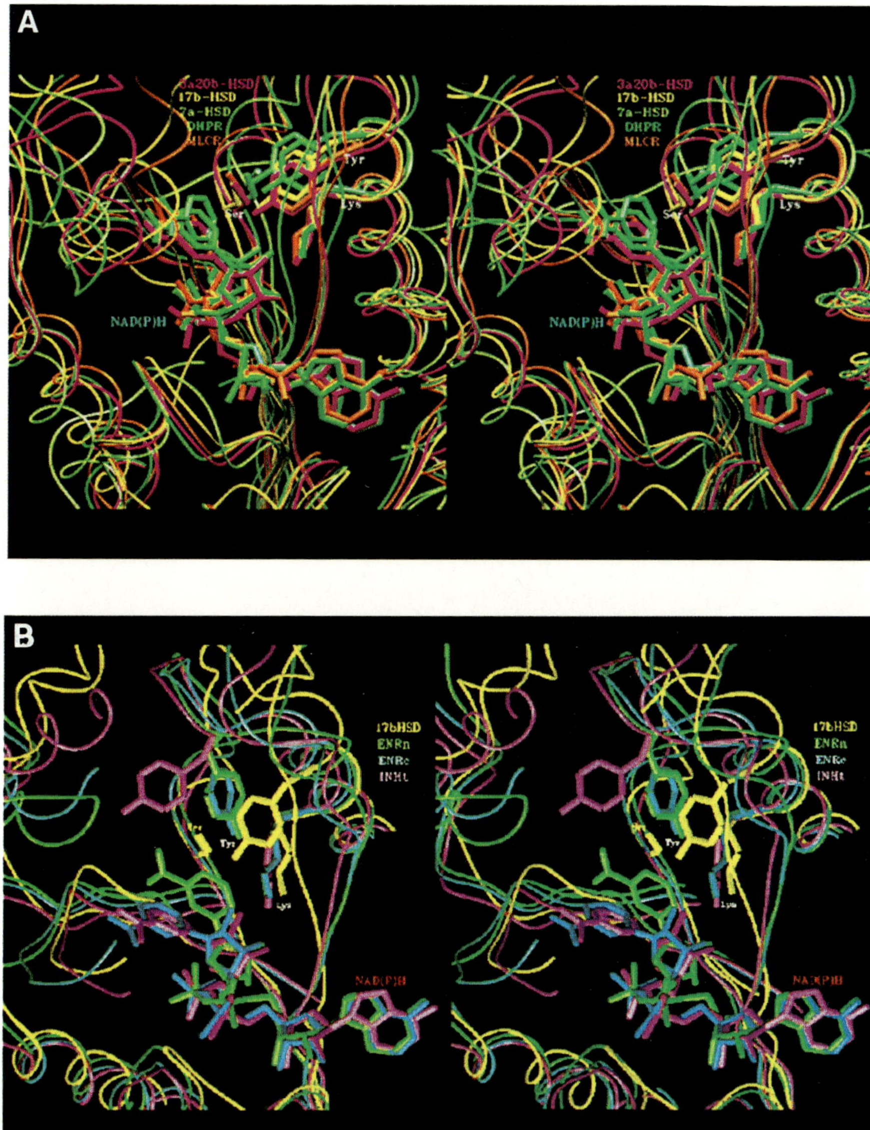

CHAPTER 5, FIG. 17. Stereo comparison of relative cofactor positions in (A) the five structures superimposed in Fig. 16A (from Duax *et al.,* 1996) and (B) the four structures superimposed in Fig. 16B.

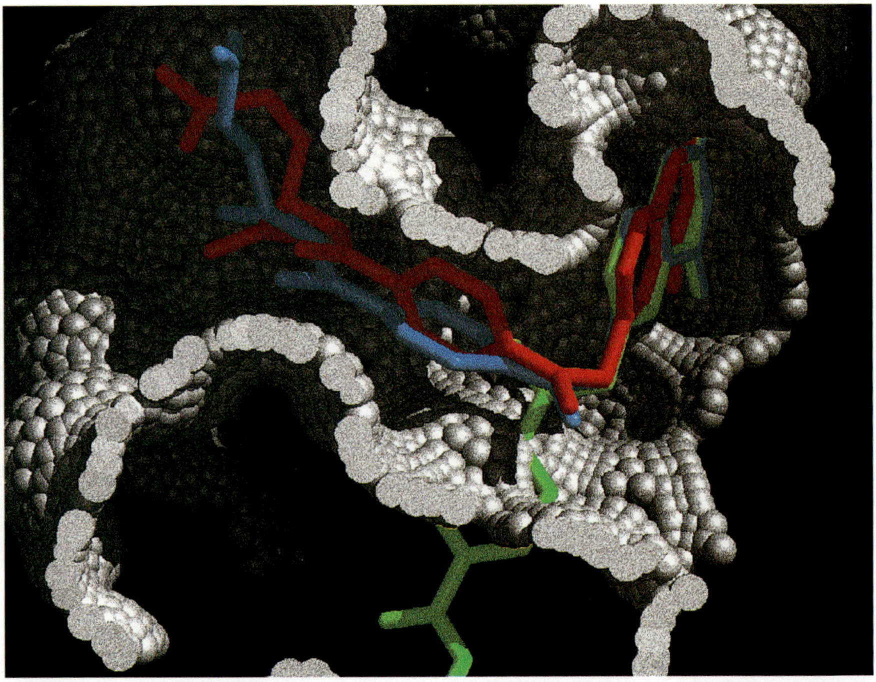

CHAPTER 6, Fig. 6. The experimental (blue) and the best fit from ensemble docking (red) orientations of methotrexate in the binding site of dihydrofolate reductase (molecular surface). The conformation of methotrexate (green) found in the Available Chemicals Directory (ACD) cannot fit into the binding site. (From Lorber and Shoichet, 1998.)